William F. Maag Library
Youngstown State University

Organic Reactions

ADVISORY BOARD

John E. Baldwin
Peter Beak
Virgil Boekelheide
George A. Boswell, Jr.
Engelbert Ciganek
Dennis Curran
David Y. Curtin
Samuel Danishefsky
Heinz W. Gschwend
Stephen Hanessian
Ralph F. Hirschmann
Herbert O. House
Robert C. Kelly

Andrew S. Kende
Steven V. Ley
James A. Marshall
Blaine C. McKusick
Jerrold Meinwald
Leo A. Paquette
Gary H. Posner
Hans J. Reich
Charles Sih
Barry M. Trost
Milán Uskokovic
James D. White

FORMER MEMBERS OF THE BOARD
NOW DECEASED

Roger Adams
Homer Adkins
Werner E. Bachmann
A. H. Blatt
Theodore L. Cairns
Arthur C. Cope
Donald J. Cram
William G. Dauben

Louis F. Fieser
John R. Johnson
Robert M. Joyce
Willy Leimgruber
Frank C. McGrew
Carl Niemann
Harold R. Snyder
Boris Weinstein

Organic Reactions

VOLUME 63

EDITORIAL BOARD

LARRY E. OVERMAN, *Editor-in-Chief*

DALE BOGER
ANDRÉ CHARETTE
SCOTT E. DENMARK
VITTORIO FARINA
LOUIS HEGEDUS
LAURA KIESSLING
MICHAEL J. MARTINELLI

STUART W. MCCOMBIE
T. V. RAJANBABU
JAMES H. RIGBY
SCOTT D. RYCHNOVSKY
AMOS B. SMITH, III
PETER WIPF

ROBERT BITTMAN, *Secretary*
Queens College of The City University of New York, Flushing, New York

JEFFERY B. PRESS, *Secretary*
JPressORxn@aol.com

LINDA S. PRESS, *Editorial Coordinator*

SUSAN CURRAN, *Editorial Assistant*

ENGELBERT CIGANEK, *Editorial Advisor*

ASSOCIATE EDITORS

R. DAVID CROUCH
ROY A. JOHNSON
C. OLIVER KAPPE

TODD D. NELSON
ALEXANDER STADLER

JOHN WILEY & SONS, INC., PUBLICATION

Published by John Wiley & Sons, Inc., Hoboken, New Jersey

Copyright © 2004 by Organic Reactions, Inc. All rights reserved.

Published simultaneously in Canada.

No part of this publication may be reproduced, stored in a retrieval system or transmitted in any form or by any means, electronic, mechanical, photocopying, recording, scanning, or otherwise, except as permitted under Sections 107 or 108 of the 1976 United States Copyright Act, without either the prior written permission of the Publisher, or authorization through payment of the appropriate per-copy fee to the Copyright Clearance Center, 222 Rosewood Drive, Danvers, MA 01923, 978-750-8400, fax 978-646-8600, or on the web at www.copyright.com. Requests for permission need to be made jointly to both the publisher, John Wiley & Sons, Inc. and the copyright holder, Organic Reactions, Inc. Requests to John Wiley & Sons, Inc. for permissions should be addressed to the Permissions Department, John Wiley & Sons, Inc., 111 River Street, Hoboken, NJ 07030, (201)748-6011, fax (201)748-6008. Requests to Organic Reactions, Inc. for permissions should be addressed to Dr. Jeffery Press, 22 Bear Berry Lane, Brewster, NY 10509, E-Mail: JPressORxn@aol.com.

Limit of Liability/Disclaimer of Warranty: While the publisher and author have used their best efforts in preparing this book, they make no representations or warranties with respect to the accuracy or completeness of the contents of this book and specifically disclaim any implied warranties of merchantability or fitness for a particular purpose. No warranty may be created or extended by sales representatives or written sales materials. The advice and strategies contained herein may not be suitable for your situation. You should consult with a professional where appropriate. Neither the publisher nor author shall be liable for any loss of profit or any other commercial damages, including but not limited to special, incidental, consequential, or other damages.

For general information on our other products and services please contact our Customer Care Department within the U.S. at 877-762-2974, outside the U.S. at 317-572-3993 or fax 317-572-4002.

Wiley also publishes its books in a variety of electronic formats. Some content that appears in print, however, may not be available in electronic format.

Library of Congress Catalog Card Number 42-20265

ISBN 0-471-44532-0

Printed in the United States of America

10 9 8 7 6 5 4 3 2 1

PREFACE TO THE SERIES

In the course of nearly every program of research in organic chemistry the investigator finds it necessary to use several of the better-known synthetic reactions. To discover the optimum conditions for the application of even the most familiar one to a compound not previously subjected to the reaction often requires an extensive search of the literature; even then a series of experiments may be necessary. When the results of the investigation are published, the synthesis, which may have required months of work, is usually described without comment. The background of knowledge and experience gained in the literature search and experimentation is thus lost to those who subsequently have occasion to apply the general method. The student of preparative organic chemistry faces similar difficulties. The textbooks and laboratory manuals furnish numerous examples of the application of various syntheses, but only rarely do they convey an accurate conception of the scope and usefulness of the processes.

For many years American organic chemists have discussed these problems. The plan of compiling critical discussions of the more important reactions thus was evolved. The volumes of *Organic Reactions* are collections of chapters each devoted to a single reaction, or a definite phase of a reaction, of wide applicability. The authors have had experience with the processes surveyed. The subjects are presented from the preparative viewpoint, and particular attention is given to limitations, interfering influences, effects of structure, and the selection of experimental techniques. Each chapter includes several detailed procedures illustrating the significant modifications of the method. Most of these procedures have been found satisfactory by the author or one of the editors, but unlike those in *Organic Syntheses* they have not been subjected to careful testing in two or more laboratories.

Each chapter contains tables that include all the examples of the reaction under consideration that the author has been able to find. It is inevitable, however, that in the search of the literature some examples will be missed, especially when the reaction is used as one step in an extended synthesis. Nevertheless, the investigator will be able to use the tables and their accompanying bibliographies in place of most or all of the literature search so often required.

Because of the systematic arrangement of the material in the chapters and the entries in the tables, users of the books will be able to find information desired by reference to the table of contents of the appropriate chapter. In the interest of economy the entries in the indices have been kept to a minimum, and, in particular, the compounds listed in the tables are not repeated in the indices.

The success of this publication, which will appear periodically, depends upon the cooperation of organic chemists and their willingness to devote time and effort to the preparation of the chapters. They have manifested their interest already by the almost unanimous acceptance of invitations to contribute to the work. The editors will welcome their continued interest and their suggestions for improvements in *Organic Reactions*.

Chemists who are considering the preparation of a manuscript for submission to *Organic Reactions* are urged to write either secretary before they begin work.

CONTENTS

CHAPTER	PAGE

1. THE BIGINELLI DIHYDROPYRIMIDINE SYNTHESIS
   C. Oliver Kappe and Alexander Stadler 1

2. MICROBIAL ARENE OXIDATIONS
   Roy A. Johnson 117

3. Cu, Ni, AND Pd MEDIATED HOMOCOUPLING REACTIONS IN BIARYL SYNTHESIS: THE ULLMANN REACTION
   Todd D. Nelson and R. David Crouch 265

CUMULATIVE CHAPTER TITLES BY VOLUME 557

AUTHOR INDEX, VOLUMES 1–63 571

CHAPTER AND TOPIC INDEX, VOLUMES 1–63 575

CHAPTER 1

THE BIGINELLI DIHYDROPYRIMIDINE SYNTHESIS

C. OLIVER KAPPE AND ALEXANDER STADLER

Institute of Chemistry, Karl-Franzens-University Graz, Graz, Austria

CONTENTS

	PAGE
ACKNOWLEDGMENTS	2
INTRODUCTION	2
MECHANISM	3
SCOPE AND LIMITATIONS	4
Building Blocks	4
Aldehydes	4
CH-Acidic Carbonyl Components	6
Urea Building Blocks	7
Solid-Phase, Fluorous-Phase, and Related Procedures	7
Asymmetric Biginelli Synthesis	9
Intramolecular (Tethered) Biginelli Reactions	10
Related Biginelli-Type Reactions	11
COMPARISON WITH OTHER METHODS	12
EXPERIMENTAL CONDITIONS	13
EXPERIMENTAL PROCEDURES	14
Ethyl 6-Methyl-2-oxo-4-phenyl-1,2,3,4-tetrahydropyrimidine-5-carboxylate [Biginelli Condensation Utilizing Ethanol/HCl as the Solvent/Catalyst System]	14
Isopropyl 4-(2-Bromo-5-nitrophenyl)-6-methyl-2-oxo-1,2,3,4-tetrahydropyrimidine-5-carboxylate [Biginelli Condensation Utilizing Acetic Acid/HCl as the Solvent/Catalyst System]	14
Ethyl 1-Benzyl-6-methyl-2-oxo-4-phenyl-1,2,3,4-tetrahydropyrimidine-5-carboxylate [Biginelli Condensation Utilizing THF/Polyphosphate Ester as the Solvent/Catalyst System]	15
Methyl 4-(3,4-Difluorophenyl)-6-ethyl-2-oxo-1,2,3,4-tetrahydropyrimidine-5-carboxylate [Biginelli Condensation Utilizing Lewis Acids as the Catalyst System]	15
Ethyl 4-(4-Hydroxy-3-methoxyphenyl)-6-methyl-2-oxo-1,2,3,4-tetrahydropyrimidine-5-carboxylate [Biginelli Condensation Utilizing Lewis Acids as the Catalyst System]	16
Ethyl 4-(3-Hydroxyphenyl)-6-methyl-2-thioxo-1,2,3,4-tetrahydropyrimidine-5-carboxylate [Microwave-Assisted Biginelli Condensation under Solvent-Free Conditions]	16
Methyl 4-(3,4-Difluorophenyl)-2-oxo-1,2,3,4-tetrahydropyrimidine-5-carboxylate [Microwave-Assisted Biginelli Condensation Utilizing Lewis Acids as the Catalyst System]	17

---

oliver.kappe@uni-graz.at

Organic Reactions, Vol. 63, Edited by Larry E. Overman et al.
ISBN 0-471-44532-0 © 2004 Organic Reactions, Inc. Published by John Wiley & Sons, Inc.

4-Phenyl-5-(thiophene-2-carbonyl)-6-trifluoromethyl-1,2,3,4-tetrahydropyrimidin-2-one
[Biginelli Condensation Utilizing Lewis Acids under Solvent-Free Conditions] . . 18
Ethyl 4-Butyl-6-methyl-2-oxo-1,2,3,4-tetrahydropyrimidine-5-carboxylate [Biginelli
Condensation Utilizing KSF-Clay as the Catalyst System] 18
Diethyl 4-(4-Bromophenyl)-6-methyl-2-oxo-1,2,3,4-tetrahydropyrimidine-5-carboxamide
[Biginelli Condensation Utilizing Ethanol/HCl as the Solvent/Catalyst System] . . 18
6-Methyl-5-nitro-4-(3,4,5-trimethoxyphenyl)-1,2,3,4-tetrahydropyrimidin-2-one [Biginelli
Condensation Utilizing Ethanol/HCl as the Solvent/Catalyst System] 19
5-Acetyl-6-methyl-4-(4-nitrophenyl)-1,2,3,4-tetrahydropyrimidin-2-one [Biginelli
Condensation Utilizing Ionic Liquids as the Reaction Medium] 19
Ethyl 6-Methyl-2-thioxo-1,2,3,4-tetrahydropyrimidine-5-carboxylate [Biginelli
Condensation Employing a Masked Aldehyde Precursor] 20
C-Galactosylpyrimidine [Biginelli Condensation Involving a Carbohydrate-Derived
Aldehyde and a β-Keto Ester Building Block] 20
Benzyl 4-(3-Methoxyphenyl)-6-methyl-2-oxo-1,2,3,4-tetrahydropyrimidine-5-carboxylate
[Biginelli Condensation Utilizing a Supported Lewis Acid Catalyst] . . . 21
Ethyl 1-(3-Carboxypropyl)-6-methyl-4-(naphthalen-2-yl)-2-oxo-1,2,3,4-tetrahydro-
pyrimidine-5-carboxylate [Biginelli Condensation Employing a Polystyrene-Supported
Urea Building Block] 21
Ethyl 1-(2-Benzoyloxyethyl)-6-methyl-4-(4-methoxyphenyl)-2-oxo-1,2,3,4-
tetrahydropyrimidine-5-carboxylate [Biginelli Condensation Employing a Fluorous
Urea Building Block 22
TABULAR SURVEY 22
Table 1. Reactions Involving Urea 24
  A. β-Keto Esters 24
  B. β-Keto Amides 62
  C. β-Diketones 65
  D. Other CH-Acidic Carbonyl Compounds 71
Table 2. Reactions Involving Substituted Ureas 73
  A. β-Keto Esters 73
  B. β-Keto Amides 82
  C. β-Diketones 83
  D. Other CH-Acidic Carbonyl Compounds 84
Table 3. Reactions Involving Thioureas 85
  A. β-Keto Esters 85
  B. β-Keto Amides 101
  C. β-Diketones 105
  D. Other CH-Acidic Carbonyl Compounds 107
Table 4. Reactions Involving Guanidine 108
Table 5. Reactions on Solid Phase 109
Table 6. Reactions in Fluorous Phase 111
REFERENCES 112

ACKNOWLEDGMENTS

The authors would like to thank the Austrian Academy of Sciences (ÖAW) and the Austrian Science Fund (FWF) for generous support of their research in this area.

INTRODUCTION

The cyclocondensation of suitable CH-acidic carbonyl compounds, aldehydes, and urea-type building blocks under acidic conditions provides multifunctionalized dihydropyrimidine derivatives (Eq. 1). The discovery of this three-component condensation process was made by Biginelli in 1893;[1] therefore, a reaction of this type is nowadays referred to as the "Biginelli reaction," "Biginelli condensation," or as the "Biginelli dihydropyrimidine synthesis." While the early examples of this cyclocondensation process typically involved a β-ketoester, aromatic aldehyde, and urea, the scope of this heterocycle synthesis has now been extended considerably by variation of all three building blocks, allowing access to a large number of multifunctionalized pyrimidine derivatives of type **1**. For this particular heterocyclic scaffold the acronym DHPM has been adopted in the literature and is also used throughout this chapter. The importance of multicomponent reactions in combinatorial chemistry has generated a renewed interest in the Biginelli reaction, and the number of publications and patents describing the synthesis of novel DHPM analogs is constantly growing.

An earlier, comprehensive review of the Biginelli reaction and of the synthetic potential of DHPMs appeared in 1993.[2] In 2000, the biological properties of DHPM derivatives were reviewed,[3] and recent trends in the Biginelli method were highlighted.[4]

In this chapter, all three-component condensations involving suitable CH-acidic carbonyl compounds, aldehydes, and urea-type building blocks following the Biginelli concept are covered. Therefore, reactions involving 1,3-diketones or nitroacetone as building blocks leading to DHPMs that follow the substitution pattern outlined in Eq. 1 are included, in contrast to earlier review articles.[2] Patents are only cited if they contain information that is otherwise not accessible.

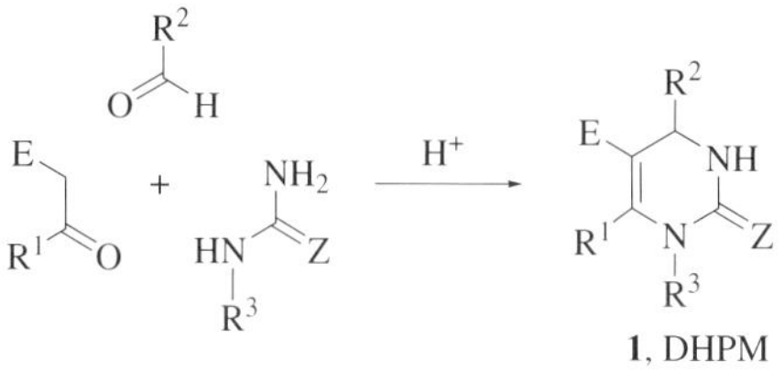

$$E = \text{ester, amide, acyl, nitro}$$
$$Z = O, S, NR$$
$$R^1\text{-}R^3 = H, \text{alkyl, (het)aryl}$$

(Eq. 1)

1, DHPM

MECHANISM

The mechanism of the Biginelli reaction has been investigated by several research groups.[5–7] Its dependence upon acidic catalysis has been experimentally established and all three possible primary reaction pathways of the three-component Biginelli system have been thoroughly studied.[5]

In the currently accepted mechanistic pathway outlined in Scheme 1,[7] the key step in the Biginelli sequence involves the acid-catalyzed formation of an *N*-acyliminium ion intermediate of type **2** from the aldehyde and urea precursors. Interception of the iminium ion **2** by the CH-acidic carbonyl component, presumably through its enol tautomer, produces an open-chain ureide **3**, which subsequently cyclizes to hexahydropyrimidine **4**. Acid-catalyzed elimination of water from **4** ultimately leads to the final DHPM product **1**. The reaction mechanism can therefore be classified as an α-amidoalkylation, or, more specifically, as an α-ureidoalkylation.[8] Consistent with this mechanistic formulation, monosubstituted ureas and thioureas furnish exclusively the *N*-1 alkylated DHPMs.[2] *N,N′*-Disubstituted ureas do not react under the reaction conditions.

Although the highly reactive *N*-acyliminium ion species **2** cannot be isolated or directly observed, evidence for the mechanism depicted in Scheme 1 is derived from the isolation of a hexahydropyrimidine analog **4** employing electron-deficient 1,3-dicarbonyl compounds (see also Eq. 13). When $R^1 = CF_3$, for example, the sequence stops at the hexahydropyrimidine stage unless forcing dehydration conditions are used.[9] In general, all DHPMs obtained by conventional Biginelli condensations are obtained as racemates (see below).

The elucidation of the mechanism of the Biginelli multicomponent reaction has prompted a renewed interest in improving the efficiency of this process. Novel catalysts, in particular Lewis acids, are now used to favor the formation and interception of the key *N*-acyliminium ion intermediates. It is proposed that these Lewis acids stabilize the *N*-acyliminium ion by coordination to the urea oxygen.[10] In some reactions, chelation of the 1,3-dicarbonyl component by suitable Lewis acids, which stabilizes the enol tautomer, has also been inferred.[10] Lewis acid conditions are discussed in detail in the Experimental Conditions section.

Scheme 1

SCOPE AND LIMITATIONS

Building Blocks

Aldehydes. Of the three building blocks in the Biginelli reaction, the aldehyde component can be varied to the greatest extent. In general, the reaction works best with aromatic aldehydes. These can be substituted in the ortho, meta, or para position with either electron-withdrawing or -donating groups. Good yields are usually obtained with meta- or para-substituted aromatic aldehydes carrying electron-withdrawing substituents. For ortho-substituted benzaldehydes having bulky substituents, yields can be significantly lower. Heterocyclic aldehydes derived from furan, thiophene, and pyridine also generally furnish acceptable yields of DHPM products.

Aliphatic aldehydes provide poor yields in the Biginelli reaction (10–40%) unless special reaction conditions are employed, i.e., Lewis acid catalysts/solvent-free methods, or using the aldehydes in protected form. For example, the 4-cyanomethyl-DHPM **5** is obtained successfully by treatment of oxazolidine-protected cyanoacetaldehyde with ethyl acetoacetate and urea (Eq. 2).[11] The C4 unsubstituted DHPM can be prepared in a similar manner employing suitable formaldehyde synthons.[11]

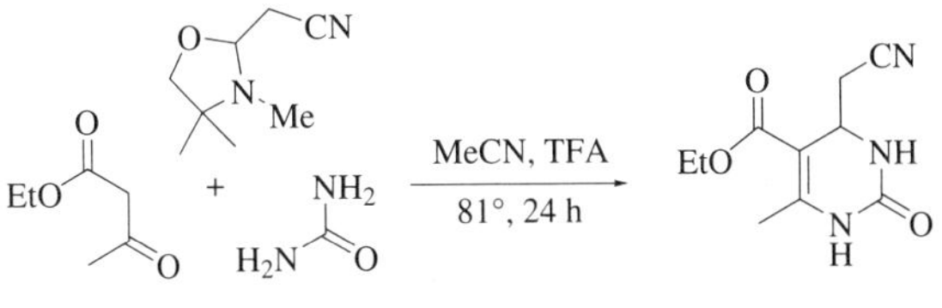

(Eq. 2)

5 (68%)

Of particular interest are reactions in which the aldehyde component is derived from a carbohydrate.[12–20] In such transformations, DHPMs having a sugar-like moiety at position 4 (C-nucleoside analogs) are obtained (Eq. 3).[20] The use of such chiral aldehydes is also of interest in the context of developing an asymmetric version of the Biginelli reaction (see below), but so far the chemical yields and diastereoselectivities that are achievable are not of practical use.

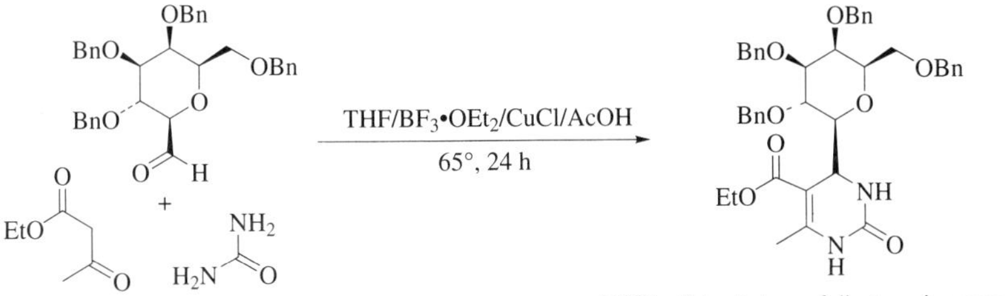

(Eq. 3)

(60%) 5:1 mixture of diastereoisomers

Bisaldehydes have been used as synthons in Biginelli reactions.[21,22] For example, the use of terephthalaldehyde under microwave irradiation provides the expected bis-DHPM product in good yields (Eq. 4).[22]

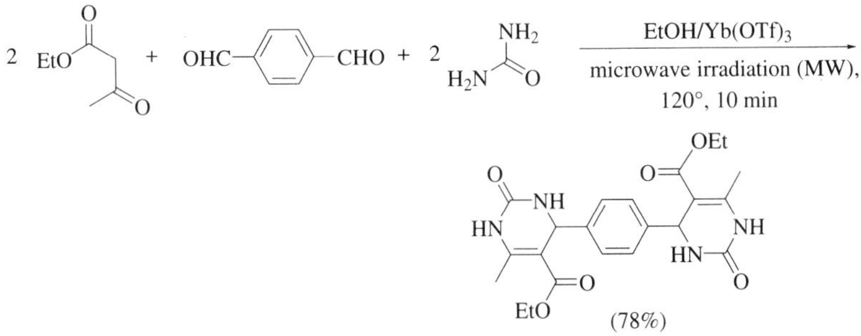

(Eq. 4)

(78%)

CH-Acidic Carbonyl Components. Traditionally, simple alkyl acetoacetates are employed as CH-acidic carbonyl building blocks, but other types of 3-oxoalkanoic esters or thioesters can also be used successfully. With ethyl 4-bromoacetoacetate, for example, the corresponding 6-bromomethyl-substituted DHPMs, which can serve as valuable templates for further synthetic transformations, are obtained (Eq. 5).[23]

Benzoylacetic esters react analogously, but yields are usually significantly lower and the overall condensation process is more sluggish. Primary, secondary, and tertiary acetoacetamides can be used in place of esters to produce pyrimidine-5-carboxamides. In addition, β-diketones serve as viable substrates in Biginelli reactions. Condensation can also be achieved using cyclic β-diketones such as cyclohexane-1,3-dione[24,25] and other cyclic β-dicarbonyl compounds (Eq. 6).[26]

For the synthesis of a C6-unsubstituted DHPM derivative, the corresponding 3-oxopropanoic ester derivative in which the aldehyde function is masked as an acetal can be employed (Eq. 7).[22,27] Apart from ester-derived CH-acidic carbonyl compounds, nitroacetone also serves as a good building block, leading to 5-nitro-substituted DHPM derivatives in generally high yields.[28]

Urea Building Blocks. The urea is the most restricted component of the Biginelli reaction in terms of allowed structural diversity. Therefore, most of the published examples involve urea itself as the building block. However, simple monosubstituted alkyl ureas generally react equally well, in a regiospecific manner, to provide good yields of N1-substituted DHPMs (Scheme 1). The N3-substituted analogs cannot be obtained by classical Biginelli condensations. Also, N,N'-disubstituted ureas do not react under Biginelli conditions. There is little published work to demonstrate that N-arylureas can take part effectively in Biginelli condensations.[29,30]

Thiourea and substituted thioureas follow the same general rules as ureas, although longer reaction times are required to achieve good conversions. Yields are typically lower compared with the corresponding urea derivatives. In a few cases, the use of unprotected guanidine has been reported in a three-component Biginelli-type condensation (Eq. 8).[31] In general, these types of cyclic guanidine derivatives need to be prepared by alternative methods (see Eq. 17).

$$\text{(Eq. 8)}$$

Solid-Phase, Fluorous-Phase, and Related Procedures

Multi-component reactions such as the Biginelli condensation, leading to interesting heterocyclic scaffolds, are particularly useful for the creation of diverse chemical libraries for biological screening. The combination of three low-molecular weight building blocks in a single operation leads to high combinatorial efficacy. As the experimental conditions for the traditional Biginelli reaction are rather straightforward, small libraries of DHPMs are readily accessible using parallel synthesis or related robotic techniques.[32–34] In addition to these conventional solution-phase methods for preparing DHPM libraries, it is also possible to employ polymer-supported reagents to aid in the product purification and reaction work-up. The use of a polymer-supported Lewis acid [Yb(III)-reagent supported on Amberlyst 15] in combination with polymer-supported scavenging resins (Amberlyst 15 and Ambersep 900 OH) that remove excess urea allows for a rapid parallel Biginelli synthesis with a simplified purification strategy.[35]

An even higher degree of throughput and automation is possible with solid-phase protocols (Merrifield type synthesis). In one example, a γ-aminobutyric acid derived urea is attached to a Wang resin using standard procedures. The resulting polymer-bound urea can be condensed with excess β-ketoester and aromatic aldehydes in the presence of a catalytic amount of hydrochloric acid to afford the corresponding immobilized DHPMs. Subsequent cleavage of product from the polystyrene resin by trifluoroacetic acid provides the free DHPMs in high yields and excellent purities (Eq. 9).[36]

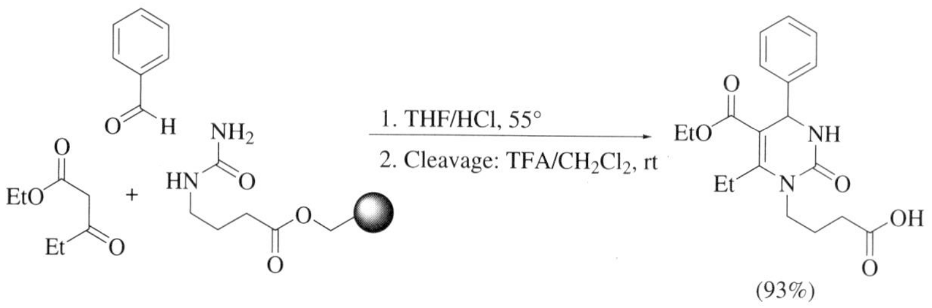

(Eq. 9)

In a variation of the above procedure, the Biginelli synthesis is adapted to fluorous-phase conditions. Here a fluorous urea derivative is prepared by attaching a suitable fluorous tag to hydroxyethylurea (Eq. 10).[37,38] The fluorous urea is then condensed with an excess of acetoacetates and aldehydes in a suitable solvent containing hydrochloric acid. After extraction of the fluorous DHPMs with a fluorous solvent (perfluorohexanes, FC-72), desilylation with tetrabutylammonium fluoride (TBAF) followed by extractive purification provides the "organic" Biginelli products in good overall yields.

In contrast to the methods described previously where the urea component is linked to the solid (or fluorous) support via the amide nitrogen, it is also possible to attach the acetoacetate building block to the solid support. Thus, Biginelli condensation of Wang-bound acetoacetates with excess aldehydes and urea/thiourea provides DHPMs on solid support. Subsequent cleavage with trifluoroacetic acid furnishes the free carboxylic acids in high overall yields (Eq. 11).[39]

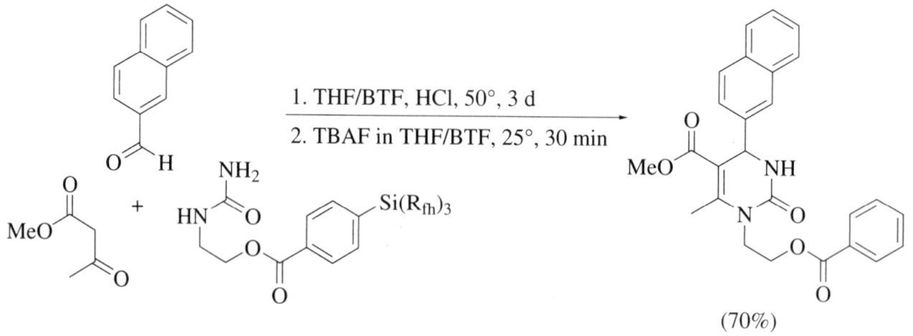

$R_{fh} = C_{10}F_{21}CH_2CH_2-$

BTF = benzotrifluoride, TBAF = tetrabutylammonium fluoride

(Eq. 10)

(Eq. 11)

There are a number of alternative solid-phase procedures described in the literature for the generation of DHPMs that use the so-called Atwal modification instead of the classical three-component Biginelli approach (see Eq. 17).[40,41] By employing any of the solid-phase synthesis methods described above, libraries of DHPMs can be generated in a relatively straightforward fashion.

Asymmetric Biginelli Synthesis

Since dihydropyrimidines (DHPMs) are inherently asymmetric molecules they are generally obtained as racemic mixtures in the traditional Biginelli reaction as well as in related processes (see below). The dramatic influence of the absolute configuration at the stereogenic center at C4 of the pyrimidine ring on the biological activity of some DHPMs is well documented.[27,42–46] Therefore, access to enantiomerically pure DHPMs is of interest. In the absence of any known general asymmetric synthesis for this heterocyclic system, resolution strategies are the methods of choice to obtain enantiomerically pure DHPM analogs. These methods include fractional crystallization techniques involving diastereomeric α-methylbenzylammonium salts[47] or covalently linked derivatives,[27,42,43,46] or rely on biocatalytic resolution.[48,49] Analytically, separation of DHPM derivatives can be readily achieved by enantioselective HPLC using a variety of different chiral stationary phases.[32,33,50] Alternatively, chiral separation can also be performed by capillary electrophoresis (CE) with chiral modifiers and buffers.[51,52] The absolute configuration of enantiomerically pure DHPMs is easily derived from circular dichroism (CD) spectra.[53,54]

Efforts to develop a practical asymmetric version of the Biginelli reaction itself have so far failed. Although chiral acetoacetates, e.g., (−)-menthyl acetoacetate, show no diastereoselectivity,[47] chiral aldehydes derived from carbohydrates can induce chirality at C4 of the pyrimidine ring (see Eq. 3). This latter approach, however, is of limited use because only moderate chemical yields and selectivities are achieved.[20] Also, the method does not allow for the preparation of the important class of 4-aryl-substituted DHPMs. Therefore, a general asymmetric access to DHPMs of the Biginelli type remains a highly desirable goal. The only known asymmetric variations of the Biginelli reaction are of an intramolecular nature and have been applied to the synthesis of natural products (see Eq. 12).

Intramolecular (Tethered) Biginelli Reactions

A special variant of the Biginelli reaction are intramolecular, or so-called tethered, Biginelli condensations where the aldehyde and urea components are linked together in one building block (Eq. 12). The "tethered Biginelli strategy" has been used in the synthesis of various polycyclic guanidinium alkaloids (e.g., batzelladine alkaloids A and D) that all have the hexahydropyrrolo[1,2-c]pyrimidine fragment **7** in common. For example, condensation of the chiral hemiaminal precursor **6** with a suitable β-ketoester leads to the desired hexahydropyrrolo[1,2-c]pyrimidine scaffold.[55] Importantly, depending on the reaction conditions (A or B), both the syn and anti stereoisomers of **7** can be obtained with high selectivities. When typical Knoevenagel conditions (morpholinium acetate) are used, cis stereoselection (4–7:1) is observed. In contrast, when the condensation is carried out in the presence of polyphosphate ester, trans stereoselection (4–20:1) is found.[55] As both types of stereoisomers are present in several alkaloids, this method provides an elegant way for the enantioselective total synthesis of natural products of this type. Tethered Biginelli strategies have been used to synthesize batzelladine B,[56] batzelladine D,[57] ptilomycalin A,[58,59] 13,14,15-isocrambescidin 800,[60,61] 13,14,15-isocrambescidin 657,[60] crambescidin 657,[59] crambescidin 800[59], and batzelladine F.[62]

(Eq. 12)

Related Biginelli-Type Reactions

For 1,3-dicarbonyl building blocks having a strong electron-withdrawing substituent such as a trifluoromethyl group, the Biginelli sequence generally stops at the hexahydropyrimidine stage (see Scheme 1).[9,63] In fact, a variety of hexahydropyrimidines can be synthesized in this way using perfluorinated 1,3-dicarbonyl compounds or β-ketoesters as building blocks (Eq. 13).[63] The stereochemistry of hexahydropyrimidine **8** was confirmed by an X-ray analysis.[9]

$$\text{(Eq. 13)}$$

8 (70%)

The steric proximity of a hydroxy substituent in the ortho position of the aromatic ring and the C6 carbon of the pyrimidine ring in DHPMs enables the formation of a six-membered ring via intramolecular Michael addition. For example, with aromatic aldehydes such as salicylaldehyde, the expected product of a Biginelli condensation is not a simple DHPM but rather an 8-oxa-10,12-diazatricyclo[7.3.1.0^{2,7}]tridecatriene derivative (Eq. 14).[64] Several examples of these unusual domino Biginelli condensation/Michael addition sequences are known.[65,66]

$$\text{(Eq. 14)}$$

(68%)

As mentioned previously (Eq. 6), cyclic β-diketones such as cyclohexane-1,3-dione and other cyclic β-dicarbonyl compounds are known to function well in the Biginelli condensation. However, for tetronic acid the reaction takes an entirely different course, following a pseudo four-component pathway to give spiro heterobicyclic products in good yields (Eq. 15).[67] The reaction proceeds by a regiospecific condensation of two molecules of aldehyde with the other reagents to afford products having the C4 and C6 substituents exclusively in cis configuration. The classical Biginelli product was not detected.

$$\text{(Eq. 15)}$$

Another interesting variation of the standard Biginelli reaction involves the use of β-keto carboxylic acids as CH-acidic carbonyl compounds. Under suitable reaction conditions, oxalacetic acid is an excellent substrate in such condensations.[68] Cyclization and in situ decarboxylation cleanly yield 5-unsubstituted 3,4-dihydropyrimidin-2(1H)-ones. By using trifluoroacetic acid (TFA) as the acidic catalyst and dichloroethane (DCE) as the solvent, excellent yields of products can be obtained (Eq. 16).

$$\text{(Eq. 16)}$$

COMPARISON WITH OTHER METHODS

Apart from the traditional Biginelli three-component condensation there are only a few other efficient synthetic methods available to prepare DHPM products. Since most of these syntheses lack the experimental and conceptual simplicity of the one-pot, one-step Biginelli procedure, few of those have any real significance today. One noticeable exception is the so-called "Atwal modification" of the Biginelli reaction.[69-71] An enone of type **9**, derived from the CH-acidic carbonyl component and the aldehyde component by Knoevenagel condensation, is first reacted with a suitable protected urea or thiourea derivative **10a,b** under almost neutral conditions. Deprotection of the resulting 1,4-dihydropyrimidine **11** using suitable reagents leads to the desired DHPMs (Eq. 17). Similar results are obtained when enone **9** is condensed with guanidine derivatives to give 2-amino-substituted pyrimidines. Although these methods require prior synthesis of enones **9**, their general reliability and broad applicability make them an attractive alternative to the traditional one-step Biginelli condensation. In some cases, a direct three-component condensation of the protected urea/thiourea with the CH-acidic carbonyl compound and the aldehyde component is also possible.[72,73] In addition, 1,4-dihydropyrimidines **11** can be acylated regiospecifically at N3, thereby making those pharmacologically relevant analogs available.[71] This sequence is important because direct regiospecific acylation or alkylation of

DHPMs is troublesome.[2] Most other procedures that lead to Biginelli-type products are very limited in their scope and are rarely used for synthetic purposes.[2,74]

$$\underset{9}{\underset{R^1}{\overset{R^2}{\text{E}}}\diagup\!\!\!\diagdown\text{O}} + \underset{10a,b}{\text{HN}\diagup\!\!\!\diagdown\text{ZR}}^{\text{NH}_2} \xrightarrow{\text{mild base}} \underset{11}{\text{(intermediate)}} \xrightarrow{\text{deprotection}} \text{product} \quad \text{(Eq. 17)}$$

a: Z = O, R = Me
b: Z = S, R = 4-methoxybenzyl

EXPERIMENTAL CONDITIONS

There is a great variety of suitable reaction conditions for carrying out Biginelli condensations. For the condensation of ethyl acetoacetate with benzaldehyde and urea, more than 40 different experimental conditions are now known (see Tabular Survey). Traditionally, Biginelli condensations are carried out in a solvent such as ethanol or methanol, but more recently aprotic solvents such as tetrahydrofuran,[10,36,75] dioxane,[39] or acetonitrile[76] have also been used successfully. In some reactions, it is necessary to use acetic acid as a solvent.[77–79] This is particularly important where condensation of an aromatic aldehyde and urea leads to precipitation of an insoluble bisureide derivative, i.e., $ArCH(NHCONH_2)_2$,[7] which might not react further along the desired pathway outlined in Scheme 1 when ethanol is used as a solvent. Biginelli reactions in water[80] and ionic liquids[81] are also known. A recent trend is to perform the condensation without any solvent, with the components either adsorbed on an inorganic support,[82] or in the presence of a suitable catalyst.[34,83]

The Biginelli condensation is strongly dependent on the amount of acidic catalyst present in the reaction medium.[5] Traditionally, strong Brønsted acids such as hydrochloric or sulfuric acid have been employed,[2] but now the use of Lewis acids such as $BF_3 \cdot OEt_2$ and $CuCl$,[10] $LaCl_3$,[84] $FeCl_3$,[85–87] $NiCl_2$,[88] $Yb(OTf)_3$,[83] $La(OTf)_3$,[89] $InCl_3$,[90] $InBr_3$,[91] $BiCl_3$,[76] $LiClO_4$,[92] $Mn(OAc)_3$,[93] or $ZrCl_4$[94] is preferred. It is also possible to use a solid acid catalyst, such as an acidic clay,[82] a zeolite,[95] or Amberlyst material.[79] In addition, amidosulfonic acid has been utilized as catalyst.[96]

Biginelli condensations generally proceed rather slowly at room temperature.[7] Therefore, it is necessary to activate these processes by heating. Apart from traditional heating methods, microwave heating employing some of the solvent/catalyst systems mentioned above has been used to shorten reaction times significantly.[22,30,34,79,86,97–102] It is also feasible to carry out Biginelli reactions using ultrasound activation.[103]

With regard to the molar ratio of the building blocks, Biginelli reactions generally employ an excess of the CH-acidic carbonyl or the urea components, rather than an excess of the aldehyde. As DHPM products are usually only sparingly soluble in solvents such as methanol or ethanol at room temperature, work-up in many cases simply involves isolation of the product by filtration. It may also be possible to precipitate the product by addition of water.

EXPERIMENTAL PROCEDURES

Ethyl 6-Methyl-2-oxo-4-phenyl-1,2,3,4-tetrahydropyrimidine-5-carboxylate [Biginelli Condensation Utilizing Ethanol/HCl as the Solvent/Catalyst System].[78] A mixture of 0.5 mol (53 g) of benzaldehyde, 0.5 mol (30 g) of urea, 0.75 mol (97.5 g) of ethyl acetoacetate, 200 mL EtOH, and 40 drops of concentrated HCl was heated under reflux for 3 hours. The reaction mixture was then cooled to 0°, and the pyrimidine was collected by filtration and dried at 50°, yielding 93.6 g of crude product. The filtrate was refluxed for an additional 2 hours and finally distilled until 155 mL of EtOH were collected. On cooling the residue an additional 21.3 g of the pyrimidine was isolated, resulting in a total yield of crude product of 88.4%. For purification, the substance was divided into two equal portions, and each was dissolved in 800 mL of 95% boiling alcohol. On cooling, the pyrimidine separated as colorless crystals, mp 202–204°. The yield of purified material was 102 g (78%); IR[22] (KBr) 3240, 3110, 1725, 1700, 1645 cm^{-1}; ^{1}H NMR[22] (200 MHz, DMSO-d_6) δ 1.12 (t, J = 7.5 Hz, 3H), 2.28 (s, 3H), 4.03 (q, J = 7.5 Hz, 2H), 5.17 (d, J = 3.5 Hz, 1H), 7.22–7.41 (m, 5H), 7.78 (br s, 1H), 9.22 (br s, 1H).

Isopropyl 4-(2-Bromo-5-nitrophenyl)-6-methyl-2-oxo-1,2,3,4-tetrahydropyrimidine-5-carboxylate [Biginelli Condensation Utilizing Acetic Acid/HCl as the Solvent/Catalyst System].[77] A mixture of 16 mmol (2.30 g) of isopropyl acetoacetate, 11 mmol (2.43 g) of 2-bromo-5-nitrobenzaldehyde, 11 mmol (0.95 g) of urea, and acetic acid (20 mL) containing 100 µL (ca. 4 drops) of concentrated HCl was heated under reflux for 24 hours; after the first 6 hours an additional quantity of concentrated HCl (100 µL) was added. After the mixture remained at room temperature overnight, the precipitate was collected by filtration and recrystallized from acetic acid to give 2.89 g (66%) of the desired pyrimidine as a colorless solid: mp 241°; IR (KBr) 3380, 3080, 2950, 1710, 1650, 1520, 1450 cm^{-1}; ^{1}H NMR (DMSO-d_6), δ 0.79, 1.12 (2d, J = 6.4 Hz, 6H), 2.36 (s, 3H), 4.75 (m, J = 6.4 Hz,

1H), 5.70 (s, 1H), 7.80–8.10 (m, 4H), 9.45 (br s, 1H); Anal. Calcd for C$_{15}$H$_{16}$BrN$_3$O$_5$: C, 45.20; H, 4.00; N, 10.60. Found: C, 45.58; H, 4.00; N, 10.24.

Ethyl 1-Benzyl-6-methyl-2-oxo-4-phenyl-1,2,3,4-tetrahydropyrimidine-5-carboxylate [Biginelli Condensation Utilizing THF/Polyphosphate Ester as the Solvent/Catalyst System].[104] A mixture of 2 mmol (260 mg) of ethyl acetoacetate, 2 mmol (212 mg) of benzaldehyde, 3 mmol (450 mg) of benzylurea, and THF (4 mL) containing 300 mg of polyphosphate ester (PPE) was heated under reflux and stirring for 15 hours. After cooling, the reaction mixture was poured onto crushed ice (10 g). Stirring was continued for several hours. The solid product was collected by filtration, washed with ice water, and dried, yielding 637 mg (91%), mp 157° (EtOH); IR (KBr) 3210, 3100, 1710, 1690, 1620 cm^{-1}; ^{1}H NMR (DMSO-d_6) δ 1.11 (t, J = 7.5 Hz, 3H), 2.38 (s, 3H), 4.03 (q, J = 7.5 Hz, 2H), 4.88, 5.11 (2d, J = 17.5 Hz, 2H), 5.25 (d, J = 3.0 Hz, 1H), 7.05–7.39 (m, 10H), 8.13 (d, J = 3.0 Hz, 1H); Anal. Calcd for C$_{21}$H$_{22}$N$_2$O$_3$: C, 71.98; H, 6.33; N, 7.99. Found: C, 72.30; H, 6.52; N, 7.95.

Methyl 4-(3,4-Difluorophenyl)-6-ethyl-2-oxo-1,2,3,4-tetrahydropyrimidine-5-carboxylate [Biginelli Condensation Utilizing Lewis Acids as the Catalyst System].[10] A 50-mL three-neck round-bottom flask fitted with a thermocouple and reflux condenser was charged under N$_2$ with sieve-dried THF (30 mL), 15.4 mmol (2.00 g) of methyl 3-oxopentanoate, 15.4 mmol (2.19 g) of 3,4-difluorobenzaldehyde, 23.1 mmol (1.39 g) of urea, 20.0 mmol (2.84 g) of BF$_3$·OEt$_2$, 1.54 mmol (0.15 g) of CuCl, and 1.54 mmol (0.1 mL) of glacial acetic acid. The mixture was heated to reflux (65°) for 8–18 hours. The solution was cooled to room temperature and quenched with 10% Na$_2$CO$_3$ solution (30 mL). Ethyl acetate

(30 mL) was added, the layers were separated, and the green aqueous solution was discarded. The organic layer was concentrated and the residue was dissolved in toluene (40 mL), cooled to room temperature, and left overnight. The resulting suspension was filtered, and the collected solid was rinsed with toluene (1 × 10 mL), and dried in vacuo at 40° to afford the desired pyrimidine as a crystalline solid in 82% yield (3.74 g), mp 182–185°; IR (KBr) 3237, 3112, 2960, 2879, 1701, 1634, 1517 cm^{-1}; ^{1}H NMR (DMSO-d_6) δ 1.11 (t, J = 7.4 Hz, 3H), 2.65 (m, 2H), 3.54 (s, 3H), 5.14 (d, J = 3.4 Hz, 1H), 6.61–6.69 (m, 3H), 7.06 (m, 1H), 7.20 (m, 1H), 7.40 (m, 1H), 7.80 (s, 1H), 9.31 (s, 1H); ^{13}C NMR (DMSO-d_6) δ 12.8, 24.0, 50.8, 52.8, 97.4, 114.9, 115.2, 117.4, 117.6, 142.2, 152.0, 154.8, 165.1; Anal. Calcd for $C_{14}H_{14}F_2N_2O_3$: C, 56.76; H, 4.76; N, 9.46. Found: C, 56.85; H, 4.70; N, 9.35.

Ethyl 4-(4-Hydroxy-3-methoxyphenyl)-6-methyl-2-oxo-1,2,3,4-tetrahydropyrimidine-5-carboxylate [Biginelli Condensation Utilizing Lewis Acids as the Catalyst System].[85] A solution of 50 mmol (6.50 g) of ethyl acetoacetate, 50 mmol (7.60 g) of vanillin (4-hydroxy-3-methoxy-benzaldehyde), 75 mmol (4.55 g) of urea, 30 mmol (8.1 g) of $FeCl_3 \cdot 6H_2O$, and 2–3 drops of concentrated HCl in EtOH (40 mL) was heated under reflux for 4 hours. The solution was cooled to room temperature and poured onto 200 g of crushed ice. Stirring was continued for several minutes. The solid product was collected by filtration, washed with ice-water and a mixture of EtOH/water (1:1), dried, and crystallized from hot EtOH, yielding 13.2 g (86%) of product.

Ethyl 4-(3-Hydroxyphenyl)-6-methyl-2-thioxo-1,2,3,4-tetrahydropyrimidine-5-carboxylate [Microwave-Assisted Biginelli Condensation under Solvent-Free Conditions].[105] A mixture of 2.3 mmol (300 mg) of ethyl acetoacetate, 2.0 mmol (244 mg) of 3-hydroxybenzaldehyde, 5 mmol (380 mg) of thiourea, and 300 mg of polyphosphate ester (PPE) was placed in a 20-mL glass beaker. The mixture was stirred for 10–20 seconds with a spatula, after which the beaker was inserted into a bed of neutral alumina (150 g) contained in a 400-mL Pyrex beaker. This set-up was irradiated in a domestic microwave oven 5 times at full power (800 W) for 10 seconds each with 1–2 minutes cooling periods between each irradiation cycle. EtOH (5 mL) was added to the hot reaction mixture, which was subsequently poured onto ice (50 g). The precipitated crude product was purified by silica gel chromatography (hexane/EtOAc 2:1) to yield 350 mg (60%) of colorless product, mp 184–186° (CH_3CN); IR (KBr) 3300, 3180, 2900–2600, 1670, 1655, 1620, 1575 cm^{-1}; ^1H NMR (200 MHz, DMSO-d_6) δ 1.14 (t, J = 7.5 Hz, 3H), 2.30 (s, 3H), 4.03 (q, J = 7.5 Hz, 2H), 5.11 (d, J = 3.5 Hz, 1H), 6.61–6.69 (m, 3H), 7.06–7.17 (m, 1H), 9.45, 9.62, 10.31 (3 br s); ^{13}C NMR (DMSO-d_6) δ 14.0, 17.2, 54.0, 59.6, 100.8, 113.3, 114.6, 117.0, 129.5, 144.8, 144.9, 157.5, 165.2, 174.2; Anal. Calcd for $C_{14}H_{16}N_2O_3S$: C, 57.52; H, 5.52; N, 9.58. Found: C, 57.33; H, 5.52; N, 9.34.

Methyl 4-(3,4-Difluorophenyl)-2-oxo-1,2,3,4-tetrahydropyrimidine-5-carboxylate [Microwave-Assisted Biginelli Condensation Utilizing Lewis Acids as the Catalyst System].[22] To a 5-mL reaction vial were added 4 mmol (592 mg) of methyl 3,3-dimethoxypropionate, 4 mmol (568 mg) of 3,4-difluorobenzaldehyde, 4 mmol (240 mg) of urea, 0.4 mmol (250 mg) of Yb(OTf)$_3$, and 2 mL of EtOH. The vessel was heated in a EmrysSynthesizer™ microwave reactor at 120° for 20 minutes, and subsequently cooled. The vessel was removed from the cavity and placed in the refrigerator for approximately 1 hour. The precipitate formed was collected by suction filtration, washed with ice-cold EtOH, and dried, yielding 377 mg (35%) of the pyrimidine, mp 229–231° (EtOH); ^1H NMR (360 MHz, DMSO-d_6) δ 3.56 (s, 3H), 5.15 (d, J = 3.0 Hz, 1H), 7.11 (br s, 1H), 7.22–7.45 (m, 3H), 7.75 (s, 1H), 9.28 (br s, 1H). Anal. Calcd for $C_{12}H_{10}F_2N_2O_3$: C, 53.74; H, 3.76; N, 10.44. Found: C, 53.36; H, 3.81; N, 10.62.

4-Phenyl-5-(thiophene-2-carbonyl)-6-trifluoromethyl-1,2,3,4-tetrahydropyrimidin-2-one [Biginelli Condensation Utilizing Lewis Acids under Solvent-Free Conditions].[83] Two and one-half mmol (555.5 mg) of 4,4,4-trifluoro-1-(thiophen-2-yl)-butane-1,3-dione, 2.5 mmol (265.3 mg) of benzaldehyde, 3.7 mmol (222 mg) of urea, and 0.125 mmol (5 mol%) of Yb(III)-triflate were heated at 100° for 20 minutes under slight stirring. Water was added, and the product was extracted with EtOAc. After the organic layer was dried (Na$_2$SO$_4$) and evaporated, the residue was recrystallized from EtOAc/hexane to afford 828 mg (94%) of the dihydropyrimidine, mp 99–102°; IR (KBr) 3200, 1670 cm^{-1}; ^{1}H NMR δ 5.25 (s, 1H), 6.61–6.69 (m, 3H), 7.11–7.34 (m, 5H), 7.52 (d, J = 3.8 Hz, 1H), 7.90 (s, 1H), 8.03 (d, J = 5.0 Hz, 1H), 9.79 (s, 1H).

Ethyl 4-Butyl-6-methyl-2-oxo-1,2,3,4-tetrahydropyrimidine-5-carboxylate [Biginelli Condensation Utilizing KSF-Clay as the Catalyst System].[82] A mixture of 0.5 g montmorillonite KSF-clay, 10 mmol (1.30 g) of ethyl acetoacetate, 10 mmol (0.86 g) of pentanal, and 15 mmol (0.90 g) of urea was heated at 130° with stirring for 48 hours. Hot MeOH (100 mL) was added and the mixture was filtered to remove the catalyst. The product crystallized after several hours and was recovered by filtration, yielding 2.05 g (86%) of pyrimidine as a white solid, mp 161–162°; IR (KBr) 3244, 3117, 1727, 1707, 1653 cm^{-1}; ^{1}H NMR (300 MHz, DMSO-d_6) δ 0.85 (t, J = 6.4 Hz, 3H), 1.18 (t, J = 7.1 Hz, 3H), 1.1–1.3 (m, 4H), 1.3–1.5 (m, 2H), 2.16 (s, 3H), 4.05 (q, J = 7.1 Hz, 2H), 4.1–4.2 (m, 1H), 7.28 (s, 1H), 8.89 (s, 1H); Anal. Calcd for C$_{12}$H$_{20}$N$_2$O$_3$: C, 60.0; H, 8.4; N, 11.7. Found: C, 60.1; H, 8.5; N, 11.9.

Diethyl 4-(4-Bromophenyl)-6-methyl-2-oxo-1,2,3,4-tetrahydropyrimidine-5-carboxamide [Biginelli Condensation Utilizing Ethanol/HCl as the Solvent/Catalyst System].[106] A mixture of 20 mmol (3.7 g) of 4-bromobenzaldehyde, 20 mmol (1.2 g) of urea, and 15 drops of concentrated HCl was added to a solution of 20 mmol (3.14 g) of N,N-diethylacetoacetamide in absolute EtOH (50 mL). The mixture was heated under reflux for 5 hours, and then cooled to yield 4.75 g (65%) of a colorless crystalline precipitate, mp 227–229° (butanol); ^{1}H NMR (360 MHz, DMSO-d_6) δ 1.05 (t, J = 7.0 Hz, 6H), 1.64 (s, 3H), 2.30 (s, 3H), 3.05 (q, J = 7.0 Hz, 4H), 4.99 (d, J = 3.0 Hz, 1H), 7.15 (d, J = 8.0 Hz, 2H), 7.32 (br s, 1H), 7.51 (d, J = 8.0 Hz, 2H), 8.48 (br s, 1H); Anal. Calcd for $C_{16}H_{20}BrN_3O_2$: C, 52.47; H, 5.50; N, 11.47. Found: C, 52.1; H, 5.7; N, 11.1.

6-Methyl-5-nitro-4-(3,4,5-trimethoxyphenyl)-1,2,3,4-tetrahydropyrimidin-2-one [Biginelli Condensation Utilizing Ethanol/HCl as the Solvent/Catalyst System].[107]

A mixture of 56.5 mmol (5.82 g) of nitroacetone, 56.5 mmol (11.08 g) of 3,4,5-trimethoxybenzaldehyde, and 113 mmol (6.78 g) of urea in absolute EtOH (100 mL) containing 20 drops of concentrated HCl was heated to reflux for 6 hours. The precipitate formed was collected by filtration, dried, and recrystallized from EtOH, yielding 14.6 g (80%) of the desired pyrimidine, mp 243–245°; ^{1}H NMR δ 2.82 (s, 3H), 3.87 (s, 3H), 3.97 (s, 6H), 5.84 (d, J = 3.5 Hz, 1H), 6.87 (s, 2H), 8.36 (d, 1H), 10.15 (s, 1H).

5-Acetyl-6-methyl-4-(4-nitrophenyl)-1,2,3,4-tetrahydropyrimidin-2-one [Biginelli Condensation Utilizing Ionic Liquids as the Reaction Medium].[81]

Twenty-five mmol (3.78 g) of 4-nitrobenzaldehyde, 25.0 mmol (2.50 g) of acetylacetone, 37.5 mmol (2.25 g) of urea, and 0.1 mmol (22.6 mg) of 1-butyl-3-methylimidazoliumtetrafluoroborate (BMImBF$_4$, ionic liquid) were successively charged into a 50-mL round-bottomed flask equipped with a magnetic stirring bar. The reaction proceeded at 100° for 30 minutes, during which time a solid product gradually formed. After the reaction was complete, the resulting pale yellow solid product was crushed, washed with water, collected by filtration, and dried in vacuo to afford the crude product in 92% yield (6.33 g). A pure product was obtained by further crystallization of the crude product from EtOAc.

Ethyl 6-Methyl-2-thioxo-1,2,3,4-tetrahydropyrimidine-5-carboxylate [Biginelli Condensation Employing a Masked Aldehyde Precursor].[11] A solution of 10 mmol (0.87 g) of 1,3-oxazinane, 10 mmol (1.30 g) of ethyl acetoacetate, and 10 mmol (0.72 g) of thiourea in anhydrous CH_3CN (30–40 mL) containing trifluoroacetic acid (TFA) (0.5 mL) was heated at reflux until the reaction was completed (TLC). The reaction mixture was made basic with cold aqueous sodium carbonate solution and extracted with $CHCl_3$ (3 × 50 mL). The extract was washed with cold water (2 × 50 mL) and dried (Na_2SO_4). The solvent was removed and the residue was crystallized to afford 1.86 g (93%) of the desired pyrimidine, mp 236°; IR (KBr) 3180, 1715 cm^{-1}; ^{1}H NMR (CDCl$_3$) δ 1.25 (t, J = 7.0 Hz, 3H), 2.23 (s, 3H), 3.98 (s, 2H), 4.11 (q, J = 7.0 Hz, 2H), 6.68 (br s, 1H), 9.58 (br s, 1H); ^{13}C NMR (CDCl$_3$) δ 13.2, 16.1, 40.2, 58.6, 94.8, 144.2, 164.3, 175.2; Anal. Calcd for $C_8H_{12}N_2O_2S$: C, 48.00; H, 6.00; N, 14.00. Found: C, 48.32; H, 5.82; N, 13.87.

C-Galactosylpyrimidine [Biginelli Condensation Involving a Carbohydrate-Derived Aldehyde and a β-Ketoester Building Block].[20] The cyclocondensation of 1.0 equivalent of ethyl 3-(3,4-bis-(benzyloxy)-5-benzyloxymethyltetrahydrofuranyl)-3-oxo-propionate, 1.0 equivalent of 3,4,5-tris-(benzyloxy)-6-benzyloxymethyltetrahydropyran-2-carboxaldehyde, and 1.5 equivalents of urea in THF (previously dried over molecular sieves) at 65° was promoted by CuCl (1.0 equivalent), $BF_3 \cdot OEt_2$ (1.3 equivalents), and acetic acid (0.2 equivalents) to give, after 24 hours, the C-galactosylpyrimidine as a mixture of diastereomers in 35% overall yield.

Benzyl 4-(3-Methoxyphenyl)-6-methyl-2-oxo-1,2,3,4-tetrahydropyrimidine-5-carboxylate [Biginelli Condensation Utilizing a Supported Lewis Acid Catalyst].[35] A screw-capped vial, equipped with a magnetic stirring bar, was charged first with 160 mg of Yb-(III)-resin [Yb(III)-reagent supported on Amberlyst 15], then with 1.5 mmol (90 mg) of urea, 0.5 mmol (68.3 mg) of 3-methoxybenzaldehyde, and 0.5 mmol (96.1 mg) of benzyl acetoacetate and heated at 120° for 5 minutes. Then another 170 mg of Yb(III)-resin was added and the reaction mixture was heated at 120° under gentle stirring for 48 hours. After cooling to 60°, MeOH (1 mL) was added. The suspension was stirred for additional 30 minutes, after which the resin was removed by filtration and washed thoroughly with EtOAc. Amberlyst 15 (400 mg) and Ambersep 900 OH (400 mg) were added to the combined filtrates. After the suspension was shaken for 2 hours, the resins were removed by filtration and washed thoroughly with MeOH. The combined filtrates were concentrated to give the desired pyrimidine in 70% yield.

Ethyl 1-(3-Carboxypropyl)-6-methyl-4-(naphthalen-2-yl)-2-oxo-1,2,3,4-tetrahydropyrimidine-5-carboxylate [Biginelli Condensation Employing a Polystyrene-Supported Urea Building Block].[36] A suspension of 0.048 mmol (50 mg) of polymer-bound urea (modified Wang-resin) in 1.5 mL of THF in a dram vial was treated with 4 equivalents (0.192 mmol, 30 mg) of 2-naphthaldehyde, 4 equivalents (0.192 mmol, 25 mg) of ethyl acetoacetate, and 50 µL of a 4:1 THF/concentrated HCl solution. The reaction mixture was stirred at 55° for 36 hours, and the resin was collected by filtration and washed with THF (3 × 5 mL), hexanes (3 × 5 mL), MeOH (3 × 5 mL), and CH_2Cl_2 (3 × 5 mL). Subsequently, the resin was washed with 3 mL of TFA, followed by 2 × 3 mL of CH$_2$Cl$_2$. The latter filtrate was concentrated in vacuo, yielding 16.5 mg (87%) of the pyrimidine.

Ethyl 1-(2-Benzoyloxyethyl)-6-methyl-4-(4-methoxyphenyl)-2-oxo-1,2,3,4-tetrahydropyrimidine-5-carboxylate [Biginelli Condensation Employing a Fluorous Urea Building Block].[38] A solution of 9.6 μmol (18 mg) of (2-((4-(tris(2-(perfluorodecyl)ethyl)silyl)benzoyl)oxy)ethyl)urea in 0.75 mL of THF/benzotrifluoride (BTF) (2:1) was treated at 25° with 96 μmol (12.5 mg) of ethyl acetoacetate, 96 μmol (13.1 mg) of 4-methoxybenzaldehyde, and 1 μL of concentrated HCl. After 3 days at 50°, the volatiles were removed in vacuo and FC-84 (fluorocarbon liquid, containing isomers of C$_7$F$_{16}$, bp 80°) and toluene (10 mL each) were added. The toluene phase was extracted with FC-84 (5 × 5 mL). The combined fluorous phases were filtered and concentrated. The resulting white solid was treated with 0.5 mL of THF/BTF (1:1) and then dropwise with a 1 M tributylammonium fluoride (TBAF) solution in THF (10 μL, 10 μmol). After the mixture was stirred for 0.5 hours at 25°, the volatiles were removed in vacuo and FC-84 and toluene (10 mL each) were added. The fluorous phase was extracted with toluene (3 × 5 mL). The combined toluene phases were extracted with saturated aqueous NaHCO$_3$ solution (3× 10 mL) and brine (3 × 10 mL), dried (Na$_2$SO$_4$), filtered, and concentrated, yielding 2.9 mg (6.6 μmol, 69%) of the desired pyrimidine, mp 75°; IR (CHCl$_3$) 3425, 3025, 1704, 1677, 1621, 1514 cm^{-1}; ^{1}H NMR (CDCl$_3$) δ 1.19 (t, J = 7.1 Hz, 3H), 2.60 (s, 3H), 3.66 (s, 3H), 3.95–4.15 (m, 3H), 4.40–4.50 (m, 3H), 5.34 (br s, 2H), 6.62 (d, J = 8.7 Hz, 2H), 7.14 (d, J = 8.7 Hz, 2H), 7.41 (t, J = 7.6 Hz, 2H), 7.56 (t, J = 7.6 Hz, 1H), 7.96 (d, J = 7.1 Hz, 2H); ^{13}C NMR (CDCl$_3$) δ 14.3, 16.5, 40.9, 53.3, 55.1, 60.3, 63.6, 105.8, 113.9, 127.4, 128.5, 129.7, 133.2, 135.5, 147.9, 153.9, 159.0, 166.2, 166.4.

TABULAR SURVEY

The tabular survey in this chapter covers all examples of the classical Biginelli three-component cyclocondensation reported in the literature from 1932 through December of 2001. Not covered are Biginelli reactions that do not lead to the expected dihydropyrimidine products, intramolecular (tethered) Biginelli reactions, and modifications that do not involve the traditional three-component condensation approach (see above). The survey begins with reactions involving urea as a building block in Table 1, continues with examples concerning the use of substituted ureas in Table 2, and thioureas in Table 3, respectively. These three parts are further organized into sub-tables according to the type of the CH-acidic carbonyl compound, i.e.: A. β-keto esters, B. β-keto amides, C. β-diketones, and D. other CH-acidic carbonyl compounds. Table 4 summarizes the examples for Biginelli reactions involving guanidine, Table 5 covers transformations on solid supports, and Table 6 presents fluorous phase Biginelli reactions.

Within each table, the entries are listed according to increasing carbon number of the respective CH-acidic carbonyl compound and, furthermore, by increasing carbon number of the aldehyde building block used. Within sub-tables, the entries are listed by increasing carbon numbers of the R-group.

Reaction conditions including solvent, temperature, and time are presented as they are available from the original references; yields are given in parentheses. A dash, "(—)," indicates that no yield is reported in the reference. All data have been reproduced as provided in the original references. Ratios of diastereomers are not reported in reactions involving chiral reactants; for none of the examples in the tables where chiral reactants were used have the absolute configurations of the products been established.

The following abbreviations have been used in the tables:

AcOH	acetic acid
$BF_3 \cdot OEt_2$	boron trifluoride etherate
$BMImBF_4$	1-butyl-3-methylimidazolium tetrafluoroborate
BMImCl	1-butyl-3-methylimidazolium chloride
$BMImPF_6$	1-butyl-3-methylimidazolium hexafluorophosphate
BTF	benzotrifluoride
BuOH	n-butanol
CAN	ceric ammonium nitrate
DCE	dichloroethane
DCM	dichloromethane
DMF	dimethylformamide
KSF	montmorillonite KSF-clay
K-10	montmorillonite K-10-clay
MeCN	acetonitrile
MW	microwave irradiation
PPA	polyphosphoric acid
PPE	polyphosphate ester (ethyl polyphosphate)
PS	polystyrene
p-TsOH	p-toluenesulfonic acid
TBAF	tetrabutylammonium fluoride
TFA	trifluoroacetic acid
THF	tetrahydrofuran
US	ultrasonification

The term "*superheating*" under microwave conditions indicates heating of the solvent above its boiling point at atmospheric pressure and "*open vessel*" under microwave conditions denotes introducing an open glass beaker in a domestic microwave oven.

TABLE 1. REACTIONS INVOLVING UREA
A. β-Keto Esters

β-Keto Ester and Aldehyde	Conditions	Product(s) and Yield(s) (%)	Refs.
C₅	EtOH/HCl, 78°, 4 h	(17)	108
	MeOH/HCl	R Temp Time	109
	Neat, MW	H 65° 3 h (65)	101
	MeOH/HCl	H — 5.7 min (40)	110
	EtOH/Yb(OTf)₃, MW	NO₂ 65° 6 h (—)	22
		NO₂ 100° 10 min (34)	
	EtOH, 78°, 5 h	(16)	13

Conditions	Temp	Time	(%)	Ref
THF/BF$_3$·OEt$_2$/CuCl/AcOH	65°	8-18 h	(88)	10
EtOH/H$_2$SO$_4$	78°	18 h	(42)	10
EtOH/LaCl$_3$	78°	5 h	(97)	11
EtOH/FeCl$_3$	78°	4 h	(86)	84, 85
Neat/Yb(OTf)$_3$	100°	20 min	(98)	83
THF/InCl$_3$	65°	7 h	(92)	90
Neat, MW	—	1 min	(—)	101
MeCN/BiCl$_3$	81°	5 h	(92)	76
MeOH/HCl	65°	4 h	(81)	112
MeCN/LiClO$_4$	81°	6 h	(90)	92
Neat/Yb(III)-resin	120°	48 h	(71)	35

Conditions	Ar	Temp	Time	(%)	Ref
Neat, MW	2-ClC$_6$H$_4$	—	0.7 min	(66)	101
Neat/Yb(III)-resin	2-BrC$_6$H$_4$	120°	48 h	(68)	35
Neat/Yb(III)-resin	4-BrC$_6$H$_4$	120°	48 h	(68)	35
Neat/Yb(OTf)$_3$	2,4-Cl$_2$C$_6$H$_3$	100°	20 min	(83)	83
Neat, MW	2,4-Cl$_2$C$_6$H$_3$	—	1 min	(48)	101
Neat, MW	2,6-Cl$_2$C$_6$H$_3$	—	13 min	(51)	101
Neat, MW	3-O$_2$NC$_6$H$_4$	—	1 min	(66)	101
Neat/Yb(OTf)$_3$	4-FC$_6$H$_4$	100°	20 min	(81)	83
AcOH/EtOH, Yb(OTf)$_3$, MW	4-FC$_6$H$_4$	120°	10 min	(81)	22
Neat/Yb(III)-resin	4-FC$_6$H$_4$	120°	48 h	(70)	35
THF/BF$_3$·OEt$_2$/CuCl/AcOH	3,4-F$_2$C$_6$H$_3$	65°	8-18 h	(88)	10, 27
EtOH/H$_2$SO$_4$	3,4-F$_2$C$_6$H$_3$	78°	18 h	(62)	10

TABLE 1. REACTIONS INVOLVING UREA (Continued)
A. β-Keto Esters (Continued)

β-Keto Ester and Aldehyde	Conditions	Product(s) and Yield(s) (%)			Refs.

C$_5$

β-Keto ester: MeO-C(=O)-CH$_2$-C(=O)-CH$_3$

Aldehyde: 4-Cl-C$_6$H$_4$-CHO

Product: 4-(4-chlorophenyl)-5-methoxycarbonyl-6-methyl-3,4-dihydropyrimidin-2(1H)-one

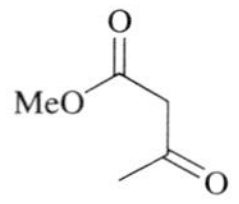

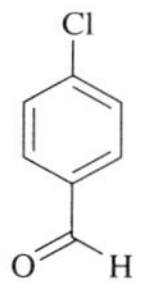

Conditions	Temp	Time	Yield	Refs.
THF/BF$_3$·OEt$_2$/CuCl/AcOH	65°	8–18 h	(95)	10
EtOH/H$_2$SO$_4$	78°	18 h	(56)	10
EtOH/LaCl$_3$	78°	5 h	(96)	111
EtOH/FeCl$_3$	78°	4 h	(96)	84, 85
THF/InCl$_3$	65°	9 h	(93)	90
Neat, MW	—	2.6 min	(68)	101
MeCN/BiCl$_3$	81°	5 h	(90)	76

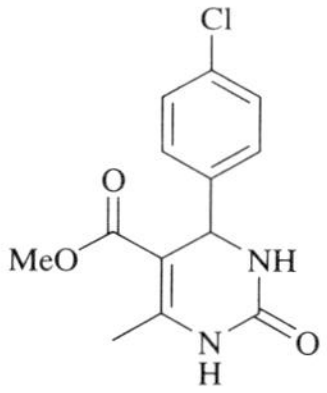

Aldehyde: 4-NO$_2$-C$_6$H$_4$-CHO

Product: 4-(4-nitrophenyl)-5-methoxycarbonyl-6-methyl-3,4-dihydropyrimidin-2(1H)-one

Conditions	Temp	Time	Yield	Refs.
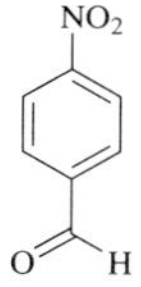 THF/BF$_3$·OEt$_2$/CuCl/AcOH	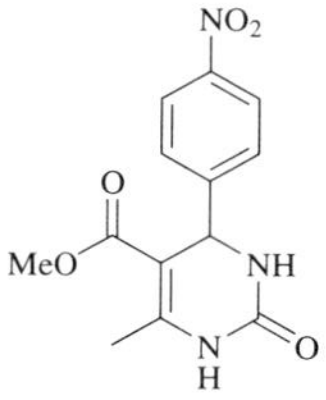 65°	8–18 h	(92)	10
EtOH/H$_2$SO$_4$	78°	18 h	(41)	10
PPE, MW	—	3 x 40 sec	(86)	34
EtOH/LaCl$_3$	78°	5 h	(68)	111
EtOH/FeCl$_3$	78°	4 h	(88)	84, 85
Neat/Yb(OTf)$_3$	100°	20 min	(91)	83
THF/InCl$_3$	65°	6 h	(91)	90
MeCN/BiCl$_3$	81°	6 h	(89)	76
THF/PPE	65°	15 h	(84)	104

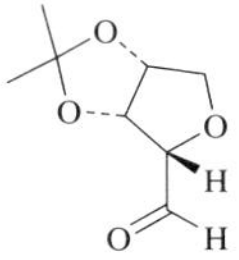

Conditions	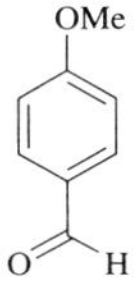 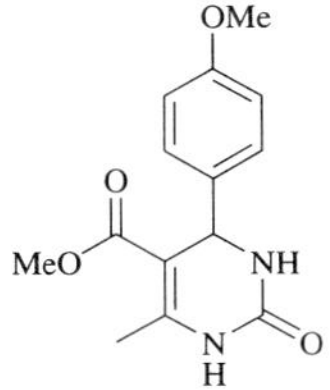 Temp	Time	(%)
THF/BF$_3$•OEt$_2$/CuCl/AcOH	65°	8-18 h	(87)
EtOH/H$_2$SO$_4$	78°	18 h	(28)
EtOH/LaCl$_3$	78°	5 h	(82)
EtOH/FeCl$_3$	78°	4 h	(88)
Neat/Yb(OTf)$_3$	100°	20 min	(99)
THF/InCl$_3$	65°	9 h	(91)
MeCN/BiCl$_3$	81°	5 h	(92)
Neat/Yb(III)-resin	120°	48 h	(71)

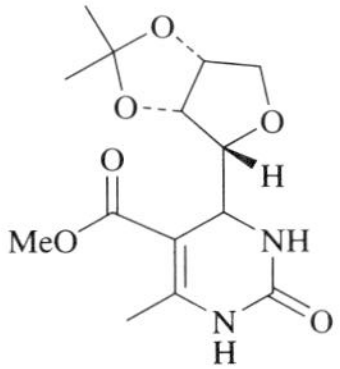

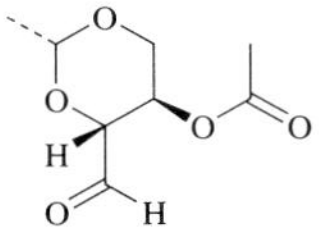

EtOH, 78°, 5 h (43) 15

EtOH, 70-80°, 55 h (38) 19

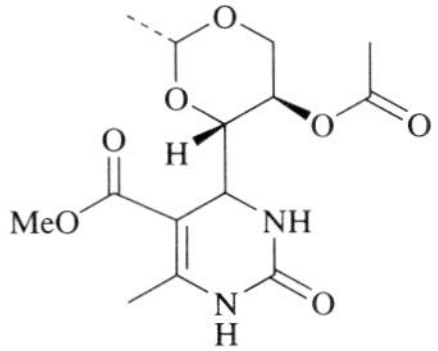

10
10
111
84, 85
83
90
76
35

TABLE 1. REACTIONS INVOLVING UREA (*Continued*)
A. β-Keto Esters (*Continued*)

β-Keto Ester and Aldehyde	Conditions	Product(s) and Yield(s) (%)				Refs.
		Ar	Temp	Time		
C_5 MeO-β-ketoester + ArCHO	PPE, MW	2-MeC$_6$H$_4$	—	3 x 40 sec	(86)	34
	THF/PPE	2-MeC$_6$H$_4$	65°	15 h	(88)	104
	EtOH/Yb(OTf)$_3$, MW	2-MeC$_6$H$_4$	120°	10 min	(73)	22
	MeOH/HCl	2-MeC$_6$H$_4$	65°	4 h	(68)	112
	MeOH/HCl	3-MeC$_6$H$_4$	65°	4 h	(78)	112
	MeCN/BiCl$_3$	4-MeC$_6$H$_4$	81°	5 h	(91)	76
	Neat/Yb(III)-resin	4-(F$_3$C)C$_6$H$_4$	120°	48 h	(65)	35
	Neat/Yb(III)-resin	3-MeOC$_6$H$_4$	120°	48 h	(71)	35
	EtOH/Yb(OTf)$_3$, MW	3-MeOC$_6$H$_3$OH-4	100°	15 min	(46)	22
	EtOH/HCl	2-(CHF$_2$O)C$_6$H$_4$	78°	3 h	(68)	113, 114
	EtOH/HCl	2-(CHF$_2$S)C$_6$H$_4$	rt	12 h	(52)	115
	THF/BF$_3$•OEt$_2$/CuCl/AcOH	2,5-Me$_2$C$_6$H$_3$	65°	8-18 h	(96)	10
	EtOH/H$_2$SO$_4$	2,5-Me$_2$C$_6$H$_3$	78°	18 h	(62)	10
sugar aldehyde	EtOH, 78°, 5 h				(49)	15

(60) EtOH/HCl, 78°, 10 h		116
(65) EtOH/HCl, 78°, 10 h; R = H, OMe		116
(4) 1. THF/BF₃·OEt₂/CuCl/AcOH, 65°, 8-18 h; 2. PPA, 100°, 30 min		27

TABLE 1. REACTIONS INVOLVING UREA (Continued)
A. β-Keto Esters (Continued)

β-Keto Ester and Aldehyde	Conditions	Product(s) and Yield(s) (%)	Refs.

C_6

β-Keto ester: MeO-C(=O)-CH2-C(=O)-Et; Aldehyde: Ar-CHO

Product: dihydropyrimidinone with MeO-C(=O), Et, Ar, NH, C=O, NH substituents

		Ar	Temp	Time		
	THF/BF$_3$•OEt$_2$/CuCl/AcOH	Ph	65°	8-18 h	(81)	10
	EtOH/H$_2$SO$_4$	Ph	78°	18 h	(42)	10
	THF/InCl$_3$	Ph	65°	6 h	(95)	90
	THF/BF$_3$•OEt$_2$/CuCl/AcOH	4-ClC$_6$H$_4$	65°	8-18 h	(89)	10
	EtOH/H$_2$SO$_4$	4-ClC$_6$H$_4$	78°	18 h	(66)	10
	THF/InCl$_3$	4-ClC$_6$H$_4$	65°	7 h	(92)	90
	THF/BF$_3$•OEt$_2$/CuCl/AcOH	4-O$_2$NC$_6$H$_4$	65°	8-18 h	(90)	10
	EtOH/H$_2$SO$_4$	4-O$_2$NC$_6$H$_4$	78°	18 h	(64)	10
	THF/InCl$_3$	4-O$_2$NC$_6$H$_4$	65°	6 h	(91)	90
	THF/BF$_3$•OEt$_2$/CuCl/AcOH	3,4-F$_2$C$_6$H$_3$	65°	8-18 h	(82)	10
	EtOH/H$_2$SO$_4$	3,4-F$_2$C$_6$H$_3$	78°	18 h	(55)	10
	PPE, MW	3,4-F$_2$C$_6$H$_3$	—	3 x 40 sec	(65)	34
	AcOH/EtOH, Yb(OTf)$_3$, MW	3,4-F$_2$C$_6$H$_3$	120°	10 min	(64)	22
	THF/BF$_3$•OEt$_2$/CuCl/AcOH	4-MeOC$_6$H$_4$	65°	8-18 h	(85)	10
	EtOH/H$_2$SO$_4$	4-MeOC$_6$H$_4$	78°	18 h	(25)	10
	THF/InCl$_3$	4-MeOC$_6$H$_4$	65°	8 h	(91)	90

β-Keto ester: MeO-C(=O)-CH2-C(=O)-CH2-OMe; Aldehyde: Ar-CHO

Product: dihydropyrimidinone with MeO-C(=O), MeOCH$_2$, Ar, NH, C=O, NH

	A. AcOH/EtOH, Yb(OTf)$_3$, MW, 120°, 10 min	Ar: A. 3-O$_2$NC$_6$H$_4$ (35)	22
	B. THF/BF$_3$•OEt$_2$/CuCl/AcOH, 65°, 8-18 h	B. 3,4-F$_2$C$_6$H$_3$ (94)	27

R	Temp	Time		
H	118°	5 h	AcOH/HCl	78, 117
H	rt	12 h	H$_2$O, pH 6.6	80
H	170°	40 min	Piperidine	118
H	81°	1 h	MeCN/TFA[b]	11
H	110°	48 h	Toluene/Zeolite	95
Me	78°	3 h	EtOH/HCl	78
Me	100°	2 h	Dioxane/HCl	78
Me	rt	12 h	H$_2$O, pH 1.0	80
Me	65°	15 h	THF/PPE	75, 104
Me	78°	5 h	EtOH/ p-toluenesulfinic acid	119
Me	81°	6 h	MeCN/TFA[b]	11
Me	65°	—	MeOH/KSF-clay	120
Me	65°	7 h	THF/InCl$_3$	90
Me	120°	20 min	EtOH/HCl, MW	22
Me	110°	48 h	Toluene/Zeolite	95

EtOH/HCl, 78°, 4 h   108

TABLE 1. REACTIONS INVOLVING UREA (*Continued*)
A. β-Keto Esters (*Continued*)

β-Keto Ester and Aldehyde	Conditions	Product(s) and Yield(s) (%)	Refs.

C_6

β-Keto ester: EtO-C(O)-CH_2-C(O)-CH_3; Aldehyde: R-CHO

Product: ethyl 6-methyl-2-oxo-4-R-1,2,3,4-tetrahydropyrimidine-5-carboxylate

		R	Temp	Time		
	H$_2$O, pH 1.0	Et	rt	12 h	(15)	80
	EtOH/*p*-toluenesulfinic acid	Et	78°	5 h	(38)	119
	MeCN/TFA[b]	Et	81°	8 h	(86)	11
	H$_2$O, pH 2.1-6	propenyl	rt	12 h	(44)	80
	EtOH/*p*-toluenesulfinic acid	Pr	78°	5 h	(35)	119
	EtOH/LaCl$_3$	Pr	78°	5 h	(60)	111
	EtOH/FeCl$_3$	Pr	78°	4 h	(73)	84, 85
	THF/InCl$_3$	Pr	65°	7 h	(85)	90
	MeCN/BiCl$_3$	Pr	81°	6 h	(72)	76
	EtOH/HCl	Pr	78°	4 h	(15)	73
	MeCN/TFA, 81°, 12 h	product with CH$_2$CN substituent	(68)	11
	MeCN/TFA, 81°, 12 h	product with CH$_2$C(O)OEt substituent	(78)	11

Aldehyde surrogate [c]: 2-(cyanomethyl)-3-methyl-4,4-dimethyloxazolidine

Aldehyde surrogate [b]: 2-(ethoxycarbonylmethyl)-4,4,6-trimethyl-tetrahydro-2H-1,3-oxazine

R	Temp	Time		
EtOH/HCl	i-Pr	78°	6 h	(32)	121
EtOH/p-toluenesulfinic acid	i-Pr	78°	5 h	(27)	119
EtOH/LaCl₃	i-Pr	78°	5 h	(56)	111
EtOH/FeCl₃	i-Pr	78°	4 h	(53)	84, 85
Neat/Yb(OTf)₃	i-Pr	100°	20 min	(83)	83
THF/InCl₃	i-Pr	65°	8 h	(83)	90
MeCN/BiCl₃	i-Pr	81°	8 h	(54)	76
EtOH/HCl	i-Pr	78°	4 h	(10)	73
Neat/KSF-clay	n-Bu	130°	48 h	(86)	82
Neat/Yb(OTf)₃	n-Bu	100°	40 min	(87)	83
EtOH/HCl	n-Bu	78°	3 h	(31)	122

R	Temp	Time		
EtOH/LaCl₃	H	78°	5 h	(67)	111
EtOH/Yb(OTf)₃, MW	H	100°	20 min	(50)	22
MeOH, US, (A) CAN	H	—	3 h	(87)	103
(B) Oxone®	H	—	4.5 h	(73)	103
MeCN/LiClO₄	H	81°	5 h	(85)	92
EtOH/HCl	NO₂	78°	3 h	(—)	123

TABLE 1. REACTIONS INVOLVING UREA (Continued)
A. β-Keto Esters (Continued)

β-Keto Ester and Aldehyde	Conditions	Product(s) and Yield(s) (%)	Refs.
C$_6$ EtOC(O)CH$_2$C(O)CH$_3$ + thiophene-2-carbaldehyde	EtOH/HCl	(thiophene-dihydropyrimidinone), Temp 78°, Time 3-4 h, (70)	73, 122
	MeOH/HCl	65°, 2 h, (—)	124
	THF/PPE	65°, 15 h, (84)	75, 104
	AcOH, MW	—, 2 min, (97)	79
	Toluene/Amberlyst 15	110°, 6 h, (83)	79
	MeCN/BiCl$_3$	81°, 6 h, (89)	76
	EtOH/AcOH, Yb(OTf)$_3$	120°, 10 min, (89)	22
	MeOH, US, (A) CAN	—, 3.5 h, (90)	103
	(B) Oxone®	—, 3 h, (78)	103
	MeCN/LiClO$_4$	81°, 5 h, (90)	92
5-R-thiophene-2-carbaldehyde	EtOH/HCl, 78°, 3 h	(thiophene-dihydropyrimidinone) R: H (83), NO$_2$ (85), Br (80)	125
pyridine-3-carbaldehyde	EtOH/AcOH, Yb(OTf)$_3$ MW, 120°, 10 min	(31)	22

R	Conditions	Temp	Time		Refs.
n-C$_5$H$_{11}$	Ionic liquid (BMImBF$_4$)	100°	30 min	(93)	81
n-C$_5$H$_{11}$	EtOH/AcOH, Yb(OTf)$_3$, MW	120°	10 min	(35)	22
n-C$_5$H$_{11}$	MeCN/LiClO$_4$	81°	7 h	(82)	92
c-C$_6$H$_{11}$	MeCN/BiCl$_3$	81°	6 h	(92)	76
c-C$_6$H$_{11}$	MeOH, US, (A) CAN	—	3 h	(90)	103
c-C$_6$H$_{11}$	(B) Oxone®	—	5 h	(76)	103
c-C$_6$H$_{11}$	MeCN/LiClO$_4$	81°	8 h	(87)	92
n-C$_6$H$_{13}$	H$_2$O, pH 1.0	rt	12 h	(12)	80
n-C$_6$H$_{13}$	EtOH/p-toluenesulfinic acid	reflux	5 h	(23)	119
n-C$_6$H$_{13}$	THF/InCl$_3$	65°	8 h	(81)	90
n-C$_6$H$_{13}$	MeCN/BiCl$_3$	81°	9 h	(50)	76
n-C$_6$H$_{13}$	MeOH, US, (A) CAN	—	3 h	(85)	103
n-C$_6$H$_{13}$	(B) Oxone®	—	4 h	(77)	103
n-C$_6$H$_{13}$	DMF/ClSiMe$_3$, urea added after 12 h	20°	14 h	(37)	126
n-C$_7$H$_{15}$	DMF/ClSiMe$_3$, urea added after 12 h	20°	14 h	(32)	126

EtOH, 78°, 5 h   (25)   13

TABLE I. REACTIONS INVOLVING UREA (*Continued*)
A. β-Keto Esters (*Continued*)

β-Keto Ester and Aldehyde	Conditions	Temp	Time	Product(s) and Yield(s) (%)	Refs.
C$_6$					
Ph-CHO, EtO-C(O)-CH$_2$-C(O)-CH$_3$	EtOH/HCl	78°	1.5-4 h	(80) [Ph/NH/N-H/Me/CO$_2$Et dihydropyrimidinone]	33, 73, 99, 117, 127-130
	H$_2$O, pH 2.1-7	rt	12 h	(40)	80
	MeOH/H$_2$O, HCl	rt	3 d	(80)	6
	Piperidine	170°	1.5 h	(68)	118
	EtOH/HCl, MW	—	3.5 min	(90)	97
	Neat on inorganic support, MW,				
	(A) SiO$_2$	—	12 min	(85)	98
	(B) Al$_2$O$_3$ (neutral)	—	10 min	(87)	98
	(C) Al$_2$O$_3$ (basic)	—	10 min	(90)	98
	(D) Al$_2$O$_3$ (acidic)	—	10 min	(97)	98
	THF/BF$_3$·OEt$_2$/CuCl/AcOH	65°	8-18 h	(94)	10
	THF/H$_2$SO$_4$	78°	18 h	(71)	10
	THF/PPE	65°	15 h	(94)	75, 104
	EtOH/*p*-toluenesulfinic acid	78°	5 h	(73)	119
	Neat/KSF-clay	130°	18 h	(82)	82
	PPE, MW	—	3 x 40 sec	(85)	34
	MeCN/TFA[b]	81°	4 h	(94)	11
	MeOH/KSF-clay	65°	—	(92)	120
	EtOH/LaCl$_3$	78°	5 h	(95)	111

EtOH/FeCl$_3$	78°	4 h	(94)	84, 85
Neat/Yb(OTf)$_3$	100°	20 min	(98)	83
THF/InCl$_3$	65°	7 h	(95)	90
EtOH/HCl, MW, (A) reflux	78°	3 h	(80)	100
(B) superheating	96°	3 h	(80)	100
(C) open vessel	—	15 x 20 sec	(78)	100
Neat/HCl, MW, open vessel	—	15 x 20 sec	(50)	100
AcOH, MW	—	2 min	(86)	79
Toluene/Amberlyst 15	110°	12 h	(80)	79
MeCN/BiCl$_3$	81°	5 h	(95)	76
Ionic liquid, (A) BMImBF$_4$	100°	30 min	(92)	81
(B) BMImPF$_6$	100°	30 min	(94)	81
(C) BMImCl	100°	30 min	(56)	81
EtOH/AcOH, Yb(OTf)$_3$, MW	120°	10 min	(92)	22
MeOH, US, (A) CAN	—	3.5 h	(92)	103
(B) Oxone®	—	5 h	(88)	103
MeCN/LiClO$_4$	81°	6 h	(89)	92
MeCN/Mn(OAc)$_3$	81°	2 h	(96)	93
Neat/Yb(III)-resin	120°	48 h	(80)	35
Toluene/Zeolite	110°	12 h	(80)	95
DMF/ClSiMe$_3$,				
(A) urea added after 12 h	20°	14 h	(80)	126
(B) all components together	20°	2 h	(61)	126

TABLE 1. REACTIONS INVOLVING UREA (Continued)
A. β-Keto Esters (Continued)

β-Keto Ester and Aldehyde	Conditions	Product(s) and Yield(s) (%)	Refs.
C₆			
EtO-C(=O)-CH₂-C(=O)-CH₃ ; 4-HO-C₆H₄-CHO	EtOH/HCl, MW	Temp —, Time 3.5 min (87)	97
	Neat/KSF-clay	130°, 48 h (88)	82
	EtOH/LaCl₃	78°, 5 h (89)	111
	EtOH/FeCl₃	78°, 4 h (84)	84, 85
	THF/InCl₃	65°, 8 h (91)	90
	AcOH, MW	—, 2 min (86)	79
	Toluene/Amberlyst 15	110°, 12 h (76)	79
	MeCN/Mn(OAc)₃	81°, 4 h (79)	93
	Toluene/Zeolite	110°, 18 h (76)	95
Ar-CHO		Ar, Temp, Time	
	EtOH/HCl	2-HOC₆H₄, 78°, 3 h (19)	122
	H₂O, pH 2.1-6.6	2-HOC₆H₄, rt, 12 h (87)	80
	EtOH/LaCl₃	2-HOC₆H₄, 78°, 5 h (70)	111
	THF/InCl₃	2-HOC₆H₄, 65°, 7 h (91)	90
	THF/InCl₃	3-HOC₆H₄, 65°, 9 h (88)	90
	EtOH/HCl	2-HOC₆H₃Br-5, 78°, 3 h (86)	131
2-Cl-C₆H₄-CHO		Temp, Time	
	EtOH/HCl	78°, 3 h (51)	132
	H₂O, pH 1.0	rt, 12 h (20)	80
	THF/PPE	65°, 15 h (83)	75, 104
	PPE, MW	—, 2 × 40 sec (95)	34
	EtOH/AcOH, Yb(OTf)₃, MW	120°, 10 min (68)	22
	MeCN/Mn(OAc)₃	81°, 2 h (76)	93

	Temp	Time		
EtOH/HCl	78°	3 h	(37)	122, 133
THF/BF$_3$•OEt$_2$/CuCl/AcOH	65°	8-18 h	(92)	10
EtOH/H$_2$SO$_4$	78°	18 h	(56)	10
Neat/KSF-clay	130°	48 h	(76)	82
MeOH/KSF-clay	65°	—	(93)	120
EtOH/LaCl$_3$	78°	5 h	(92)	111
EtOH/FeCl$_3$	78°	4 h	(90)	84, 85
Neat/Yb(OTf)$_3$	100°	20 min	(97)	83
THF/InCl$_3$	65°	6.5 h	(92)	90
AcOH, MW	—	3 min	(84)	79
Toluene/Amberlyst 15	110°	15 h	(79)	79
MeCN/BiCl$_3$	81°	5 h	(90)	76
Ionic liquid, (A) BMImBF$_4$	100°	30 min	(96)	81
(B) BMImPF$_6$	100°	30 min	(98)	81
MeOH, US, (A) CAN	—	4 h	(89)	103
(B) Oxone®	—	6.5 h	(85)	103
MeCN/Mn(OAc)$_3$	81°	3.5 h	(78)	93
Neat/Yb(III)-resin	120°	48 h	(68)	35

Ar		Temp	Time		
3-FC$_6$H$_4$	EtOH/HCl, MW	—	12 x 30 sec	(80)	30
4-FC$_6$H$_4$	EtOH/HCl	78°	3 h	(63)	122, 134
4-FC$_6$H$_4$	EtOH/HCl, MW	—	12 x 30 sec	(86)	30
4-FC$_6$H$_4$	Neat/Yb(OTf)$_3$	100°	20 min	(94)	83
4-FC$_6$H$_4$	Neat/Yb(III)-resin	120°	48 h	(68)	35
3-ClC$_6$H$_4$	EtOH/LaCl$_3$	78°	5 h	(87)	111
3-ClC$_6$H$_4$	EtOH/FeCl$_3$	78°	4 h	(82)	84, 85
2-BrC$_6$H$_4$	Neat/Yb(OTf)$_3$	100°	20 min	(97)	83
2-BrC$_6$H$_4$	Neat/Yb(III)-resin	120°	48 h	(68)	35
3-BrC$_6$H$_4$	EtOH/LaCl$_3$	78°	5 h	(97)	111
3-BrC$_6$H$_4$	EtOH/FeCl$_3$	78°	4 h	(83)	84, 85

TABLE 1. REACTIONS INVOLVING UREA (Continued)
A. β-Keto Esters (Continued)

β-Keto Ester and Aldehyde	Conditions	Product(s) and Yield(s) (%)	Refs.

C_6 (ethyl acetoacetate + 2-nitrobenzaldehyde)

Product: ethyl 4-(2-nitrophenyl)-6-methyl-2-oxo-1,2,3,4-tetrahydropyrimidine-5-carboxylate

Conditions	Temp	Time	(Yield)	Refs.
H₂O, pH 1-3	rt	12 h	(50)	80
THF/PPE	65°	5-15 h	(84)	75, 104
MeCN/TFA[b]	81°	4 h	(88)	11
EtOH/AcOH, Yb(OTf)₃, MW	120°	10 min	(54)	22
MeCN/Mn(OAc)₃	81°	4 h	(77)	93

(ethyl acetoacetate + 3-nitrobenzaldehyde)

Product: ethyl 4-(3-nitrophenyl)-6-methyl-2-oxo-1,2,3,4-tetrahydropyrimidine-5-carboxylate

Conditions	Temp	Time	(Yield)	Refs.
EtOH/HCl, MW	—	4 min	(88)	97
THF/PPE	65°	5-15 h	(87)	75, 104
PPE, MW	—	3 x 40 sec	(93)	34
MeCN/TFA[b]	81°	5 h	(90)	11
AcOH, MW	—	4 min	(90)	79
Toluene/Amberlyst 15	110°	14 h	(84)	79
MeCN/BiCl₃	81°	6 h	(85)	76
MeCN/Mn(OAc)₃	81°	4 h	(75)	93

Solvent/Catalyst	Temp	Time		
EtOH/HCl, MW	—	4 min	(70)	97
THF/BF$_3$·OEt$_2$/CuCl/AcOH	65°	8-18 h	(91)	10
EtOH/H$_2$SO$_4$	78°	18 h	(—)	10
THF/PPE	65°	15 h	(77)	75, 104
MeOH/KSF-clay	65°	—	(89)	120
EtOH/LaCl$_3$	78°	4.5 h	(80)	111
EtOH/FeCl$_3$	78°	4 h	(83)	84, 85
Neat/Yb(OTf)$_3$	100°	20 min	(94)	83
THF/InCl$_3$	65°	6 h	(93)	90
AcOH, MW	—	4 min	(88)	79
Toluene/Amberlyst 15	110°	15 h	(79)	79
MeCN/BiCl$_3$	81°	6 h	(90)	76
Ionic liquid, (A) BMImBF$_4$	100°	30 min	(90)	81
(B) BMImPF$_6$	100°	30 min	(92)	81
MeOH, US, (A) CAN	—	7 h	(85)	103
(B) Oxone®	—	9.5 h	(73)	103
MeCN/LiClO$_4$	81°	5 h	(90)	92
Neat/Yb(III)-resin	120°	48 h	(72)	35
Toluene/Zeolite	110°	12 h	(68)	95

TABLE 1. REACTIONS INVOLVING UREA (Continued)
A. β-Keto Esters (Continued)

β-Keto Ester and Aldehyde	Conditions		Product(s) and Yield(s) (%)			Refs.
		Ar	Temp	Time		
C₆	THF/BF₃·OEt₂/CuCl/AcOH	3,4-F₂C₆H₃	65°	8-18 h	(81)	10
	EtOH/H₂SO₄	3,4-F₂C₆H₃	78°	18 h	(66)	10
	THF/PPE	3,4-F₂C₆H₃	65°	15 h	(84)	75, 104
	PPE, MW	3,4-F₂C₆H₃	—	3 x 40 sec	(87)	34
	EtOH/Yb(OTf)₃, MW	3,4-F₂C₆H₃	120°	10 min	(61)	22
	EtOH/HCl	2-ClC₆H₃F-6	78°	3 h	(—)	133
	EtOH/HCl	2-ClC₆H₃OH-5	78°	3 h	(30)	132
	EtOH/HCl, MW	2,3-Cl₂C₆H₃	—	3.5 min	(85)	97
	THF/PPE	2,3-Cl₂C₆H₃	65°	15 h	(79)	75, 104
	PPE, MW	2,3-Cl₂C₆H₃	—	3 x 40 sec	(91)	34
	EtOH/LaCl₃	2,4-Cl₂C₆H₃	78°	5 h	(93)	111
	Neat/Yb(OTf)₃	2,4-Cl₂C₆H₃	100°	20 min	(89)	83
	Toluene/Zeolite	2,4-Cl₂C₆H₃	110°	12 h	(74)	95
	MeOH, US, (A) CAN	3,4-Cl₂C₆H₃	—	5 h	(90)	103
	(B) Oxone®	3,4-Cl₂C₆H₃	—	7.5 h	(83)	103
	MeCN/LiClO₄	3,4-Cl₂C₆H₃	81°	10 h	(85)	92
	EtOH/HCl	3,4-Cl₂C₆H₃	78°	3 h	(59)	134
	AcOH/HCl	2,6-Cl₂C₆H₃	118°	8 h	(65)	135
	AcOH, MW	2,6-Cl₂C₆H₃	—	4 min	(82)	79
	Toluene/Amberlyst 15	2,6-Cl₂C₆H₃	110°	15 h	(71)	79
	EtOH/HCl	3,5-(O₂N)₂C₆H₃	78°	3 h	(—)	33

PhCH2CHO	MeOH, US, (A) CAN, 6 h (B) Oxone®, 6 h	(91) (87) 103
terephthalaldehyde	EtOH/Yb(OTf)3, MW, 120°, 10 min	(78) 22
sugar-aldehyde	EtOH, 78°, 5 h	(—) 16

TABLE 1. REACTIONS INVOLVING UREA (Continued)
A. β-Keto Esters (Continued)

β-Keto Ester and Aldehyde	Conditions	Product(s) and Yield(s) (%)			Refs.
			Temp	Time	
C₆ (EtO-β-ketoester + 4-OMe-benzaldehyde)	H₂O, pH 2.1-6.6	(product: 4-(4-methoxyphenyl)-dihydropyrimidinone ethyl ester)	rt	12 h	(75)	80
	EtOH/HCl, MW		—	3 min	(98)	97
	Neat/inorganic support, MW,					
	(A) SiO₂		—	12 min	(80)	98
	(B) Al₂O₃ (neutral)		—	10 min	(85)	98
	(C) Al₂O₃ (basic)		—	10 min	(85)	98
	(D) Al₂O₃ (acidic)		—	9 min	(92)	98
	THF/BF₃•OEt₂/CuCl/AcOH		65°	8-18 h	(85)	10
	EtOH/H₂SO₄		78°	18 h	(37)	10
	Neat/KSF-clay		130°	18 h	(78)	82
	MeCN/TFA^b		81°	4 h	(92)	11
	MeOH/KSF-clay		65°	—	(82)	120
	EtOH/LaCl₃		78°	5 h	(93)	111
	EtOH/FeCl₃		78°	4 h	(94)	84, 85
	Neat/Yb(OTf)₃		100°	20 min	(96)	83
	THF/InCl₃		65°	9 h	(90)	90
	AcOH, MW		—	2 min	(88)	79
	Toluene/Amberlyst 15		110°	12 h	(71)	79
	MeCN/BiCl₃		81°	6 h	(90)	76
	Ionic liquid, (A) BMImBF₄		100°	30 min	(95)	81
	(B) BMImPF₆		100°	30 min	(98)	81
	Toluene/Zeolite		110°	12 h	(71)	95

	Ar	Temp	Time		
MeOH/KSF-clay	4-MeC$_6$H$_4$	65°	—	(88)	120
AcOH, MW	4-MeC$_6$H$_4$	—	2 min	(85)	79
Toluene/Amberlyst 15	4-MeC$_6$H$_4$	110°	8 h	(73)	79
MeCN/BiCl$_3$	4-MeC$_6$H$_4$	81°	5 h	(97)	76
THF/PPE	4-MeC$_6$H$_4$	65°	15 h	(86)	104
MeCN/LiClO$_4$	4-MeC$_6$H$_4$	81°	7 h	(89)	92
THF/PPE	2-(F$_3$C)C$_6$H$_4$	65°	15 h	(68)	75, 104
PPE, MW	2-(F$_3$C)C$_6$H$_4$	—	3 x 40 sec	(76)	34
EtOH/AcOH Yb(OTf)$_3$, MW	2-(F$_3$C)C$_6$H$_4$	120°	20 min	(49)	22
Neat/Yb(OTf)$_3$	4-(F$_3$C)C$_6$H$_4$	100°	20 min	(87)	83
Neat/Yb(III)-resin	4-(F$_3$C)C$_6$H$_4$	120°	48 h	(70)	35
THF/PPE	2-MeOC$_6$H$_4$	65°	15 h	(83)	104
THF/InCl$_3$	3-MeOC$_6$H$_4$	65°	9 h	(90)	90
EtOH/HCl	2-(F$_2$HCO)C$_6$H$_4$	78°	3 h	(—)	113, 114
EtOH/HCl	3,4-(OCH$_2$O)C$_6$H$_3$	78°	3 h	(49)	78
EtOH/HCl, MW	3,4-(OCH$_2$O)C$_6$H$_3$	—	3.5 min	(85)	97
EtOH/LaCl$_3$	3,4-(OCH$_2$O)C$_6$H$_3$	78°	5 h	(91)	111
EtOH/FeCl$_3$	3,4-(OCH$_2$O)C$_6$H$_3$	78°	4 h	(82)	84, 85
AcOH, MW	3,4-(OCH$_2$O)C$_6$H$_3$	—	2 min	(87)	79
Toluene/Amberlyst 15	3,4-(OCH$_2$O)C$_6$H$_3$	110°	10 h	(72)	79
MeCN/LiClO$_4$	3,4-(OCH$_2$O)C$_6$H$_3$	81°	7 h	(90)	92
EtOH/HCl, MW	3-MeOC$_6$H$_3$OH-4	—	4 min	(90)	97
EtOH/LaCl$_3$	3-MeOC$_6$H$_3$OH-4	78°	5 h	(92)	111
EtOH/FeCl$_3$	3-MeOC$_6$H$_3$OH-4	78°	4 h	(86)	84, 85
MeCN/Mn(OAc)$_3$	3-MeOC$_6$H$_3$OH-4	81°	4 h	(76)	93
EtOH/HCl	3-MeOC$_6$H$_2$OH-4-I-5	78°	3.5 h	(85)	131
EtOH/HCl	2,4-(MeO)$_2$C$_6$H$_3$	78°	4.5 h	(—)	33

TABLE 1. REACTIONS INVOLVING UREA (Continued)
A. β-Keto Esters (Continued)

β-Keto Ester and Aldehyde	Conditions	Product(s) and Yield(s) (%)				Refs.
		Ar	Temp	Time		
C₆ (EtO-CO-CH₂-CO-CH₃ + Ar-CHO)	MeCN/TFA[b]	3,4,5-(MeO)₃C₆H₂	81°	2 h	(84)	11
	MeCN/BiCl₃	3,4,5-(MeO)₃C₆H₂	81°	3.5 min	(95)	76
	MeOH, US, (A) CAN (B) Oxone®	3,4,5-(MeO)₃C₆H₂	—	12 h	(90)	103
			—	3 h	(85)	103
	EtOH/HCl	3,4,5-(MeO)₃C₆H₂	78°	8 h	(65)	134
	EtOH/HCl, MW	2,4,6-(MeO)₃C₆H₂	—	14 min	(70)	97
	Toluene/Zeolite	2,4,6-(MeO)₃C₆H₂	110°	12 h	(71)	95
	EtOH/HCl	3,4-(EtO)₂C₆H₃	78°	12 min	(55)	132
	MeCN/LiClO₄	3,4-(EtO)₂C₆H₃	81°	10 min	(90)	92

Conditions	Temp	Time	(%)	Refs
AcOH/HCl	118°	8 h	(60)	135
EtOH/HCl, MW	—	3 min	(96)	97
THF/PPE	65°	15 h	(75)	75, 104
MeCN/TFA[b]	81°	6 h	(82)	11
EtOH/HCl, MW, (A) reflux	78°	3 h	(54)	100
(B) superheating	96°	3 h	(75)	100
(C) open vessel	—	15 x 20 sec	(78)	100
Neat/HCl, open vessel, (A) MW	—	15 x 20 sec	(53)	100
(B) thermal	120°	30 min	(50)	100
AcOH, MW	—	3 min	(85)	79
Toluene/Amberlyst 15	110°	13 h	(78)	79
EtOH/HCl, MW	120°	10 min	(52)	22
MeOH, US, (A) CAN	—	3 h	(90)	103
(B) Oxone®	—	4.5 h	(92)	103
MeCN/LiClO₄	81°	8 h	(87)	92
MeCN/Mn(OAc)₃	81°	3.5 h	(94)	93

Conditions				
THF/PPE, 65°, 15 h			(92)	75, 104

TABLE 1. REACTIONS INVOLVING UREA (Continued)
A. β-Keto Esters (Continued)

β-Keto Ester and Aldehyde	Conditions	Product(s) and Yield(s) (%)	Refs.
C₆ EtOC(O)CH₂C(O)CH₃ + PhCH=CHCHO	EtOH/HCl, 78°, 3-6 h	[styryl dihydropyrimidinone] (—)	1, 121
	Neat/KSF-clay, 130°, 48 h	(70)	82
	Neat/Yb(OTf)₃, 100°, 20 min	(81)	83
	THF/InCl₃, 65°, 9 h	(90)	90
	MeCN/BiCl₃, 6 h	(90)	76
	MeOH, US, (A) CAN	(85)	103
	(B) Oxone®	(82)	103
	MeCN/LiClO₄, 81°, 7 h	(83)	92
PhCH₂CH₂CHO	MeCN/LiClO₄, 81°, 7 h	[phenethyl dihydropyrimidinone] (81)	92
2-R-C₆H₄CHO	MeOH/HCl, 65°, 15 h	R: vinyl (74); allyl (39)	136, 137; 136

R	Temp	Time			
Me	78°	3 h	(65)		138, 139
Me	rt	12 h	(10)		80
Me	—	4 h	(88)		103
Me	—	6 h	(70)		103
Me	110°	12 h	(63)		95
Et	78°	3 h	(72)		132
CH₂CH₂Cl	78°	3 h	(33)		122

(27) 122

R		
Me	(34)	
CH₂CH₂Cl	(39)	122

R		
H	(38)	
MeO	(—)	140

TABLE 1. REACTIONS INVOLVING UREA (Continued)
A. β-Keto Esters (Continued)

β-Keto Ester and Aldehyde	Conditions	Product(s) and Yield(s) (%)	Refs.
C_6 EtO-β-ketoester + 3-(oxiranylmethoxy)-4-methoxybenzaldehyde	EtOH/HCl, 78°, 15-18 h	(product, —)	140

Second aldehyde: 2,4,5-trisubstituted benzaldehyde (R^1, R^2, R^3)

Product table:

Conditions	R^1	R^2	R^3	Temp	Time	Refs.
EtOH/HCl	MeO	MeO	Et	78°	13-15 h (49)	141
EtOH/HCl	MeO	EtO	Et	78°	13-15 h (50)	141
EtOH/HCl	MeO	MeO	Pr	78°	13-15 h (53)	141
EtOH/HCl	EtO	MeO	Et	78°	13-15 h (49)	141
Neat/inorganic support, MW,	EtO	MeO	Et			
(A) SiO_2				—	14 min (80)	98
(B) Al_2O_3 (neutral)				—	12 min (82)	98
(C) Al_2O_3 (basic)				—	12 min (85)	98
(D) Al_2O_3 (acidic)				—	10 min (89)	98
EtOH/HCl	MeO	EtO	Pr	78°	13-15 h (40)	141
EtOH/HCl	EtO	MeO	Pr	78°	13-15 h (48)	141
Neat/inorganic support, MW,	EtO	MeO	Pr			
(A) SiO_2				—	14 min (75)	98
(B) Al_2O_3 (neutral)				—	12 min (80)	98
(C) Al_2O_3 (basic)				—	12 min (82)	98
(D) Al_2O_3 (acidic)				—	10 min (90)	98
EtOH/HCl	EtO	EtO	Et	78°	13-15 h (44)	141
EtOH/HCl	EtO	EtO	Pr	78°	13-15 h (54)	141

Aldehyde	Product	Conditions	Temp	Time	(Yield)	Refs.
1-naphthaldehyde	4-(1-naphthyl)-DHPM	EtOH/HCl	78°	3 h	(—)	33
		MeCN/BiCl₃	81°	5 h	(91)	76
		MeCN/LiClO₄	81°	8 h	(87)	92
2-naphthaldehyde	4-(2-naphthyl)-DHPM	EtOH/HCl	78°	3 h	(—)	33
		AcOH, MW	—	3 min	(88)	79
		Toluene/Amberlyst 15	110°	11 h	(77)	79
		MeOH, US, (A) CAN	—	4.5 h	(84)	103
		(B) Oxone®	—	6 h	(78)	103
2-(diethylphosphoryloxy)benzaldehyde		MeOH/H₂O, 2M HCl, rt, 3 d			(—)	142
4-methoxy-1-naphthaldehyde		EtOH/HCl, 78°, 3 h			(—)	33

TABLE 1. REACTIONS INVOLVING UREA (*Continued*)

A. β-Keto Esters (*Continued*)

β-Keto Ester and Aldehyde	Conditions	Product(s) and Yield(s) (%)	Refs.
	MeCN/LiClO₄, 81°, 6 h	(90)	92
	EtOH/HCl, 78°, 10 h	R: H (65), MeO (65)	116
	MeCN/LiClO₄, 81°, 12 h	Ar = 9-anthracenyl (81)	92
	EtOH/HCl, 78°, 3 h	(28)	122

EtOH/HCl, 78°, 5 h		(58)	12
EtOH, 78°, 20 h		R: H (—), EtO (—)	143
EtOH, 78°, 20 h		R: H (—), NO₂ (—), MeO (—), EtO (—)	143
EtOH, 78°, 20 h		(—)	143

Glu = 2,3,4,6-tetra-O-acetyl glucosyl (AcO, OAc groups shown)

TABLE 1. REACTIONS INVOLVING UREA (*Continued*)

A. β-Keto Esters (*Continued*)

β-Keto Ester and Aldehyde	Conditions	Product(s) and Yield(s) (%)	Refs.
C₆	EtOH, 78°, 5 h	(—)	16
	THF/BF₃•OEt₂/CuCl/AcOH, 65°, 24 h	(63)	20
	EtOH, 78°, 15 h	(—)	15

Ar	
2-furyl	(43)
Ph	(62)
4-HOC$_6$H$_4$	(52)
4-MeOC$_6$H$_4$	(57)

Ar		Time	
4-MeC$_6$H$_4$	(A)	5 h	(88)
	(B)	8 h	(81)
4-MeOC$_6$H$_4$	(A)	4 h	(88)
	(B)	5 h	(80)
3,4-(OCH$_2$O)C$_6$H$_3$	(A)	4.5 h	(90)
	(B)	4 h	(85)

Ar	
Ph	(49)
4-O$_2$NC$_6$H$_4$	(60)
4-BrC$_6$H$_4$	(44)

TABLE I. REACTIONS INVOLVING UREA (*Continued*)
A. β-Keto Esters (*Continued*)

β-Keto Ester and Aldehyde	Conditions	Product(s) and Yield(s) (%)			Refs.

C₇

β-Keto Ester: EtO-C(=O)-CH₂-C(=O)-Et
Aldehyde: Ar-CHO

Product structure: EtO₂C-C(=C(Et)-NH-C(=O)-NH-)-CH(Ar) (dihydropyrimidinone with Et, Ar substituents)

		Ar	Temp	Time	
	THF/BF₃·OEt₂/CuCl/AcOH	Ph	65°	8-18 h (83)	10
	EtOH/H₂SO₄	Ph	78°	18 h (41)	10
	THF/InCl₃	Ph	65°	7 h (89)	90
	THF/BF₃·OEt₂/CuCl/AcOH	4-ClC₆H₄	65°	8-18 h (84)	10
	EtOH/H₂SO₄	4-ClC₆H₄	78°	18 h (56)	10
	THF/InCl₃	4-ClC₆H₄	65°	6 h (92)	90
	THF/BF₃·OEt₂/CuCl/AcOH	4-O₂NC₆H₄	65°	8-18 h (90)	10
	EtOH/H₂SO₄	4-O₂NC₆H₄	78°	18 h (44)	10
	THF/InCl₃	4-O₂NC₆H₄	65°	6 h (90)	90
	THF/BF₃·OEt₂/CuCl/AcOH	3,4-F₂C₆H₃	65°	8-18 h (82)	10
	EtOH/H₂SO₄	3,4-F₂C₆H₃	78°	18 h (61)	10
	THF/BF₃·OEt₂/CuCl/AcOH	4-MeOC₆H₄	65°	8-18 h (79)	10
	EtOH/H₂SO₄	4-MeOC₆H₄	78°	18 h (40)	10
	THF/InCl₃	4-MeOC₆H₄	65°	8 h (85)	90

β-Keto Ester: EtO-C(=O)-CH₂-C(=O)-H(CF₂)₂
Aldehyde: Ph-CHO

Product: EtO₂C-C(=C(H(CF₂)₂)-NH-C(=O)-NH-)-CH(Ph) dihydropyrimidinone

| | 1. EtOH/HCl, 78°, 6 h
2. Toluene/*p*-TsOH, 6 h | | | (64) | 63 |

C8

[Structure: Ar-CHO + i-PrO-C(O)-CH2-C(O)-CH3 → dihydropyrimidinone with Ar, NH, C=O, N-H, methyl, i-PrO-C(O)]

Ar	Temp	Time		
Ph	65°	15 h	(84)	THF/PPE	75, 104
Ph	120°	10 min	(50)	EtOH/Yb(OTf)₃, MW	22
Ph	120°	48 h	(78)	Neat/Yb(III)-resin	35
3-O₂NC₆H₄	—	3 x 40 sec	(94)	PPE, MW	34
3-O₂NC₆H₄	65°	15-24 h	(88)	THF/PPE	48, 104
3-O₂NC₆H₄	120°	10 min	(73)	EtOH/Yb(OTf)₃, MW	22
4-O₂NC₆H₄	120°	48 h	(73)	Neat/Yb(III)-resin	35
2-BrC₆H₃NO₂-5	118°	24 h	(66)	AcOH/HCl	77
2-BrC₆H₄	120°	48 h	(70)	Neat/Yb(III)-resin	35
4-ClC₆H₄	120°	48 h	(64)	Neat/Yb(III)-resin	35
4-FC₆H₄	120°	48 h	(71)	Neat/Yb(III)-resin	35
3-MeOC₆H₄	120°	48 h	(73)	Neat/Yb(III)-resin	35
4-MeOC₆H₄	120°	48 h	(75)	Neat/Yb(III)-resin	35
2-(CHF₂O)C₆H₄	78°	3 h	(—)	EtOH/HCl	113, 114
4-(F₃C)C₆H₄	120°	48 h	(68)	Neat/Yb(III)-resin	35

4-O₂N-C₆H₄-CHO + EtO-C(O)-CH₂-C(O)-Pr

EtOH/Yb(OTf)₃, MW, 120°, 10 min

[Product: dihydropyrimidinone with 4-NO₂C₆H₄, NH, C=O, Pr, EtO-C(O)] (41) 22

TABLE 1. REACTIONS INVOLVING UREA (Continued)
A. β-Keto Esters (Continued)

β-Keto Ester and Aldehyde	Conditions	Product(s) and Yield(s) (%)	Refs.

C$_8$

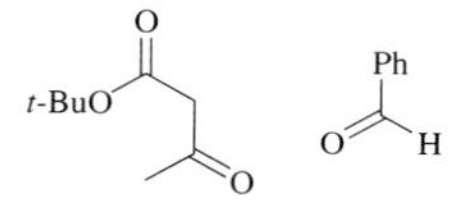

β-keto ester: t-BuO-C(=O)-CH$_2$-C(=O)-CH$_3$; aldehyde: PhCHO

Conditions:
- EtOH/HCl
- PPE, MW
- THF/PPE
- EtOH/Yb(OTf)$_3$, MW

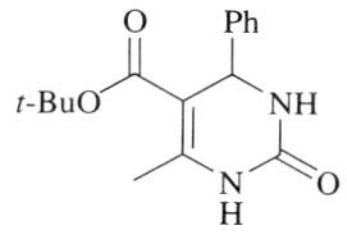

Product: dihydropyrimidinone with Ph, C(=O)O-t-Bu, CH$_3$ substituents

Temp	Time	Yield
78°	3 h	(51)
—	3 × 40 sec	(81)
65°	15 h	(77)
120°	10 min	(51)

Refs: 145, 34, 104, 22

C$_9$

β-keto ester: EtO-C(=O)-CH$_2$-C(=O)-t-Bu; aldehyde: PhCHO

Conditions:
(A) THF/BF$_3$·OEt$_2$/CuCl/AcOH, 65°, 8–18 h
(B) EtOH/H$_2$SO$_4$, 78°, 18 h

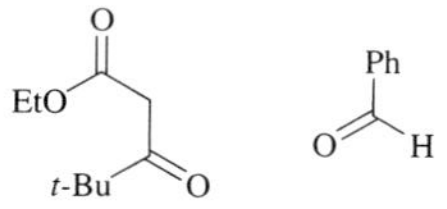

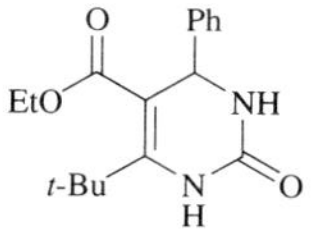

Product: A (85); B (32)

Refs: 10, 10

β-keto ester: EtO-C(=O)-CH$_2$-C(=O)-F(CF$_2$)$_4$; aldehyde: PhCHO

Conditions:
1. EtOH/HCl, 78°, 6 h
2. Toluene/p-TsOH, 6 h

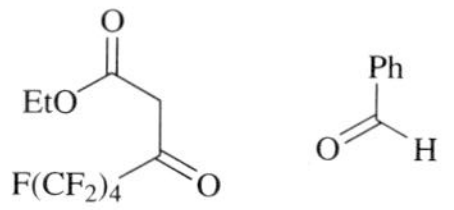

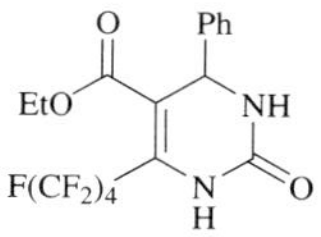

Product: dihydropyrimidinone (56)

Refs: 63

β-keto ester: i-Pr-CH$_2$-O-C(=O)-CH$_2$-C(=O)-CH$_3$; aldehyde: ArCHO, Ar = C$_6$H$_4$OCHF$_2$-2

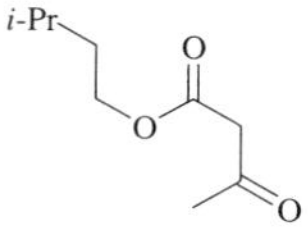

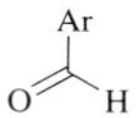

Conditions: EtOH/HCl, 78°, 3 h

Product: dihydropyrimidinone (—)

Refs: 113, 114

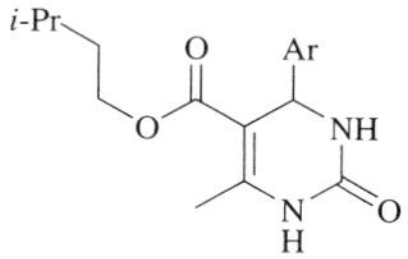

R	Temp	Time	
AcOH/HCl	Me	118°	44 h (21)	146
EtOH/HCl	Me	50°	48 h (46)	145
EtOH/HCl	Ph	78°	3 h (31)	145
THF/BF$_3$•OEt$_2$/CuCl/AcOH	Ph	65°	8-18 h (70)	10
EtOH/H$_2$SO$_4$	Ph	78°	18 h (10)	10
MeOH/KSF-clay	Ph	65°	— (80)	120
THF/LaCl$_3$	Ph	65°	9 h (84)	90
MeCN/LiClO$_4$	Ph	81°	10 h (75)	92
AcOH/EtOH, Yb(OTf)$_3$, MW	4-O$_2$NC$_6$H$_4$	120°	20 min (40)	22

AcOH/HCl, 118°, 5 h   (—)   147

EtOH/HCl, 78°, 18 h   (2)   108

TABLE 1. REACTIONS INVOLVING UREA (Continued)
A. β-Keto Esters (Continued)

β-Keto Ester and Aldehyde	Conditions	Product(s) and Yield(s) (%)	Refs.
C₁₁ (BnO-acetoacetate + RCHO)	AcOH/HCl; EtOH/HCl; EtOH/HCl	Dihydropyrimidinone (BnO₂C, Me, R, NH-C(=O)-NH): R=H, 118°, 5 h (25); R=Me, 78°, 5 h (42); R=Ph, 78°, 3 h (68)	145
	Neat/Yb(III)-resin, 120°, 48 h	Ar product: Ph (80); 4-FC₆H₄ (73); 3-MeOC₆H₄ (70); 4-MeOC₆H₄ (75)	35
C₁₆ (BnO-benzoylacetate + RCHO)	EtOH/HCl	Ph-substituted dihydropyrimidinone: R=Me, 50°, 48 h (40); R=Ph, 78°, 8 h (42)	145
C₃₁ (sugar-derived β-keto ester + PhCHO)	THF/CuCl/BF₃·OEt₂/AcOH, 65°, 24 h	(92)	20

	THF/CuCl/BF₃·OEt₂/AcOH, 65°, 24 h	(42)	20
	THF/CuCl/BF₃·OEt₂/AcOH, 65°, 24 h	(35)	20
	THF/CuCl/BF₃·OEt₂/AcOH, 65°, 24 h	(75)	20

[a] This structure is a protected aldehyde.
[b] The aldehyde was protected as an oxazinane.
[c] The aldehyde was protected as an oxazolidine.

TABLE 1. REACTIONS INVOLVING UREA (*Continued*)
B. β-Keto Amides

β-Keto Amide and Aldehyde	Conditions	Product(s) and Yield(s) (%)	Refs.
C₄ H₂N-CO-CH₂-CO-CH₃ + RCHO	MeOH/HCl, 65°	(product: dihydropyrimidinone with R, CONH₂, Me) R: Pr (—); *i*-Pr (—)	148
H₂N-CO-CH₂-CO-CH₃ + ArCHO	EtOH/HCl, 78°, 5 h EtOH/HCl, 78°, 5 h EtOH/HCl, 78°, 5 h EtOH/HCl, 78°, 5 h EtOH/HCl, MW, 120°, 15 min	(product: dihydropyrimidinone with Ar, CONH₂, Me) Ar: Ph (41); 4-BrC₆H₄ (46); 3-O₂NC₆H₄ (39); 4-O₂NC₆H₄ (52); 4-O₂NC₆H₄ (59)	106, 148 106 106 106 22
C₅ MeHN-CO-CH₂-CO-CH₃ + ArCHO	EtOH/HCl, 78°, 3 h	(product: dihydropyrimidinone with Ar, CONHMe, Me) Ar: Ph (90); 4-HOC₆H₄ (60); 4-ClC₆H₄ (60); 2-O₂NC₆H₄ (80); 3-O₂NC₆H₄ (90); 4-O₂NC₆H₄ (60); 2,4-Cl₂C₆H₃ (90); 3,4-Cl₂C₆H₃ (70); 4-MeOC₆H₄ (60); 3-MeOC₆H₃OH-2 (89); 3,4-(OCH₂O)C₆H₃ (74); 3,4-(MeO)₂C₆H₃ (60); 3,4,5-(MeO)₃C₆H₂ (65)	149

C8

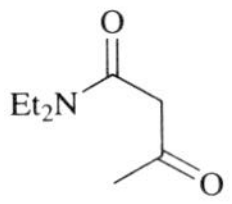

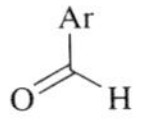

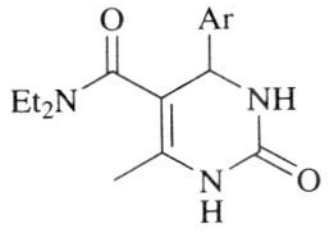

Ar-CHO + Et₂N-CO-CH₂-CO-CH₃

			Ar	Temp	Time	
AcOH			Ph	118°	10 h	(73)	106
AcOH			3-O₂NC₆H₄	118°	10 h	(81)	106
EtOH/HCl			4-O₂NC₆H₄	78°	5 h	(75)	106
EtOH/HCl			4-BrC₆H₄	78°	5 h	(65)	106
EtOH/AcOH, Yb(OTf)₃, MW			4-BrC₆H₄	120°	10 min	(66)	22

Product: 5-(Et₂N-CO)-4-Ar-6-methyl-3,4-dihydropyrimidin-2(1H)-one

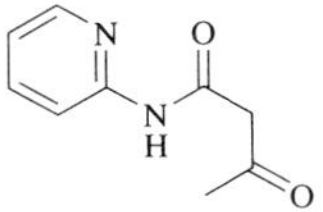

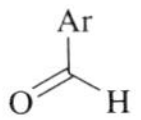

C9

Ar-CHO + 2-pyridyl-NH-CO-CH₂-CO-CH₃

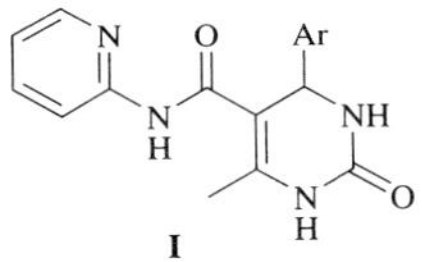

	Ar		(83)	106
EtOH/HCl, 78°, 5 h	4-O₂NC₆H₄		(65)	
	4-BrC₆H₄		(51)	
	4-Me₂NC₆H₄		(61)	
	3,4,5-(MeO)₃C₆H₂		(66)	

Product: 5-(2-pyridyl-NH-CO)-4-Ar-6-methyl-3,4-dihydropyrimidin-2(1H)-one **I**

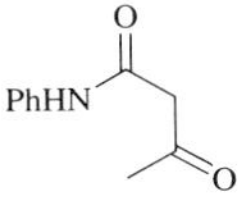

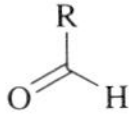

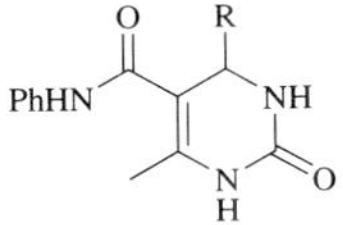

I Ar = 3-O₂NC₆H₄ (49)

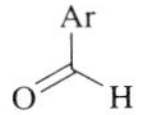

AcOH, 118°, 10 h

C10

R-CHO + PhHN-CO-CH₂-CO-CH₃

	R	Temp	Time		
MeOH/HCl	Pr	65°	—	(—)	148
MeOH/HCl	i-Pr	65°	—	(—)	148
AcOH	Ph	118°	10 h	(50)	106
MeOH/HCl	Ph	65°	20 min	(—)	148
EtOH/AcOH, MW	Ph	120°	20 min	(55)	22
AcOH	4-O₂NC₆H₄	118°	10 h	(64)	106
AcOH	4-BrC₆H₄	118°	10 h	(63)	106
AcOH	3-MeOC₆H₃OH-4	118°	10 h	(61)	106
EtOH/HCl, MW	3-MeOC₆H₃OH-4	100°	15 min	(28)	22
AcOH	4-Me₂NC₆H₄	118°	10 h	(89)	106

TABLE I. REACTIONS INVOLVING UREA (Continued)
B. β-Keto Amides (Continued)

β-Keto Amide and Aldehyde	Conditions	Product(s) and Yield(s) (%)	Refs.
C₁₅ [pyrazolone β-keto amide] + ArCHO	EtOH/HCl, 78°, 3-5 h	[dihydropyrimidinone-carboxamide product] Ar / Yield Ph (71) 4-FC₆H₄ (69) 4-ClC₆H₄ (66) 4-BrC₆H₄ (78) 4-O₂NC₆H₄ (63) 3,4-Cl₂C₆H₃ (70) 4-MeOC₆H₄ (60) 3,4,5-(MeO)₃C₆H₂ (62)	134

TABLE 1. REACTIONS INVOLVING UREA (Continued)
C. β-Diketones

β-Diketone and Aldehyde	Conditions	Product(s) and Yield(s) (%)	Refs.
	Piperidine, 170°, 40 min	R / H (72)	118
	EtOH/HCl, 50°, 6 h	Me (4)	150
	EtOH/HCl, 78°, 5 h	Et (32)	151
	AcOH, 118°, 3 h	(—)	151
	THF/InCl₃, 65°, 8 h	(90)	90
	THF/InCl₃, 65°, 6 h	(93)	90
	EtOH/HCl, 78°, 3 h	(—)	13

TABLE 1. REACTIONS INVOLVING UREA (Continued)
C. β-Diketones (Continued)

β-Diketone and Aldehyde	Conditions	Product(s) and Yield(s) (%)			Refs.

β-Diketone: pentane-2,4-dione (C$_5$); Aldehyde: PhCHO

Product: 5-acetyl-6-methyl-4-phenyl-3,4-dihydropyrimidin-2(1H)-one

Conditions	Temp	Time	Yield (%)	Refs.
EtOH/HCl	78°	3.5 h	(55)	151-153
Piperidine	170°	40 min	(64)	118
Neat/KSF-clay	130°	48 h	(74)	82
Neat/Yb(OTf)$_3$	100°	20 min	(94)	83
THF/InCl$_3$	65°	7 h	(94)	90
Ionic liquid (BMImBF$_4$)	100°	30 min	(99)	81
EtOH/Yb(OTf)$_3$	120°	10 min	(53)	22
MeCN/LiClO$_4$	81°	7 h	(88)	92
Neat/Yb(III)-resin	120°	48 h	(71)	35
DMF/ClSiMe$_3$;				
(A) urea added after 12 h	20°	14 h	(62)	126
(B) all components together	20°	2 h	(57)	126

Aldehyde: ArCHO

Product: 5-acetyl-6-methyl-4-aryl-3,4-dihydropyrimidin-2(1H)-one

Conditions	Ar	Temp	Time	(%)	Refs.
EtOH/HCl	2-HOC$_6$H$_4$	78°	2 h	(49)	151
EtOH/HCl	2-ClC$_6$H$_4$	78°	5 h	(32)	153
EtOH/HCl	3-ClC$_6$H$_4$	78°	5 h	(31)	153
EtOH/HCl	4-ClC$_6$H$_4$	78°	5 h	(37)	151, 153
EtOH/HCl	2-BrC$_6$H$_4$	78°	5 h	(25)	153
EtOH/HCl	3-BrC$_6$H$_4$	78°	5 h	(27)	153
Neat/Yb(III)-resin	4-FC$_6$H$_4$	120°	48 h	(70)	35
EtOH/HCl	3-O$_2$NC$_6$H$_4$	78°	3 h	(70)	151
Neat/Yb(OTf)$_3$	4-O$_2$NC$_6$H$_4$	100°	20 min	(90)	83
Ionic liquid (BMImBF$_4$)	4-O$_2$NC$_6$H$_4$	100°	30 min	(92)	81
EtOH/HCl	2-MeC$_6$H$_4$	78°	5 h	(34)	153
EtOH/HCl	3-MeC$_6$H$_4$	78°	5 h	(39)	153
EtOH/HCl	4-MeC$_6$H$_4$	78°	5 h	(38)	153
EtOH/HCl	2-MeOC$_6$H$_4$	78°	3 h	(40)	151, 153
EtOH/HCl	3-MeOC$_6$H$_4$	78°	5 h	(29)	153
THF/InCl$_3$	3-MeOC$_6$H$_4$	65°	9 h	(92)	90
Neat/Yb(III)-resin	3-MeOC$_6$H$_4$	120°	48 h	(65)	35
Neat/Yb(OTf)$_3$	4-MeOC$_6$H$_4$	100°	20 min	(91)	83
THF/InCl$_3$	4-MeOC$_6$H$_4$	65°	9 h	(91)	90
Neat/Yb(III)-resin	4-MeOC$_6$H$_4$	120°	48 h	(71)	35
EtOH/HCl	3-MeOC$_6$H$_3$OH-4	78°	3.5 h	(64)	151
EtOH/HCl	3,4-(MeO)$_2$C$_6$H$_3$	78°	1 h	(44)	151
EtOH/HCl	4-Me$_2$NC$_6$H$_4$	78°	3 h	(40)	151

EtOH/HCl, 78°, 5 h				(78)	21

TABLE 1. REACTIONS INVOLVING UREA (Continued)
C. β-Diketones (Continued)

β-Diketone and Aldehyde	Conditions	Product(s) and Yield(s) (%)	Refs.
C_5: pentane-2,4-dione + PhCH=CHCHO	AcOH, 118°, 3 h	[product: 4-(2-phenylvinyl)-6-methyl-5-acetyl-3,4-dihydropyrimidin-2(1H)-one] (—)	151
C_6: cyclohexane-1,3-dione + ArCHO	EtOH/HCl, 78°, 20 h	[product: 4-Ar-quinazolinone-type bicycle]	
		Ar	
		Ph (46)	24, 25
		2-FC_6H_4 (—)	24
		2-ClC_6H_4 (80)	24, 154
		3-ClC_6H_4 (62)	24, 25
		4-ClC_6H_4 (54)	24, 25, 155
		2-BrC_6H_4 (44)	24, 25
		3-BrC_6H_4 (45)	24, 25
		4-BrC_6H_4 (44)	24, 25
		3-O_2NC_6H_4 (50)	25
		4-O_2NC_6H_4 (46)	25
		2,3-Cl_2C_6H_3 (56)	24
		2,4-Cl_2C_6H_3 (50)	24
		2,6-Cl_2C_6H_3 (20)	24
		3,4-Cl_2C_6H_3 (61)	24
		2-MeC_6H_4 (45)	24, 25, 155
		3-MeC_6H_4 (57)	24
		4-MeC_6H_4 (43)	24, 25
		2-MeOC_6H_4 (30)	24, 25
		3-MeOC_6H_4 (42)	24, 25
		4-MeOC_6H_4 (48)	24, 25

Reactants	Conditions	Product	Yield	Refs.
PhCHO + Et-C(O)CH2C(O)CH3	EtOH/HCl, 78°, 3.5 h	(Ph, Et-CO, Me DHPM)	(47)	152
PhCHO + thienyl-C(O)CH2C(O)CF3	Neat/Yb(OTf)3, 100°, 20 min	(Ph, thienyl-CO, CF3 DHPM)	(94)	83
ArCHO + PhC(O)CH2C(O)CH3				
Ar = Ph	EtOH/HCl, 78°, 3.5 h		(—)	152
Ar = Ph	Neat/KSF-clay, 130°, 48 h		(74)	82
Ar = Ph	THF/InCl3, 65°, 9 h		(88)	90
Ar = 4-MeOC6H4	THF/InCl3, 65°, 9 h		(90)	90
OHC-C6H4-CHO + PhC(O)CH2C(O)CH3	EtOH/HCl, 78°, 5 h	bis-DHPM	(71)	21

TABLE I. REACTIONS INVOLVING UREA (*Continued*)
C. β-Diketones (*Continued*)

β-Diketone and Aldehyde	Conditions	Product(s) and Yield(s) (%)	Refs.
C₁₀ (Ph-CO-CH₂-CO-CF₃ + PhCHO)	Neat/Yb(OTf)₃, 100°, 20 min	(96)	83
C₁₅ (Ph-CO-CH₂-CO-Ph + R-C₆H₄-CHO)	BuOH/HCl, 118°, 3.5 h [a]	R: H (60), Me (40)	152

[a] These conditions are applied to the aldehyde and the bisureide of the starting β-diketone.

TABLE 1. REACTIONS INVOLVING UREA (Continued)
D. Other CH-Acidic Carbonyl Compounds

CH-Acidic Carbonyl Compound and Aldehyde	Conditions	Product(s) and Yield(s) (%)	Refs.
C₃	EtOH/LaCl₃, MW, 100°, 15 min	(83)	22
	EtOH/HCl, 78°, 6 h	Ar: Ph (91)	107, 156, 157
		4-HOC₆H₄ (82)	157
		2-O₂NC₆H₄ (26)	157
		3-O₂NC₆H₄ (67)	157
		2-ClC₆H₄ (79)	157
		4-ClC₆H₄ (82)	157
		4-MeOC₆H₄ (84)	107, 157
		2-(F₃C)C₆H₄ (65)	157
		2-(CHF₂O)C₆H₄ (43)	157
		2-(CHF₂S)C₆H₄ (61)	157
		3,4-(OCH₂O)C₆H₃ (45)	157
		3,4,5-(MeO)₃C₆H₂ (80)	107, 157
C₄	(A) EtOH/H₂SO₄, 78°, 12 h (B) TFA/DCE, 82°, 12 h	R (A/B) 2-Thienyl (51/72) Ph (55/88) cyclohexyl (50/75) 3-O₂NC₆H₄ (73/99) 2,3-Cl₂C₆H₃ (34/96) 3,5-(MeO)₂C₆H₃ (62/75)	68

TABLE 1. REACTIONS INVOLVING UREA (*Continued*)
D. Other CH-Acidic Carbonyl Compounds (*Continued*)

CH-Acidic Carbonyl Compound and Aldehyde	Conditions	Product(s) and Yield(s) (%)	Refs.
C_6 EtS-COCH_2-COCH_3 ; PhCHO	EtOH/HCl, 78°	dihydropyrimidinone (Ph, EtS-CO-) (85)	158
MeO-CO-CH(OMe)-... ; ArCHO	EtOH/Yb(OTf)$_3$, MW, 120°, 20 min	dihydropyrimidinone **I** (MeO-CO-): Ar = Ph (65); 3,4-F$_2$C$_6$H$_3$ (70); 3,4,5-(MeO)$_3$C$_6$H$_2$ (35)	22
Ar = 3,4-F$_2$C$_6$H$_3$; ArCHO	THF/BF$_3$·OEt$_2$/CuCl/AcOH, 65°, 8-24 h	**I** (—)	27
C_9 4-hydroxycoumarin ; ArCHO	MeOH/HCl, 65°, 4 h	chromeno-pyrimidinone: Ar = Ph (59); 4-MeC$_6$H$_4$ (64); 4-MeOC$_6$H$_4$ (51); 3,4-(MeO)$_2$C$_6$H$_3$ (53)	26
C_{10} 4-hydroxy-8-methylcoumarin ; ArCHO	MeOH/HCl, 65°, 4 h	methyl-chromeno-pyrimidinone: Ar = Ph (66); 4-MeC$_6$H$_4$ (70); 4-MeOC$_6$H$_4$ (62); 3,4-(MeO)$_2$C$_6$H$_3$ (58)	26
C_{13} 4-hydroxy-benzo[h]coumarin ; ArCHO	MeOH/HCl, 65°, 4 h	benzo-chromeno-pyrimidinone: Ar = Ph (46); 4-MeC$_6$H$_4$ (51); 4-MeOC$_6$H$_4$ (48); 3,4-(MeO)$_2$C$_6$H$_3$ (50)	26

TABLE 2. REACTIONS INVOLVING SUBSTITUTED UREAS

A. β-Keto Esters

β-Keto Ester, Aldehyde, and Urea	Conditions	Product(s) and Yield(s) (%)	Refs.
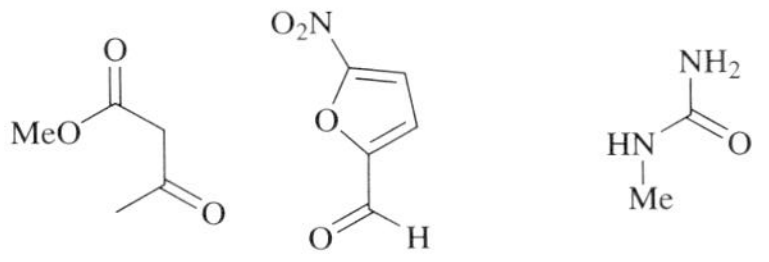 C5	MeOH/HCl, 65°, 3 h	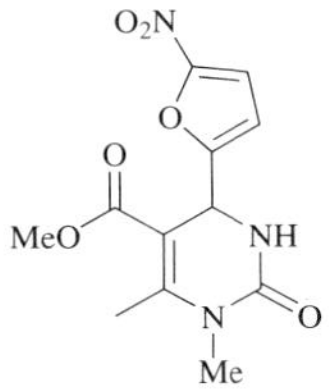 (—)	159
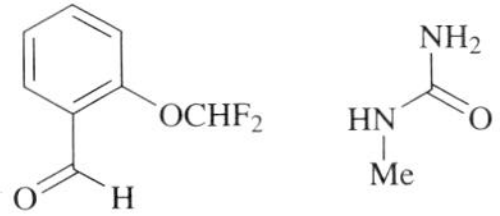	EtOH/HCl, 78°, 8 h	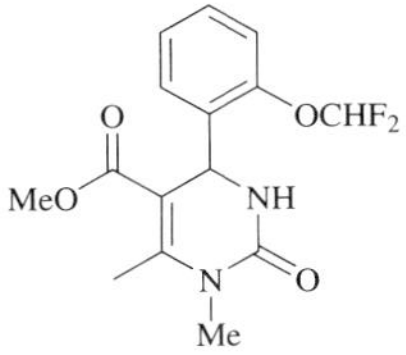 (—)	113, 114
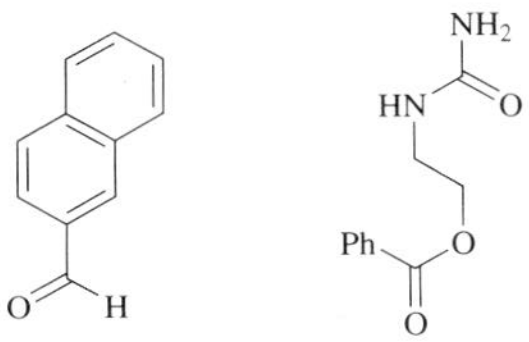	THF/HCl, 25°, 1-2 d	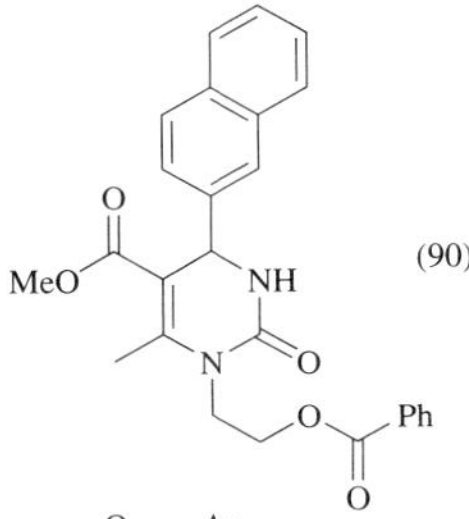 (90)	38
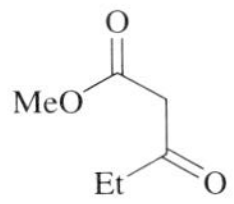 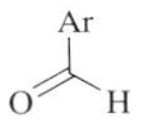 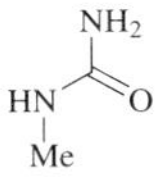 C6	EtOH/Yb(OTf)$_3$, MW, 120°, 10 min	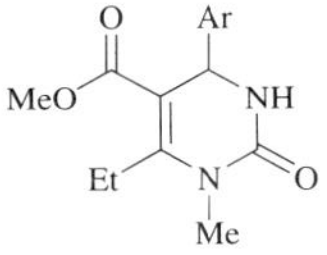 (26)	22

Ar = 2,3-Cl$_2$C$_6$H$_3$

TABLE 2. REACTIONS INVOLVING SUBSTITUTED UREAS (Continued)

A. β-Keto Esters (Continued)

β-Keto Ester, Aldehyde, and Urea	Conditions	Product(s) and Yield(s) (%)	Refs.
C₆ (EtO-acetoacetate + PhCHO-type aldehyde + methylurea)	EtOH/HCl, 78°, 2 h	(—)	160
(1,3-dichloro-2-ethoxypropanal[a] + methylurea)	EtOH/HCl, 78°, 6 h	(17) [CH₂Cl substituent]	108
(furfural + N-R urea)	EtOH/HCl, 78°, 3 h	R: Me (75), Et (70)	125
(thiophene-2-carbaldehyde + methylurea)	EtOH/HCl, 78°, 3 h	(—)	33
(benzaldehyde + methylurea)		(—)	

	Temp	Time		
EtOH/HCl	78°	3 h	(74)	33, 127, 161, 162
MeOH/H₂O, HCl	rt	3 d	(—)	6
THF/PPE	65°	15-24 h	(95)	48, 75, 104
PPE, MW	—	3 × 40 sec	(89)	34
DMF/ClSiMe₃:				
(A) urea added after 12 h	20°	14 h	(73)	48
(B) all components together	20°	2 h	(51)	126

	Ar	
THF/PPE, 65°, 15-24 h	3-O₂NC₆H₄	(86)	48, 104
	2,3-Cl₂C₆H₃	(93)	48

	Ar	
EtOH/HCl, 78°, 3 h	I (—)	33
	2-HOC₆H₄	
	3-O₂NC₆H₄	
	2,4-Cl₂C₆H₃	
	2,4-(O₂N)₂C₆H₃	
	3,5-(O₂N)₂C₆H₃	
	4-(NC)C₆H₄	
	3,4-(OCH₂O)C₆H₃	
	4-(F₃C)C₆H₄	
	3-MeOC₆H₄	
	4-MeOC₆H₄	
	2,4-Me₂C₆H₃	
	2,3-(MeO)₂C₆H₃	
	2,4-(MeO)₂C₆H₃	
	4-i-PrC₆H₄	
	1-naphthyl	
	2-naphthyl	

TABLE 2. REACTIONS INVOLVING SUBSTITUTED UREAS (Continued)
A. β-Keto Esters (Continued)

β-Keto Ester, Aldehyde, and Urea	Conditions	Product(s) and Yield(s) (%)	Refs.

C₆

EtO-C(O)-CH₂-C(O)-CH₃ ; Ar-CHO ; H₂N-C(O)-NH-Me

| 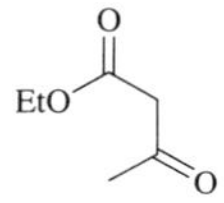 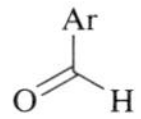 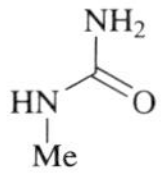 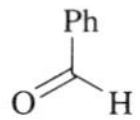 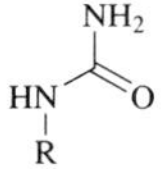 | EtOH/HCl, 78°, 3 h | 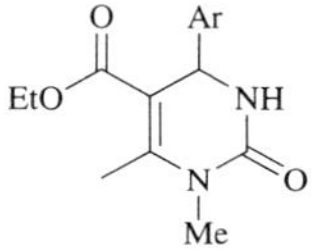 (—) Ar: 5-O₂N-naphthyl (49); 2-MeO-naphthyl (18); 4-MeO-naphthyl (55); 9-phenanthryl (47); 9-anthryl (82); 1-pyrenyl 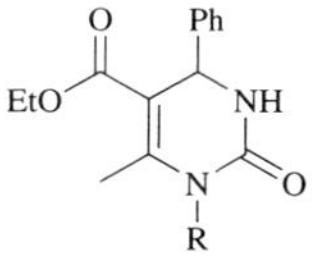 | 33 |

Ph-CHO ; H₂N-C(O)-NH-R

Product: 6-Ph, 5-CO₂Et, 4-Me, 3-N(R), 2-oxo, 1-NH dihydropyrimidinone

		R / Temp / Time	
	EtOH/HCl	Et / 78° / 3 h	161
	EtOH/AcOH, Yb(OTf)₃, MW	Et / 120° / 10 min	22
	EtOH/HCl	n-Bu / 78° / 3 h	161
	EtOH/HCl	n-hexyl / 78° / 3 h	161
	EtOH/HCl, MW	4-FC₆H₄ / — / 12 × 30 sec	30

Ar-CHO (Ar = 1-naphthyl) ; H₂N-C(O)-NH-R

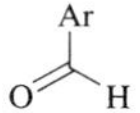

Product: 6-Ar, 5-CO₂Et, 4-Me, 3-N(R), 2-oxo dihydropyrimidinone

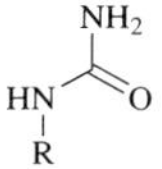

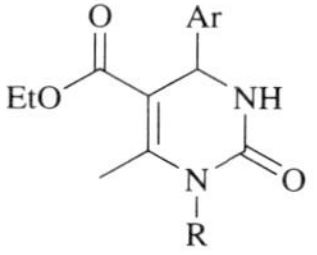

		R / Temp / Time	
	EtOH/Yb(OTf)₃, MW	allyl / 120° / 10 min (41)	22
	EtOH/HCl	allyl / 78° / 3 h (—)	33
	EtOH/HCl	Ph / 78° / 3 h (—)	33
	EtOH/HCl	3,5-Me₂C₆H₃ / 78° / 3 h (—)	33

R	Temp	Time	
H	65°	15 h	(91)
H	120°	10 min	(43)
vinyl	65°	24 h	(47)

Ar	
Ph	(74)
4-MeOC$_6$H$_4$	(78)
2-naphthyl	(62)

TABLE 2. REACTIONS INVOLVING SUBSTITUTED UREAS (Continued)
A. β-Keto Esters (Continued)

β-Keto Ester, Aldehyde, and Urea	Conditions	Product(s) and Yield(s) (%)	Refs.
C₇			
i-PrO β-ketoester; 3-NO₂-C₆H₄-CHO; allyl urea (NH₂C(O)NH-allyl)	EtOH/AcOH, Yb(OTf)₃, MW, 120°, 10 min	3,4-dihydropyrimidinone with 3-NO₂-phenyl, i-PrO ester, N-allyl (34)	22
i-PrO β-ketoester; 2-OCHF₂-C₆H₄-CHO; N-Me urea	EtOH/HCl, 78°, 3 h	3,4-dihydropyrimidinone with 2-OCHF₂-phenyl, i-PrO ester, N-Me (—)	113, 114
EtO β-ketoester (Et substituent); ArCHO; urea with CH₂CH₂OC(O)Ph	THF/HCl, 25°, 1–2 d	dihydropyrimidinone product: Ar = Ph (73); 4-MeOC₆H₄ (38); 2-naphthyl (66)	41
C₈			
t-BuO β-ketoester; PhCHO; N-Me urea	EtOH/HCl, 78°, 3 h	3,4-dihydropyrimidinone with Ph, t-BuO ester, N-Me (—)	145

C₉	aldehyde (2-OCHF₂-C₆H₄-CHO)	urea (NH₂-C(O)-NHMe)	EtOH/HCl, 78°, 3 h	product (i-Pr-CH₂CH₂-O-C(O)- dihydropyrimidinone with 2-OCHF₂-C₆H₄), (—)	113, 114
	Ar-CHO	urea (NH₂-C(O)-NHMe)	EtOH/HCl, 78°, 3 h	product with Ar, (—)	33

Ar
1-naphthyl
9-phenanthryl

	Ph-CHO	urea (NH₂-C(O)-NHMe)	EtOH/HCl, 78°	furfuryl ester dihydropyrimidinone, Ph, (74)	158
C₁₁	R-CHO (with EtO-C(O)-CH₂-C(O)-Ph)	urea (NH₂-C(O)-NHMe)	EtOH/HCl, 50°, 48 h; EtOH/HCl, 50°, 20 h; Neat/KSF-clay, 130°, 48 h	EtO₂C, Ph-substituted dihydropyrimidinone	145; 145, 162; 82

R	
Me	(—)
Ph	(—)
Ph	(75)

TABLE 2. REACTIONS INVOLVING SUBSTITUTED UREAS (Continued)
A. β-Keto Esters (Continued)

β-Keto Ester, Aldehyde, and Urea	Conditions	Product(s) and Yield(s) (%)	Refs.
C₁₁	EtOH/HCl	R: H, 78°, 3.5 h (—); Me, 50°, 48 h (—)	145
	EtOH/HCl	Ar: Ph 78° 3 h (64)	145
	PPE/THF	Ph 65° 15-24 h (93)	48, 104
	PPE/THF	2,3-Cl₂C₆H₃ 65° 15 h (93)	75, 104
	EtOH/AcOH, Yb(OTf)₃, MW	2,3-Cl₂C₆H₃ 120° 10 min (25)	22
	EtOH/HCl	3,4-F₂C₆H₃ 78° 5 h (79)	27
	MeOH/tartaric acid	2-naphthyl 65° 20 h (70)	47
C₁₂	THF/HCl, 25°, 1-2 d	Ar: Ph (56); 4-MeOC₆H₄ (18); 2-naphthyl (33)	38
C₁₄	MeOH/tartaric acid, 65°, 20 h	(53)	47

80

R	Temp	Time
Me	50°	48 h (45)
Ph	50°	24 h (50)

a This structure is a protected aldehyde.

TABLE 2. REACTIONS INVOLVING SUBSTITUTED UREAS (*Continued*)
B. β-Keto Amides

β-Keto Amide, Aldehyde, and Urea			Conditions	Product(s) and Yield(s) (%)	Refs.

C₄: H₂N-CO-CH₂-CO-CH₃ ; Ar-CHO ; H₂N-CO-NH-Me

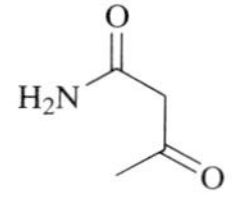

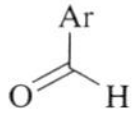

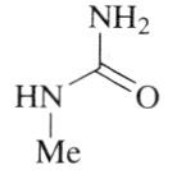

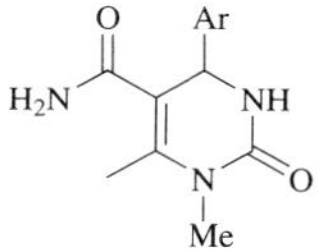

EtOH/HCl, 78°, 5 h — Ar = Ph (85) — 163
EtOH/HCl, 78°, 3 h — Ar = 1-naphthyl (—) — 33

C₆: Me₂N-CO-CH₂-CO-CH₃ ; Ar-CHO (Ar = 1-naphthyl) ; H₂N-CO-NH-Me

EtOH/HCl, 78°, 3 h — (—) — 33

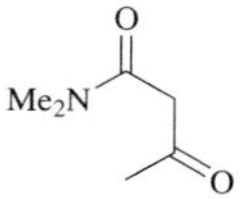

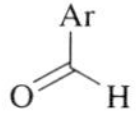

C₈: Et₂N-CO-CH₂-CO-CH₃ ; Ph-CHO ; H₂N-CO-NH-R

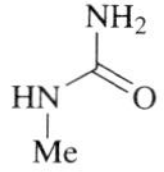

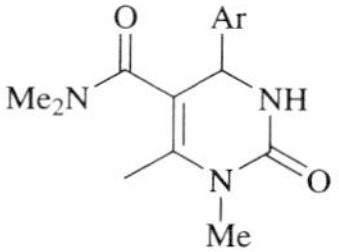

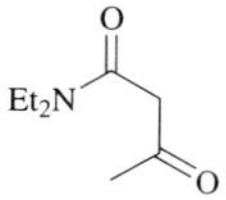

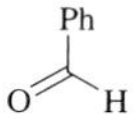

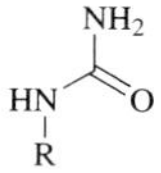

EtOH/HCl, 78°, 3 h

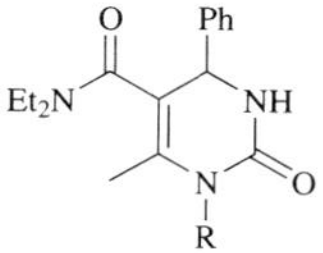

R	
Me	(75)
Et	(52)
n-Bu	(45)
n-hexyl	(41)

161

C₁₁: o-MeC₆H₄-NH-CO-CH₂-CO-CH₃ ; Ar-CHO (Ar = 1-naphthyl) ; H₂N-CO-NH-Me

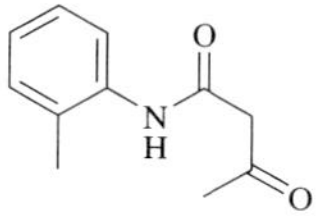

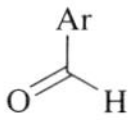

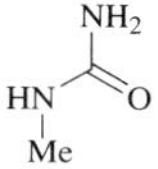

EtOH/HCl, 78°, 3 h — (—) — 33

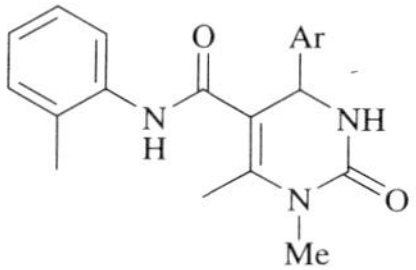

TABLE 2. REACTIONS INVOLVING SUBSTITUTED UREAS (Continued)

C. β-Diketones

β-Diketone, Aldehyde, and Urea	Conditions	Product(s) and Yield(s) (%)	Refs.
C$_5$	EtOH/HCl, 50°, 6 h	(3)	150
C$_6$	EtOH/HCl, 78°, 20 h	(57)	24, 25

TABLE 2. REACTIONS INVOLVING SUBSTITUTED UREAS (*Continued*)
D. Other CH-Acidic Carbonyl Compounds

CH-Acidic Carbonyl Compound, Aldehyde, and Urea	Conditions	Product(s) and Yield(s) (%)	Refs.
C_3 O_2N-CH$_2$-C(O)-Me ; Ar-CHO ; H$_2$N-C(O)-N(H)Me	EtOH/HCl, 78°, 6 h	Ar–NH–C(O)–N(Me) ring with O$_2$N, Me substituents Ar = Ph (64) 2-O$_2$NC$_6$H$_4$ (59) 3-O$_2$NC$_6$H$_4$ (79) 2-ClC$_6$H$_4$ (75) 2-(CF$_3$)C$_6$H$_4$ (49) 3,4,5-(MeO)$_3$C$_6$H$_2$ (75)	156, 157 157 157 157 157 157
C_6 MeO-C(O)-CH$_2$-CH(OMe)$_2$; 3,4-F$_2$C$_6$H$_3$-CHO ; H$_2$N-C(O)-N(H)Me	THF/BF$_3$·OEt$_2$/CuCl/AcOH, 65°, 8-24 h	dihydropyrimidinone with 3,4-F$_2$C$_6$H$_3$, CO$_2$Me, N-Me (—)	27

TABLE 3. REACTIONS INVOLVING THIOUREAS
A. β-Keto Esters

β-Keto Ester, Aldehyde, and Urea			Conditions	Product(s) and Yield(s) (%)	Refs.
C₅					
MeO-C(O)-CH₂-C(O)-Me	MeCHO	H₂N-C(S)-NH₂	2-Propanol/EtOH/HCl, rt	dihydropyrimidinethione (45)	164
	sugar-CHO (with OH)	H₂N-C(S)-NH₂	EtOH, 70-80°, 55 h	(27)	19
	acetonide-CHO	H₂N-C(S)-NH₂	EtOH, 78°, 12 h	(—)	18
	PhCHO	MeHN-C(S)-NH₂	AcOH/HCl, 118°, 6-8 h	(71)	165
	PhCHO	MeHN-C(S)-NHMe	AcOH/HCl, 118°, 6-8 h	(60)	165

TABLE 3. REACTIONS INVOLVING THIOUREAS (Continued)
A. β-Keto Esters (Continued)

β-Keto Ester, Aldehyde, and Urea	Conditions	Product(s) and Yield(s) (%)			Refs.
		Temp	Ar	Time	
C₅	EtOH/HCl	78°	Ph	3-8 h (88)	165-170
	2-Propanol/HCl	rt	Ph	24 h (—)	164
	EtOH/HCl	78°	2-FC₆H₄	3-4 h (—)	177
	EtOH/HCl	78°	2-ClC₆H₄	8 h (50)	166, 168, 169
	EtOH/HCl	78°	3-ClC₆H₄	8 h (76)	166-169
	EtOH/HCl	78°	4-ClC₆H₄	8 h (79)	167-169
	EtOH/HCl	78°	2-BrC₆H₄	8 h (75)	167
	EtOH/HCl	78°	4-BrC₆H₄	3-10 h (79)	167-169, 171
	2-Propanol/HCl	rt	4-BrC₆H₄	24 h (—)	164
	EtOH/HCl	78°	2-HOC₆H₃Br-5	12 h (48)	168
	2-Propanol/HCl	rt	2-ClC₆H₃F-6	24 h (—)	164
	2-Propanol/HCl	rt	3,4-Cl₂C₆H₃	24 h (—)	164
	EtOH/HCl	78°	2-O₂NC₆H₄	8-48 h (73)	166-168
	EtOH/HCl	78°	3-O₂NC₆H₄	3-12 h (73)	166-168
	2-Propanol/HCl	rt	3-O₂NC₆H₄	24 h (—)	164
	EtOH/HCl	78°	4-O₂NC₆H₄	9 h (68)	168, 169
	EtOH/HCl	78°	2-MeC₆H₄	— (—)	166
	EtOH/HCl	78°	3-MeC₆H₄	— (—)	166
	EtOH/HCl	78°	4-MeC₆H₄	3-9 h (73)	167-169, 171, 172

2-Propanol/HCl	rt	4-MeC$_6$H$_4$	24 h	(—)	164
EtOH/LaCl$_3$, MW	120°	4-MeC$_6$H$_4$	10 min	(58)	22
EtOH/HCl	78°	2-MeOC$_6$H$_4$	1-3 h	(87)	166, 168, 169
2-Propanol/HCl	rt	2-MeOC$_6$H$_4$	24 h	(—)	164
EtOH/HCl	78°	3-MeOC$_6$H$_4$	—	(—)	166
EtOH/HCl	78°	4-MeOC$_6$H$_4$	3-9 h	(75)	167-169, 171, 172
2-Propanol/HCl	rt	4-MeOC$_6$H$_4$	24 h	(—)	164
EtOH/HCl	78°	4-(MeCONH)C$_6$H$_4$	3 h	(77)	168, 169
2-Propanol/EtOH/HCl	rt	4-Me$_2$NC$_6$H$_4$	24 h	(—)	164
2-Propanol/HCl	rt	3,4,5-(MeO)$_3$C$_6$H$_2$	24 h	(—)	164

EtOH/HCl, 78°, 10 h   (95)   116

TABLE 3. REACTIONS INVOLVING THIOUREAS (Continued)

A. β-Keto Esters (Continued)

β-Keto Ester, Aldehyde, and Urea	Conditions	Product(s) and Yield(s) (%)	Refs.
C₅	EtOH/HCl, 78°, 10 h	R: H (70); MeO (70)	116
C₆	EtOH/HCl, 78°, 3 h	Ar: 2-MeC₆H₄ (32); 3-MeC₆H₄ (27); 2-MeOC₆H₄ (71); 3-MeOC₆H₄ (38)	173; 173; 167, 173; 173
	MeCN/TFA, reflux	R: H, 1.5 h (93); Me, 8 h (84)	11
	EtOH/HCl, 78°, 3 h	R: Me (22); i-Pr (21)	121, 164; 121

	Temp	Time	Ar		
PPE/EtOH, MW	—	5 x 10 sec	3-HOC$_6$H$_4$	(60)	105
EtOH/Yb(OTf)$_3$, MW	120°	20 min	3-HOC$_6$H$_4$	(45)	22
EtOH/HCl	78°	—	4-HOC$_6$H$_4$	(—)	139
EtOH/HCl, MW	—	12 x 30 sec	4-FC$_6$H$_4$	(81)	30
EtOH/HCl	78°	3 h	4-ClC$_6$H$_4$	(73)	99, 133, 139, 174, 175
2-Propanol/HCl	rt	24 h	4-BrC$_6$H$_4$	(—)	164
EtOH/HCl	78°	3 h	4-BrC$_6$H$_4$	(76)	175
EtOH/HCl, MW	—	4 min	2-O$_2$NC$_6$H$_4$	(50)	97
EtOH/HCl	78°	8 h	3-O$_2$NC$_6$H$_4$	(24)	170
EtOH/HCl, MW	—	8 x 30 sec	3-O$_2$NC$_6$H$_4$	(78)	30
PPE, MW	—	3 x 40 sec	3-O$_2$NC$_6$H$_4$	(71)	34
EtOH/HCl, MW	—	15 x 20 sec	3-O$_2$NC$_6$H$_4$	(50)	100
Neat/HCl, (A) MW	—	15 x 20 sec	3-O$_2$NC$_6$H$_4$	(53)	100
(B) thermal	120°	30 min		(50)	100
EtOH/HCl	78°	3 h	2,4-F$_2$C$_6$H$_3$	(75)	130
EtOH/HCl, MW	—	3.5 min	2,3-Cl$_2$C$_6$H$_3$	(90)	97
EtOH/HCl	78°	3 h	3,4-Cl$_2$C$_6$H$_4$	(71)	134, 175
2-Propanol/EtOH/HCl	rt	24 h	3,4-Cl$_2$C$_6$H$_4$	(—)	164
2-Propanol/HCl	rt	24 h	3-O$_2$NC$_6$H$_3$Cl-4	(—)	164
2-Propanol/HCl	rt	24 h	2-FC$_6$H$_3$Cl-6	(—)	164
EtOH/HCl	78°	—	2-FC$_6$H$_3$Cl-6	(—)	99, 133

TABLE 3. REACTIONS INVOLVING THIOUREAS (*Continued*)
A. β-Keto Esters (*Continued*)

β-Keto Ester, Aldehyde, and Urea	Conditions	Product(s) and Yield(s) (%)			Refs.
			Temp	Time	
C$_6$ EtO-CO-CH$_2$-CO-CH$_3$, Ph-CHO, H$_2$N-C(=S)-NH$_2$	EtOH/HCl	Ph, NH, S, EtO-CO, N-H, CH$_3$ (6-methyl dihydropyrimidinethione)	78°	2-8 h (76)	33, 99, 127, 139, 160, 170, 174, 176
	MeOH/H$_2$O, HCl		rt	3 d (70)	177
	MeOH/H$_2$O, 2M HCl		rt	24 h (80)	178
	EtOH/HCl, MW		—	3 min (90)	97
	2-Propanol/HCl		rt	18 h (60)	164
	Neat/inorganic support, MW,				
	(A) SiO$_2$		—	12 min (62)	98
	(B) Al$_2$O$_3$ (neutral)		—	10 min (85)	98
	(C) Al$_2$O$_3$ (basic)		—	10 min (85)	98
	(D) Al$_2$O$_3$ (acidic)		—	9 min (92)	98
	PPE, MW		—	3 x 40 sec (82)	34
	MeCN/TFA[b]		reflux	4 h (95)	11
	EtOH/LaCl$_3$		78°	5 h (96)	111
	THF/InCl$_3$		65°	9 h (91)	90
	EtOH/HCl, MW,				
	(A) reflux		78°	3 h (33)	100
	(B) superheating		96°	3 h (35)	100
	(C) open vessel		—	15 x 20 sec (58)	100
	Neat/HCl, open vessel,				
	(A) MW		—	15 x 20 sec (62)	100
	(B) thermal		120°	30 min (67)	100
	EtOH/AcOH, LaCl$_3$, MW		120°	20 min (56)	22

R	Conditions	Temp	Time		Ref
2-furyl	EtOH/HCl	78°	8 h	(53)	179
3-MeC₆H₄	MeOH/HCl	rt	3 d	(70)	177
3-MeC₆H₄	EtOH/HCl, MW	—	3.5 min	(88)	97
4-MeC₆H₄	2-Propanol/EtOH/HCl	rt	24 h	(—)	164
4-MeC₆H₄	EtOH/HCl	78°	3 h	(—)	176
2-MeOC₆H₄	AcOH	118°	—	(—)	180
3-MeOC₆H₄	THF/InCl₃	65°	9 h	(90)	90
4-MeOC₆H₄	MeOH/HCl (1:1)	rt	3 d	(70)	177
4-MeOC₆H₄	EtOH/HCl	78°	3 h	(75)	174
4-MeOC₆H₄	EtOH/HCl, MW	—	3 min	(99)	97
4-MeOC₆H₄	2-Propanol/HCl	rt	24 h	(—)	164
	Neat/inorganic support, MW,				
4-MeOC₆H₄	(A) SiO₂	—	12 min	(85)	98
4-MeOC₆H₄	(B) Al₂O₃ (neutral)	—	13 min	(86)	98
4-MeOC₆H₄	(C) Al₂O₃ (basic)	—	10 min	(90)	98
4-MeOC₆H₄	(D) Al₂O₃ (acidic)	—	10 min	(98)	98
4-MeOC₆H₄	MeCN/TFA[b]	reflux	4 h	(87)	11
4-MeOC₆H₄	EtOH/LaCl₃	78°	5 h	(85)	111
2,5-(MeO)₂C₆H₃	EtOH/HCl	78°	3 h	(70)	181
4-Me₂NC₆H₄	EtOH/HCl	78°	3 h	(65)	132, 139, 174,
4-Me₂NC₆H₄	2-Propanol/HCl	rt	24 h	(—)	175
4-Et₂NC₆H₄	EtOH/HCl	78°	3 h	(54)	164
β-styryl	EtOH/HCl	78°	6 h	(78)	132
					121

TABLE 3. REACTIONS INVOLVING THIOUREAS (Continued)
A. β-Keto Esters (Continued)

β-Keto Ester, Aldehyde, and Urea	Conditions	Product(s) and Yield(s) (%)			Refs.

C$_6$ EtO$_2$C-CH$_2$-C(O)-CH$_3$; Ar-CHO ; H$_2$N-C(=S)-NH$_2$

Product: EtO$_2$C—[6-methyl-4-Ar-2-thioxo-1,2,3,4-tetrahydropyrimidine]

Conditions	Ar	Temp	Time	(Yield)	Refs.
EtOH/HCl	3-MeO-4-HOC$_6$H$_2$Br-5	78°	3 h	(85)	131
EtOH/HCl	3,4-(MeO)$_2$C$_6$H$_3$	78°	3 h	(48)	132
MeOH/H$_2$O, HCl	3,4-(MeO)$_2$C$_6$H$_3$	rt	3 d	(—)	177
EtOH/HCl, MW	3,4-(MeO)$_2$C$_6$H$_3$	—	3 min	(90)	97
EtOH/HCl	3,4-(EtO)$_2$C$_6$H$_3$	78°	—	(69)	132, 175
2-Propanol/HCl	3,4,5-(MeO)$_3$C$_6$H$_2$	rt	24 h	(—)	164
EtOH/HCl	3,4,5-(MeO)$_3$C$_6$H$_2$	78°	3 h	(77)	134
EtOH/HCl	2-EtO-4-MeOC$_6$H$_2$Et-5	78°	13-15 h	(54)	141
Neat/inorganic support, MW,	2-EtO-4-MeOC$_6$H$_2$Et-5				
(A) SiO$_2$		—	13 min	(73)	98
(B) Al$_2$O$_3$ (neutral)		—	12 min	(79)	98
(C) Al$_2$O$_3$ (basic)		—	12 min	(80)	98
(D) Al$_2$O$_3$ (acidic)		—	11 min	(87)	98
EtOH/HCl	2-EtO-4-MeOC$_6$H$_2$Pr-5	78°	13-15 h	(48)	141
Neat/inorganic support, MW.	2-EtO-4-MeOC$_6$H$_2$Pr-5				
(A) SiO$_2$		—	14 min	(75)	98
(B) Al$_2$O$_3$ (neutral)		—	12 min	(80)	98
(C) Al$_2$O$_3$ (basic)		—	12 min	(82)	98
(D) Al$_2$O$_3$ (acidic)		—	10 min	(90)	98
EtOH/HCl	2,4-(MeO)$_2$C$_6$H$_2$Et-5	78°	13-15 h	(48)	141
EtOH/HCl	2,4-(MeO)$_2$C$_6$H$_2$Pr-5	78°	13-15 h	(47)	141

	EtOH/HCl	2,4-(EtO)₂C₆H₃Et-5	78°	13-15 h	(44)	141
	EtOH/HCl	2,4-(EtO)₂C₆H₃Pr-5	78°	13-15 h	(50)	141
	EtOH/HCl	2-MeO-4-EtOC₆H₂Et-5	78°	13-15 h	(52)	141
	EtOH/HCl	2-MeO-4-EtOC₆H₂Pr-5	78°	13-15 h	(44)	141

Ar-CHO + thiourea (NH₂-C(=S)-NH-Me) →

Ar—[dihydropyrimidine with EtO₂C, Me, N-Me, C=S, NH]

	Ar	Temp	Time		
EtOH/HCl	Ph	78°	3 h	(83)	33, 127, 182
PPE, MW	Ph	—	3 x 40 sec	(78)	34
K-10-clay, MW	Ph	—	20 min	(38)	183
EtOH/LaCl₃, MW	Ph	120°	10 min	(41)	22
MeOH/H₂O, 2M HCl	Ph	rt	24 h	(70)	178
K-10-clay, MW	4-BrC₆H₄	—	20 min	(38)	183
K-10-clay, MW	3,4-Cl₂C₆H₃	—	20 min	(64)	183
MeOH/H₂O, 2M HCl	4-MeOC₆H₄	rt	24 h	(72)	178
MeOH/H₂O, 2M HCl	3,4-(OCH₂O)C₆H₃	rt	24 h	(75)	178
MeOH/H₂O, 2M HCl	3,4-(MeO)₂C₆H₃	rt	24 h	(70)	178

4-F-C₆H₄-CHO + thiourea (NH₂-C(=S)-NH-Ph) →

[4-fluorophenyl dihydropyrimidine with EtO₂C, Me, N-Ph, C=S, NH]

EtOH/HCl, MW, 12 x 30 sec				(88)	30

TABLE 3. REACTIONS INVOLVING THIOUREAS (Continued)
A. β-Keto Esters (Continued)

β-Keto Ester, Aldehyde, and Urea	Conditions	Product(s) and Yield(s) (%)	Refs.
C₆			
EtO-CO-CH₂-CO-CH₃; Ar-CHO; HN(Ar¹)-C(=S)-NH₂, Ar¹ = 4-FC₆H₄	EtOH/HCl, MW	Ar / Time / 3-O₂NC₆H₄ 14 × 30 sec (70); 4-MeOC₆H₄ 12 × 30 sec (75)	30
Ar-CHO; HN(Ar¹)-C(=S)-NH₂, Ar¹ = 2-CF₃C₆H₄	EtOH/HCl, MW	Ar / Time / Ph 22 × 30 sec (86); 4-FC₆H₄ 20 × 30 sec (65)	30
2-(EtO)₂P(O)O-C₆H₄-CHO; H₂N-C(=S)-NH₂	MeOH/H₂O, HCl, rt, 3 d	(—)	142
4-R-naphthyl-CHO; H₂N-C(=S)-NH₂	EtOH/HCl, 78°, 3 h	R / H (80); MeO (—)	33, 130; 33

94

Aldehyde	Thiourea	Conditions	Product	(%)	Ref.
(R=naphthyl-CHO, R=H, OMe)	HN(Me)C(S)NH₂	EtOH/HCl, 78°, 3 h	naphthyl-dihydropyrimidinethione (N-Me, CO₂Et, Me)	(—)	33
5-OMe-7-OH-2-Me-chromone-6-CHO	H₂NC(S)NH₂	EtOH/HCl, 78°, 10 h	chromonyl-dihydropyrimidinethione (CO₂Et, Me)	(60)	116
4-OMe-furochromone-3-CHO	H₂NC(S)NH₂	EtOH/HCl, 78°, 10 h	furochromonyl-dihydropyrimidinethione (CO₂Et, Me)	(70)	116

TABLE 3. REACTIONS INVOLVING THIOUREAS (Continued)
A. β-Keto Esters (Continued)

β-Keto Ester, Aldehyde, and Urea	Conditions	Product(s) and Yield(s) (%)	Refs.
C₆	EtOH/HCl, 78°, 10 h	(70)	116
	EtOH/HCl, 78°, 3 h	(—)	33
	EtOH/HCl, 78°, 3 h	(—)	33

R = H, Me

C₇	furfural + 3-chloropropyl acetoacetate	thiourea (H₂N-C(=S)-NH₂)	EtOH/HCl, 78°, 8 h	product: 4-(2-furyl)-6-methyl-5-(3-chloropropoxycarbonyl)-3,4-dihydropyrimidine-2(1H)-thione (58)	179
	PhCHO + EtO₂C-CH₂-CO-CH(CF₂)₂	thiourea	1. EtOH/HCl, 78°, 6 h; 2. Toluene/p-TsOH, 110°, 6 h	4-Ph-5-EtO₂C-6-H(CF₂)₂-dihydropyrimidine-2-thione (40)	63
	RCHO + PrO-CO-CH₂-CO-CH₃	thiourea	EtOH/HCl, 78°, 8 h	4-R-5-PrO₂C-6-methyl-dihydropyrimidine-2-thione	
				R: 2-furyl (90); Ph (38); 3-O₂NC₆H₄ (30)	179; 170; 170
	RCHO + i-PrO-CO-CH₂-CO-CH₃	thiourea	EtOH/HCl, 78°, 8 h	4-R-5-(i-PrO₂C)-6-methyl-dihydropyrimidine-2-thione	
				R: 2-furyl (47); Ph (37); 3-O₂NC₆H₄ (24)	179; 170; 170
C₈	furfural + RO-CO-CH₂-CO-CH₃	thiourea	EtOH/HCl, 78°, 8 h	4-(2-furyl)-5-RO₂C-6-methyl-dihydropyrimidine-2-thione	
				R: n-Bu (41); i-Bu (65); s-Bu (44); t-Bu (39)	179

TABLE 3. REACTIONS INVOLVING THIOUREAS (Continued)
A. β-Keto Esters (Continued)

β-Keto Ester, Aldehyde, and Urea			Conditions	Product(s) and Yield(s) (%)	Refs.
C₈					
			AcOH, 118°	(—) Ar: 2-FC₆H₄, 2-ClC₆H₄, 2-MeC₆H₄, 2-MeOC₆H₄, 3-MeOC₆H₄, 4-MeOC₆H₄, 2-EtOC₆H₄, 2-i-PrOC₆H₄	180
C₉					
			EtOH/HCl, 78°, 8 h	(54)	179
			1. EtOH/HCl, 78°, 6 h 2. Toluene/p-TsOH, 110°, 6 h	(48)	63
		Ar = 2-MeOC₆H₄	AcOH, 118°	(—)	180
		Ar = 2-MeOC₆H₄	AcOH, 118°	(—)	180

| C10 | aldehyde: 2-MeOC6H4CHO | thiourea | AcOH, 118° | product (dihydropyrimidinethione with CO2CH2CH2NMe2, i-Pr) | (—) | 180 |

| C11 | RCHO | thiourea | 2-Propanol/EtOH/HCl, rt | product with CO2Me, Ph | (—) | 164 |

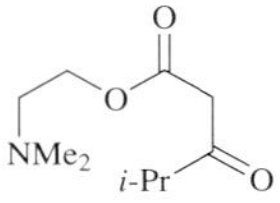

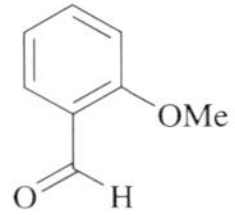

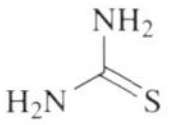

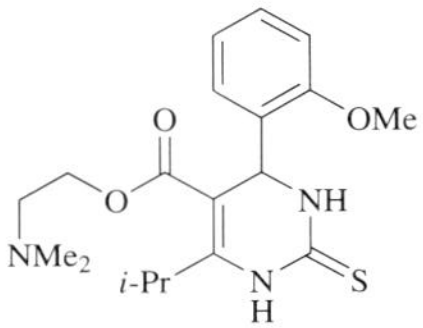

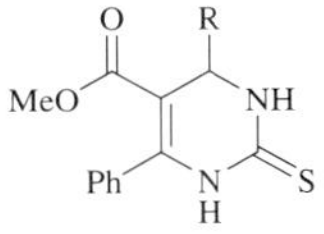

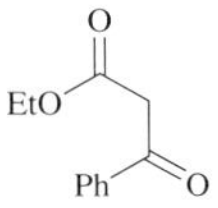

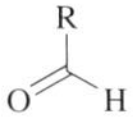

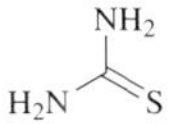

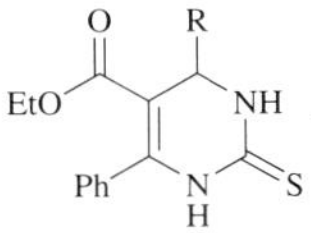

R
Me
Et
4-MeOC6H4
4-O2NC6H4
2-ClC6H3F-6

RCHO	thiourea	conditions	product with CO2Et, Ph	yield	ref
		2-Propanol/EtOH/HCl, rt	R = Me	(30)	164
		2-Propanol/EtOH/HCl, rt	R = Et	(—)	164
		EtOH/HCl, 78°, 3 h	R = Ph	(70)	174
		2-Propanol/EtOH/HCl, rt	R = 4-MeOC6H4	(—)	164
		2-Propanol/EtOH/HCl, rt	R = 4-O2NC6H4	(—)	164
		2-Propanol/EtOH/HCl, rt	R = 2-ClC6H3F-6	(—)	164

| furfural | thiourea | EtOH/HCl, 78°, 8 h | product (furyl, Me, CO2Bn) | (46) | 179 |

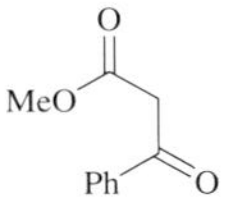

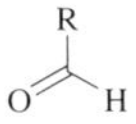

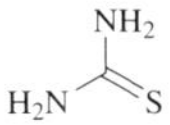

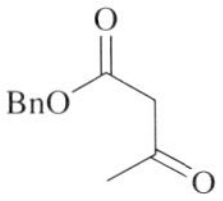

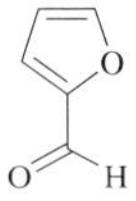

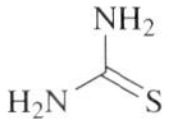

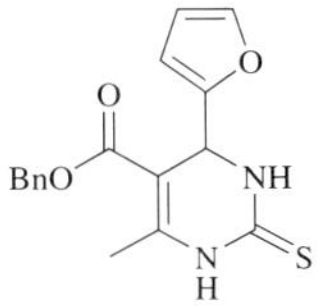

TABLE 3. REACTIONS INVOLVING THIOUREAS (*Continued*)
A. β-Keto Esters (*Continued*)

β-Keto Ester, Aldehyde, and Urea	Conditions	Product(s) and Yield(s) (%)	Refs.
C₁₁	EtOH/HCl, 78°, 3 h	(—)	33
C₁₄	AcOH, 118°	(—)	180

a This structure is an aldehyde protected as an oxazinane.
b The aldehyde was protected as an oxazinane.

TABLE 3. REACTIONS INVOLVING THIOUREAS (*Continued*)

B. β-Keto Amides

β-Keto Amide, Aldehyde, and Urea			Conditions	Product(s) and Yield(s) (%)	Refs.
C₄					
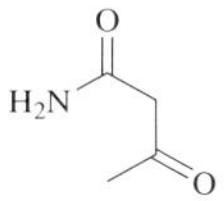 H₂N-C(=O)-CH₂-C(=O)-CH₃	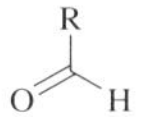 R-CHO	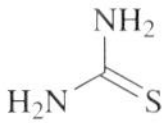 H₂N-C(=S)-NH₂	MeOH/HCl, 65° 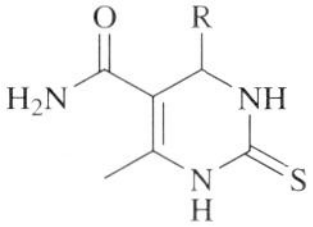	[pyrimidine-thione with R, CONH₂, Me] (—) R: Pr, i-Pr, 2-furyl	148, 148, 184
	Ar-CHO	H₂N-C(=S)-NH₂	A. MeOH/HCl, 65° B. EtOH/HCl, 78°	[pyrimidine-thione with Ar] (—) Ar: A. Ph; B. 2-ClC₆H₄; B. 4-HOC₆H₄; B. 4-O₂NC₆H₄; B. 4-MeOC₆H₄	148, 184, 184, 184, 184
	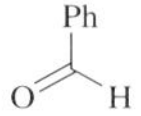 Ph-CHO	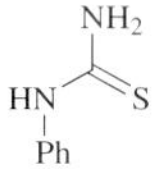 H₂N-C(=S)-NH-Ph	EtOH/HCl, MW, 120°, 15 min 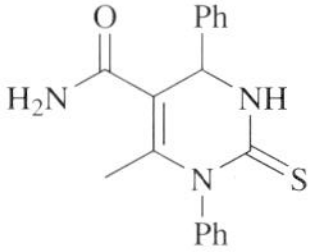	[N-Ph pyrimidine-thione with Ph] (21)	22
	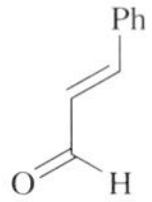 Ph-CH=CH-CHO	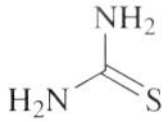 H₂N-C(=S)-NH₂	EtOH/HCl, 78° 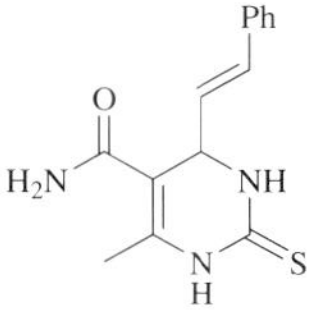	[pyrimidine-thione with styryl] (—)	184

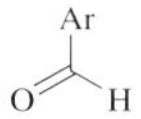

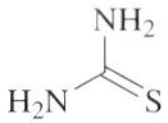

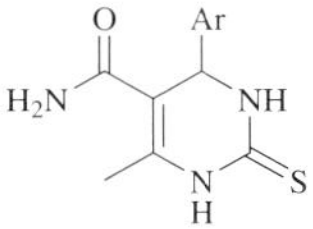

101

TABLE 3. REACTIONS INVOLVING THIOUREAS (Continued)
B. β-Keto Amides (Continued)

	β-Keto Amide, Aldehyde, and Urea		Conditions	Product(s) and Yield(s) (%)	Refs.
C₅	MeHN-CO-CH₂-CO-CH₃ ; Ar-CHO ; H₂N-CS-NH₂		EtOH/HCl, 78°, 3 h	**I** with MeHN-CO- group, Ar, methyl, NH, NH, S (structure I) Ar / Yield: Ph (85) 4-HOC₆H₄ (89) 4-ClC₆H₄ (68) 2-O₂NC₆H₄ (90) 3-O₂NC₆H₄ (91) 4-O₂NC₆H₄ (65) 2,4-Cl₂C₆H₃ (80) 3,4-Cl₂C₆H₃ (64) 4-MeOC₆H₄ (62) 3-MeOC₆H₃OH-2 (80) 3,4-(OCH₂O)C₆H₃ (68) 3,4-(MeO)₂C₆H₃ (62) 3,4,5-(MeO)₃C₆H₂ (79)	149
			EtOH/AcOH, Yb(OTf)₃, MW, 120°, 10 min	**I** (66) Ar = 2-O₂NC₆H₄	22
C₈	Me₂N-(CH₂)₃-NH-CO-CH₂-CO-CH₃ ; Ar-CHO (Ar = 2-MeOC₆H₄) ; H₂N-CS-NH₂		AcOH, 118°	structure with Ar, NMe₂-propyl-NH-CO, methyl, NH, NH, S (—)	180
C₁₀	PhHN-CO-CH₂-CO-CH₃ ; R-CHO ; H₂N-CS-NH₂	R = Me, Et R = Pr, i-Pr	2-Propanol/EtOH/HCl, rt MeOH/HCl, 65°	structure with R, PhHN-CO, methyl, NH, NH, S (—) (—)	164 148

| Ar-CHO | H2N-C(=S)-NH2 | | Product (Ar, R substituent) | |

Reaction scheme: ArCHO + thiourea + PhHN-CO-CH2-CO-CH3 → 4-Ar-6-methyl-2-thioxo-1,2,3,4-tetrahydropyrimidine-5-carboxanilide

R	Conditions	(%)	Refs.
Ph	MeOH/HCl, 65°	(—)	148
Ph	EtOH/HCl, 78°, 3 h	(71)	185
Ph	2-Propanol/HCl, 78°, 3 h	(—)	164
2-ClC$_6$H$_4$	EtOH/HCl, 78°, 3 h	(84)	185
2-ClC$_6$H$_4$	EtOH/LaCl$_3$, MW, 120°, 10 min	(89)	22
4-ClC$_6$H$_4$	EtOH/HCl, 78°, 3 h	(70)	185
4-BrC$_6$H$_4$	2-Propanol/HCl, rt	(—)	164
3-O$_2$NC$_6$H$_4$	2-Propanol/HCl, rt	(—)	164
4-O$_2$NC$_6$H$_4$	EtOH/HCl, 78°, 3 h	(72)	185
3-O$_2$NC$_6$H$_3$Cl-4	2-Propanol/HCl, rt	(—)	164
3,4-Cl$_2$C$_6$H$_3$	2-Propanol/HCl, rt	(—)	164
2-FC$_6$H$_3$Cl-6	2-Propanol/HCl, rt	(—)	164
4-MeC$_6$H$_4$	2-Propanol/HCl, rt	(—)	164
2-MeOC$_6$H$_4$	EtOH/HCl, 78°, 3 h	(82)	185
4-MeOC$_6$H$_4$	EtOH/HCl, 78°, 3 h	(65)	185
4-Me$_2$NC$_6$H$_4$	EtOH/HCl, 78°, 3 h	(65)	185
3,4,5-(MeO)$_3$C$_6$H$_2$	2-Propanol/HCl, rt	(—)	164

Reaction scheme: ArCHO + thiourea + 3-chloroanilide of acetoacetic acid → N-(3-chlorophenyl)-4-Ar-6-methyl-2-thioxo-1,2,3,4-tetrahydropyrimidine-5-carboxamide; EtOH, 78°, 6 h — 186

Ar	(%)
Ph	(63)
4-O$_2$NC$_6$H$_4$	(60)
4-ClC$_6$H$_4$	(57)
3,4,5-(MeO)$_3$C$_6$H$_2$	(54)

TABLE 3. REACTIONS INVOLVING THIOUREAS (*Continued*)
B. β-Keto Amides (*Continued*)

β-Keto Amide, Aldehyde, and Urea	Conditions	Product(s) and Yield(s) (%)	Refs.
C₁₂ (β-keto amide with MeO, OMe substituted anilide; ArCHO; H₂N-C(=S)-NH₂)	EtOH, 78°, 6 h	Dihydropyrimidine-thione product with Ar variation: Ar: Ph (62); 2-O₂NC₆H₄ (52); 4-ClC₆H₄ (48); 4-MeOC₆H₄ (61); 3,4-(MeO)₂C₆H₃ (39); 3,4,5-(MeO)₃C₆H₂ (56)	186
C₁₅ (PhHN-CO-CH₂-CO-Ph; RCHO; thiourea)	2-Propanol/HCl, rt	Dihydropyrimidine-thione product (—) R: Me; 4-O₂NC₆H₄; 2-ClC₆H₃F-6; 4-MeOC₆H₃	164
C₁₅ (pyrazolone β-keto amide; ArCHO; thiourea)	EtOH/HCl, 78°, 3–5 h	Dihydropyrimidine-thione product with Ar variation: Ar: Ph (66); 4-FC₆H₄ (54); 4-ClC₆H₄ (73); 4-BrC₆H₄ (75); 4-O₂NC₆H₄ (58); 3,4-Cl₂C₆H₃ (79); 4-MeOC₆H₄ (55); 3,4,5-(MeO)₃C₆H₂ (69)	134

TABLE 3. REACTIONS INVOLVING THIOUREAS (Continued)
C. β-Diketones

	β-Diketone, Aldehyde, and Urea			Conditions	Product(s) and Yield(s) (%)	Refs.
C_5	(pentane-2,4-dione)	PhCHO	thiourea (HN-R, NH₂, S)	EtOH/HCl, 78°	R = H (—); R = Me (72) (4-Ph-5-acetyl-6-methyl-3,4-dihydropyrimidine-2(1H)-thione)	99
				MeOH/H₂O, 2 M HCl, rt, 24 h		178
C_6	(cyclohexane-1,3-dione)	ArCHO	thiourea (H_2N, NH_2, S)	EtOH/HCl, 78°	(octahydroquinazoline-thione, Ar variations)	
					Ar = Ph, 20 h (35)	187
					Ar = 2-ClC₆H₄, 20 h (42)	187
					Ar = 3-ClC₆H₄, 20 h (49)	187
					Ar = 4-ClC₆H₄, 20 h (51)	187
					Ar = 2-BrC₆H₄, 20 h (46)	187
					Ar = 3-BrC₆H₄, 20 h (25)	187
					Ar = 4-BrC₆H₄, 20 h (56)	187
					Ar = 4-FC₆H₄, 3-4 h (84)	188, 189
					Ar = 2-MeC₆H₄, 20 h (53)	187
					Ar = 3-MeC₆H₄, 20 h (41)	187
					Ar = 4-MeC₆H₄, 20 h (41)	187
					Ar = 2-MeOC₆H₄, 20 h (49)	187
					Ar = 3-MeOC₆H₄, 20 h (37)	187
					Ar = 4-MeOC₆H₄, 20 h (45)	187
C_{10}	(1-phenylbutane-1,3-dione)	PhCHO	thiourea (H_2N, NH_2, S)	THF/InCl₃, 65°, 8 h	(90)	90

TABLE 3. REACTIONS INVOLVING THIOUREAS (*Continued*)
C. β-Diketones (*Continued*)

β-Diketone, Aldehyde, and Urea	Conditions	Product(s) and Yield(s) (%)	Refs.
C$_{10}$ PhCOCH$_2$COCH$_3$ + terephthalaldehyde + H$_2$N-C(=S)-NH$_2$	EtOH/HCl, 78°, 5 h	bis-dihydropyrimidinethione product (88)	21

TABLE 3. REACTIONS INVOLVING THIOUREAS (Continued)
D. Other CH-Acidic Carbonyl Compounds

CH-Acidic Carbonyl Compound, Aldehyde, and Urea	Conditions	Product(s) and Yield(s) (%)	Refs.
C9	MeOH/HCl, 65°, 4 h	Ar / Ph (54) / 4-MeC6H4 (59) / 4-MeOC6H4 (50) / 3,4-(MeO)2C6H3 (47)	26
C10	MeOH/HCl, 65°, 4 h	Ar / Ph (61) / 4-MeC6H4 (59) / 4-MeOC6H4 (55) / 3,4-(MeO)2C6H3 (49)	26
C13	MeOH/HCl, 65°, 4 h	Ar / Ph (45) / 4-MeC6H4 (50) / 4-MeOC6H4 (47) / 3,4-(MeO)2C6H3 (46)	26

TABLE 4. REACTIONS INVOLVING GUANIDINE

β-Keto Ester and Aldehyde	Conditions	Product(s) and Yield(s) (%)	Refs.
C₁₁			
EtO−C(O)−CH₂−C(O)−Ph ; thiophene-2-carbaldehyde	DMF/NaHCO₃, 70°, 3 h	4-(thien-2-yl)-5-ethoxycarbonyl-6-phenyl-2-amino-1,4-dihydropyrimidine (85)	31
EtO−C(O)−CH₂−C(O)−Ph ; 4-R-C₆H₄−CHO	DMF/NaHCO₃, 70°, 3 h	4-(4-R-C₆H₄)-5-ethoxycarbonyl-6-phenyl-2-amino-1,4-dihydropyrimidine; R: H (75), Cl (85), Me (85), MeO (75)	31

TABLE 5. REACTIONS ON SOLID PHASE

β-Keto Ester, Aldehyde, and Urea	Conditions	Product(s) and Yield(s) (%)	Refs.
C_5 (MeO β-ketoester, PhCHO, urea-resin)	Urea on PS-Wang resin 1. THF/HCl, 55° 2. Cleavage: TFA/DCM, rt	dihydropyrimidinone product with MeO ester, Ph, and butyric acid chain (93)	36
C_6 (EtO β-ketoester, ArCHO, urea-resin)	Urea on PS-Wang resin 1. THF/HCl, 55° 2. Cleavage: TFA/DCM, rt	dihydropyrimidinone product with EtO ester, Ar, and butyric acid chain Ar Ph (80) 3-$O_2NC_6H_4$ (98) 2-ClC_6H_4 (87) 4-HOC_6H_4 (87) 4-$MeOC_6H_4$ (93) 2-naphthyl (87)	36
C_7 (EtO β-ketoester with Et, PhCHO, urea-resin)	Urea on PS-Wang resin 1. THF/HCl, 55° 2. Cleavage: TFA/DCM, rt	dihydropyrimidinone product with EtO ester, Ph, Et, and butyric acid chain (93)	36
C_{11} (BnO β-ketoester, PhCHO, urea-resin)	Urea on PS-Wang resin 1. THF/HCl, 55° 2. Cleavage: TFA/DCM, rt	dihydropyrimidinone product with BnO ester, Ph, and butyric acid chain (93)	36

TABLE 5. REACTIONS ON SOLID PHASE (*Continued*)

β-Keto Ester, Aldehyde, and Urea	Conditions	Product(s) and Yield(s) (%)	Refs.
C₁₁ (resin-bound acetoacetate, PhCHO, H₂N-C(=Y)-NH₂)	Ester on PS-Wang resin 1. Dioxane/HCl, 70°, 18 h 2. Cleavage: TFA/DCM, rt	Y / O (81) / S (44)	39
(resin-bound acetoacetate, ArCHO, urea)	Ester on PS-Wang resin 1. Dioxane/HCl, 70°, 18 h 2. Cleavage: TFA/DCM, rt	Ar 3-O₂NC₆H₄ (86) 4-O₂NC₆H₄ (80) 2-ClC₆H₄ (77) 2,3-Cl₂C₆H₃ (86) 3,4-F₂C₆H₃ (73) 4-MeC₆H₄ (83) 2-(CF₃)C₆H₄ (70)	39
C₁₂ (resin-bound β-keto ester with Et, ArCHO, urea)	Ester on PS-Wang resin 1. Dioxane/HCl, 70°, 18 h 2. Cleavage: TFA/DCM, rt	Ar 3-O₂NC₆H₄ (88) 3,4-F₂C₆H₃ (73)	39
C₁₃ (resin-bound urea with NH₂, EtO-β-ketoester with Ph, PhCHO)	Urea on PS-Wang resin 1. THF/HCl, 55° 2. Cleavage: TFA/DCM, rt	(67)	36

TABLE 6. REACTIONS IN FLUOROUS PHASE

β-Keto Ester, Aldehyde, and Urea	Conditions	Product(s) and Yield(s) (%)	Refs.
C5 (methyl acetoacetate; 2-naphthaldehyde; urea with R_{fh} tag, $R_{fh} = C_{10}F_{21}CH_2CH_2$)	1. THF/BTF, HCl, 50°, 3 d 2. TBAF in THF/BTF, 25°, 30 min	(2-naphthyl dihydropyrimidinone, MeO ester) (70)	38
C6 (ethyl acetoacetate; ArCHO; urea with R_{fh} tag)	1. THF/BTF, HCl, 50°, 3 d 2. TBAF in THF/BTF, 25°, 30 min	(Ar dihydropyrimidinone, EtO ester) Ar Ph (71) 4-MeOC6H4 (69) 2-naphthyl (55)	38
C7 (ethyl 2-ethylacetoacetate; ArCHO; urea with R_{fh} tag)	1. THF/BTF, HCl, 50°, 3 d 2. TBAF in THF/BTF, 25°, 30 min	(Ar dihydropyrimidinone with Et, EtO ester) Ar Ph (47) 2-naphthyl (60)	37, 38

REFERENCES

[1] Biginelli, P. *Gazz. Chim. Ital.* **1893**, *23*, 360.
[2] Kappe, C. O. *Tetrahedron* **1993**, *49*, 6937.
[3] Kappe, C. O. *Eur. J. Med. Chem.* **2000**, *35*, 1043.
[4] Kappe, C. O. *Acc. Chem. Res.* **2000**, *33*, 879.
[5] Folkers, K.; Johnson, T. B. *J. Am. Chem. Soc.* **1933**, *55*, 3784.
[6] Sweet, F.; Fissekis, J. D. *J. Am. Chem. Soc.* **1973**, *95*, 8741.
[7] Kappe, C. O. *J. Org. Chem.* **1997**, *62*, 7201.
[8] Petersen, H. *Synthesis* **1973**, 243.
[9] Kappe, C. O.; Falsone, S. F.; Fabian, W. M. F.; Belaj, F. *Heterocycles* **1999**, *51*, 77.
[10] Hu, E. H.; Sidler, D. R.; Dolling, U. H. *J. Org. Chem.* **1998**, *63*, 3454.
[11] Singh, K.; Singh, J.; Deb, P. K.; Singh, H. *Tetrahedron* **1999**, *55*, 12873.
[12] Lopez Sastre, J. A.; Molina Molina, J. *An. Quim.* **1978**, *74*, 353; *Chem. Abstr.* **1978**, *89*, 163889.
[13] Lopez Aparicio, F. J.; Lopez Herrera, F. J. *Carbohydr. Res.* **1979**, *69*, 243.
[14] Lopez Aparicio, F. J.; Lopez Sastre, J. A.; Molina Molina, J. *Carbohydr. Res.* **1981**, *95*, 113.
[15] Lopez Aparicio, F. J.; Lopez Sastre, J. A.; Molina Molina, J.; Lopez Herrera, F. J. *An. Quim., Ser. C* **1981**, *77*, 147; *Chem. Abstr.* **1982**, *97*, 72696.
[16] Lopez Aparicio, F. J.; Lopez Sastre, J. A.; Molina Molina, J.; Romero-Avila Garcia, M. C. *An. Quim., Ser. C* **1981**, *77*, 348; *Chem. Abstr.* **1982**, *97*, 39304.
[17] Molina Molina, J.; Abad Lorenzo, J. P.; Lopez Sastre, J. A. *An. Quim., Ser. C* **1982**, *78*, 250; *Chem. Abstr.* **1982**, *97*, 145217.
[18] Valpuesta Fernandez, M.; Lopez Herrera, F. J.; Lupion Cobos, T. *Heterocycles* **1986**, *24*, 679.
[19] Valpuesta Fernandez, M.; Lopez Herrera, F. J.; Lupion Cobos, T. *Heterocycles* **1988**, *27*, 2133.
[20] Dondoni, A.; Massi, A.; Sabbatini, S. *Tetrahedron Lett.* **2001**, *42*, 4495.
[21] Shaker, R. M.; Abdel-Latif, F. F. *J. Chem. Res. (S)* **1997**, 294.
[22] Stadler, A.; Kappe, C. O. *J. Comb. Chem.* **2001**, *3*, 624.
[23] Chiba, T.; Sato, H.; Kato, T. *Heterocycles* **1984**, *22*, 493.
[24] Mayer, K. K.; Dove, S.; Pongratz, H.; Ertan, M.; Wiegrebe, W. *Heterocycles* **1998**, *48*, 1169.
[25] Sarac, S.; Yarim, M.; Ertan, M.; Boydag, S.; Erol, K. *Pharmazie* **1998**, *53*, 91; *Chem. Abstr.* **1998**, *128*, 230344.
[26] Brahmbhatt, D. I.; Raolji, G. B.; Pandya, S. U.; Pandya, U. R. *Indian J. Chem.* **1999**, *38B*, 839; *Chem. Abstr.* **1999**, *132*, 64235.
[27] Barrow, J. C.; Nantermet, P. G.; Selnick, H. G.; Glass, K. L.; Rittle, K. E.; Gilbert, K. F.; Steele, T. G.; Homnick, C. F.; Freidinger, R. M.; Ransom, R. W.; Kling, P.; Reiss, D.; Broten, T. P.; Schorn, T. W.; Chang, R. S. L.; O'Malley, S. S.; Olah, T. V.; Ellis, J. D.; Barrish, A.; Kassahun, K.; Leppert, P.; Nagarathnam, D.; Forray, C. *J. Med. Chem.* **2000**, *43*, 2703.
[28] Remennikov, G. Y. *Chem. Het. Compounds (New York)* **1997**, *33*, 1369; *Chem. Abstr.* **1998**, *129*, 216523.
[29] Namazi, H.; Mirzaei, Y. R.; Azamat, H. *J. Heterocycl. Chem.* **2001**, *38*, 1051.
[30] Dandia, A.; Saha, M.; Taneja, H. *J. Fluorine Chem.* **1998**, *90*, 17.
[31] Vanden Eynde, J. J.; Hecq, N.; Kataeva, O.; Kappe, C. O. *Tetrahedron* **2001**, *57*, 1785.
[32] Lewandowski, K.; Murer, P.; Svec, F.; Frechet, J. M. J. *Chem. Commun.* **1998**, 2237.
[33] Lewandowski, K.; Murer, P.; Svec, F.; Frechet, J. M. J. *J. Comb. Chem.* **1999**, *1*, 105.
[34] Kappe, C. O.; Kumar, D.; Varma, R. S. *Synthesis* **1999**, 1799.
[35] Dondoni, A.; Massi, A. *Tetrahedron Lett.* **2001**, *42*, 7975.
[36] Wipf, P.; Cunningham, A. *Tetrahedron Lett.* **1995**, *36*, 7819.
[37] Studer, A.; Hadida, S.; Ferritto, R.; Kim, S.-Y.; Jeger, P.; Wipf, P.; Curran, D. P. *Science* **1997**, *275*, 823.
[38] Studer, A.; Jeger, P.; Wipf, P.; Curran, D. P. *J. Org. Chem.* **1997**, *62*, 2917.
[39] Valverde, M. G.; Dallinger, D.; Kappe, C. O. *Synlett* **2001**, 741.
[40] Robinett, L. D.; Yager, K. M.; Phelan, J. C. In *211th National Meeting of the American Chemical Society*; Organic Division, Abstract of Papers: New Orleans, LA, 1996.
[41] Kappe, C. O. *Bioorg. Med. Chem. Lett.* **2000**, *10*, 49.

[42] Atwal, K. S.; Swanson, B. N.; Unger, S. E.; Floyd, D. M.; Moreland, S.; Hedberg, A.; O'Reilly, B. C. *J. Med. Chem.* **1991**, *34*, 806.

[43] Rovnyak, G. C.; Atwal, K. S.; Hedberg, A.; Kimball, S. D.; Moreland, S.; Gougoutas, J. Z.; O'Reilly, B. C.; Schwartz, J.; Malley, M. F. *J. Med. Chem.* **1992**, *35*, 3254.

[44] Grover, G. J.; Dzwonczyk, S.; McMullen, D. M.; Normandin, D. E.; Parham, C. S.; Sleph, P. G.; Moreland, S. *J. Cardiovasc. Pharmacol.* **1995**, *26*, 289.

[45] Rovnyak, G. C.; Kimball, S. D.; Beyer, B.; Cucinotta, G.; DiMarco, J. D.; Gougoutas, J.; Hedberg, A.; Malley, M.; McCarthy, J. P.; Zhang, R. A.; Moreland, S. *J. Med. Chem.* **1995**, *38*, 119.

[46] Nagarathnam, D.; Miao, S. W.; Lagu, B.; Chiu, G.; Fang, J.; Dhar, T. G. M.; Zhang, J.; Tyagarajan, S.; Marzabadi, M. R.; Zhang, F. Q.; Wong, W. C.; Sun, W. Y.; Tian, D.; Wetzel, J. M.; Forray, C.; Chang, R. S. L.; Broten, T. P.; Ransom, R. W.; Schorn, T. W.; Chen, T. B.; O'Malley, S.; Kling, P.; Schneck, K.; Bendesky, R.; Harrell, C. M.; Vyas, K. P.; Gluchowski, C. *J. Med. Chem.* **1999**, *42*, 4764.

[47] Kappe, C. O.; Uray, G.; Roschger, P.; Lindner, W.; Kratky, C.; Keller, W. *Tetrahedron* **1992**, *48*, 5473.

[48] Schnell, B.; Krenn, W.; Faber, K.; Kappe, C. O. *J. Chem. Soc., Perkin Trans. 1* **2000**, 4382.

[49] Schnell, B.; Strauss, U. T.; Verdino, P.; Faber, K.; Kappe, C. O. *Tetrahedron: Asymmetry* **2000**, *11*, 1449.

[50] Kleidernigg, O. P.; Kappe, C. O. *Tetrahedron: Asymmetry* **1997**, *8*, 2057.

[51] Wang, F.; Loughlin, T.; Dowling, T.; Bicker, G.; Wyvratt, J. *J. Chromatogr. A* **2000**, *872*, 279.

[52] Lecnik, O.; Schmid, M. G.; Kappe, C. O.; Gübitz, G. *Electrophoresis* **2001**, *22*, 3198.

[53] Krenn, W.; Verdino, P.; Uray, G.; Faber, K.; Kappe, C. O. *Chirality* **1999**, *11*, 659.

[54] Uray, G.; Verdino, P.; Belaj, F.; Kappe, C. O.; Fabian, W. M. F. *J. Org. Chem.* **2001**, *66*, 6685.

[55] McDonald, A. I.; Overman, L. E. *J. Org. Chem.* **1999**, *64*, 1520.

[56] Franklin, A. S.; Ly, S. K.; Mackin, G. H.; Overman, L. E.; Shaka, A. J. *J. Org. Chem.* **1999**, *64*, 1512.

[57] Cohen, F.; Overman, L. E.; Sakata, S. K. L. *Org. Lett.* **1999**, *1*, 2169.

[58] Overman, L. E.; Rabinowitz, M. H.; Renhowe, P. A. *J. Am. Chem. Soc.* **1995**, *117*, 2657.

[59] Coffey, D. S.; McDonald, A. I.; Overman, L. E.; Rabinowitz, M. H.; Renhowe, P. A. *J. Am. Chem. Soc.* **2000**, *122*, 4893.

[60] Coffey, D. S.; Overman, L. E.; Stappenbeck, F. *J. Am. Chem. Soc.* **2000**, *122*, 4904.

[61] Coffey, D. S.; McDonald, A. I.; Overman, L. E.; Stappenbeck, F. *J. Am. Chem. Soc.* **1999**, *121*, 6944.

[62] Cohen, F.; Overman, L. E. *J. Am. Chem. Soc.* **2001**, *123*, 10782.

[63] Saloutin, V. I.; Burgart, Y. V.; Kuzueva, O. G.; Kappe, C. O.; Chupakhin, O. N. *J. Fluorine Chem.* **2000**, *103*, 17.

[64] Svetlik, J.; Hanus, V.; Bella, J. *J. Chem. Res. (S)* **1991**, 4.

[65] Baldwin, J. J. U.S. Patent 4,609,494 (1986); *Chem. Abstr.* **1987**, *106*, 18636.

[66] Rehani, R.; Shah, A. C. *Indian J. Chem.* **1994**, *33B*, 775; *Chem. Abstr.* **1994**, *122*, 9985.

[67] Byk, G.; Gottlieb, H. E.; Herscovici, J.; Mirkin, F. *J. Comb. Chem.* **2000**, *2*, 732.

[68] Bussolari, J. C.; McDonnell, P. A. *J. Org. Chem.* **2000**, *65*, 6777.

[69] O'Reilly, B. C.; Atwal, K. S. *Heterocycles* **1987**, *26*, 1185.

[70] Atwal, K. S.; O'Reilly, B. C.; Gougoutas, J. Z.; Malley, M. F. *Heterocycles* **1987**, *26*, 1189.

[71] Atwal, K. S.; Rovnyak, G. C.; O'Reilly, B. C.; Schwartz, J. *J. Org. Chem.* **1989**, *54*, 5898.

[72] Sár, C. P.; Hankovszky, O. H.; Jerkovich, G.; Pallagi, I.; Hideg, K. *ACH-Models in Chemistry* **1994**, *131*, 363.

[73] Vanden Eynde, J. J.; Audiart, N.; Canonne, V.; Michel, S.; Van Haverbeke, Y.; Kappe, C. O. *Heterocycles* **1997**, *45*, 1967.

[74] Shutalev, A. D.; Kishko, E. A.; Sivova, N. V.; Kuznetsov, A. Y. *Molecules* **1998**, *3*, 100.

[75] Kappe, C. O.; Falsone, S. F. *Synlett* **1998**, 718.

[76] Ramalinga, K.; Vijayalakshmi, P.; Kaimal, T. N. B. *Synlett* **2001**, 863.

[77] Jauk, B.; Pernat, T.; Kappe, C. O. *Molecules* **2000**, *5*, 227.

[78] Folkers, K.; Harwood, H. J.; Johnson, T. B. *J. Am. Chem. Soc.* **1932**, *54*, 3751.

[79] Yadav, J. S.; Reddy, B. V. S.; Reddy, E. J.; Ramalingam, T. *J. Chem. Res. (S)* **2000**, 354.

[80] Ehsan, A.; Karimullah *Pakistan J. Sci. Ind. Res.* **1967**, *10*, 83; *Chem. Abstr.* **1967**, *68*, 78231.

[81] Peng, J.; Deng, Y. *Tetrahedron Lett.* **2001**, *42*, 5917.

[82] Bigi, F.; Carloni, S.; Frullanti, B.; Maggi, R.; Sartori, G. *Tetrahedron Lett.* **1999**, *40*, 3465.

[83] Ma, Y.; Qian, C.; Wang, L. M.; Yang, M. *J. Org. Chem.* **2000**, *65*, 3864.

[84] Lu, J.; Bai, Y. J.; Wang, Z. J.; Yang, B.; Ma, H. R. *Tetrahedron Lett.* **2000**, *41*, 9075.
[85] Lu, J.; Ma, H. R. *Synlett* **2000**, 63.
[86] Tu, S. J.; Zhou, J. F.; Cai, P. J.; Wang, H.; Feng, J. C. *Synth. Commun.* **2002**, *32*, 147.
[87] Lu, J.; Bai, Y.-J.; Guo, Y.-H.; Wang, Z.-J.; Ma, H.-R. *Chinese J. Chem.* **2002**, *20*, 681; *Chem. Abstr.* **2002**, *138*, 4503.
[88] Lu, J.; Bai, Y. *Synthesis* **2002**, 466.
[89] Chen, R. F.; Qian, C. T. *Chinese J. Chem.* **2002**, *20*, 427; *Chem. Abstr.* **2002**, *137*, 310885.
[90] Ranu, B. C.; Hajra, A.; Jana, U. *J. Org. Chem.* **2000**, *65*, 6270.
[91] Fu, N. Y.; Yuan, Y. F.; Cao, Z.; Wang, S. W.; Wang, J. T.; Peppe, C. *Tetrahedron* **2002**, *58*, 4801.
[92] Yadav, J. S.; Reddy, B. V. S.; Srinivas, R.; Venugopal, C.; Ramalingam, T. *Synthesis* **2001**, 1341.
[93] Kumar, K. A.; Kasthuraiah, M.; Reddy, C. S.; Reddy, C. D. *Tetrahedron Lett.* **2001**, *42*, 7873.
[94] Reddy, C. V.; Mahesh, M.; Raju, P. V. K.; Babu, T. R.; Reddy, V. V. N. *Tetrahedron Lett.* **2002**, *43*, 2657.
[95] Rani, V. R.; Srinivas, N.; Kishan, M. R.; Kulkarni, S. J.; Raghavan, K. V. *Green Chem.* **2001**, *3*, 305.
[96] Jin, T. S.; Zhang, S. L.; Zhang, S. Y.; Guo, J. J.; Li, T. S. *J. Chem. Res. (S)* **2002**, 37.
[97] Gupta, R.; Gupta, A. K.; Paul, S.; Kachroo, P. L. *Indian J. Chem.* **1995**, *34B*, 151; *Chem. Abstr.* **1995**, *122*, 160598.
[98] Gupta, R.; Paul, S.; Gupta, A. K. *Indian J. Chem. Technology* **1998**, *5*, 340.
[99] Mallakpour, S. E.; Hajipour, A.-R.; Faghihi, K.; Foroughifar, N.; Bagheri, J. *J. Appl. Polym. Sci.* **2001**, *80*, 2416.
[100] Stadler, A.; Kappe, C. O. *J. Chem. Soc., Perkin Trans. 2* **2000**, 1363.
[101] Stefani, H. A.; Gatti, P. M. *Synth. Commun.* **2000**, *30*, 2165.
[102] Xue, S.; Shen, Y.-C.; Li, Y.-L.; Shen, X.-M.; Guo, Q.-X. *Chinese J. Chem.* **2002**, *20*, 385; *Chem. Abstr.* **2002**, *137*, 169482.
[103] Yadav, J. S.; Reddy, B. V. S.; Reddy, K. B.; Raj, K. S.; Prasad, A. R. *J. Chem. Soc., Perkin Trans. 1* **2001**, 1939.
[104] Falsone, F. S.; Kappe, C. O. *Arkivoc* **2001**, *2* (2), 122; *Chem. Abstr.* **2001**, *137*, 154900.
[105] Kappe, C. O.; Shishkin, O. V.; Uray, G.; Verdino, P. *Tetrahedron* **2000**, *56*, 1859.
[106] Duburs, G.; Khanina, E. L. *Khim. Geterotsikl. Soedin.* **1976**, 220; *Chem. Abstr.* **1976**, *85*, 32946.
[107] Remennikov, G. Y.; Shavaran, S. S.; Boldyrev, I. V.; Kurilenko, L. K.; Klebanov, B. M.; Kukhar, V. P. *Khim.-Farm. Zh.* **1991**, *25*, 35; *Chem. Abstr.* **1991**, *115*, 71570.
[108] Ashby, J.; Griffiths, D. *J. Chem. Soc., Perkin Trans. 1* **1975**, 657.
[109] Taguchi, H.; Yazawa, H.; Arnett, J. F.; Kishi, Y. *Tetrahedron Lett.* **1977**, 627.
[110] Hull, R.; Swain, G. GB Patent 868,030 (1958); *Chem. Abstr.* **1958**, *56*, 7729.
[111] Lu, J.; Ma, H. R.; Li, W. H. *Chinese J. Org. Chem.* **2000**, *20*, 815.
[112] Kappe, C. O.; Fabian, W. M. F.; Semones, M. A. *Tetrahedron* **1997**, *53*, 2803.
[113] Kastron, V. V.; Vitolina, R. O.; Khanina, E. L.; Duburs, G.; Kimenis, A. A. *Khim.-Farm. Zh.* **1987**, *21*, 948; *Chem. Abstr.* **1988**, *108*, 16014.
[114] Vitolina, R.; Kimenis, A. *Khim.-Farm. Zh.* **1989**, *23*, 285; *Chem. Abstr.* **1989**, *111*, 188.
[115] Kastron, V. V.; Vitolina, R.; Khanina, E. L.; Duburs, G.; Kimenis, A.; Kondratenko, N. V.; Popov, V. I.; Yagupol'skii, L. M.; Kolomeitsev, A. A.; U.S.S.R. Patent SU 1,433,958 (1988); *Chem. Abstr.* **1989**, *111*, 7423.
[116] Fawzy, N. M.; Mandour, A. H.; Zaki, M. A. *Egypt. J. Chem.* **2000**, *43*, 401; *Chem. Abstr.* **2000**, *135*, 257209.
[117] Khromov-Borisov, N. V.; Savchenko, A. M. *Zh. Obshch. Khim.* **1952**, *22*, 1680; *Chem. Abstr.* **1953**, *47*, 54900.
[118] Bakibaev, A. A.; Filimonov, V. D. *Zh. Org. Khim.* **1991**, *27*, 854; *Chem. Abstr.* **1991**, *115*, 158931.
[119] Shutalev, A. D.; Sivova, N. V. *Chem. Het. Compounds (New York)* **1998**, *34*, 848; *Chem. Abstr.* **1999**, *130*, 267397.
[120] Lin, H.; Ding, J.; Chen, X.; Zhang, Z. *Molecules* **2000**, *5*, 1240.
[121] Buzueva, A. M. *Khim. Geterotsikl. Soedin.* **1969**, 345; *Chem. Abstr.* **1969**, *71*, 30439.
[122] Konyukhov, V. N.; Sakovich, G. S.; Krupnova, L. V.; Pushkareva, Z. V. *Zh. Org. Khim.* **1965**, *1*, 1487; *Chem. Abstr.* **1966**, *64*, 35878.
[123] Hirao, I.; Kato, Y.; Hujimoto, T. *Nippon Kagaku Zasshi* **1964**, *85*, 52; *Chem. Abstr.* **1964**, *61*, 76556.

[124] Zigeuner, G.; Knopp, C. *Monatsh. Chem.* **1970**, *101*, 1541.
[125] Zigeuner, G.; Hamberger, H.; Blaschke, H.; Sterk, H. *Monatsh. Chem.* **1966**, *97*, 1408.
[126] Zavyalov, S. I.; Kulikova, L. B. *Khim.-Farm. Zh.* **1992**, *26*, 116; *Chem. Abstr.* **1993**, *119*, 160222.
[127] Kappe, C. O.; Roschger, P. *J. Heterocycl. Chem.* **1989**, *26*, 55.
[128] Holden, M. S.; Crouch, R. D. *J. Chem. Ed.* **2001**, *78*, 1104.
[129] Khlebnikov, A. I.; Akhmedzhanov, R. R.; Naboka, O. I.; Novozheeva, T. P.; Saratikov, A. S. *Pharm. Chem. J.* **1999**, *33*, 644.
[130] El-Gaby, M. S. A.; Abdel-Hamide, S. G.; Ghorab, M. M.; El-Sayed, S. *Acta Pharm.* **1999**, *49*, 149.
[131] Jani, M. K.; Undavia, N. K.; Trivedi, P. B. *J. Indian Chem. Soc.* **1990**, *67*, 847; *Chem. Abstr.* **1990**, *115*, 8721.
[132] McKinstry, D. W.; Reading, E. H. *J. Franklin Inst.* **1944**, *237*, 203.
[133] Foroughifar, N.; Shariatzadeh, S. M. *Oriental J. Chem.* **2000**, *16*, 427; *Chem. Abstr.* **2000**, *134*, 326493.
[134] El-Ashmawy, M. B. *Saudi Pharm. J.* **1997**, *5*, 156; *Chem. Abstr.* **1997**, *128*, 102050.
[135] George, T.; Tahilramani, R.; Mehta, D. V. *Synthesis* **1975**, 405.
[136] Jauk, B.; Belaj, F.; Kappe, C. O. *J. Chem. Soc., Perkin Trans. 1* **1999**, 307.
[137] Kappe, C. O.; Peters, K.; Peters, E.-M. *J. Org. Chem.* **1997**, *62*, 3109.
[138] Folkers, K.; Johnson, T. B. *J. Am. Chem. Soc.* **1934**, *56*, 1374.
[139] Foroughifar, N.; Shariatzadeh, S. M.; Khaledi, A. M.; Khasnavi, E.; Masoudnia, M. *Ultra Sci.* **2000**, *12*, 277; *Chem. Abstr.* **2000**, *135*, 242198.
[140] Gupta, R.; Sudan, S.; Kachroo, P. L.; Jain, S. M. *Indian J. Chem.* **1996**, *35B*, 985.
[141] Sudan, S.; Gupta, R.; Bani, S.; Singh, G. B.; Jain, S. M.; Kachroo, P. L. *J. Indian Chem. Soc.* **1996**, *73*, 431.
[142] Kryukov, L. N.; Lebedeva, N. Y.; Kostrova, S. M. *Zh. Obshch. Khim.* **1990**, *60*, 1066; *Chem. Abstr.* **1990**, *113*, 172204.
[143] Jain, S. M.; Khajuria, R. K.; Dhar, K. L.; Singh, S.; Singh, G. B. *Indian J. Chem.* **1991**, *30B*, 805.
[144] Khanina, E. L.; Kastrons, V. In *6th Sint. Issled. Biol. Soedin., Tezisy Dokl. Konf. Molodykh Uch.*; Romadan, Y. P., Ed.; Zinatne: Riga, USSR, 1978, p 16; *Chem. Abstr.* **1978**, *92*, 163928.
[145] Zigeuner, G.; Knopp, C.; Blaschke, H. *Monatsh. Chem.* **1976**, *107*, 587.
[146] Folkers, K.; Johnson, T. B. *J. Am. Chem. Soc.* **1933**, *55*, 1140.
[147] Fabian, W. M. F.; Semones, M. A.; Kappe, C. O. *J. Mol. Struct. (Theochem)* **1998**, *432*, 219.
[148] Kato, T.; Chiba, T.; Sasaki, M. *Yakugaku Zasshi* **1981**, *101*, 182.
[149] Sadanandam, Y. S.; Shetty, M. M.; Diwan, P. V. *Eur. J. Med. Chem.* **1992**, *27*, 87.
[150] Zigeuner, G.; Nischk, W.; Juraszovits, B. *Monatsh. Chem.* **1966**, *97*, 1611.
[151] Chi, Y.-F.; Ling, Y.-C. *Sc. Sinica* **1957**, *VI*, 247.
[152] Ivanovskaya, L. Y.; Dubovenko, Z. D.; Mamaev, V. P. *Izv. Sib. Otd. Akad. Nauk SSSR, Ser. Khim. Nauk* **1969**, 132; *Chem. Abstr.* **1970**, *72*, 66892.
[153] Yarim, M.; Sarac, S.; Ertan, M.; Batu, Ö.; Erol, K. *Il Farmaco* **1999**, *54*, 359.
[154] Chi, Y.-F.; Wu, Y.-L. *Hua Hsüeh Hsüeh Pao* **1956**, *22*, 188; *Chem. Abstr.* **1958**, *52*, 35294.
[155] Kendi, E.; Sarac, S.; Yarim, M.; Ertan, M.; Läge, M.; Krebs, B. *Cryst. Res. Technol.* **1997**, *32*, 857.
[156] Remennikov, G. Y.; Boldyrev, I. V.; Kapran, N. A.; Kurilenko, L. K. *Khim. Geterotsikl. Soedin.* **1993**, 388; *Chem. Abstr.* **1993**, *120*, 77251.
[157] Remennikov, G. Y.; Shavaran, S. S.; Boldyrev, I. V.; Kapran, N. A.; Kurilenko, L. K.; Schevchuk, V. G.; Klebanov, B. M. *Khim.-Farm. Zh.* **1994**, 25; *Chem. Abstr.* **1995**, *122*, 9988.
[158] Kadis, V.; Stradins, J.; Khanina, E. L.; Duburs, G.; Muceniece, D. *Khim. Geterotsikl. Soedin.* **1985**, 117; *Chem. Abstr.* **1985**, *102*, 166142.
[159] Hull, R. GB Patent 984,365 (1965); *Chem. Abstr.* **1965**, *62*, 74268.
[160] Folkers, K.; Johnson, T. B. *J. Am. Chem. Soc.* **1933**, *55*, 2886.
[161] Khanina, E. L.; Andaburskaya, M. B.; Duburs, G.; Zolotoyabko, R. M. *Latv. PSR Zinat. Akad. Vestis, Kim. Ser.* **1978**, 197; *Chem. Abstr.* **1978**, *89*, 43319.
[162] Khanina, E. L.; Liepins, E.; Muceniece, D.; Duburs, G. *Khim. Geterotsikl. Soedin.* **1987**, 668; *Chem. Abstr.* **1988**, *108*, 112372.
[163] Khanina, E. L.; Duburs, G. *Khim. Geterotsikl. Soedin.* **1982**, 535; *Chem. Abstr.* **1982**, *97*, 55766.
[164] Bózsing, D.; Sohár, P.; Gigler, G.; Kovács, E. *Eur. J. Med. Chem.* **1996**, *31*, 663.

[165] Khanina, E. L.; Zolotoyabko, R. M.; Muceniece, D.; Duburs, G. *Khim. Geterotsikl. Soedin.* **1989**, 1076; *Chem. Abstr.* **1990**, *112*, 198292.
[166] Tozkoparan, B.; Ertan, M.; Krebs, B.; Läge, M.; Kelicen, P.; Demirdamar, R. *Arch. Pharm. (Weinheim)* **1998**, *331*, 201.
[167] Balkan, A.; Tozkoparan, B.; Ertan, M.; Sara, Y.; Ertekin, N. *Boll. Chim. Farmaceutico* **1996**, *135*, 648.
[168] Ertan, M.; Balkan, A.; Sarac, S.; Uma, S.; Renaud, J. F.; Rolland, Y. *Arch. Pharm. (Weinheim)* **1991**, *324*, 135.
[169] Ertan, M.; Balkan, A.; Sarac, S.; Rübseman, K.; Renaud, J. F.; Uma, S. *Arzneim.-Forsch./Drug Res.* **1991**, *41*, 725; *Chem. Abstr.* **1991**, *115*, 114455.
[170] Tan, R.; Sun, P. *Chinese J. Med. Chem.* **1997**, *7*, 283; *Chem. Abstr.* **1997**, *130*, 66458q.
[171] Tozkoparan, B.; Ertan, M.; Kelicen, P.; Demirdamar, R. *Il Farmaco* **1999**, *54*, 588.
[172] Balkan, A.; Ertan, M.; Burgemeister, T. *Arch. Pharm. (Weinheim)* **1992**, *325*, 499.
[173] Tozkoparan, B.; Yarim, M.; Sarac, S.; Ertan, M.; Kelicen, P.; Altinok, G.; Demirdamar, R. *Arch. Pharm. (Weinheim)* **2000**, *333*, 415.
[174] Sherif, S. M.; Youssef, M. M.; Mobarak, K. M.; Abdel Fatah, A. S. M. *Tetrahedron* **1993**, *49*, 9561.
[175] Ghorab, M. M.; Mohamed, Y. A.; Mohamed, S. A.; Ammar, Y. A. *Phosphorus, Sulfur and Silicon* **1996**, *108*, 249.
[176] Mohamed, N. K.; Aly, A. A.; Hassan, A. A.; Mourad, A.-F. E.; Hopf, H. *J. Prakt. Chem.* **1996**, *338*, 745.
[177] Akhtar, M. S.; Seth, M.; Bhaduri, A. P. *Indian J. Chem.* **1987**, *26B*, 556; *Chem. Abstr.* **1988**, *108*, 150408b.
[178] Sharma, S. D.; Kaur, V.; Bhutani, P.; Khurana, J. P. S. *Bull. Chem. Soc. Jpn.* **1992**, *65*, 2246.
[179] Hu, C.; Ding, L.; Xing, G.; Xin, Y.; Wang, S. *Chinese J. Med. Chem.* **2001**, *11*, 255.
[180] Wichmann, J.; Adam, G.; Kolczewski, S.; Mutel, V.; Woltering, T. *Bioorg. Med. Chem. Lett.* **1999**, *9*, 1573.
[181] Ismail, M. M. F. *Az. J. Pharm. Sci.* **1999**, *23*, 1; *Chem. Abstr.* **1999**, *134*, 252308.
[182] Khanina, E. L.; Muceniece, D.; Kadysh, P. V.; Duburs, G. *Khim. Geterotsikl. Soedin.* **1986**, 1223; *Chem. Abstr.* **1987**, *107*, 39737.
[183] Krstenansky, J. L.; Khmelnitsky, Y. *Bioorg. Med. Chem.* **1999**, *7*, 2157.
[184] Kato, T. Japanese Patent JP 59,190,974 (1984); *Chem. Abstr.* **1985**, *102*, 132067.
[185] Sharaf, M. A. F.; Aal, F. A. A.; Fatah, A. M. A.; Khalik, A. M. R. A. *J. Chem. Res. (S)* **1996**, 354.
[186] Parmar, J. M.; Parikh, A. R. *Indian J. Heterocycl. Chem.* **2001**, *10*, 205.
[187] Sarac, S.; Yarim, M.; Ertan, M.; Erol, K.; Aktan, Y. *Boll. Chim. Farmaceutico* **1997**, *136*, 657.
[188] Abdel-Gawad, S. M.; El-Gaby, M. S. A.; Ghorab, M. M. *Il Farmaco* **2000**, *55*, 287.
[189] Ghorab, M. M.; Abdel-Gawad, S. M.; El-Gaby, M. S. A. *Il Farmaco* **2000**, *55*, 249.

CHAPTER 2

MICROBIAL ARENE OXIDATIONS

ROY A. JOHNSON

67 Marin Ave., Sausalito, California 94965

CONTENTS

	PAGE
ACKNOWLEDGMENTS	118
INTRODUCTION	118
MECHANISM AND STEREOCHEMISTRY	120
SCOPE AND LIMITATIONS	125
Toluene Dioxygenase (TDO)	128
Naphthalene Dioxygenase (NDO)	130
Biphenyl Dioxygenase (BPDO)	131
Benzoate Dioxygenase (BZDO)	133
Other Dioxygenases	133
Other Oxidations by Arene Dioxygenases	134
Site-Directed Mutagenesis of Dioxygenases	137
Scope of Method	138
APPLICATIONS TO SYNTHESIS	139
Other Synthetic Intermediates by Modification of Microbially Produced Dihydrodiols	139
Use of Dihydrodiols in Synthesis	141
COMPARISON WITH OTHER METHODS	145
EXPERIMENTAL CONDITIONS	146
Comments about Yields	146
Stability and Isolation of Dihydrodiols	146
Microorganisms (Cultures)	148
Inducers	148
EXPERIMENTAL PROCEDURES	149
Handling the Microorganisms	149
Preparation of Media	150
Mineral salt broth (MSB) medium	151

---

rajohns4@pacbell.net

Organic Reactions, Vol. 63, Edited by Larry E. Overman et al.
ISBN 0-471-44532-0 © 2004 Organic Reactions, Inc. Published by John Wiley & Sons, Inc.

Mineral salt broth (MSB) agar plates	151
Mineral salts mixture (MSM) medium	151
Minimal salts medium	151
M9 medium	151
Luria-Bertani (LB) broth	152
Luria-Bertani (LB) agar	152
FERMENTATIONS	152
(1S-cis)-3-Chloro-3,5-cyclohexadiene-1,2-diol	152
(1S-cis)-3-Methyl-3,5-cyclohexadiene-1,2-diol	153
(1S-cis)-3-Methyl-3,5-cyclohexadiene-1,2-diol via a Fermentation with Immobilized E. coli JM109(pKST11)	153
(1S-cis)-1,6-Dihydroxy-2,4-cyclohexadiene-1-carboxylic acid	154
1. From benzoic acid with *Pseudomonas putida* U103	154
2. From benzoic acid with *Alcaligenes eutrophus* B9	155
(1R-cis)-1,2-Dihydro-1,2-naphthalenediol	156
(1S-cis)-3-Iodo-3,5-cyclohexadiene-1,2-diol	156
(1S-cis)-3-Phenyl-3,5-cyclohexadiene-1,2-diol	157
(1S-cis)-3-Bromo-4,5-difluoro-3,5-cyclohexadiene-1,2-diol	157
GLOSSARY	158
TABULAR SURVEY	159
Table 1. Microbiological Oxygenations of Benzenes	160
Table 2. Microbiological Oxygenations of Benzoic and Naphthoic Acids and Esters	183
Table 3. Microbiological Oxygenations of Biphenyls	187
Table 4. Microbiological Oxygenations of Naphthalenes	192
Table 5. Microbiological Oxygenations of Polycyclic Aromatics	201
Table 6. Microbiological Oxygenations of Heterocycles	205
Table 7. Microbiological Oxygenations of Olefins	213
Table 8. Microbiological Benzylic Oxygenations	217
Table 9. Microbiological Oxygenation of Thiols	225
Table 10. Dihydrodiols That Have Not Been Isolated	235
Table 11. Chemical Transformations of Dihydrodiols	253
REFERENCES	258

ACKNOWLEDGMENTS

The author extends warmest thanks to David T. Gibson, Derek R. Boyd, and Gregg M. Whited for their gracious responses to inquires and requests for assistance. The author also acknowledges, with thanks, access to the library facilities available at the University of California, Berkeley, the University of California, San Francisco, and Stanford University.

INTRODUCTION

The metabolism of organic molecules by living organisms is of fundamental interest to biologists, microbiologists, and biochemists. The primary avenue of metabolism in most living organisms is via oxidative pathways. The aromatic hydrocarbons (arenes) are subject to such oxidative degradation; their metabolism by both mammalian and microbial systems has been extensively studied. It is the capacity of certain microorganisms to oxidatively convert arenes into arene cis-dihydrodiols that provides the foundation of this chapter. Arenes are subject to a variety of metabolic oxidations but when the expression "microbial arene oxidation" is used within the field of organic chemistry, it generally refers to the oxidation exemplified by Eq. 1. The focus of this chapter is the oxidative transformation and the valuable molecules that are generated by the process.

$$\text{benzene} \xrightarrow{\textit{Pseudomonas putida } F1} \text{cis-3,5-cyclohexadiene-1,2-diol (1)} \tag{Eq. 1}$$

Throughout this chapter the shortened term, "dihydrodiol," is used for the array of functional groups shown in the structure of cis-3,5-cyclohexadiene-1,2-diol (**1**). Also, note that the words "oxidation" and "oxygenation" are used interchangeably as are "microbial" and "microbiological." Please refer to the Glossary for definitions of unfamiliar terminology.

When considering microbial arene oxidations, it is important to be aware of the distinction between bacteria and fungi. The cis-dihydrodiols are produced only by bacteria. As early as 1953, an optically active dihydrodiol was isolated from fermentation of naphthalene with a bacterium.[1] However, the stereochemical relationship of the hydroxy groups was not determined; rather, the stereochemical relationship of these groups was assigned as trans by analogy to the trans-dihydrodiol produced by mammalian metabolism of naphthalene. Arene oxidations performed by fungi also produce trans-dihydrodiols. In the following years several other dihydrodiols obtained from the fermentations of arenes with bacteria were likewise interpreted to have the trans-diol configuration. A seminal study by Gibson and coworkers in 1968 placed the chemistry and stereochemistry of microbial arene oxidations on solid footing. These investigators demonstrated a fundamental difference in the first step of arene metabolism in mammalian and bacterial systems. Using the bacterium *Pseudomonas putida* F1, they showed conclusively that cis-3,5-cyclohexadien-1,2-diol is produced from benzene by this microorganism, as shown in Eq. 1.[2] The early history of arene oxidations and of these discoveries are included in an excellent review by Gibson.[3]

The next crucial step in the development of the microbial arene oxidation was a mutagenesis experiment with *P. putida* F1. Treatment of the microorganism with *N*-methyl-*N*′-nitro-*N*-nitrosoguanidine (MNNG) gave a number of mutant strains of the organism. One of these mutant strains, designated *P. putida* 39/D, was devoid of the enzyme cis-dihydrodiol dehydrogenase normally present in the natural strain. This enzyme carries out the dehydrogenation of the dihydrodiol, efficiently converting it into a catechol. The absence of this enzyme allows the accumulation of the dihydrodiol during the fermentation, a factor essential to isolation of useful quantities of these compounds.[3,4]

The assignment of dihydrodiol configuration and the use of mutant strains of microorganisms laid the groundwork for the development of microbial arene oxidations as a process useful to organic synthesis. Other important developments that have contributed toward achievement of that goal include: (a) determination of enantiomeric purities and absolute configurations of the dihydrodiols, (b) discovery of other organisms that can produce dihydrodiols, (c) cloning of the genes for the dioxygenase enzyme complex and their expression in *Escherichia coli*, and (d) a demand for the dihydrodiols as a consequence of their potential value in organic synthesis.

These advances have led to the development of microbial arene oxidations that are suitable for organic synthesis. The attractive features of these oxidations are three-fold. First, the process is one of a very few that disrupts the aromatic system of arenes. Second, the array of functional groups generated in the dihydrodiol products is useful. Third, the process is highly enantioselective, affording enantiomerically pure products in most cases.

For this chapter, the literature has been reviewed through 2001. As noted above, the early literature was reviewed in 1971.[3] Progress from 1971 to 1984,[5] and the generation of dihydrodiols from halogenated aryls,[6] from polychlorinated biphenyls,[7] and from phthalates[8] are all included in a book devoted to microbial degradation of organic molecules (see ref. 5). The microbial degradation of polycyclic aromatic hydrocarbons, including formation of dihydrodiols, has been reviewed[9] as have the reactions catalyzed by naphthalene dioxygenases.[10] The oxygenations of polycyclic aromatic hydrocarbons are the subject of a recent review article.[11] Enzymic structure-function relationships of the dioxygenases have been reviewed[12] and the role of Rieske centers in dioxygenases is included as part of a larger review.[13] Several articles combine summaries of dihydrodiol preparation with applications to organic syntheses[14–16] including a detailed review of the stereochemistry of the process. A fascinating personal account of his involvement in the development of the field has been written by Gibson.[17] Finally, references to numerous reviews devoted primarily to synthetic applications based on the use of dihydrodiols are enumerated later in this Chapter.

MECHANISM AND STEREOCHEMISTRY

The "reagents" used in microbial arene oxidations are oxygen and an enzyme complex. The oxygenation reaction takes place within the cells of the living microorganisms where the enzyme complex is found. A key component of the enzyme complex is a dioxygenase that catalyzes the addition of oxygen to an arene, generating the dihydrodiol. The dioxygenase requires the support of electron transport and oxygen transport systems, with which it is coupled, in order to achieve the oxygenation. Not surprisingly, there are small differences in the dioxygenases found in different microorganisms with the consequence that there are several versions of the reagent. Distinctions between the different enzymes are made on the basis of the substrate that the microorganism was originally found to oxidize. More than a dozen classes of dioxygenases are known but the four most widely used for arene oxygenations will be emphasized in this chapter. They are: (a) toluene dioxygenase (TDO), (b) naphthalene dioxygenase (NDO), (c) biphenyl dioxygenase (BPDO), and (d) benzoic acid (or benzoate) dioxygenase (BZDO). Fortunately, the specificity of each of these dioxygenases is not nearly as stringent as the names imply. Each of the four classes of enzymes catalyzes the dioxygenation of a broad range of substrates. There also is considerable overlap between the four classes so that some arenes are oxygenated by more than one of the enzyme classes. More than one microorganism may carry the dioxygenase of any given class. These may be different natural microorganisms, mutant strains of the natural microorganisms, or recombinant strains expressing the dioxygenase of the natural microorganisms. Further discussion of these microorganisms and how they are related may be found in the "Scope and Limitations" section.

The unequivocal demonstration of a cis-diol configuration in the dihydrodiols clearly differentiates the mechanism of arene oxidations by bacteria from the mechanism in mammalian and fungal systems. Whereas mammalian and fungal oxidations occur via an arene oxide intermediate, the cis-configuration of the diols suggests addition of both atoms of the oxygen molecule in bacterial oxidations. Several studies confirm this mode of oxygenation with experiments in which the fermentations were done using mixtures of $^{16}O_2$ and $^{18}O_2$. Analysis of the dihydrodiols produced under these conditions clearly shows that both atoms of the oxygen molecule are incorporated into the same dihydrodiol. Although these results are consistent with the intermediacy of an arene-dioxetane, such an intermediate has yet to be isolated or detected.

The dramatic progress in recent years in the X-ray crystallographic study of protein structure has greatly enhanced understanding of the mechanisms of enzyme-catalyzed reactions. This technique has been successfully applied to bacterial dioxygenases. The first X-ray crystallographic studies of an aryl dioxygenase are of a naphthalene dioxygenase[18,19] (NDO) from *Pseudomonas* sp. 9816-4 expressed in a recombinant strain of *E. coli*.[20] Three different versions of the structure have been reported; they include: (a) enzyme with no substrate,[18] (b) enzyme with indole bonded via dioxygen to iron in the catalytic site,[21] and (c) the enzyme after crystals were soaked with indole.[21] Indole is known to be a substrate of NDO and is oxidized to indoxyl, which, in turn, undergoes further air oxidation to indigo.[22]

The crystallographic studies indicate that the catalytic site of the enzyme is a flattened and elongated cavity having the approximate dimensions of 6 Å by 8 Å by 10 Å. Computational docking studies with the enzyme structure suggest that the cavity can nicely accommodate substrates such as naphthalene and biphenyl. The cavity is mostly lined by hydrophobic amino acids with the exception of a polar area at the bottom located near a mononuclear iron center.[21]

Further development and refinement of X-ray structural data combined with computational analysis are very likely to lead to models useful for the prediction of regio- and stereochemical outcomes of microbial arene oxidations. In the meantime, enough empirical data is available from the structure assignments already in the literature to generate a predictive model for the stereochemical course of many dioxygenation reactions.[23] The scheme shown in Figure 1 is useful for predicting the regio- and stereochemical outcome for the dioxygenation of most benzenes as well as for many bicyclic and polycyclic arenes.

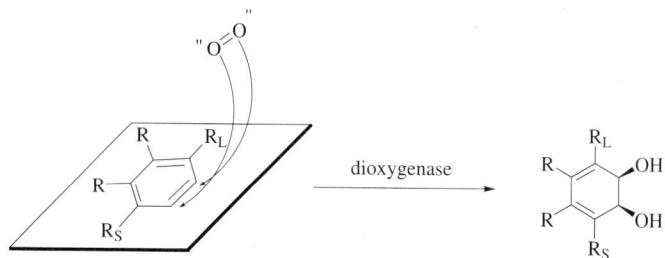

Figure 1. An empirical model for predicting the sense of enantioselectivity in the dioxygenation of arenes by TDOs, NDOs, and BPDOs. R_L is a large (in terms of space occupied) substituent group; R_S is a small substituent group.

To use the model shown in Figure 1, mono-substituted benzenes are oriented so that the substituent is in the position of R_L. The model then predicts that dioxygenation will produce a dihydrodiol having the constitution and absolute configuration shown. All monosubstituted dihydrodiols for which complete stereochemical data is available are consistent with the model. Nearly all dioxygenations of monosubstituted benzenes give dihydrodiols with >95% ee. One exception is fluorobenzene for which the derived dihydrodiol has an ee of ~60%,[24,23,25] reflecting the approximately equal volume of space occupied by the hydrogen and fluorine atoms. These observations are consistent with the steric effect, as opposed to an electronic effect, of the substituent(s) being dominant in direction of the oxygenation process.

Di- and polysubstituted benzenes are likewise oriented in the model with the largest substituent in the position of R_L.[23] Where there is ambiguity about which group to rank as the largest, some guidance may be found from the results summarized in Table 1. These results provide the following preferences in order of decreasing van der Waals radii of the substituent, $CF_3 > I > Br > CH_3 \geq Cl > F > H$. This ranking is derived from a very limited sample set, however, and should be used with caution in making predictions.

Naphthalene is oriented in the model so that the second ring occupies the positions of R_L and the adjacent R. This orientation predicts oxygenation at C-1 and C-2 rather than at C-2 and C-3 of the naphthalene ring, consistent with observed results. Substituted naphthalenes likewise are oriented so that the ring to which the substituent is attached occupies the positions of R_L and the adjacent R. Now, however, the model is not sufficiently detailed to allow a choice to be made between the two different orientations that are possible for the substituent-bearing ring. As illustrated in Figure 2, a 2-substituted naphthalene can be aligned in the model in the two orientations shown. Each orientation predicts a different regioselectivity of the oxygenation process. In practice, two or three dihydrodiols are obtained from the oxygenation (Eq. 2).[26]

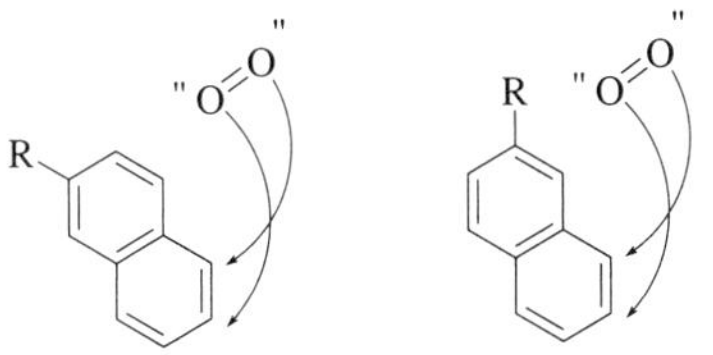

Figure 2. Two orientations of a 2-substituted naphthalene in the model of Figure 1.

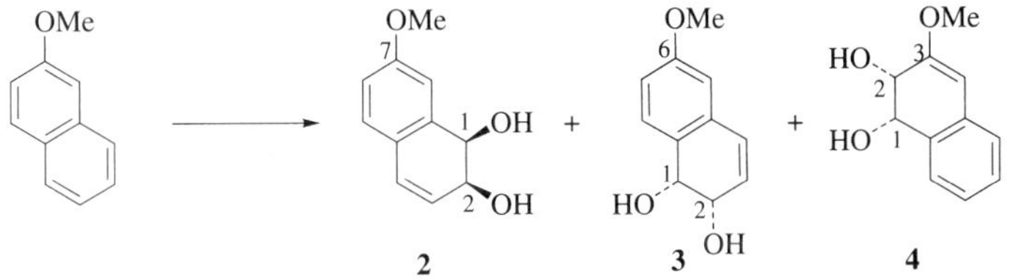

(Eq. 2)

Microorganism	Dioxygenase Type	(Yield of crude diols), 2:3:4
Pseudomonas putida 39D	TDO	(29%), 12:73:15
Escherichia coli JM109(pDTG601)	TDO	(64%), 17:69:14
Pseudomonas putida NCIB 9816	NDO	(6%), 93:7:0
Escherichia coli JM109(pDTG141)	NDO	(57%), 93:7:0
Escherichia coli C534(ProR/Sac)	NDO	(61%), 92:8:0
Beijerinckia sp. B8/36	BPDO	(36%), 74:26:0

Although the names given the classes of dioxygenases, e.g., toluene dioxygenase (TDO), suggest highly selective reagents, the dioxygenases are quite broad in their substrate specificity. Naphthalenes, for example, may be oxygenated with TDOs or BPDOs as well as with NDOs. This permits some latitude in choosing a microorganism to use for a dioxygenation process. It is important, then, to consider if regio- and stereoselectivity of dioxygenation is sensitive to (a) the class of dioxygenase, and (b) to different microorganisms within the same class of dioxygenase (see Eq. 2). Oxygenation of 2-methoxynaphthalene can generate the three constitutionally different dihydrodiols, (1*R*-*cis*)-7-methoxy-, (1*R*-*cis*)-6-methoxy-, and (1*S*-*cis*)-3-methoxy-1,2-dihydronaphthalene-1,2-diol (**2**, **3**, and **4**) shown in the equation. The results from oxygenation of 2-methoxynaphthalene with six different microorganisms are summarized. Note that the six include two TDO organisms (one mutant strain and one recombinant strain, see TDO section below), three NDO organisms (one mutant and two recombinant strains, see NDO section below), and one BPDO organism (a mutant strain, see BPDO section below). In these results, the regioselectivity of dioxygenation varies considerably between microorganisms having different classes of dioxygenase but only slightly between microorganisms of the same dioxygenase class. The absolute configurations of all three dihydrodiols are those that would be predicted by the model shown in Figure 1.

In another example, the oxygenation of 4-iodotoluene with two TDO-producing organisms, one the mutant strain *P. putida* UV4 (derived from the natural strain *P. putida* NCIMB 11767) and the other the recombinant strain *Escherichia coli* JM109(pDTG601) (derived from *P. putida* F1), is compared (Eq. 3).[23,27] Both organisms convert the substrate into (1*S-cis*)-3-iodo-6-methyl-3,5-cyclohexadiene-1,2-diol (**5**) but the product from the mutant strain has 80% ee whereas that from the recombinant strain has >98% ee. This difference in ee may reflect slight differences between the two TDOs, since each is derived from a different natural strain of *P. putida*. It should be added that the transformations from a wild-type organism to a mutant or to a recombinant strain are not expected to alter the amino acid sequence of the dioxygenase and its catalytic selectivity.

$$\text{4-iodotoluene} \xrightarrow{\textit{Pseudomonas putida} \text{ UV4 or } \textit{Escherichia coli} \text{ JM109(pDTG601)}} \mathbf{5} \quad \text{(Eq. 3)}$$

The use of the model to predict the stereoselectivity of oxygenation of polycyclic aromatic hydrocarbons must also be done with care. Orientation of anthracene in the model so that the fused rings take the positions of R_L and the adjacent R correctly predicts the structure of the oxygenation product (Eq. 4).[28,29] Orientation of phenanthrene presents the same dilemma that was faced with the 2-substituted naphthalenes. Two orientations of phenanthrene are possible and dihydrodiols from both are obtained from the oxygenation process (Eq. 5).[28,30,31] Similar considerations hold when the model is used for predicting oxygenation of polycyclic hydrocarbons having more than three rings.

$$\text{anthracene} \xrightarrow{\textit{Beijerinckia} \text{ sp. B8/36}} \text{product (25%)} \quad \text{(Eq. 4)}$$

$$\text{phenanthrene} \xrightarrow{\textit{Beijerinckia} \text{ sp. B8/36}} (4R,3S)\text{-diol (32%)} + (1R,2S)\text{-diol (8%)} \quad \text{(Eq. 5)}$$

4:1; both >95% ee

The model may also be applied, with care, in the oxygenations of heteroaromatic compounds. Oxygenation of phenazine with *S. yanoikuyae* B8/36 produces a dihydrodiol having constitution and absolute configuration consistent with the model of Figure 1 (Eq. 6).[32] Also produced in the fermentation is another, more polar product that was determined to be the product of bis-dihydroxylation of phenazine. In a separate experiment, the mono-dihydrodiol was shown to be a substrate for the second dioxygenation process.[32]

A different regiochemistry is found for BZDO (Eq. 7).[33–36] The enantioselectivity of BZDO oxidations has not been sufficiently studied to develop an empirical model for predicting stereochemistry. In fact, the absolute configuration of only one product of BZDO oxidations, (1S-cis)-1,6-dihydroxy-2,4-cyclohexadiene-1-carboxylic acid (**6**), has been determined (Eq. 7).[33]

Several general methods have been developed for assignment of absolute configuration to new chiral dihydrodiols. These include preparation of chiral boronate derivatives coupled with NMR analysis,[37] preparation of 4-phenyl-1,2,4-triazoline-3,5-dione (PTAD) adducts followed by derivatization as the Mosher esters and NMR analysis,[25] and circular dichroism correlations[29] of the dihydrodiols and their derivatives and of cyclic η^4-diene complexes of tricarbonyliron.[38]

Further aspects of the regio- and stereochemistry of dioxygenations catalyzed by the different dioxygenases are included in the next section.

SCOPE AND LIMITATIONS

When used in the context of organic chemistry, the expression "microbial arene oxidations" describes the oxidative conversion of arenes into cis-dihydrodiols by microorganisms. This transformation is attractive to chemists because, in one reaction, the aromatic system of the arene is disrupted in a controlled manner and a useful array of functional groups is generated. In addition, the oxygenation process usually is highly enantioselective as described in the preceding section.

The range of arenes that can be converted into dienediols by this method is broad. Benzene, numerous substituted benzenes, and a variety of naphthalenes all are good substrates for the reaction. (A "good substrate" is one that is converted in good yield into a dienediol.) Polycyclic aromatic hydrocarbons also are subject to dioxygenation by microorganisms although, at the present stage of process development, the yields of dienediols are not synthetically useful. Heteroaromatic compounds also have been widely studied as substrates for dioxygenation and many are converted into dienediols. The list of heteroaromatic compounds that have been converted into dienediols includes: thiophenes, benzo- and dibenzothiophenes, benzo- and dibenzofurans, indole (the final product is indigo), quinolines, isoquinolines, quinoxalines, quinazolines, acridines, phenazines, chromenes, and thiachromenes as well as several other polycyclic heteroaromatic compounds.

Oxygenations of substituted arenes by dioxygenases are not limited to the π-electron system. Depending on the groups attached or fused to the arene nucleus, other oxidation products may be formed. In some reactions, these become the major products of the oxygenation process. Indenes and indanes are prominent among such substrates. With indenes, oxidations of both the cyclopentene double bond and of the benzylic position are observed, and with indanes, benzylic oxidations predominate. Another oxidative "side-reaction" is the oxidation of thioethers, usually to sulfoxides. These oxidations are described further in a later section of this chapter.

Another limitation of microbial arene oxidations as a route to enantiomerically pure diols is that, in most cases, only one enantiomer can be accessed by this method. In a few cases (see "Applications to Synthesis"), the dihydrodiol can be manipulated chemically to produce the opposite enantiomer. In these cases, however, the additional chemical steps detract somewhat from the convenience of the one-step microbial generation of a dihydrodiol. A promising approach to production of both enantiomers of a dihydrodiol is in the very early stages of development. The combination of genetic engineering with site-directed mutagenesis of the dioxygenase can alter dramatically the stereochemical outcome of the oxygenation process. This approach is described further in the section on "Site-Directed Mutagenesis of Dioxygenases."

Many microorganisms catalyze the dioxygenation of arene molecules, requiring a choice to be made when planning to perform such an experiment. Most of the work leading to practical preparations of dihydrodiols has been done with microorganisms of the four classes already outlined, i.e., TDO, NDO, BPDO, and BZDO. Benzene dioxygenase (BDO) is sometimes considered a fifth class of dioxygenase but in this review it is included as part of the TDO class. These four classes of dioxygenases and the microorganisms that produce them are discussed below.

Within each of the four classes of dioxygenases, two or three microbial strains have been studied extensively for biotransformation of the arene substrates. A listing of the most frequently used microorganisms, and their mutant and recombinant strains, for each of the four classes of enzymes is given in Table A. Coordination of this information with the examples given in the Tables at the end of the chapter may be helpful in making the choice of an organism to use for a desired dioxygenation process.

TABLE A. MICROORGANISMS FREQUENTLY USED IN OXYGENATION OF ARENES

Dioxygenase	Microorganism (type); inducer(s)[a]	Source	Reference
TDO	*P. putida* F1 (wild type)	ATCC 700007; DSM 6899	2
	P. putida 39/D (mutant of *P. putida* F1); toluene	ATCC 700008; DSM 6414	39, 40
	E. coli JM109(pDTG601) (clone of *P. putida* F1 dioxygenase); IPTG[b]		41
TDO	*P. putida* NCIMB 11767 (wild type); toluene		42, 43
	P. putida UV4 (mutant of *P. putida* NCIMB 11767); no inducer required	Proprietary	42
	E. coli JM109(pKST11) (clone of *P. putida* NCIMB 11767 dioxygenase); IPTG[b]		44
NDO	*Pseudomonas* sp. NCIB 9816-4		45, 10
	Pseudomonas sp. NCIB 9816-11 (mutant of *P.* NCIB 9816-4); salicylate		46, 10
	E. coli JM109(DE3)(pDTG141) (clone of *P.* NCIB 9816-4 dioxygenase); IPTG[b]		45, 10
NDO	*P. fluorescens* N3 (wild type)		47, 48
	P. fluorescens N3 TCC1 (mutant of *P. fluorescens* N3); salicylate, grown on succinate	NCIMB 40605	47, 48
BPDO	*Beijerinckia* sp. B1 ≡ *S. yanoikuyae* B1 (wild type)	ATCC 51230; DSM 6900	49, 11
	Beijerincka sp. B8/36 ≡ *S. yanoikuyae* B8/36 (mutant of *S. yanoikuyae* B1 wild type); biphenyl, *m*-xylene		49, 11
BPDO	*Pseudomonas* sp. LB400 ≡ *Burkholderia* sp. LB400		50, 51
	E. coli BL21(DE3)/pLys (clone of *B.* sp. LB400)		52
	E. coli FM4560(pGEM410) (clone of *B.* sp. LB400); grown on succinate		50
BZDO	*P. putida* JT103 (wild type)		53
	P. putida U103 (mutant) ; benzoate		53, 33
BZDO	*A. eutrophus* 335; wild type	ATCC 17697; DSM 531	35
	A. eutrophus 335 strain B9 (mutant of *A. eutrophus* 335); benzoate		34

[a] A discussion of inducers may be found in the section "Experimental Conditions."
[b] IPTG = isopropyl-β-D-thiogalactopyranoside

The mutant and recombinant versions of two *P. putida* strains, F1 and NCIMB 11767, are widely used for TDO oxygenations. In this case, a choice of which organism to use can be made on the basis of the availability of the strains. The mutant *P. putida* 39/D, from *P. putida* F1, can be obtained from a commercial repository such as the American Type Culture Collection (ATCC). However, the mutant *P. putida* UV4, from NCIMB 11767, is a proprietary strain that at the present time is not freely available for general use. The other natural and mutant microorganism strains given in the table are generally available from central repositories or from the research groups working with the strains. The recombinant strains of the microorganisms listed in Table A may, in most cases, be requested from the research group in which the strain was constructed. [See also the later subsection titled "Microorganisms (Cultures)."]

Some knowledge of the nomenclature of microbiology is essential to chemists when reading the literature of microbial arene oxidations for the first time. Subtle differences in names of microorganisms, particularly in designations of strains, may at first be confusing. It is also important to be able to follow the trail of names created as the microorganism containing a specific dioxygenase is modified from the natural (wild-type) microorganism to a mutant organism and to a recombinant organism. Typically, a natural—or wild-type—microorganism capable of metabolizing an arene molecule is first discovered. This is followed by development of a mutant strain of the organism that is capable of accumulating the dihydrodiol product. Until the past decade, the use of such mutant strains has been the primary means for generating large quantities of dihydrodiols. More recently, the development of recombinant strains of microorganisms into which the dioxygenase is cloned has found increasing application and holds much promise for further improvements in the production of dihydrodiols in practical quantities. Occasionally, the taxonomic classification of a microorganism is changed with the consequence that the name is changed (see both the NDO and BPDO sections below for examples). Detailed examples of microbial nomenclature are included below for the four major classes of dioxygenases.

Because scientists from a variety of disciplines have reported microbial arene oxidations, a wide range of reporting styles is found in the experimental details. The data range from detailed descriptions (including complete stereochemical assignment) of processes that yield multi-gram quantities (per liter of culture) of product to processes for which the dihydrodiol has not been isolated or characterized. Results of the latter type are placed in Table 10 to provide the reader with references that point to the potential for preparation of these dihydrodiols by microbial arene oxidation.

Toluene Dioxygenase (TDO)

Most dioxygenations of substituted benzenes are done with microorganisms expressing TDO. Among such microorganisms, two have been used for the majority of these oxidations. The first of the two is *P. putida* F1,[2,54] the microorganism that played a key role in the discovery and development of arene oxidations (see Introduction). The designation "F1" is a descriptor given to the natural strain of this microorganism. A typical dioxygenation catalyzed by *P. putida* F1 is the conversion of toluene into (1*S-cis*)-3-methyl-3,5-cyclohexadiene-1,2-diol (**7**, Eq. 8). However, this natural strain contains the enzyme, *cis*-1,2-dihydrodiol dehydrogenase, which converts the dihydrodiol into a catechol, reducing the amount of dihydrodiol that can be isolated. When *P. putida* F1 was exposed to chemical mutation with MNNG, a mutant strain lacking the *cis*-1,2-dihydrodiol dehydrogenase was obtained.[39] Using the mutant strain, dihydrodiol **7** can be produced at the rate of almost 1g/L of fermentation broth and with greater than 98% ee.[39] The mutant strain is named *P. putida* 39/D (sometimes, F39/D).[39] The genes of *P. putida* F1 responsible for dioxygenase activity have been cloned into *E. coli*. The resulting recombinant organism is named *E. coli* JM109(pDTG601).[41] In this name, JM109 refers to a strain of *E. coli* that frequently is used for cloning and (pDTG601) refers to the genetic sequence (plasmid) cloned from *P. putida* F1 into *E. coli* JM109. The *E. coli* JM109(pDTG601) also is free of *cis*-1,2-dihydrodiol dehydrogenase, permitting the dihydrodiol to accumulate.

$$\text{PhMe} \xrightarrow[\text{or}]{\begin{array}{c}\textit{Pseudomonas putida } 39/\text{D } (0.93 \text{ g/L}) \\ \text{or} \\ \textit{Pseudomonas putida } \text{UV4 } (60\%) \\ \text{or} \\ \textit{Escherichia coli } \text{JM109(pDTG601)}\end{array}} \text{Me-cyclohexadiene-diol } \mathbf{7} \quad \text{(Eq. 8)}$$

The second strain of *P. putida* that has been widely used for TDO activity is NCIMB 11767.[42] Mutation of this microorganism again produces a strain capable of accumulating dihydrodiol. This mutant strain is named *P. putida* UV4 and produces (1*S-cis*)-3-methyl-3,5-cyclohexadiene-1,2-diol in 60% yield and >98% ee.[55] A thorough description of the development of the mutant from the wild-type organism is available.[56] However, the mutant strain is proprietary and available only by request from the owners.[57] The dioxygenase genes from *P. putida* NCIMB 11767 have been cloned into *E. coli* and the resultant organism is *E. coli* JM109(pKST 11).[58] Immobilization of either *P. putida* UV4 or *E. coli* JM109(pKST 11) cells in barium alginate beads has been shown to improve the yield of dihydrodiols as measured on a mol/g of dry cell weight basis.[59] This immobilization technology has not as yet been used in preparation of dienediols for synthetic applications. Work to improve the fermentation processes for this microorganism continues.[60] *P. putida* UV4 has also been used for oxygenation of naphthalenes and various heteroaromatic compounds. Regioselectivity is lower with heteroaromatic compounds than with arenes but the level of enantioselectivity remains high. The isolated yields of the heterocyclic dienediols are low.

The sense of enantioselectivity of TDO oxidations is very consistent and allows formulation of the model shown in Figure 1. This figure summarizes existing data and can be used as an empirical model to predict the stereochemical outcome of new oxidations performed with either the mutants or the clones of these two TDO microorganisms (see the section on "Mechanism and Stereochemistry").

Some benzoic acids and esters are substrates for TDO-containing organisms and are oxygenated to give dihydro-2,3-dihydroxy-1-carboxylic acids having regiochemistry and enantioselectivity (Eq. 9)[61] consistent with the model shown in Figure 1. A second class of microorganisms, containing the dioxygenase, BZDO, also may be used for dioxygenation of benzoic acids and esters. The regiochemistry of these dioxygenations differs from the TDO oxidations (Eq. 7). As a class, the BZDO oxygenations are discussed further in a separate section below.

$$\text{PhCO}_2\text{Me} \xrightarrow{\textit{Pseudomonas putida } \text{UV4}} \text{methyl dihydrodiol benzoate } (42\%) \quad \text{(Eq. 9)}$$

As noted above, oxidation products other than dihydrodiols may be produced by TDO-containing organisms. These are discussed below in the section that addresses the issue of by-products formed in microbial arene oxidations.

Naphthalene Dioxygenase (NDO)

The groundwork for NDO and the oxygenation of naphthalene was laid with the discovery[62] of a strain of *P. putida* capable of growing on naphthalene. The wild-type strain was then transformed by mutation into *P. putida* 119, a strain that was able to accumulate the dienediol.[63] A large-scale fermentation with *P. putida* 119 produced a sufficient amount of the product for complete characterization as (1*R*-*cis*)-1,2-dihydro-1,2-naphthalenediol (**8**, Eq. 10).[63] The mutant strain, *P. putida* 119, subsequently was found to revert during fermentation to the wild strain and so has not been widely used in subsequent dioxygenation studies [These unpublished results of Gibson, D. T. and Klečka, G.M. are referenced (#52) in this chapter's reference 15.]

$$\text{naphthalene} \xrightarrow[\text{Escherichia coli JM109(DE3)(pDTG141)}]{\textit{Pseudomonas} \text{ sp. strain 9816/11 or}} \textbf{8} \text{ (1R-cis-1,2-dihydronaphthalenediol)} \quad \text{(Eq. 10)}$$

Another *Pseudomonas* strain that catalyzes the dioxygenation of naphthalene is *Pseudomonas* sp. NCIB 9816 (also known as NCIB 9816-4; or simply as 9816-4).[63] A mutant of this organism, *Pseudomonas* sp. 9816-11 (also: 9816 strain 11; 9816-11; or 9816/11)[46] that lacks the cis-dihydrodiol dehydrogenase enzyme, has been used in much subsequent work. The structural genes for the NDO in NCIB 9816-4 have been cloned and expressed in *E. coli* giving the strains JM109(DE3)(pDTG121)[64] and JM109(DE3)(pDTG141),[65] which can be used for performing dioxygenations. An excellent review summarizing oxygenations catalyzed by the NDO common to *Pseudomonas* sp. strain NCIB 9816-4, mutant strain NCIB-11, and the recombinant *E. coli* strain JM109(DE3)(pDTG141) has been published.[10]

Both *Pseudomonas* sp. NCIB 9816-11 and *E. coli* JM109(DE3)(pDTG141) oxygenate a variety of substrates other than naphthalenes. Indene and indanes are oxygenated by these organisms at benzylic positions (Eq. 11).[45] Several tricyclic (Eq. 12)[66] and heterocyclic arenes (Eq. 13)[67] are substrates for these organisms.

$$\text{indene} \xrightarrow[\text{Escherichia coli JM109-(DE3)(pDTG141)}]{\textit{Pseudomonas putida} \text{ 9816-11 or}} \text{diol (53\%) 90\% ee} + \text{hydroxyindene (43\%) 94\% ee} \quad \text{(Eq. 11)}$$

$$\text{dihydrophenanthrene} \xrightarrow[\text{Escherichia coli JM109(DE3)(pDTG141)}]{\textit{Pseudomonas putida} \text{ 9816/11 or}} \text{diol (8\%) >95\% ee} + \text{hydroxyl product (3\%) >98\% ee} \quad \text{(Eq. 12)}$$

[Eq. 13 scheme: dibenzofuran → Pseudomonas sp. NCIB 9816-11 or Escherichia coli JM109(DE3)(pDTG141) → two dihydrodiol products (34%, >95% ee) and (22%, >95% ee)]

Another microorganism that has been used in the oxygenation of an extensive series of substituted naphthalenes (for example, Eq. 14)[47] is *P. fluorescens* N3 TTC1, a mutant of the wild-type *P. fluorescens* N3.[47] The dioxygenase of this organism has not been biochemically characterized as an NDO but the structural pattern of dihydrodiols produced is similar to that of *Pseudomonas* sp. NCIB 9816-11.

[Eq. 14 scheme: 1-chloronaphthalene → *Pseudomonas fluorescens* N3 TTC1 (65%) → two dihydrodiol products]

The model shown in Figure 1 can be used with care to predict the sense of regioselectivity and enantioselectivity of oxygenations catalyzed by the NDO enzymes. A later section provides descriptions of by-products of microbial arene oxidations by NDO microorganisms (see "Other Oxidations by Arene Dioxygenases").

Biphenyl Dioxygenase (BPDO)

A search for microorganisms that could be used in removal of polychlorinated biphenyls from the environment led to discovery of an organism with BPDO activity.[49] The microorganism was initially identified as a *Beijerinckia* species and was designated as species B1.[11] Mutation of the wild type gave *Beijerinckia* B8/36 in which the dihydrodiol dehydrogenase activity is blocked, allowing accumulation of the dihydrodiol.[49] In 1996 *Beijerinckia* B1 was reclassified as *Sphingomonas yanoikuyae* B1[68] and, accordingly, *Beijerincka* B8/36 became *Sphingomonas yanoikuyae* B8/36. In this chapter, name usage corresponds to the nomenclature used in the literature references. Cloning of the genes for the BPDO activity of *S. yanoikuyae* B1 into another microorganism has so far been unsuccessful.[11] A thorough account of the history of this microorganism in the production of dihydrodiols has been published.[11]

The fermentation of biphenyl with *Beijerincka* B8/36 gives a product characterized as (1*S-cis*)-3-phenyl-3,5-cyclohexadiene-1,2-diol (**9**, Eq. 15).[49] Although the biphenyls are a structural subclass of benzenes, the oxygenation products obtained from them are listed separately in Table 3. The structurally related dibenzocyclobutane (Eq. 16)[69,70] and fluorenes (e.g., Eq. 17)[67] are included with the biphenyls. It should be noted that biphenyl also is subject to oxygenation by NDO- and TDO-producing organisms.

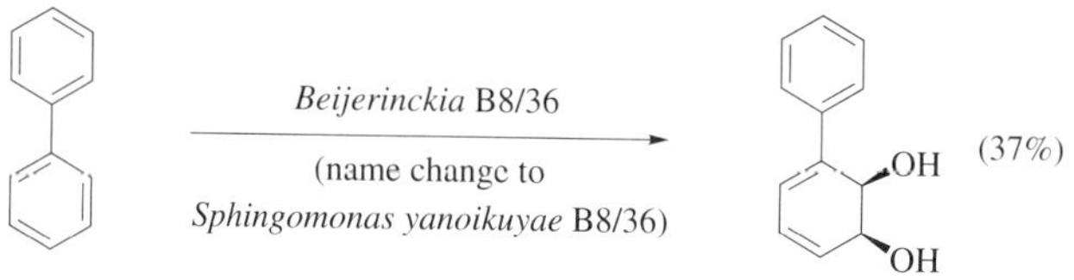

(Eq. 15)

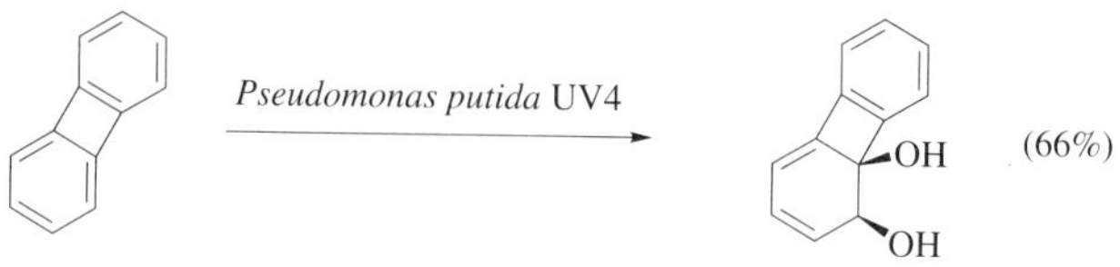

(Eq. 16)

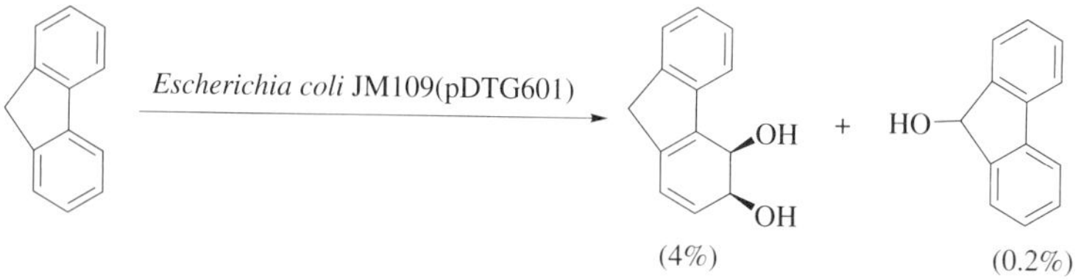

(Eq. 17)

The BPDO enzyme is relatively non-specific with regard to substrate structure and has found considerable application in the dioxygenation of a variety of arenes, including a number of polycyclic aromatics (e.g., Eqs. 4 and 5) and many heteroaromatic compounds (Eq. 6).[11] Several chlorinated biphenyls are reported to be oxygenated by *Beijerinckia* B8/36 to dihydrodiols but the results remain unpublished.[11]

Another organism capable of oxygenating biphenyl and related compounds is *Pseudomonas* sp. strain LB400.[51] This microorganism was isolated from polychlorinated biphenyl (PCB) contaminated soil.[71] *Pseudomonas* sp. LB400 has recently been reclassified as *Burkholderia* sp. strain LB400.[72] The BPDO enzyme complex from cells of *Pseudomonas* sp. LB400 has been isolated, purified and used for small scale oxygenation of chlorinated biphenyls (Eq. 18).[73] The small scale of these oxidations allowed for identification of the constitution of several dihydrodiols but not for a complete analysis of absolute configuration and enantiomeric excess.

(Eq. 18)

In another approach, the genes for the BPDO activity of *Burkholderia* sp. LB400 were cloned and expressed into *E. coli* giving strain BL21(DE3)/pLys.[52] Oxidations with the recombinant strain were done on a small scale and only limited characterization of dihydrodiols was therefore possible. Two dihydrodiols from this study are included in Table 3 and the remainder are included in Table 10. Yet another recombinant strain containing genes for BPDO is *E. coli* FM4560(pGEM410). This recombinant strain contains enzymes for the complete metabolic degradation of PCBs and is not useful for isolation of dihydrodiols.[50]

Benzoate Dioxygenase (BZDO)

The regiochemical differences in the dioxygenation of benzoic acids and esters by TDO and BZDO has already been noted. Oxygenations with TDOs give dihydro-2,3-dihydroxy-1-carboxylic acids (Eq. 9) whereas oxygenations with BZDOs occur at carbons 1 and 2 of the benzoic acid ring, giving dihydro-1,2-dihydroxy-1-carboxylic acids (also called ipso-cis-dihydrodiols).

Benzoic acid is converted into (1*S-cis*)-1,2-dihydroxy-3,5-cyclohexadiene-1-carboxylic acid (**6**, Eq. 7) by fermentation with either *Alcaligenes eutrophus* mutant strain B9[35,34] or with *P. putida* strain U103.[33] This dihydrodiol carboxylic acid is the only compound of this class for which the absolute configuration has been reported. Detailed descriptions have been published for carrying out the fermentation of benzoic acid with *P. putida* U103 (20–30 g dihydrodiol in 12 L)[33] or with *A. eutrophus* B9 (38 g of dihydrodiol from 6 L and 270 g from 80 L fermentations).[34]

The relative rates of dioxygenation of a series of seventeen benzoic acids by three microorganisms, *A. eutrophus* B9, *Pseudomonas* sp. B13, and *P. putida* mt-2, have been compared but no products were isolated (the results are not included in the Tables).[74] In a second report from the same laboratory,[36] the oxygenation of nine of these carboxylic acids with *A. eutrophus* B9 is described. The cis-relationships of the nine dihydrodiols are confirmed with spectral and chemical data. Circular dichroism (CD) spectral data suggest that all nine have the same absolute configuration as that of dihydrodiol **6** (Eq. 7).[36]

The mutant organism, *P. putida* PpJT103, has been used to oxidize several difluorobenzoic acids to 1,2-dihydroxy-1-carboxylic acids for which all stereochemistry except absolute configuration has been assigned.[75,53]

Dioxygenation of naphthalene-2-carboxylic acid by *P. testosteroni* A3 produces the cis-1,2-diol in 74% yield (Eq. 19).[76]

Other Dioxygenases

The cloning of the genes for the chlorobenzene dioxygenase (CDO) activity of *Pseudomonas* sp. strain P51 and their expression in *E. coli* DH5$_\alpha$(pTCB144) has been described and the capacity of the recombinant strain for dioxygenation activity has been explored.[77] Of fifty-six compounds screened with this strain, thirty-five were substrates and were converted into one or more dihydrodiol products. Many of these dihydrodiols were not thoroughly characterized but enough analytical data was presented to allow the conclusion that both regioselectivity and enantioselectivity in these oxygenations parallels that of organisms that produce TDO. In some cases the ee's of the dihydrodiols exceed those of the analogous compound produced by a TDO. An example is the dihydrodiol obtained from 4-chlorotoluene. From fermentation with the TDO producing organism, *P. putida* UV4, the dihydrodiol has 15% ee but from *E. coli* DH5$_\alpha$(pTCB144), the dihydrodiol has 77% ee. This recombinant strain offers promise for the dihydroxylation of a broad spectrum of arene molecules.

The marine bacterium *Nocardioides* sp. strain KP7 contains an arene dioxygenase that has been described as a phenanthrene dioxygenase.[78] The gene cluster (phdABCD) from this bacterium that codes for the dioxygenase has been introduced into *Streptomyces lividans*. The recombinant organism, *S. lividans* pIJ6021-phdABCD, converts 1-methoxynaphthalene into a mixture of 8-methoxy-1,2-dihydro-1,2-naphthalenediol and 8-methoxynapthol in a ratio of about 2:1. The organism also converts phenanthrene into *cis*-3,4-dihydroxy-3,4-dihydrophenanthrene in a nearly quantitative transformation.[78] The general usefulness of this organism for dioxygenation processes remains to be determined.

Other Oxidations by Arene Dioxygenases

A consequence of the broad substrate selectivity of the dioxygenases is that they also catalyze other types of oxidations, including: monooxygenations (usually at benzylic carbon), dihydroxylation of olefins, sulfoxidations, dehydrogenations, and *N*- and *O*-dealkylations. In some cases, these oxidations represent the major or even the exclusive transformation of a substrate. With enantiomeric substrates, the oxidations often proceed with high enantiospecificity. However, since oxidations of these types often can be achieved efficiently with chemical oxidants, the use of fermentations for these processes is of less impact. In order to place the oxidative scope of dioxygenases into larger perspective, the oxidations of olefins, of benzylic carbon, and of sulfur are included in Tables 7–9, respectively, and are discussed below.

A variety of olefinic bonds are oxygenated by dioxygenases. With styrene and substituted styrenes, oxygenation of the arene nucleus usually predominates over olefin oxygenation, a selectivity that cannot be achieved with chemical oxidants. Oxygenation of styrene with the TDO microorganisms, *P. putida* 39D and *P. putida* UV4, gives the dihydrodiol as the main product (Eq. 20).[79–81,25] However, when the styrene olefin is confined in a second ring, such as in dihydronaphthalene or indene, oxygenation of the styrene double bond is favored over that of the arene system. Also of note in the oxygenation of indene is a reversal of enantioselectivity depending on whether a TDO or an NDO organism is used for the fermentation. Dioxygenation of indene with *P. putida* 9816-11[45] or *E. coli* JM109(DE3)(pDTG141),[45] both NDO microorganisms, occurs at the olefinic bond giving the (1*R*,2*S*)-diol and is accompanied by benzylic hydroxylation of the indene methylene group (Eq. 21). Oxygenation of indene with *P. putida* UV4[82] or *E. coli* D160-1,[83] TDO microorganisms, also occurs on the alkene but now gives the (1S,2R)-diol together with products of benzylic oxidation. The *E. coli* D160-1 was engineered to contain (1R,2S)-*cis*-dihydrodiol dehydrogenase in addition to TDO so that, in effect, it is able to carry out a kinetic resolution of the mixture generated by dioxygenation. The dihydroxylation of indene has received special attention because of the potential of the (1S,2R)-diol as a precursor to an anti-viral agent.[83]

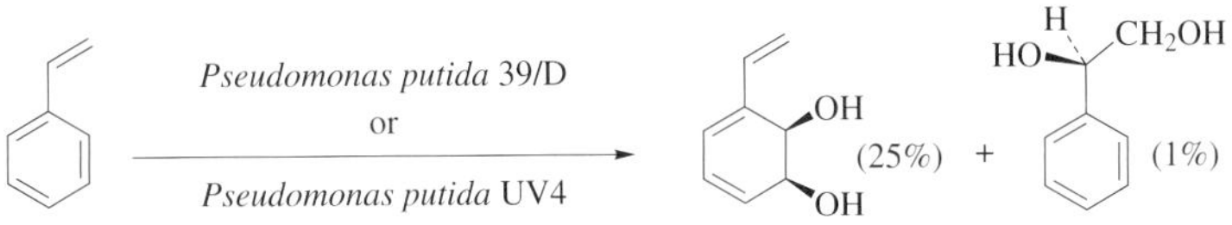

(Eq. 20)

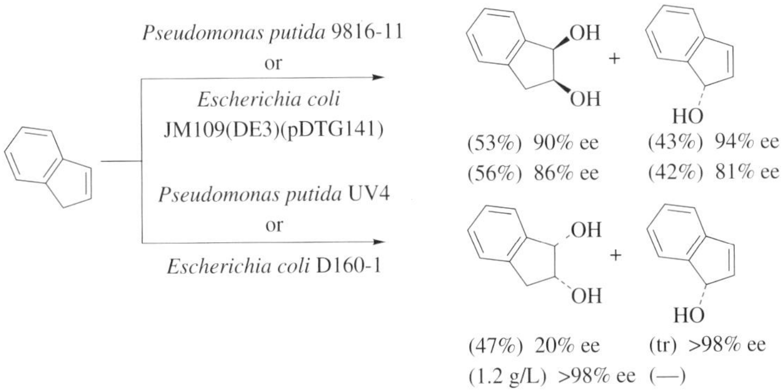

(Eq. 21)

Exploration of the oxygenation of the olefinic bond has been extended to both acyclic and cyclic mono-, di-, and polyenes. 2-Methyl-1,3-butadiene is one of a series of four acyclic olefins whose oxygenation with dioxygenases has been studied.[84] Oxygenation of this substrate with *P. putida* ML 2 produced two glycols in a ratio of 4:1 (Eq. 22). Each of the two products is the result of dioxygenation of one or the other of the two olefinic bonds in the substrate. Yields were not reported in the preliminary communication but absolute configurations and enantiomeric excesses were determined.

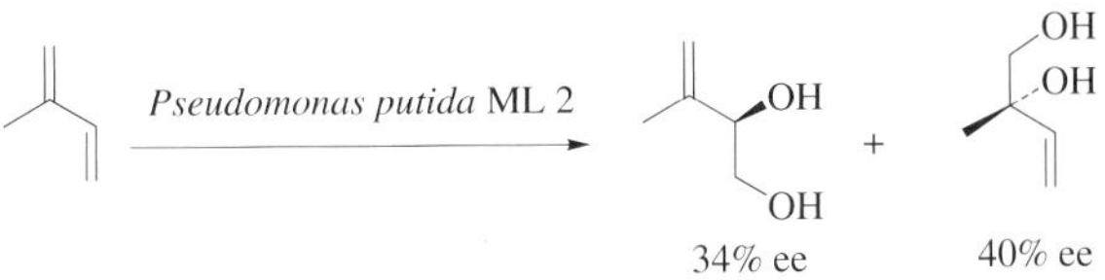

(Eq. 22)

A variety of cyclic mono-, di-, and trienes ranging from 1-methylcyclohexa-2,3-diene to norbornadiene to azulene has been subjected to oxygenation by various dioxygenase-containing microorganisms, usually with good yields of cis-diols produced. For example, oxygenation of dimethylfulvene with either *P. putida* RE213[85] or *P. putida* UV4[86] gives the cis-(1*R*,2*S*)-diol (Eq. 23) with greater than 98% ee.

$$\text{dimethylfulvene} \xrightarrow[\textit{Pseudomonas putida} \text{ UV4 (ca. 30\%)}]{\textit{Pseudomonas putida} \text{ RE213 (95\%)} \atop \text{or}} \text{cis-diol} \quad \text{(Eq. 23)}$$

Monooxygenation at benzylic carbon is widely observed (see Table 8). Even in the dioxygenation of toluene by *P. putida* UV4, which gives the dihydrodiol in good yield, careful examination of the total product mixture reveals the presence of a small amount of triol wherein the methyl group is oxidized to a benzyl alcohol.[55] Such side reactions, usually minor, are observed in the dioxygenations of a variety of alkyl substituted benzenes with this microorganism.[55] Benzylic oxygenations also are observed in cyclic substrates such as indanes (Eq. 24)[87], indenes, dihydronaphthalenes, and a series of dimethylnaphthalenes where they often represent the major pathway of oxidation.

$$\text{indane-N}_3 \xrightarrow{\textit{Pseudomonas putida} \text{ UV4}} \text{hydroxyindane-N}_3 \quad \text{(Eq. 24)}$$
(61%), >98 ee

The oxidation of aryl sulfides by dioxygenases generally occurs at sulfur in preference to the arene ring (see Table 9). Sulfides are oxidized to sulfoxides and, when the sulfide is unsymmetrical, often with a high degree of enantioselectivity. Methyl phenyl sulfide, for example, is oxidized to methyl phenyl sulfoxide and, depending on the choice of organism, either enantiomer of the sulfoxide may be formed (Eq. 25).[44,88,58] Thiophenes and benzthiophenes are subject to both dioxygenation and sulfoxidation leading to mixtures of four to six products.[89]

(Eq. 25)

PhSMe → (via *Pseudomonas putida* UV4 or TDO from *Escherichia coli* JM109(pDTG601A)) → Me-S(=O)-Ph (95%) >98% ee; (−) >98% ee

PhSMe → (via *Pseudomonas putida* NCIMB 8859) → Me-S(=O)-Ph (33%) 91% ee

Site-Directed Mutagenesis of Dioxygenases

As noted earlier, a limitation of microbial arene oxidation is that, in most cases, only one enantiomer can be generated by this method. Site-directed mutagenesis of the microbial oxygenases offers promise as a method for obtaining either enantiomer of a dihydrodiol and changing the regiochemical outcome of the process. Biphenyl has been used as a substrate to study the effect that site-directed mutagenesis can have on the dioxygenation process. The result of oxygenation of biphenyl with *Beijerinckia* sp. B8/36 (name changed to *Sphingomonas yanoikuyae* B8/36), a BPDO producing organism, is shown in Eq. 15. Biphenyl may also be oxygenated with *E. coli* JM109(DE3)(pDTG141), an NDO producing organism, in which case the dihydrodiols, (2R,3S-cis)-dihydrodiol (>98% ee) and (3R,4S-cis)-dihydrodiol (>98% ee) (**10** and **11**, Eq. 26), are generated in a ratio of 87:13.[90,31] When valine is substituted in place of phenylalanine-352 (code for this substitution: F352V) in the active site of this NDO, both the regiochemistry and the enantioselectivity of the oxygenation process are altered. Now the 2,3- and 3,4-dihydrodiols are obtained in a 4:96 ratio and the sense of enantioselectivity is changed such that the (3S, 4R-cis)-dihydrodiol **12** is obtained with 75% ee.[31] Fourteen other NDO variants were generated by substitution of from one to five amino acids near the active site of the native enzyme and were screened for dioxygenation of biphenyl.[91] The results shown (Eq. 26, top two lines) are the most divergent with regard to alteration of regio- and enantioselectivity.

(Eq. 26)

In a further step, genes encoding both the NDO F352V enzyme variant and a toluene dihydrodiol dehydrogenase were expressed in an *E. coli* strain, JM109(DE3)(pDTG141-F352V)(pDTG511). Addition of the dihydrodiol dehydrogenase permits a kinetic resolution of the enantiomerically impure dihydrodiol oxygenation product to proceed. Fermentation of biphenyl with this *E. coli* strain gives enantiomerically pure (3S,4R-cis)-dihydrodiol **13** (Eq. 26).[31]

The oxygenation of phenanthrene with the same set of variant organisms used for biphenyl has been examined. The "baseline" dioxygenation of phenanthrene with *Beijerinckia* sp. B8/36 has been discussed previously (Eq. 5). Oxygenation of phenanthrene with *E. coli* JM109(DE3)(pDTG141-F352V)[31] gives (3*S-cis*)-3,4-dihydro-3,4-phenanthrenediol (**14**, >95% ee) and (1*S-cis*)-1,2-dihydro-1,2-phenanthrenediol (**15**, 91% ee) (see Eq. 27) in a ratio of 17:83. Here again, a dramatic shift of regio- and enantioselectivity of oxygenation is seen with the organism altered by site-directed mutagenesis. Oxygenation of phenanthrene with all of the fourteen variants used for biphenyl gave various ratios of 1,2- and 3,4-dihydrodiols and in several cases the 9,10-dihydrodiol also was produced (Eq. 27).[91] Finally, as with biphenyl, oxygenation of phenanthrene with *E. coli* JM109(DE3)(pDTG141-F352V)(pDTG511) gave enantiomerically pure (1*R-cis*)-1,2-dihydro-1,2-phenanthrenediol (shown in Eq. 5).[31]

(Eq. 27)

Another approach to modification of the oxygenation enzymes is directed evolution employing either DNA shuffling[92] or mutation of single genes[93] to generate libraries of modified organisms. These libraries are then screened for enhanced oxygenation of arenes. In one case, this has led to a strain that oxidizes 4-picoline to 3-hydroxy-4-picoline.[93] Although the product is not a dihydrodiol, the oxidation is assumed to proceed through the highly reactive dihydrodiol and illustrates that this approach may have potential for generating new arene dioxygenases.

Scope of Method

Because microbial arene oxidations developed from studies of metabolism, much of the initial work has examined the oxygenation of relatively simple molecules. Most substrates were simple benzenes, naphthalenes, and biphenyls having relatively few functional groups or additional complex ring structures. As the field has evolved, the method has been extended to increasingly larger aromatic hydrocarbons and their heterocyclic congeners. Such work now encompasses four- and five-ring polycyclic aromatic and heteroaromatic hydrocarbons. The conversion of the smaller arenes into dihydrodiol synthons is, of course, what has brought the method to the attention of chemists. What remains largely unexplored is application of the method to more elaborate substrates such as is exemplified in the dioxygenation of 7-oxodehydroabietic acid (**16**) by *P. abietaniphila* BKME-9 (Eq. 28).[94] Also unexplored is use of the method for introduction of a cis-diol at an intermediate or final step of a synthetic sequence. Applications to synthesis have all been based on the use of a dihydrodiol at the outset of the synthesis.

$$\underset{\mathbf{16}}{\text{[structure with } HO_2C, O, Pr\text{-}i\text{]}} \xrightarrow{\textit{Pseudomonas abietaniphila } \text{BKME-9 or } \textit{Escherichia coli } \text{XL1 Blue MR(pVM20)}} \underset{Pr\text{-}i}{\text{[structure with } HO_2C, O, OH, OH, Pr\text{-}i\text{]}} \quad \text{(Eq. 28)}$$

In addition, questions about the suitability of unusual arenes as substrates for different oxygenases are unanswered. For example, the capability of dioxygenases to oxygenate such unusual molecules as [2.2]-paracyclophane, bullvalene, semibullvalene, triphenylene, and biphenyls of restricted rotation should be explored. The question of whether a fullerene might be a substrate for a dioxygenase also remains to be examined.

Several attempts have been made to utilize microbial arene oxidations in industrial scale syntheses, including the use of *cis*-3,5-cyclohexadiene-1,2-diol in polymer production[42] and of (1*R*-*cis*)-1,2-dihydronaphthalene-1,2-diol in the production of indinavir.[95] The fermentation step in the latter synthesis was shown to be practical but development into a full-scale industrial process failed because of the lack of proprietary control of the process. Still, there is no inherent reason that industrial arene dioxygenations cannot be implemented. Other microbial biotransformations, including monooxygenations, are performed routinely as industrial processes. Arene dioxygenation awaits another synthetic target derived from a dihydrodiol needed on a production scale to demonstrate usefulness as an industrial process.

APPLICATIONS TO SYNTHESIS

Representative examples of the use of dihydrodiols in synthesis are presented in this section, which is divided into two parts. In the first part, some synthetic modifications of microbially produced dihydrodiols that generate other synthetic intermediates are presented. Chemical modifications of the dihydrodiols have also been reviewed elsewhere.[15,96,97] In the second part, representative examples in which dihydrodiols have been elaborated into more complex molecules are summarized. The work related to the second part has been extensively reviewed,[98–100] both from the point of view of natural product classes[101,102] and of the contributions of individual research groups.[103,104] The approach used here is to show one or more examples of syntheses derived from each of a variety of different dihydrodiols. Further, rather than showing the entire multi-step synthesis, only the starting dihydrodiol and the final natural product are shown.

Other Synthetic Intermediates by Modification of Microbially Produced Dihydrodiols

The simple, one-step modification of dihydrodiols by replacement of a reactive group provides a route to new dihydrodiols. Most modifications of this kind have been done using (1*S*-*cis*)-3-iodo-3,5-cyclohexadiene-1,2-diol (**17**). For example, the palladium-catalyzed reaction of this iodide with phenyl tributyltin sulfide gives

($1S$-cis)-3-phenylthio-3,5-cyclohexadiene-1,2-diol (**18**, Eq. 29).[105,25,27] A similar palladium-catalyzed Sonogashira coupling with trimethylsilylacetylene gives ($1S$-cis)-3-(trimethylsilyl)ethynyl-3,5-cyclohexadiene-1,2-diol.[105,106]

(Eq. 29)

Another one-step approach to new dihydrodiols is the catalytic removal of iodine from a dihydrodiol obtained by oxygenation of a disubstituted benzene substrate. In the case of 4-fluoroiodobenzene, the microbially produced ($1S$-cis)-6-fluoro-3-iodo-3,5-cyclohexadiene-1,2-diol (**19**) has an 88% ee.[27,23,107] Catalytic removal of the iodine gives ($1R$-cis)-3-fluoro-3,5-cyclohexadiene-1,2-diol (**20**), still with an 88% ee (Eq. 30).[27] Fluorodihydrodiol **20** produced by this procedure is enantiomeric to that obtained from the direct oxygenation of fluorobenzene.[24,23,25]

(Eq. 30)

Another example employing removal of iodine illustrates routes to both enantiomers of a new dihydrodiol. Oxygenations of 2-fluoroiodobenzene and 3-fluoroiodobenzene give ($1S$-cis)-4-fluoro-3-iodo-3,5-cyclohexadiene-1,2-diol (**21**) and ($1S$-cis)-5-fluoro-3-iodo-3,5-cyclohexadiene-1,2-diol (**23**), respectively, in yields of 75% and with ee >98%. Catalytic removal of iodine from each of these dihydrodiols produces the enantiomers, ($1S$-cis)-4-fluoro-3,5-cyclohexadiene-1,2-diol (**22**) and ($1R$-cis)-3-fluoro-3,5-cyclohexadiene-1,2-diol (**24**), respectively (Eqs. 31 and 32).[27]

(Eq. 31)

(Eq. 32)

In another approach to the generation of both enantiomers of a dihydrodiol, an enzymatic kinetic resolution step is used. (1S-cis)-6-Bromo-3-iodo-3,5-cyclohexadiene-1,2-diol (**25**) is obtained with an ee of 22% from oxygenation of 4-bromoiodobenzene with *P. putida* UV4 (Eq. 33).[23] Hydrogenolysis of this dihydrodiol to remove the iodine gives the (1R-cis)-3-bromo-3,5-cyclohexadiene-1,2-diol (**26**, 22% ee) which, when placed in fermentation with the dienediol dehydrogenase-bearing organism, *P. putida* NCIMB 8859, is "upgraded" to (1R-cis)-3-bromo-dienediol (**27**) with ≥ 98% ee.[108] The enantiomeric (1S-cis)-3-bromo-3,5-cyclohexadiene-1,2-diol is produced in the oxygenation of bromobenzene.[106,25] The pathway from 4-bromoiodobenzene to the (1R)-enantiomer is of relatively low overall yield.

(Eq. 33)

Use of Dihydrodiols in Synthesis

The use of dihydrodiols in synthesis developed very slowly. In 1983, the preparation of polyphenylene in three steps from cis-3,5-cyclohexadiene-1,2-diol (**1**, Eq. 34) was described.[109]

(Eq. 34)

The use of this dihydrodiol as a monomer for the production of polymers became feasible with the economic production of dihydrodiols by industrial scale fermentations.[42,56] Work to improve the process and the quality of the polymer produced has been reported.[110,111]

The application of dihydrodiols to the synthesis of natural products was pioneered in 1987 when dihydrodiol **1** was used in a six-step synthesis of (±)-pinitol (Eq. 35).[112]

(Eq. 35)

Following the synthesis of (±)-pinitol, reports describing the synthesis of both unnatural and natural products from dihydrodiol precursors rapidly increased in number. Shown below are examples that illustrate the use of a variety of dihydrodiols in synthesis. Where not immediately obvious, the disposition of the atoms of the dihydrodiol in the final product is indicated by numbering of carbon atoms.

Among the many natural products syntheses based on **1** (see Figure 3) are an eight-step synthesis of 3,4-dihydroxy-α-tropolone (**28**),[113] an eleven-step synthesis of (±)-*myo*-inositol 1,4,5-triphosphate (**29**),[114] and two-step syntheses of conduritols A (**30**) and D (**31**).[115] Dihydrodiol **1** also has been used as a precursor for several "unnatural" products. The compound, *anti-o,o'*-dibenzene (**32**) is derived from the dihydrodiol in three steps, the first of which is a photochemical dimerization.[116] Benzobarrelene (**33**) is prepared from **1** in a four-step sequence that includes addition of benzyne to the diene.[117]

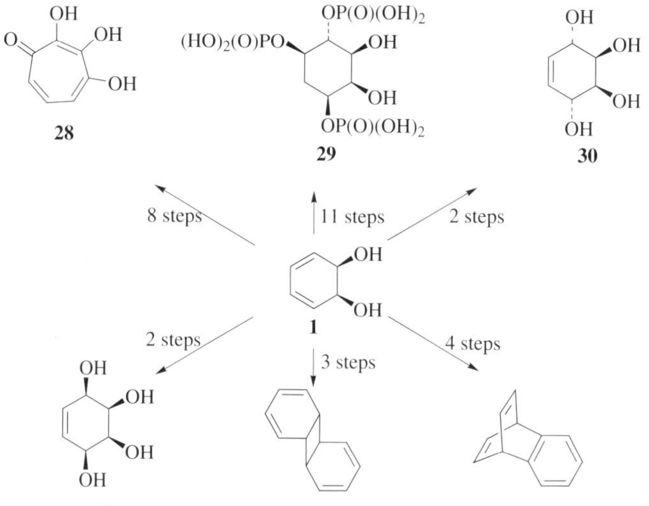

Figure 3. *cis*-3,5-Cyclohexadiene-1,2-diol (**1**) serves as the starting point for synthesis of natural and "unnatural" products.

In the acid-catalyzed dehydration of 3-substituted dihydrodiols, the predominant product in most cases is the *o*-phenol (Eq. 36).[118] An exception is found when the 3-substituent is a ketal. Acid-catalyzed dehydration of these dihydrodiols results in formation of the *m*-phenol as well as hydrolysis of the ketal (Eq. 37). The 3-acetyl or 3-formylphenols are useful precursors in the synthesis of 3-ethynylphenol.[81]

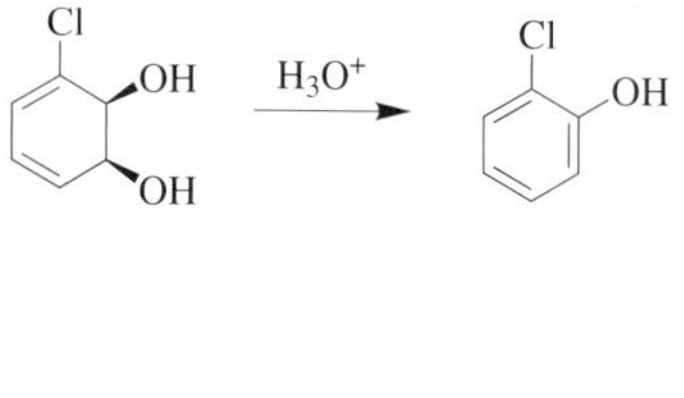

(Eq. 36)

$$\text{(Eq. 37)}$$

(1S-cis)-3-Methyl-3,5-cyclohexadiene-1,2-diol serves as the starting point for a four-step synthesis of pseudo-α-L-fucopyranose (Eq. 38).[119] Diels-Alder cycloadditions and anionic oxy-Cope rearrangements are used as key steps in syntheses of several cis-decalins[120] and bicyclo[5.3.1]undecanes from the 3-methyldihydrodiol.[120a] The cis-decalins have potential use as intermediates leading to the synthesis of natural products such as mevinolin and artemisinic acid while the bicyclo[5.3.1]undecanes are envisioned as taxol™ precursors.

$$\text{(Eq. 38)}$$

pseudo-α-L-fucopyranose

(1S-cis)-3-Chloro-3,5-cyclohexadiene-1,2-diol is the dihydrodiol that has most frequently been used as a synthetic precursor. The compound serves as a starting point for a nine-step synthesis of L-ascorbic acid (Eq. 39). In this synthesis, the C(1) hydroxyl group of the dihydrodiol directs epoxidation of the electron rich C(5)-C(6) double bond affording the syn isomer. Subsequent manipulation of the molecule results in rehybridization of the two original carbinol atoms.[121] Among other syntheses, the 3-chlorodihydrodiol serves as the starting material for a seventeen-step synthesis of the amaryllidaceae alkaloid (+)-trianthine.[122]

$$\text{(Eq. 39)}$$

L-ascorbic acid

As a synthetic intermediate, (1S-cis)-3-bromo-3,5-cyclohexadiene-1,2-diol closely parallels (1S-cis)-3-chloro-3,5-cyclohexadiene-1,2-diol. Several compounds from the amaryllidaceae family have been constructed from the bromodihydrodiol including (+)-pancratistatin, prepared in a thirteen-step synthesis (Eq. 40),[123] and the related (+)-7-deoxypancratistatin, obtained in nine steps.[124] In a seven-step synthesis of cyclopropane intermediate **34**, a compound proposed for use in preparation of pyrethrins, the atoms of the starting bromodihydrodiol are barely recognizable in the final product (Eq. 40).[125]

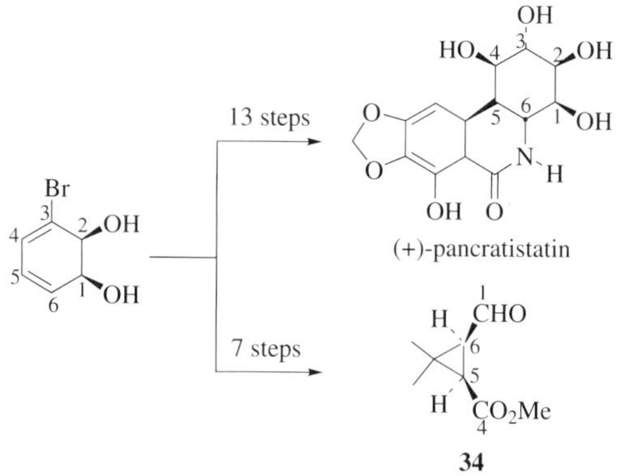

(Eq. 40)

Several dihydrodiols have been used in new and different approaches to morphine. The most advanced of these uses (1S-cis)-3-(2-bromoethyl)-4-bromo-3,5-cyclohexadiene-1,2-diol as a key intermediate. Thirteen synthetic steps are used to convert the dihydrodiol into *ent*-C14-*epi*-morphinan (Eq. 41).[126,127] (1S-cis)-3-(2-Azidoethyl)-3,5-cyclohexadiene-1,2-diol,[128] (1S-cis)-3-(2-bromoethyl)-3,5-cyclohexadiene-1,2-diol,[129] and (1S-cis)-3-(2,3-dimethoxyphenyl)-3,5-cyclohexadiene-1,2-diol[130] all have been used to synthesize intermediates on potential pathways to morphine.

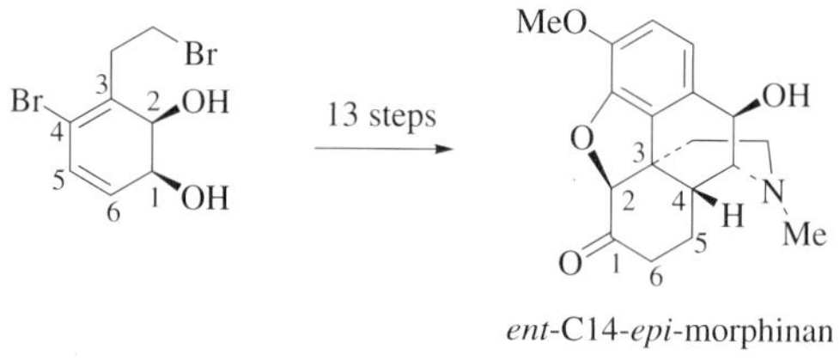

(Eq. 41)

(1S-cis)-3,5-Dibromo-3,5-cyclohexadiene-1,2-diol serves as a chiral starting material in an eleven-step route to the Amaryllidaceae alkaloid (+)-narciclasine (Eq. 42).[131]

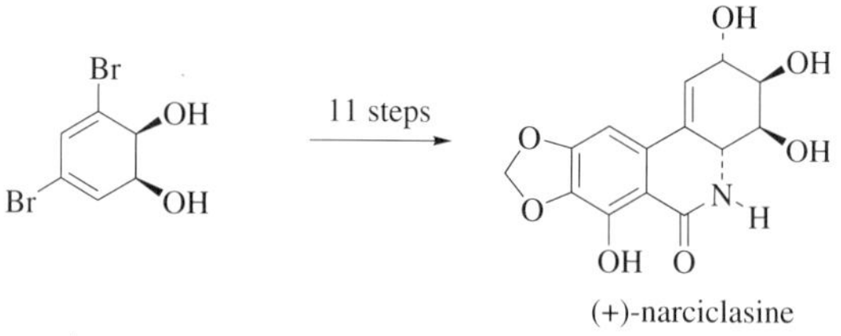

(Eq. 42)

(1*S-cis*)-3-Vinyl-3,5-cyclohexadiene-1,2-diol is converted in eleven steps to the tricyclic natural product (−)-zeylena (Eq. 43).[132]

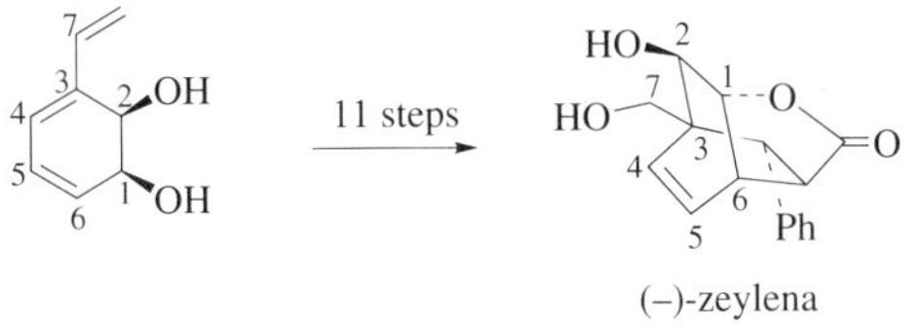

(Eq. 43)

(−)-zeylena

(1*R-cis*)-1,2-Dihydronaphthalene-1,2-diol is converted, in five steps, to polyhydroxylated tetrahydronaphthalene ethers (Eq. 44). Both of the tetrahydronaphthalene units of the ether product are derived from the dienediol.[133]

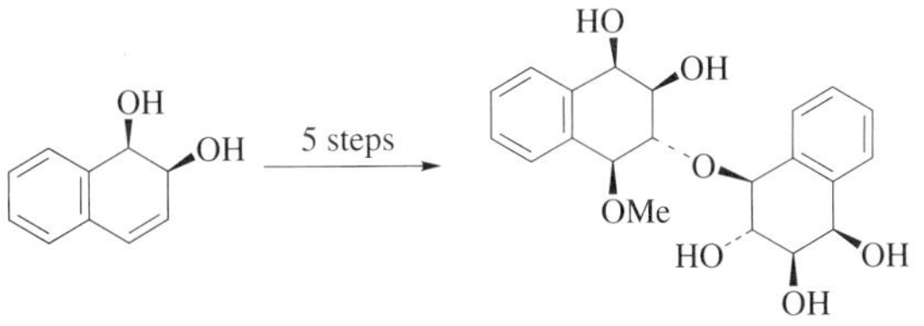

(Eq. 44)

The diol moiety of the dihydrodiols is frequently protected during synthetic manipulations as an acetonide. By carrying out the ketalization reaction with a resin-linked acetonide, the dihydrodiol grouping has been used in solid-phase synthesis for the preparation[134] of small combinatorial sortiments.[135] Both (1*S-cis*)-3-chloro- and (1*S-cis*)-3-bromo-3,5-cyclohexadiene-1,2-diol are readily incorporated into the resin[134] and are used in coupling reactions (Eq. 45).[134,136]

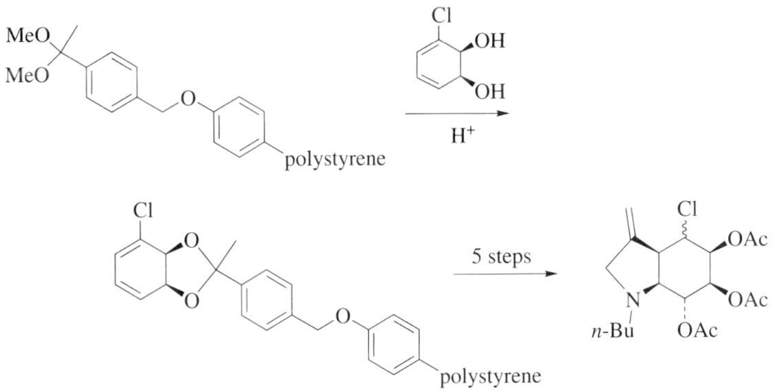

(Eq. 45)

COMPARISON WITH OTHER METHODS

A synthesis of *cis*-3,5-cyclohexadiene-1,2-diol from benzene described in 1959[137] is lengthy and not easily applicable to substituted dihydrodiols. A second synthesis (of the acetonide) from 1,4-cyclohexadiene described in 1982[138] likewise is not of general use. A photochemically induced reaction of benzene with osmium tetroxide described in 1995 is postulated to proceed through an osmate ester of cis-3,5-cyclohexadiene-1,2-diol but the isolated products of the reaction were a hexaacetoxycyclohexane and three isomers of chloropentaacetoxycyclohexane.[139]

EXPERIMENTAL CONDITIONS

Comments about Yields

The yield of a reaction is of paramount importance to the synthetic chemist. Because the determination of yield is a requirement of a good experimental procedure, it usually is reported in chemistry journals. Chemical yield often is of less concern to microbiologists and therefore is not always to be found in the experimental details reported in microbiology journals. In the Tables of this chapter, yields based on starting material are provided wherever possible. Indeed, when the reported data permit, yields have been calculated and are included in the Tables.

A frequent practice in reports of microbial arene oxidations is to report g/L of product isolated from the fermentation of a given substrate. It may be that in such experiments, substrate was supplied to the fermentation in excess until product concentration was observed to reach a maximum. Volatile substrates may be partially lost from the fermentation medium by evaporation or via the air stream if the experiment is not properly designed. The yield data reported in the Tables for such experiments are given simply as g/L. On occasion, an impressive weight of product may be reported for these experiments but this value may still represent a low chemical yield.

The expression "relative yield" also is found in some literature reports and usually expresses the yield of a given product as a percentage of the total weight (or total area under GC peaks) of product(s) isolated or measured in the experiment. Clearly, such values are not true chemical yields but do serve to give a measure of the relative amount of a product formed in a fermentation.

Stability and Isolation of Dihydrodiols

The key to isolation and storage of dihydrodiols is to realize that the compounds are sensitive to acid-catalyzed dehydration. The corollary of this observation is that dihydrodiols are stable at or above pH ~ 9. The following approach to the isolation and storage of dihydrodiols is recommended.[140] (The method is varied slightly depending on the molecular weight of the dihydrodiol being produced.) With lower molecular weight compounds, once the oxygenation process has reached a maximum level of dihydrodiol, the fermentation mixture is centrifuged in order to separate the mixture into a solid pellet and a supernatant solution. The supernatant is separated, adjusted to pH ~ 9, and concentrated approximately ten-fold by evaporation of water under partially reduced pressure. At this point, the remaining aqueous solution is relatively concentrated with inorganic salts, and the dihydrodiol is stable and can be stored in this condition. When the dihydrodiol is needed, it can be extracted easily into an organic solvent and then isolated by the usual methods. For the higher molecular weight dihydrodiols (starting with the dihydrodiols from naphthalenes and biphenyls), the *addition* of water to the fermentation mixture before performing the centrifugation and separation step is recommended to provide the aqueous volume necessary for solubilization of the product.

The stability and ease of isolation of dihydrodiols are often discussed anecdotally in the literature. For example, attempts to isolate *cis*-3,5-dimethyl-3,5-cyclohexadiene-1,2-diol from the oxygenation of *m*-xylene were unsuccessful, a result that has been attributed to the instability of the compound.[141] On the other hand, *cis*-3,6-dimethyl-3,5-cyclohexadiene-1,2-diol, the oxidation product from *p*-xylene, was isolated and characterized. The latter dihydrodiol was observed to be unstable and was converted into 2,5-dimethylphenol within six hours at room temperature.[141] The dihydrodiols derived from oxygenation of methoxy- and ethoxybenzene are characterized as unstable and subject to rapid aromatization, although derivatives from which structure and absolute configuration are determined have been prepared.[142,25,38] The half-life of (3*S-cis*)-3-chloro-3,5-cyclohexadiene-1,2-diol in a CDCl$_3$ solution was measured by NMR spectroscopy. At room temperature, the half-life was found to be four days but the diol could be stored at $-20°$ to $-80°$ for several months without decomposition.[40] The room temperature NMR experiment may be questioned since deuterochloroform is not a good solvent for storage of dihydrodiols because it has a propensity to generate traces of DCl.

In detailed studies leading to the production of *cis*-3,5-cyclohexadiene-1,2-diol from benzene on a ton scale, the dihydrodiol is stored following purification as a solution in ethyl acetate containing a small amount of triethylamine.[56]

Dihydrodiols are subject to acid-catalyzed elimination of water, resulting in aromatization and formation of a phenol. A study of the acid-catalyzed aromatization of a series of 3-substituted dihydrodiols has been reported.[118] The results are consistent with dehydration via a benzonium ion like intermediate. Rate constants for the process are highest for electron-releasing substituents such as alkoxy and alkyl groups and lowest for electron-withdrawing groups such a trifluoromethyl, sulfinyl, and sulfonyl. A difference in rate of 10^7 is observed between the most- and least-stabilized dihydrodiols in this study.

A practical consequence of the above results is that care should be taken to avoid acidic conditions in the isolation and handling of dihydrodiols. The precautions needed are often evident in descriptions of experimental conditions wherein pH is adjusted to >7, solvents are given alkaline washes, and chromatography is conducted with neutral adsorbents.

The sensitivity of the dihydrodiols to acid gives pause when considering oxygenation of substrates of the benzoic acid class. In several cases, the dihydrodiol-carboxylic acid products, while stated to be "unstable," have nevertheless been isolated and characterized. In the isolation of (1*S-cis*)-1,6-dihydroxy-2,4-cyclohexadiene-1-carboxylic acid from the fermentation broth, the pH of the mixture is kept at 4 during the extraction process. The product is isolated in good yield and is stable up to a year when stored at $-20°$. The sodium salt of the acid also was prepared and characterized.[33] In general, dihydrodiol-carboxylic acids may be isolated and stored as salts. The closely related (1*S-cis*)-1,6-dihydroxy-4-methyl-2,4-cyclohexadiene-1-carboxylic acid is described as unstable and is characterized as the sodium salt.[143] The free acid is generated as needed by acidification and rapid extraction of an aqueous solution of the salt. Alternately, the carboxylic acids can be converted into methyl esters, stabilizing the compounds.

Microorganisms (Cultures)

Culture (microorganism) procurement and handling are aspects of performing microbial arene oxidations that are likely to be unfamiliar to most organic chemists. Herewith are a few comments concerning the procurement of the specific microorganism desired for a given dioxygenation reaction. There are various sources for traditional, wild-type (natural) microorganisms. There are many central repositories throughout the world to which microbiologists submit pure cultures of new microorganisms or microbial strains as they are discovered and characterized. Several examples of such repositories are the American Type Culture Collection (10801 University Boulevard, Manassas, VA 20110; http://www.atcc.org); Deutsche Sammlung von Mikroorganismen und Zellkulturen GmbH (DSM; Mascheroder Weg 1b, D-38124 Braunschweig, Germany; http://www.dsmz.de); National Collections of Industrial, Food and Marine Bacteria (NCIMB; 23 St. Machar Drive, Aberdeen, AB24 3RY, Scotland, UK; http://www.ncimb.co.uk). These repositories maintain the pure cultures and provide "samples" or "seed cultures" of most microorganisms for a reasonable fee. If a microorganism strain is not available through such a central collection, it then is necessary to obtain the culture from the research group that has reported working with the microorganism.

As is evident from the previous sections of this chapter, the microorganisms most desirable for arene oxidations either are mutants of the wild-type strain or, increasingly, are microorganisms (usually *E. coli*) that contain plasmids encoded with the genes expressing the enzymes necessary for the dioxygenation reaction. These mutant and genetically engineered microbial strains often are not available from the aforementioned depositories. Consequently, it is necessary to request and obtain these strains from the laboratories in which they were originated or may now be in use. Publication of results in the journals of the American Society for Microbiology (ASM) implies that microorganisms described in the work will be supplied to other research groups upon request (the *Journal of Bacteriology* and *Applied and Environmental Microbiology* are the two ASM journals in which most arene oxidation work is published). Such requests require documentation of the capability to work with the microorganism. Occasionally, certain mutants or clones may be proprietary and, therefore, difficult to obtain for general use in carrying out arene oxidations.

Once obtained, transfer of the microbial culture is necessary in order to perform the fermentation that will produce the desired dihydrodiol. Basic microbiology techniques are required for such transfers. While not difficult, these techniques and standard microbiological equipment may not be familiar to the chemist. A description of some of these techniques is given in a following section (see "Experimental Procedures"). In addition, the assistance of a colleague familiar with basic microbiological manipulations can be most helpful and should be sought.

Inducers

The dioxygenases of mutant and recombinant strains of microorganisms often require that an inducer compound be added at an early phase of the fermentation to stimulate production of the enzyme. The inducer may be a natural substrate of the dioxygenase or another compound known to have the desired effect. For example, the TDO of *P. putida* 39/D is induced by toluene while the NDO of *Pseudomonas* sp. NCIB 9816-11 is induced by salicylate or succinate ions. Inducers used with some of the more frequently used microorganisms are included in Table A. The ability of a new substrate to induce the dioxygenase of a microorganism may be evaluated by using the indigo test.[40]

EXPERIMENTAL PROCEDURES

Handling the Microorganisms

The storage, transfer, and growth of microbial cells requires use of several fundamental techniques of microbiology. These techniques are not particularly difficult to perform but they require specialized equipment and adherence to strict standards. It is important at the outset of a fermentation that one be confident that the microbial cells are of high quality. An attempt has been made to include enough information to permit the chemist to proceed on his/her own. However, at this stage, consultation with a scientist familiar with basic microbiological manipulations can be helpful.

Stock cultures of microorganisms may be stored by several different methods. A traditional method is on an agar slant (or slope) in a screw-capped vial or test tube. The agar slant is inoculated with cells of the pure culture, the inoculum is allowed to establish growth on the surface of the agar, and then the closed tube is stored at low temperature. In another method of storage, cells of the pure culture are grown in a medium, such as Luria-Bertani broth, after which aliquots of the medium and culture are transferred to cryovials. Sterile glycerol containing the same nutrients as the broth is added to the vials, which are then closed and stored at low temperatures. In a third method of storage, the medium in which the culture has been allowed to grow is lyophilized in a vial or tube. The residual dry pellet is closed in the container and stored at low temperature.

The seed culture of the microorganism will most likely arrive or be made available growing on an agar slant, i.e., a test tube in which nutrient agar has been allowed to solidify to give a slanted surface. Seed cultures also may be obtained as glycerol solutions or lyophilized powders.

The transfer of cells of a microorganism requires a strong laboratory burner (as a Meeker burner), an inoculating loop (either a wire loop or a supply of disposable loops), and either a laminar flow hood or a work area that is relatively free of drafts. The wire loop is sterilized before and after each use by heating to redness in the burner flame. Mouths of test tubes and flasks used in any transfer process are flamed before and after use to kill contaminating, adventitious microorganisms and to keep air currents moving out the container. In preparation for fermentation, the microorganism is transferred from its storage medium to the surface of an agar medium in a Petri dish. Using a sterile loop for each step, a loopful of the cells, either scraped from the surface of a slant or dipped from a glycerol storage solution is streaked along one side of an imaginary square on the surface of the agar plate. Then, the next loop is passed through one end of the first streak and the loop is used to make a second streak along the adjoining side of the square. A third loop is likewise passed through the end of the second streak and this loop is used to make a third streak. The process is repeated one more time. The goal of this procedure is to dilute out microorganism cells such that a single colony can be isolated for use in further manipulations. The dish is covered and the cells are allowed to grow until heavy growth is observed (~24–48 hours). The fewest colonies of the microorganism should be observed to grow where the fourth streak was made and these colonies should be useful for inoculating flasks of nutrient media either for small-scale transformations of arene molecules, for further inoculation of larger fermentors, or for inoculation of an agar slant for storage of seed cells in case the working culture becomes contaminated.[40]

In the case where a large-scale fermentation is being performed, the inoculated flask is plugged with cotton and placed on a rotary shaker at 30° for ~24 hours. The contents of the flask are added to the sterile contents of the larger fermentor, flaming the mouth of the flask before the transfer.

A detailed description of the storage and handling of *P. putida* UV4 is given in the literature.[144] A detailed discussion, designed for chemists, of the experimental aspects of fermentations provides a good orientation to the subject.[145] The experimental description for the laboratory scale production of (1*S-cis*)-3-chloro-3,5-cyclohexadiene-1,2-diol is given in an *Organic Syntheses* procedure.[40]

Preparation of Media

Several forms of media will be necessary for handling of the culture. A medium is an appropriate mixture of nutrients in which the microorganism can grow and multiply. A liquid medium is needed for the fermentation and biotransformation of substrate while a solid medium is used for maintenance and storage of the culture.

Media may differ in the details of the ingredients and, consequently, different media are often used in different experimental descriptions. In reality, the media used for the microorganisms that carry out arene oxidations are quite similar to one another. The greatest difference among experimental descriptions will be in the carbon source used to support growth of the microorganism. Most of the microorganisms used for arene oxidations will grow on any of the standard media.

Listed below are components and recipes for the procedures that are found in the following section. Listed first are the stock solutions used in the preparation of the media. Listed second are the basic recipes for the media used in the procedures; variations are found in the individual procedures.

Stanier's stock solution A.[146] A 1 M aqueous buffer (pH 6.8) solution of $Na_2HPO_4 \cdot 7H_2O$ (268.1 g/L) and KH_2PO_4 (136 g/L).

Stanier's stock solution B.[146] An aqueous solution of $(NH_4)_2SO_4$ (1.0 g/L).

Stanier's stock solution C.[146] (based on Hutner's vitamin-free mineral base[147]) For 1 L, nitrilotriacetic acid (NTA; 10 g), $MgSO_4 \cdot 7H_2O$ (14.45 g), $CaCl_2 \cdot 2H_2O$ (3.335 g), $(NH_4)_6Mo_7O_{24} \cdot 4H_2O$ (9.3 mg), $FeSO_4 \cdot 7H_2O$ (99.0 mg), and Metals "44" solution (see below) (50 mL). The following procedure for mixing the components of Solution C is recommended.[40] First, the NTA is dissolved with stirring in 150 mL of distilled water. Next, a solution of the $MgSO_4 \cdot 7H_2O$ in 150 mL of distilled water is added to the NTA solution with stirring at a rate to avoid clouding of the mixture. Then a solution of the $CaCl_2 \cdot 2H_2O$ in 150 mL of distilled water is likewise added slowly so as to avoid any cloudiness of the solution (cloudiness will result in formation of insoluble precipitates). Next a solution containing both the $(NH_4)_6Mo_7O_{24} \cdot 4H_2O$ and the $FeSO_4 \cdot 7H_2O$ in 150 mL of distilled water is added with stirring. At this point, the combined solution should have a pale yellow color. The Metals 44 solution is added and the total volume is brought to 1.0 L with distilled water. The pH of the total solution should be carefully adjusted to 6.8 with 10 M aqueous NaOH solution—preferably in 0.2-mL aliquots—otherwise insoluble precipitates will form.

Metals "44" solution.[147] The Metals 44 solution is prepared as follows (for 100 mL): Ethylenediaminetetraacetic acid (EDTA; 250 mg), $ZnSO_4 \cdot 7H_2O$ (1.095 g), $FeSO_4 \cdot 7H_2O$ (500 mg), $MnSO_4 \cdot H_2O$ (154 mg), $CuSO_4 \cdot 5H_2O$ (39.2 mg), $Co(NO_3)_2 \cdot 6H_2O$ (24.8 mg), and $Na_2B_4O_7 \cdot 10H_2O$ (17.7 mg) are dissolved one at a time in 100 mL of distilled water followed by ~1–2 drops of 1 M H_2SO_4 (to retard precipitation). The resulting solution should be aquamarine blue in color.

Vishniac and Santer's trace metal solution.[148] For 1.0 L in H_2O, EDTA (50.0 g), $ZnSO_4 \cdot 7H_2O$ (22.0 g), $CaCl_2$ (5.54 g), $MnCl_2 \cdot 4H_2O$ (5.06 g), $FeSO_4 \cdot 7H_2O$ (4.99 g), $(NH_4)_6Mo_7O_{24} \cdot 4H_2O$ (1.10 g), $CuSO_4 \cdot 5H_2O$ (1.57 g), $CoCl_2 \cdot 6H_2O$ (1.61 g).

Ribbons' trace metal solution.[149,33] For 1 L of H_2O, $ZnSO_4 \cdot 7H_2O$ (0.2 g), $CaCl_2 \cdot 2H_2O$ (4.38 g), $MnSO_4 \cdot 7H_2O$ (0.4 g), $FeSO_4 \cdot 7H_2O$ (8.0 g), $CuSO_4 \cdot 5H_2O$ (0.4 g), $CoCl_2 \cdot 6H_2O$ (0.04 g), H_3BO_4 (0.004 g), citric acid (100 g).

Mineral salt broth (MSB) medium.[146] This medium is made up of, for 1 L, 40 mL of Stanier's stock solution A, 20 mL of Stanier's stock solution B, and 1.0 g of $(NH_4)_2SO_4$ brought to the final volume of 1 L with distilled water. When this medium is autoclaved, a precipitate will form. This precipitate will re-dissolve upon cooling.

Mineral salt broth (MSB) agar plates.[40] In a 1-L flask equipped with a magnetic stir bar are placed Stanier's stock solution A (20 mL), Stanier's stock solution B (10 mL), a solution containing $(NH_4)_2SO_4$ (200 mg/L; 7.5 mL), and L-arginine (2.5 g) and the total volume of the solution is brought to 250 mL by the addition of distilled water. In a second 1-L flask, Bacto-Agar (10 g) and 250 mL of distilled water are mixed. Both solutions are sterilized in an autoclave. A precipitate in the MSB flask at this point will dissolve upon cooling and stirring of the solution. When both solutions are at about 50°, they are combined, quickly mixed, and poured into 100 mm diameter Petri dishes (approximately 20 mL of solution per dish). The solution gels and solidifies upon cooling.

Mineral salts mixture (MSM) medium.[33] To prepare this medium (1 L), glucose (10 g), $MgSO_4 \cdot 7H_2O$ (0.25 g), KH_2PO_4 (3 g), $(NH_4)_2SO_4$ (1 g), and polypropyleneglycol (1 g; as an antifoam agent) are dissolved in distilled water and the volume is brought to 1 L. The solution is sterilized (121°, 25 minutes) and then Ribbons' trace metal solution (10 mL) is added.

Minimal salts medium.[150] This medium (1.0 L) is made up of KH_2PO_4 (2 g), NH_4Cl (3 g), $MgSO_4 \cdot 7H_2O$ (0.4 g), and Vishniac and Santer's[148] trace element solution (2 mL).

M9 medium.[151] For 1.0 L, the following are added to 750 mL of sterile deionized H_2O (cooled to 50° or lower): "5xM9" salts (200 mL); 20% appropriate carbon source (e.g., 20% glucose) (20 mL); and sterile deionized H_2O to 1 L. If necessary, the M9 medium is supplemented with stock solutions of the appropriate amino acids.

The "5xM9" salts solution is made by dissolving the following salts in deionized water to a final volume of 1 L: $Na_2HPO_4 \cdot 7H_2O$ (64 g), KH_2PO_4 (15 g), NaCl (2.5 g), and NH_4Cl (5.0 g). The salt solution may be divided into 200-mL aliquots and sterilized by autoclaving for 15 minutes at 15 lb/sq. in. on a liquid cycle.

Luria-Bertani (LB) broth. This medium (1.0 L) is made up of tryptone (10 g; Difco 0123), yeast extract (5.0 g; Difco 0127), and NaCl (10.0 g) dissolved by stirring in 1 L of distilled or deionized water. The solution may be placed as needed in appropriate containers and sterilized for 20 minutes at 121°.

Luria-Bertani (LB) agar. This medium (1.0 L) is prepared as for Luria-Bertani broth, above, except that agar (15 g/L) is included in the aqueous solution before sterilization.

FERMENTATIONS

The following experimental procedures describe the production of dihydrodiols with a variety of microorganisms. Among the references from which these procedures are taken, references 40, 34 (see the Supporting Information), and 149 are especially thorough in description of experimental techniques.

(1S-cis)-3-Chloro-3,5-cyclohexadiene-1,2-diol.[40] A flask of inoculum (50 mL in a 250-mL flask) of *Pseudomonas putida* 39/D was prepared in advance of the fermentation. Mineral salt broth (MSB) medium (0.5 L) and L-arginine hydrochloride (10 g) were placed in a 2.8-L Fernbach flask fitted with an air inlet tube and a vapor bulb extended through the closure of the flask mouth. The flask, fittings, and contents were sterilized in an autoclave. After cooling, the previously prepared flask of inoculum was transferred to the Fernbach flask using aseptic technique. The vapor bulb was charged with chlorobenzene (10 mL), and the flask was shaken on a rotary shaker at 150 rpm and 30° for 48 hours. The vapor bulb with excess chlorobenzene was removed, and the pH of the aqueous contents of the flask was measured and adjusted to pH ~ 9 if necessary. The aqueous mixture was divided equally into centrifuge tubes and the solids were separated by centrifugation for 30 minutes at ~8,000 rpm. The aqueous supernatant was decanted, combined, saturated with NaCl, and extracted with EtOAc (4 × 100 mL). The combined organic extracts were dried (Na_2SO_4 or $MgSO_4$), filtered, and concentrated under reduced pressure, giving 190 mg of a tan colored solid. Centrifugation was used to aid breaking of any emulsions observed with the organic extracts. Recrystallization of the solid from CH_2Cl_2-hexane gave an off-white solid, 0.160 g, mp 82–84°; $[\alpha]_D^{25}$ +54° (c 0.59, $CHCl_3$); ^{1}H NMR ($CDCl_3$) δ 6.12 (m, 1H), 5.87 (m, 2H), 4.48 (m, 1H), 4.19 (t, J = 7.3 Hz, 1H), 2.74 (d, J = 7.3 Hz, 1H), 2.63 (d, J = 8.4 Hz, 1H); ^{13}C NMR ($CDCl_3$) δ 134.9 (C), 128.0 (CH), 123.4 (CH), 122.7 (CH), 71.4 (CH), 69.1 (CH).

(1S-cis)-3-Methyl-3,5-cyclohexadiene-1,2-diol.[39,149] A highly detailed description of the procedure that follows may be found in reference 149. Four 2-L flasks, each containing glucose medium[146] (500 mL), were inoculated with *P. putida* 39/D. Toluene was supplied to each fermentation flask by placing 2 mL in a glass tube suspended above the medium by a neoprene stopper. The open end of the tube above the stopper was plugged with cotton. A hole in the glass tube below the stopper allowed toluene to diffuse into the flask. The flasks were shaken on a reciprocal shaker at 30° for 30 hours. The contents of the flasks were placed in centrifuge bottles and centrifuged at ~15,000g for 30 minutes. The clear supernatant liquid was decanted and evaporated to dryness under reduced pressure while warming to no higher than 40°. The residue was extracted with MeOH. Removal of the MeOH left 2.42 g of yellow oil that was taken up in $CHCl_3$ and subjected to silica gel chromatography (column, 3 × 50 cm). The column was eluted first with $CHCl_3$ and then with 0.5% (v/v) CH_3OH in $CHCl_3$ (800 mL) to give the dihydrodiol. The dihydrodiol was crystallized two times from petroleum ether (30–60°), giving 1.94 g (83%) of (1S-cis)-3-methyl-3,5-cyclohexadiene-1,2-diol, mp 59°;[39] $[\alpha]_D$ +25° (*c* 0.4, CH_3OH);[39] ^1H NMR $(CDCl_3)$[149] δ 5.91–5.86 (m, 1H), 5.77–5.69 (m, 2H), 4.27 (m, 1H), 2.85–2.78 (m, 2H, exchangeable with D_2O), 1.90 (s, 3H).

(1S-cis)-3-Methyl-3,5-cyclohexadiene-1,2-diol via a Fermentation with Immobilized *E. coli* JM109(pKST 11).[59]

1. Preparation of the inoculum. *E. coli* JM109(pKST 11) was maintained on Luria-Bertani agar plates containing 0.2% of a 50 mg/mL solution of ampicillin in water. A 250-mL flask containing Luria-Bertani broth (50 mL) and 0.2% of a 50 mg/mL solution of ampicillin in water was inoculated with *E. coli* JM109(pKST11) from the culture plates. The flask was shaken on a rotary shaker at 30° for 12 hours. Culture medium from the flask (20 mL) was used to inoculate a 2-L flask containing Luria-Bertani broth (500 mL) and 0.2% of a 50 mg/mL solution of ampicillin in water. After four hours of growth on an orbital shaker at 30°, a solution of 0.1 M isopropyl-β-D-thiogalactopyranoside in water (5 mL) was added to the flask for induction of the dioxygenase. The bacteria were harvested (see next step) after 1.5 hours of further incubation at 30° on the orbital shaker and were used for immobilization and biotransformation of the substrate.

2. Immobilization of the cells. Cells from the inoculum culture (40 mL) were collected by centrifugation at 2,500 g for 6 minutes. The supernatant was removed and the bacterial pellet mixed with an autoclaved solution of 3% (w/v) sodium alginate

(from *Laminaria hyperborea* supplied by Fluka Chemicals) in water (60 mL). The mixture was added dropwise to a chilled solution of 0.05 M $BaCl_2$ (500 mL). The resulting gel beads (diameter, 2–3 mm; cell load, 6–13 × 10^9 cells per gram bead) were hardened for 15 minutes at 4°. The biocatalyst mixture was filtered through a No. 1 Whatman filter and washed with distilled water (500 mL).

3. Biotransformation. Fifteen grams of beads were packed in a 250-mL glass jacketed column (3 cm × 30 cm) maintained at 30°. The reaction medium was composed of 0.1 M Tris-HCl buffer pH 7.0, 0.5% (w/v) glucose, and 10% (v/v) minimal salts medium[150] and was pumped from the bottom to the top of the column with a peristaltic pump at a flow rate of 20 mL/ minute. The beads were fluidized by an oxygen flow of 100 mL/minute. The medium was recirculated through a reservoir where the pH and temperature were controlled at 7.0 and 30°, respectively. The total volume of medium in circulation was 1.5 L. Toluene (30 mL) was mixed with 300 mL of tetradecane and added directly to the reaction medium. The production of diol was measured by HPLC. Isolation of the dihydrodiol from the reaction medium was not described.

(1S-*cis*)-1,6-Dihydroxy-2,4-cyclohexadiene-1-carboxylic acid.

1. From benzoic acid with *Pseudomonas putida* U103.[33] In advance of the large-scale fermentation, inoculum was prepared, first by inoculating mineral salts medium[33] (50 mL) in a shaker flask (250-mL) with *Pseudomonas putida* U103 and shaking the flask at 30° for 24 hours on a rotary shaker. The contents of this flask were then used to inoculate two flasks (500 mL) each containing mineral salts medium (200 mL). These flasks were shaken at 30° for 24 hours before being used to inoculate the large fermentation tank.

A 14-L New Brunswick Microferm (Model MF-114) fermentor was charged with glucose (10 g/L), $MgSO_4$ (0.25 g/L), KH_2PO_4 (3.0 g/L), $(NH_4)_2SO_4$ (1 g/L), and polypropylene glycol antifoam agent (1.0 g/L, 11.4 L), and was sterilized at 121° for 25 minutes before Ribbons' trace metal solution (114 mL) was added. After cooling, the pH of the medium was adjusted to 7.0 ± 0.1 with 10 M aqueous NH_4OH solution. The pH was controlled automatically throughout the fermentation with the use of 10 M aqueous NH_4OH solution and 4 M aqueous H_3PO_4 solution. The inocula from the two flasks prepared in advance were transferred to the fermentor and the culture was left to grow for 24 hours at 30°, pH 7, and a controlled oxygen content of 40%. With a glucose feed of 13 mmol/hour, sodium benzoate (5 mmol) was added to induce the oxygenase. Once oxidation was initiated, sodium benzoate was fed to the fermentation at a rate of 5 mmol/hour until unoxidized benzoic acid began to accumulate in the medium.

The contents of the fermentor were centrifuged in order to separate solids. The supernatant was concentrated to between 1/20th and 1/50th of the original volume by evaporation under reduced pressure at a temperature of 40° or less. The pH of the concentrated solution was adjusted to 4.0 by the addition of dilute aqueous HCl solution. The solution was repeatedly extracted with EtOAc until UV analysis of the solution indicated complete extraction of the product. Sodium sulfate was added to aid extraction of the product and the pH of the solution was maintained at ~4. The combined EtOAc extracts were dried (MgSO$_4$), filtered, and concentrated under reduced pressure leaving a yellow solid. Much of the yellow color was removed by carefully washing the solid with CH$_2$Cl$_2$, after which (1S-cis)-1,6-dihydroxy-2,4-cyclohexadiene-1-carboxylic acid (20–30 g) was obtained as an off-white solid; [α]$_D$ −106° (c 0.5, EtOH); λ_{max} (EtOH) 261 (ε 3,400); ^{1}H NMR (CDCl$_3$ + DMSO-d$_6$) δ 4.1 (br s, 1H), 5.2 (br d, 2H), 5.2 (d m, 1H), 5.4 (br m, 1H).

2. From benzoic acid with *Alcaligenes eutrophus* B9.[34] The authors of this procedure have described two sets of conditions for medium-scale fermentations of benzoic acid with *A. eutrophus* B9. The larger scale of the two procedures is described here. The yield of product from the smaller scale procedure was higher (38 g, 74%) than from the larger scale (270 g, 39%).

A sterile pipette tip was streaked across the surface of a frozen glycerol stock solution of *A. eutrophus* B9 to produce small shards (ca. 10 mg). The frozen shards were added to a sterile, baffled 2-L Erlenmeyer flask containing mineral salt broth medium (500 mL; note: the mineral salt broth medium used in this experiment contains the same components as described in "Preparation of Media" with the exception that the concentration of Metals "44" was doubled) and aqueous sodium succinate solution (1.67 mL of a 1.5 M stock solution, 5 mM final concentration). The flask was shaken at 250 rpm for 24 hours at 30° on a rotary shaker. The white, heterogeneous mixture was added to a large-scale fermentor[34] containing mineral salt broth medium (80 L) and aqueous sodium succinate solution (267 mL of a 1.5 M stock solution, 5 mM final concentration). The solution was warmed to an internal temperature of 30° by circulating warm water through a 30′ coil of Tygon tubing (1/2″ diameter). Air filtered through cotton was sparged through the medium. After 24 hours, aqueous sodium benzoate (240 mL of a 1.0 M solution) and aqueous sodium succinate (1.33 mL of a 1.5 M solution) were added to the white heterogeneous mixture. The resulting mixture was aerated vigorously for 6 hours at an internal temperature of 30°. After induction, sufficient aqueous sodium benzoate (160 to 400 mL of a 1.0 M solution, depending on the rate of consumption) was added hourly to maintain a concentration of 10–20 mM (determined by absorbance at 225 nm). Aqueous sodium succinate (135 mL of a 1.5 M solution) was added when the rate of oxidation decreased as determined by UV absorbance. The fermentation mixture was maintained at pH 6.8 (monitored every hour) by periodic additions of aqueous NaH$_2$PO$_4$ (2.0 M solution). These additions were continued for a period of 22 hours, then the fermentation mixture was aerated overnight at an internal temperature of 30° to maximize conversion. The fermentation mixture was centrifuged, in portions, at 2,000 rpm (bench top centrifuge) to remove the majority of solid material. The supernatant was concentrated to a volume of 20 L using a large-scale rotary evaporator (bath temperature <45°). The concentrate was centrifuged, in portions, at 6,000 rpm (Sorvall GS-3 rotor, model SLA-3000) to remove remaining solids. The supernatant was concentrated to 6 L using a large-scale rotary evaporator (<45°) and the concentrate was divided into three 2-L portions. The light gray solutions were cooled to 0° and acidified to pH 3.0 using concentrated HCl. The acidified solutions were each extracted repeatedly with EtOAc (~60 L) until less than 50 mg of material was isolated per 1 L of EtOAc extract. The organic extracts were dried (Na$_2$SO$_4$) and concentrated (large-scale rotary evaporator, <45°), yielding a light brown residue. Trituration of the residue with CH$_2$Cl$_2$ (2 L) followed by drying in vacuo of the solids gave pure (1S-cis)-1,6-dihydroxy-2,4-cyclohexadiene-1-carboxylic acid as a white powder, mp 95–96° dec (210 g, 30%). The CH$_2$Cl$_2$ wash was concentrated, leaving a brown residue. The residue was dissolved in a minimal amount of EtOAc and the resulting solution was cooled to −20° to precipitate a light yellow solid, which was collected by filtration. The solid was triturated with CH$_2$Cl$_2$ (2 × 250 mL) followed by drying of the solid in vacuo to give additional product (60 g, 9%) as an off-white powder.

(1R-cis)-1,2-Dihydro-1,2-naphthalenediol.[47] Flasks containing M9 medium (600 mL total, also containing 20 mM sodium succinate, and 10 mg of salicylic acid as an inducer) were inoculated with cells of *Pseudomonas fluorescens* TCC1 (NCIMB 40605). The flasks were incubated at 30° for 24 hours after which the cells were separated from the liquid medium by centrifugation and the clear supernatant was decanted. The cells were washed with distilled water, collected again by centrifugation and transferred to a water-jacketed stirred reactor containing M9 medium (1 L), sodium succinate (5 mM), naphthalene (6.4 g, 0.050 mol; finely ground), and polyethyleneglycol 8000 [0.1% (w/v); as an antifoaming agent]. The fermentation was maintained at 30°, stirred at 1,600 rpm, and aerated with a flow of 9 L/minute of air through a sparger controlled by an in-line flow meter. The fermentation was continued for 24 hours after which the solids were separated by centrifugation. The supernatant was extracted with EtOAc (4 × 150 mL), the extracts were dried (Na$_2$SO$_4$), filtered, and concentrated, giving crude product. The crude material was crystallized from hexane, giving (1R-cis)-1,2-dihydronaphthalene-1,2-diol (6.51 g, 80%); [α]$_D$ +220° (MeOH);[47] mp 115–116°;[63] [α]$_D$ +220° (c 0.05–0.1, MeOH);[63] ^{1}H NMR[63] (CDCl$_3$) δ 4.36 (dd, J = 3.8, 5.1 Hz, 1H), 4.67 (d, J = 5.1 Hz, 1H), 6.03 (dd, 1H, J = 3.8, 9.9 Hz, 1H), 6.53 (d, J = 9.9 Hz, 1H), 7.0–7.6 (m, 4H).

(1S-cis)-3-Iodo-3,5-cyclohexadiene-1,2-diol.[25,105] A minimal salts medium[150] (500 mL) containing gluconate (12% w/v) in a 2-L flask was inoculated with *Pseudomonas putida* UV4 and shaken at 30° for 24 hours.[152] The cells were separated from the liquid medium by centrifugation. The clear supernatant was discarded and the cells were resuspended in minimal salts medium (500 mL) containing pyruvate (12% w/v) in a 2-L flask. Iodobenzene (2 g, 9.8 mmol) was added to the flask and the fermentation carried out for 50 hours at 30° using an orbital shaker (400 rev/min). After the contents of the flask were centrifuged, the clear supernatant was separated and saturated with solid NaCl, and then extracted with CH_2Cl_2 (5 × 100 mL). The extract was dried (Na_2SO_4), filtered, and concentrated. The residue was purified by chromatography, giving (1*S-cis*)-3-iodo-3,5-cyclohexadiene-1,2-diol (1.88 g, 80%); mp 64–81° (dec.); $[\alpha]_D$ +41° (*c* 0.5, MeOH); ^{1}H NMR (CDCl$_3$, TMS) δ 6.69 (d, *J* = 5.5 Hz, 1H), 6.03 (dd, *J* = 4.2, 9.4 Hz, 1H), 5.72 (m, 1H), 4.43 (m, 1H), 4.28 (d, *J* = 6.1 Hz, 1H).

(1*S-cis*)-3-Phenyl-3,5-cyclohexadiene-1,2-diol.[49,153] In preparation for a 10-L fermentation, five 500-mL flasks of glucose medium (100 mL) containing 0.2% sodium succinate and 0.1% (w/v) biphenyl were inoculated with *Beijerinckia* B8/36 (now classified as *Sphingomonas yanoikuyae* B8/36), then incubated at 27° on a rotary shaker at 150 rpm for 12 hours. This inoculum was transferred to a New Brunswick Model M14 Microferm fermentor containing glucose medium (10 L), sodium succinate (0.2%), and biphenyl (0.1% w/v). The fermentation was carried out for five hours after which the contents of the fermentor were centrifuged and the supernatant decanted. The supernatant was extracted with EtOAc (total, 3 L), and the extract was dried (Na_2SO_4), filtered, and concentrated, giving 4.6 g (37%) of white solid. Crystallization from hexane gave (1*S-cis*)-3-phenyl-3,5-cyclohexadiene-1,2-diol; mp 93°; λ_{max} (MeOH) 303 (ε, 13,600) and 223 nm (ε, 9,200); ^{1}H NMR (CDCl$_3$) δ 7.4 (m, 5H), 6.35 (m, 1H), 5.9 (m, 2H), 4.5 (dd, 2H).

(1*S-cis*)-3-Bromo-4,5-difluoro-3,5-cyclohexadiene-1,2-diol.[154] *E. coli* JM109(pDTG601) was grown overnight at 35° in a 2.8-L Fernbach flask containing 500 mL of MSB medium supplemented with glucose (0.2%), thiamine (1mM), isopropyl-β-D-thiogalactoside (10 mg/L), and ampicillin (100 mg/L).[26,154] This culture was transferred to a 12-L fermentor containing 8 L of the same medium and the cells were grown to an optical density of 70 at 660 nm. The substrate, 1-bromo-2,3-difluorobenzene (unspecified amount), was added dropwise to the culture and the progress of the biotransformation was followed by observing oxygen consumption and CO_2 production by the culture. Diol formation was monitored by measuring absorbance at 270 nm typical of the cyclohexadiene moiety. After all metabolic activity ceased (or no further diol formation was observed), the fermentation was stopped and the pH of the contents of the fermentor was adjusted to pH 8.4 with aqueous NaOH solution. The contents of the fermentor were centrifuged and the clear supernatant was decanted, saturated with solid NaCl, and extracted with EtOAc (previously washed with saturated aqueous $NaHCO_3$ solution). The extracts were dried (Na_2SO_4), filtered, and concentrated. The crude product residue was purified by flash chromatography over deactivated (10% H_2O) silica gel using hexane-EtOAc, 7:3, for elution. (1S-cis)-3-Bromo-4,5-difluoro-3,5-cyclohexadiene-1,2-diol was obtained in the quantity of 0.7 g/L of fermentation volume; mp 104.5–105.5°; $[\alpha]_D$ +34.6° (c 0.49, MeOH); ^{1}H NMR ($CDCl_3$) δ 5.12 (m, 1H), 4.60 (br s, 1H), 4.47 (br s, 1H); ^{13}C NMR (C_3D_6O, TMS) δ 148.5 (C, dd, J = 259.0, 26.3 Hz), 148.4 (C, dd, J = 258.6, 28.2 Hz), 109.2 (CH, dd, J = 9.9, 1.5 Hz), 106.8 (C, dd, J = 13.3, 3.4 Hz), 73.0 (CH, d, J = 2.3 Hz), 67.7 (CH, d, J = 8.4 Hz); ^{19}F NMR ($CDCl_3$) δ −124.2 (br s), −132.8 (br s).

GLOSSARY

Aerobic: A procedure (growth, biotransformation, fermentation, etc.) carried out in the presence of air or oxygen.

Agar: (1) A polysaccharide from seaweed, commonly used as a base (gelling agent) for solid media. (2) A solid medium (jargon). The gelling temperature is between 25–35° and the gel remains solid to about 90°.

Aseptic Techniques: Procedures that minimize accidental entry of undesired organisms into a culture or fermentation.

Autoclave: An instrument in which equipment and media are sterilized at elevated temperature and high pressure.

Bacteria: A Kingdom of cellular organisms. Bacteria are single-celled and lack a nucleus.

Culture: A population of microorganisms that is growing and/or alive, confined in an environment in which viability is retained.

Fungus: A member of the Fungi Kingdom, which consists of unicellular or multicellular eukaryotic organisms that lack chlorophyll. They are frequently involved in the decay of dead organic matter.

Fermentation: The transformation of one molecule type (substrate) to another (product) by microorganisms.

Incubation: The time during which a culture is kept under a given set of conditions for a defined time.

Inoculate: To deliberately introduce microorganisms (usually of a single species) into a culture medium.

Inoculum: The microorganism(s) used to inoculate.

Medium (Culture Medium): (1) An appropriate mixture of nutrients capable of supporting the growth and multiplication of a microorganism. (2) A liquid culture medium (jargon).

Nutrient: A material that the microorganism uses as food or growth stimulant.

Slant (or Slope): Agar placed in a tube and allowed to solidify at a slant.

Species: A taxonomically distinct kind of microorganism. Individual cells within a species may differ in biochemical properties.

Spores: Reproductive bodies, of one or more cells, at rest (non-growing) until introduced into a nutrient medium; the "seeds" of microorganisms.

Sterilize: To kill all undesired microorganisms.

Strain: A pure culture descended from a single individual of a species and thus presumably more biochemically homogeneous than the aggregate of individuals of a species. A strain often arises from a single individual cell.

Substrate: The material or compound to be acted on by an organism or an enzyme to produce a product chemically related to the substrate. For arene oxygenations, the substrates are preferably not nutrients.

Transfer: The introduction of a small amount of an organism into a virgin nutrient environment in order to increase the supply of the organism.

Yeast: Single celled fungi that reproduce asexually by budding or fission and sexually by haploid spores.

TABULAR SURVEY

Entries are arranged according to increasing carbon and hydrogen count of the substrate. Yields are given by the numbers in parentheses. A "(—)" indicates that no yield was provided in the original reference. Other expressions of yield may be encountered in microbiology literature and these are given as described, e.g., g/L, moles/g dry cell weight. The enantiomeric excess (ee) is given by the number following the yield parentheses. A "—" indicates that no measure of ee was provided in the original reference.

The structures of arene substrate and predominant dihydrodiol product are drawn in a consistent manner throughout the Tables. This is intended to provide the reader with a recognizable pattern related to the stereoselectivities observed in arene dioxygenations. Where the absolute configuration of a dihydrodiol has been reported, the information is reflected in the drawing of the compound. When absolute configurations are unknown, bonds are drawn with a thin line.

The entries of Table 10, "Dihydrodiols That Have Not Been Isolated," are included to give the reader a full scope of arenes that have been subjected to microbial arene oxidation. The compounds in this table have been used as substrates for fermentations, but for which product structures are assigned primarily by analogy to related examples. Biological and/or analytical data have been obtained that are consistent with the indicated dihydroxylation reactions, but the products have not been isolated or further characterized. The reader contemplating use of a microbial arene oxidation reaction should take encouragement if precedent is found in this table.

The following abbreviations are used in the tables:

AIBN	2,2′-azobis(isobutyronitrile)
THF	tetrahydrofuran
Ts	tosyl, toluenesulfonyl

TABLE 1. MICROBIOLOGICAL OXYGENATIONS OF BENZENES

Substrate		Microorganism, Conditions	Product(s), Yield(s) (% or g/L) and Enantiomeric Excess %		Refs.
C₆					
	C₆H₃Cl₃	Pseudomonas sp. PS12		(—), —	155
	C₆H₃BrF₂	Pseudomonas putida 39/D		(50 mg/L), >98	154
		Escherichia coli JM109(pDTG601)	"	(0.7 g/L), >98	154
	C₆H₄FI	Pseudomonas putida UV4		(—), >98	27
	C₆H₄FI	Pseudomonas putida UV4		(—), >98	27
	C₆H₄FI	Pseudomonas putida UV4		(60), 88	27, 23
		Escherichia coli JM109(pDTG601)	"	(—), 88	107

Substrate	Formula	Organism	Product	Yield	Ref
1,4-dichlorobenzene	$C_6H_4Cl_2$	*Xanthobacter flavus* 14p1	3,6-dichlorocyclohexa-3,5-diene-1,2-diol	(—)	156
2-chloroiodobenzene	C_6H_4ClI	*Pseudomonas putida* F1	"	(—)	157
		Pseudomonas sp.	"	(—)	158
	C_6H_4ClI	*Pseudomonas putida* UV4	3-chloro-2-iodocyclohexa-3,5-diene-1,2-diol	(—), >98	27
	C_6H_4ClI	*Escherichia coli* JM109(pDTG601)	"	(—), >98	27
4-chloroiodobenzene	C_6H_4ClI	*Pseudomonas putida* UV4	3-chloro-6-iodocyclohexa-3,5-diene-1,2-diol	(25), 15	23
		Escherichia coli DH5α(pTCB144)	"	(—), 67	77
1,3-dibromobenzene	$C_6H_4Br_2$	*Escherichia coli* JM109(pDTG601)	3,5-dibromocyclohexa-3,5-diene-1,2-diol	(4 g/L), >99	131
2-bromoiodobenzene	C_6H_4BrI	*Pseudomonas putida* UV4	3-bromo-2-iodocyclohexa-3,5-diene-1,2-diol	(—), >98	27
3-bromoiodobenzene	C_6H_4BrI	*Pseudomonas putida* UV4	5-bromo-2-iodocyclohexa-3,5-diene-1,2-diol	(—), >98	27

TABLE 1. MICROBIOLOGICAL OXYGENATIONS OF BENZENES (*Continued*)

Substrate	Microorganism, Conditions	Product(s), Yield(s) (% or g/L) and Enantiomeric Excess %	Refs.
C₆H₄BrI (4-I-C₆H₄-Br)	*Pseudomonas putida* UV4	(I)(Br)-cis-diol-OH,OH (22), 22	23
	Escherichia coli JM109(pTG601)	" (—), 20	107
C₆H₅NO₂ (NO₂-C₆H₅)	*Pseudomonas putida* 39/D	*cis* (NO₂)-diol-OH,OH (—), —	159
	Pseudomonas putida TB 103	" (—), —	160
C₆H₅F (F-C₆H₅)	*Pseudomonas putida* UV4	(F)-diol-OH,OH (—), *ca.* 60	24, 23, 25
	Pseudomonas putida UV4 immobilized in barium alginate beads	" (0.7 mol/g dry cell weight)	59
	Pseudomonas mutant	" (9.5 g/L)	161
	Escherichia coli DH5α(pTCB144)	" (—), —	77
C₆H₅Cl (Cl-C₆H₅)	*Pseudomonas putida* 39/D	(Cl)-diol-OH,OH (1g/L), —	80, 40
	Pseudomonas putida UV4	" (80), >98	24, 25

Substrate	Formula	Organism	Yield (ee)	Ref.
(Cl-C₆H₄-D, ortho)	C₆H₄DCl	Pseudomonas putida UV4 immobilized in barium alginate beads	(3.0 mol/g dry cell weight)	59
"		A bacterium, strain WR1306	(—), —	162
"		Pseudomonas mutant A	(9.5 g/L)	163
3-Cl-C₆H₄-D structure with OH, D, OH (cis-diol)	C₆H₄DCl	Pseudomonas putida UV4	(—), —	164
3-Cl structure with OH, D, OH	C₆H₄DCl	Pseudomonas putida UV4	(—), —	164
Br-C₆H₅ with cis-diol OH, OH	C₆H₅Br	Pseudomonas putida UV4	(77), >98	105, 25
"		Pseudomonas putida 39/D	(—), —	81
"		Escherichia coli DH5α(pTCB144)	(—), —	77
I-C₆H₅ with cis-diol OH, OH	C₆H₅I	Pseudomonas putida UV4	(85), >98	105, 25
"		Escherichia coli JM109(pDTG601)	(—), >98	27
"		Escherichia coli DH5α(pTCB144)	(—), —	77

TABLE 1. MICROBIOLOGICAL OXYGENATIONS OF BENZENES (*Continued*)

Substrate	Microorganism, Conditions	Product(s), Yield(s) (% or g/L) and Enantiomeric Excess %	Refs.
C_6H_6	*Pseudomonas putida* F1	*cis*-dihydrodiol (OH, OH) (1 g/L)	2, 4
	Pseudomonas putida 11767 mutant	" (40–50 g/L)	42, 56
	Pseudomonas mutant D	" (40 g/L)	163
	Pseudomonas putida UV4 immobilized in barium alginate beads	" 4.1 mol/g dry cell weight	59
	Moraxella sp. cell-free extracts	" (—)	165
	Escherichia coli DH5α(pTCB144)	" (—)	77
	Pseudomonas putida F1/$^{18}O_2$	(^{18}OH, ^{18}OH) (0.6 g/L)	4
$C_7H_4BrF_3$ (CF_3, Br)	*Pseudomonas putida* 39/D	(Br, OH, OH, CF_3) (48 mg/L, —) + (OH, OH, CF_3) (2 mg/L)	154
	Escherichia coli JM109(pDTG601)	" (0.7 g/L), >98 + " (20 mg/L)	154
$C_7H_4F_3I$ (CF_3, I)	*Pseudomonas putida* UV4	(CF_3, OH, OH, I) (50), >98	23

Substrate	Formula	Organism	Product	Yield	Ref.
3-cyanobenzene (CN-C6H5)	C7H5N	*Pseudomonas putida* UV4	cis-diol with CN, OH, OH	(3.9 g/L), >98	61, 25
(trifluoromethyl)benzene (CF3-C6H5)	C7H5F3	*Pseudomonas putida* UV4	cis-diol with CF3, OH, OH	(ca. 65), >98	24, 25
"		*Pseudomonas* mutant	"	(1.25 g/L)	166
"		*Pseudomonas* mutant D	"	(>6 g/L)	167
2-D-(trifluoromethyl)benzene	C7H4DF3	*Pseudomonas putida* UV4	cis-diol with CF3, D, OH, OH	(—), —	164
3-D-(trifluoromethyl)benzene	C7H4DF3	*Pseudomonas putida* UV4	cis-diol with CF3, D, OH, OH	(—), —	164
2,4-dichlorotoluene	C7H6Cl2	*Escherichia coli* DH5α(STE7)	cis-diol with CH3, Cl, Cl, OH, OH	(—), —	168
2,5-dichlorotoluene	C7H6Cl2	*Escherichia coli* DH5α(STE7)	cis-diol with Cl, CH3, Cl, OH, OH	(—), —	168

165

TABLE 1. MICROBIOLOGICAL OXYGENATIONS OF BENZENES (*Continued*)

Substrate	Microorganism, Conditions	Product(s), Yield(s) (% or g/L) and Enantiomeric Excess %	Refs.
$C_7H_6Cl_2$ (3,4-dichlorotoluene)	*Escherichia coli* DH5$_\alpha$(STE7)	dichloromethylcatechol (—), —	168
$C_7H_6Cl_2$ (2,6-dichlorotoluene)	*Escherichia coli* DH5$_\alpha$(STE7)	dichloromethylcatechol (—), —	168
C_7H_8O (benzaldehyde)	*Pseudomonas putida* UV4	cis-dihydrodiol with CH$_2$OH (8), >98	55
C_7H_7F (4-fluorotoluene)	*Pseudomonas putida* UV4	cis-dihydrodiol (21), 83	23
"	*Pseudomonas putida* 39/D	" (—), —	153
"	*Escherichia coli* DH5$_\alpha$(pTCB144)	" (—), 49	77
C_7H_7Cl (4-chlorotoluene)	*Pseudomonas putida* UV4	cis-dihydrodiol (20), 15	23
"	*Pseudomonas putida* F1	" (1.5 mg/L), —	54

C₇H₇Br	[structure: Br-phenyl-methyl]		*Pseudomonas putida* 39/D	(—), —	153
			Pseudomonas putida JS6	(—), —	169
			Escherichia coli DH5α(pTCB144)	(—), 77	77
		[structure: Br, OH, OH, methyl cyclohexadiene]	*Pseudomonas putida* UV4	(13), 37	23
			Escherichia coli DH5α(pTCB144)	(—), 77	77
C₇H₇I	[structure: I-phenyl-methyl]	[structure: I, OH, OH, methyl cyclohexadiene]	*Pseudomonas putida* UV4	(24), 80-88	23, 27
			Escherichia coli JM109(pDTG601)	(—), >98	27
			Escherichia coli DH5α(pTCB144)	(—), 98	77
C₇H₇I	[structure: ortho-I-methylbenzene]	[structure: I, methyl, OH, OH cyclohexadiene]	*Pseudomonas putida* UV4	(—), >98	27
C₇H₈	[structure: toluene]	[structure: methyl, OH, OH cyclohexadiene]	*Pseudomonas putida* 39/D	(0.93 g/L), —	39, 170, 171, 80
			Escherichia coli JM109(pDTG601)	(—), —	41

TABLE 1. MICROBIOLOGICAL OXYGENATIONS OF BENZENES (Continued)

Substrate	Microorganism, Conditions	Product(s), Yield(s) (% or g/L) and Enantiomeric Excess %	Refs.
C_7H_8 (toluene)	Escherichia coli DH5α(pDTG927)	(—), —	172
	Pseudomonas putida UV4	(ca. 60), >98 + (4), >98 [CH$_2$OH / OH,OH diol structure]	24, 25, 55
	Pseudomonas putida UV4 immobilized in beads	(6.1 mol/g dry cell weight)	59
	Escherichia coli JM109(pKST11) immobilized in beads	(1.2 mol/g dry cell weight)	59
	Pseudomonas putida NG1	(18-24 g/L), >98	173
	Escherichia coli TG2(pTAC365)	(0.4 g/L), —	174, 175
	Pseudomonas mutant A	(16 g/L), —	163
	Rhodococcus rhodochrous strain OFS	(—), —	176
	Escherichia coli DH5α(pTCB144)	(—), —	77
C_7H_7D (2-D-toluene)	Pseudomonas putida UV4	(—), — [cis-diol with D]	164
C_7H_7D (3-D-toluene)	Pseudomonas putida UV4	(—), — [cis-diol with D]	164

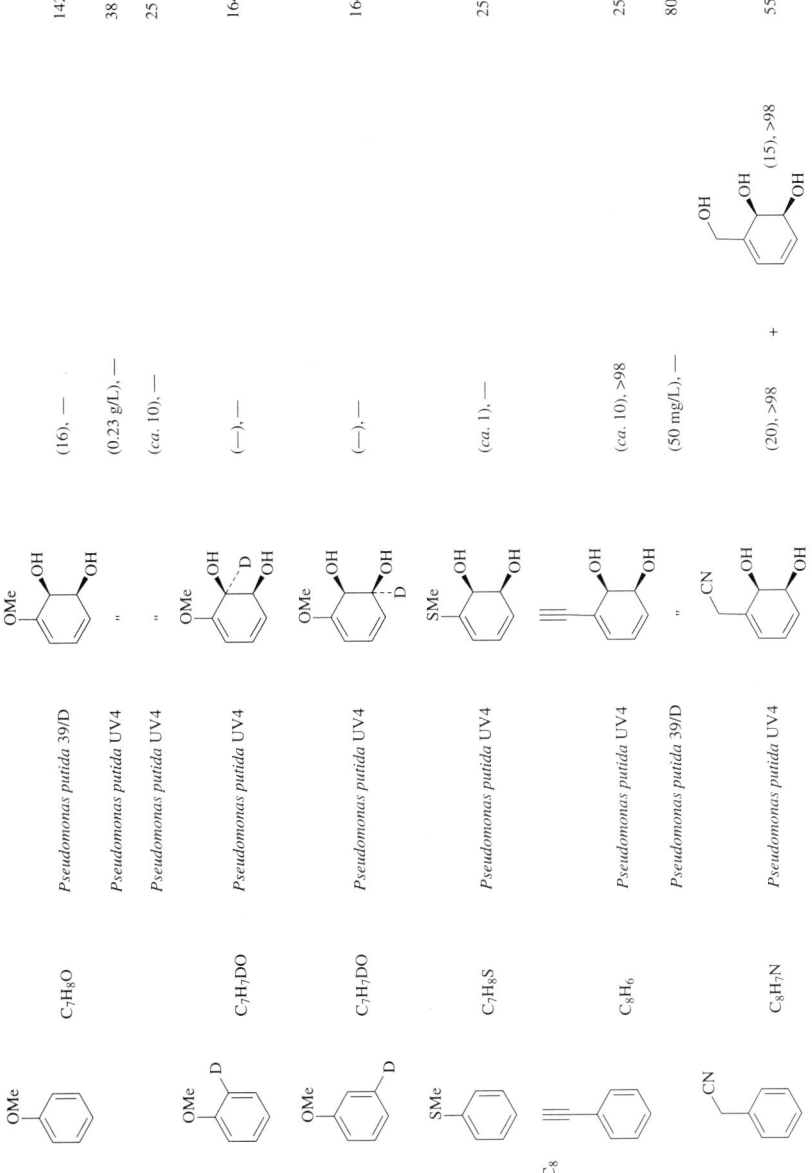

TABLE 1. MICROBIOLOGICAL OXYGENATIONS OF BENZENES (Continued)

Substrate	Microorganism, Conditions	Product(s), Yield(s) (% or g/L) and Enantiomeric Excess %	Refs.
C_8H_7NO	Pseudomonas putida UV4	(18), >98	55
$C_8H_7F_3$	Pseudomonas putida UV4	(28), >98	23
C_8H_7Cl	Pseudomonas putida 39/D	(1.4), >98 + (2.6), 73 + (tr)	79, 177
C_8H_7Cl	Pseudomonas putida 39/D	(1), 54 + (2), >98 + (tr)	79
C_8H_7Cl	Pseudomonas putida 39/D	(17), 15 + (4), 79	79
C_8H_7Br	Pseudomonas putida 39/D	(1), 91 + (5), —	178

170

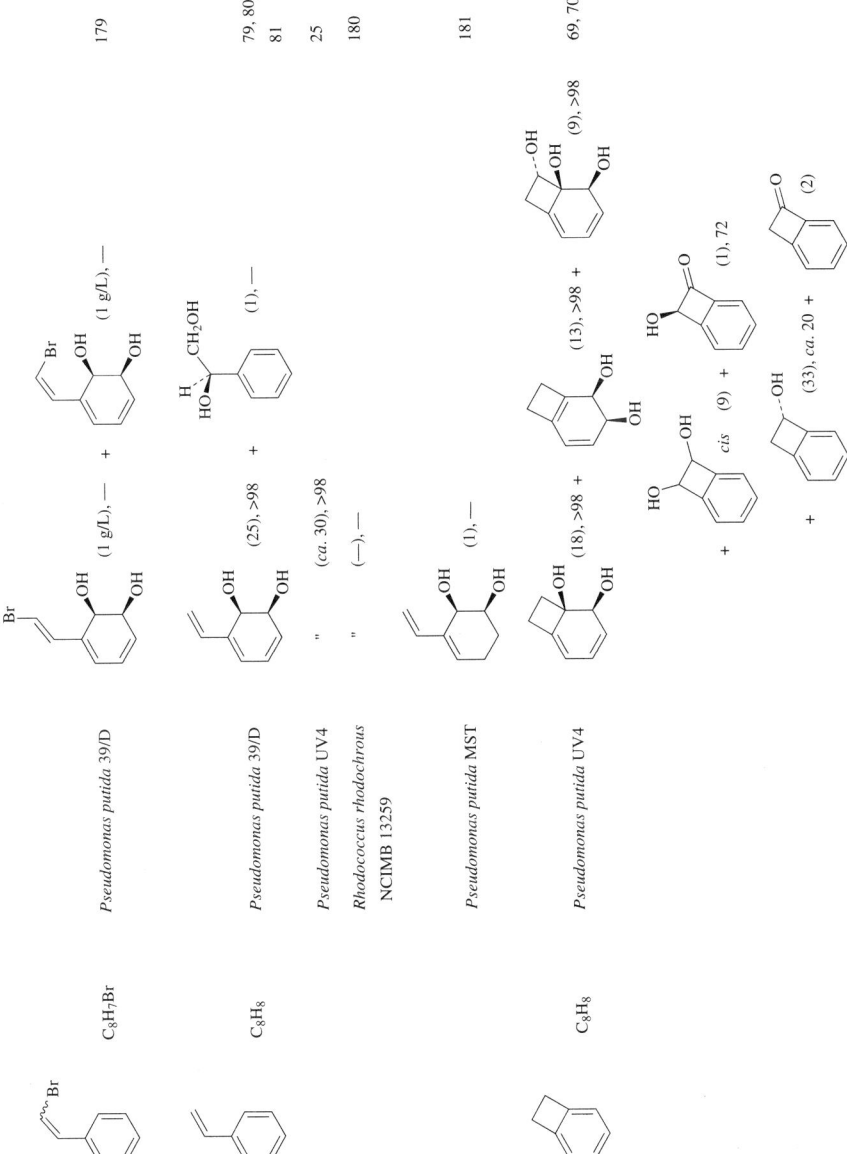

TABLE 1. MICROBIOLOGICAL OXYGENATIONS OF BENZENES (Continued)

Substrate	Microorganism, Conditions	Product(s), Yield(s) (% or g/L) and Enantiomeric Excess %	Refs.
C_8H_8	Pseudomonas fluorescens 127-68 XVII	(—), 0 + (—)	182
C_8H_8O	Pseudomonas putida UV4	(60) + (16), >98	69
C_8H_8O	Pseudomonas putida 39/D	(5), —	183, 81
	Pseudomonas putida UV4	"	38
	Pseudomonas putida ICI strain 11767	(4), —	184
$C_8H_8Br_2$	Pseudomonas putida 39/D or Escherichia coli JM109(pDTG601)	(ca. 0.3 g/L), >95 + (ca. 0.15 g/L)	185
$C_8H_9N_3$	Escherichia coli JM109(pDTG601)	(ca. 42), >98	186

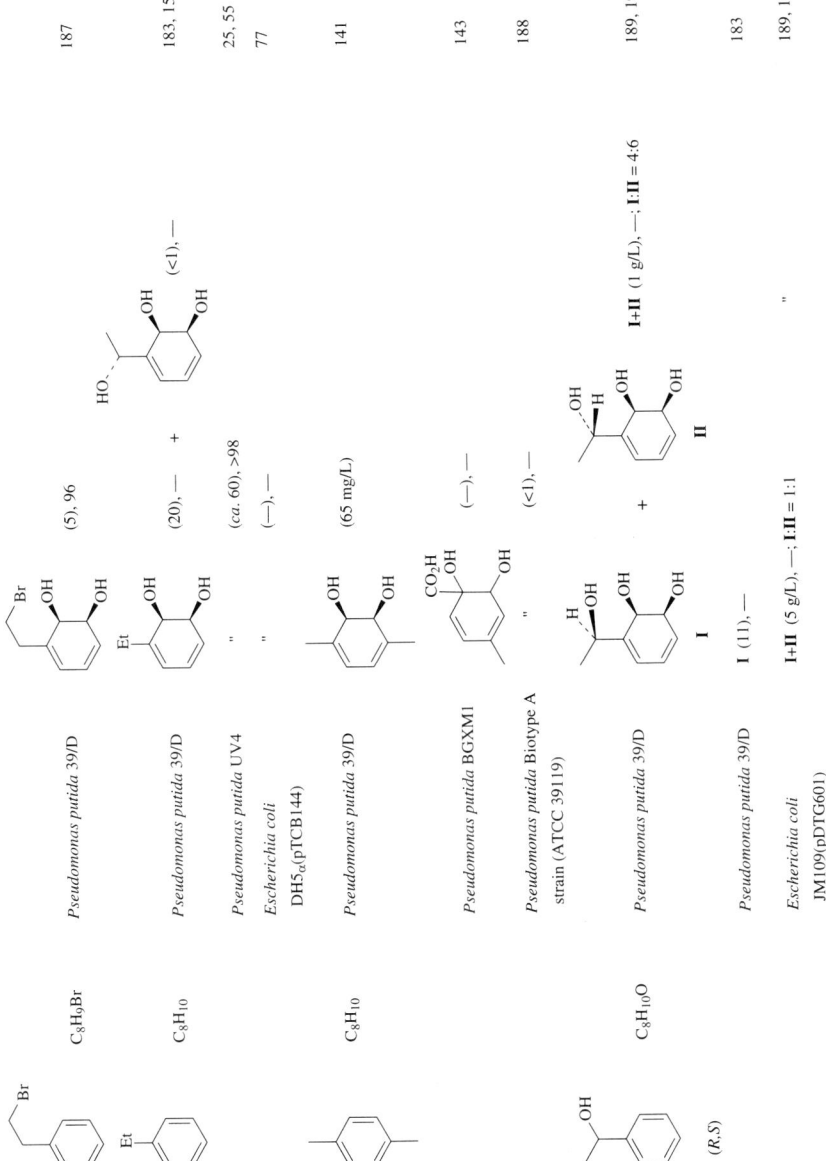

TABLE 1. MICROBIOLOGICAL OXYGENATIONS OF BENZENES (*Continued*)

Substrate		Microorganism, Conditions	Product(s), Yield(s) (% or g/L) and Enantiomeric Excess %	Refs.
(1-phenylethanol, H/OH)	$C_8H_{10}O$	*Pseudomonas putida* 39/D	(diol product) (15), —	183
		Pseudomonas putida 39/D	" (1 g/L), —	189, 190
		Pseudomonas putida UV4	" (20), >98	55
		Escherichia coli JM109(pDTG601)	" (1 g/L), —	189, 190
(1-phenylethanol, OH/H)	$C_8H_{10}O$	*Pseudomonas putida* 39/D	(diol product) (7), —	183
		Pseudomonas putida 39/D	" (7), —	189
		Pseudomonas putida UV4	" (8), >98	55
		Escherichia coli JM109(pDTG601)	" (1 g/L), —	189
(2-phenylethanol)	$C_8H_{10}O$	*Escherichia coli* JM109(pDTG601)	(diol product) (4), 94	186
(phenetole, OEt)	$C_8H_{10}O$	*Pseudomonas putida* UV4	(diol product) (1.15 g/L), >98	191, 25
		Pseudomonas putida 39/D	" (—), —	142

174

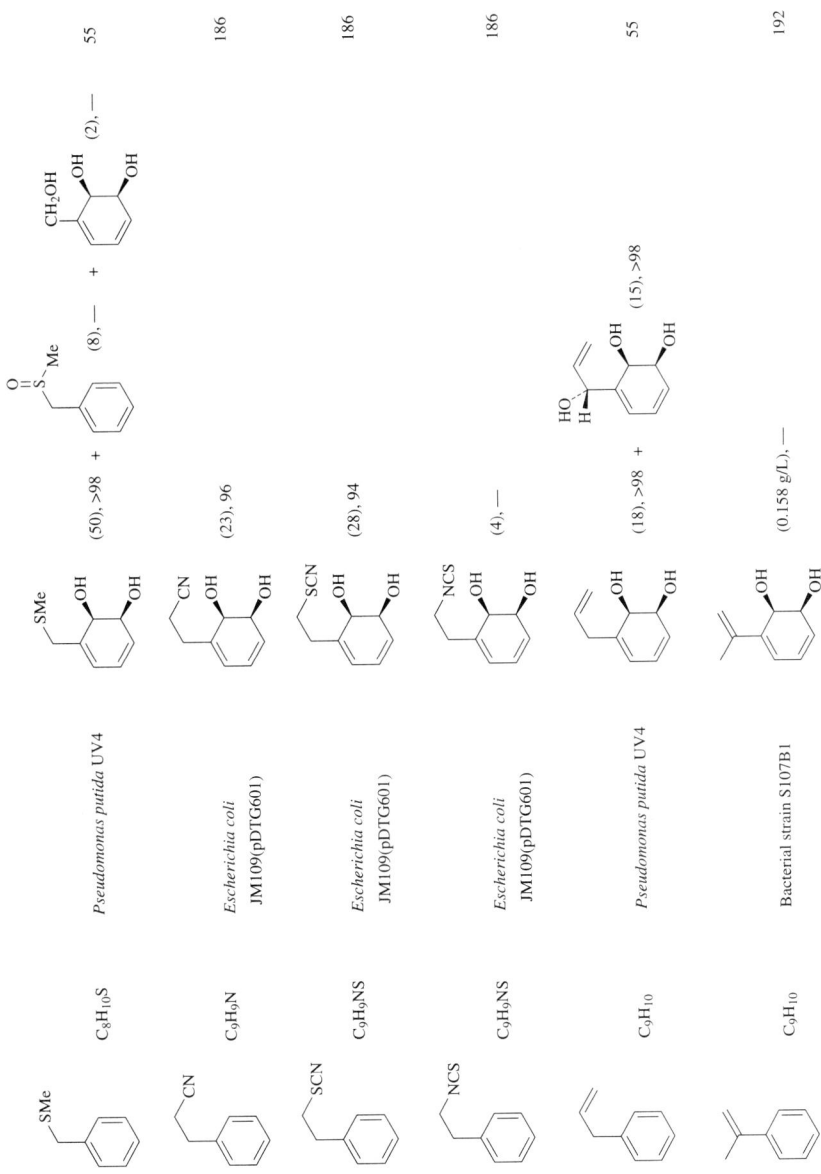

TABLE 1. MICROBIOLOGICAL OXYGENATIONS OF BENZENES (*Continued*)

Substrate		Microorganism, Conditions	Product(s), Yield(s) (% or g/L) and Enantiomeric Excess %		Refs.
(isopropenylbenzene)	C_9H_{10}	*Pseudomonas putida* MST	(isopropenyl diol), (3), —	+ (2-phenylallyl alcohol), (tr)	181
(benzyl acetate, OAc)	$C_9H_{10}O_2$	*Pseudomonas putida* UV4	(OAc diol), (*ca.* 20), >98		25, 24
(2-phenyl-1,3-dioxolane)	$C_9H_{10}O_2$	*Pseudomonas* mutant D	" (9.3 g/L), —		193
		Pseudomonas putida 39/D	(dioxolane diol), (—), —		81
(4-ethyltoluene)	C_9H_{12}	*Pseudomonas putida* UV4	(Et, Me diol), (16), >98		55
(propylbenzene)	C_9H_{12}	*Pseudomonas putida* UV4	(propyl diol), (15), >98	+ (Et, HO, H diol), (26), >98	55
(1-phenyl-1-propanol)	$C_9H_{12}O$	*Pseudomonas putida* UV4	(Et, HO, H diol), (8), >98		55

Substrate	Formula	Organism	Product (yield %, ee %)	Product 2 (yield %, ee %)	Ref
PhCH(OH)Et	$C_9H_{12}O$	*Pseudomonas putida* UV4	(53), >98		55
4-amino-3-chloro-1-phenylpyridazinone	$C_{10}H_8ClN_3O$	Bacterium, strain E	(—), —		194
2-methyl-3-phenyl-1-propene	$C_{10}H_{12}$	*Pseudomonas putida* UV4	(9), >98	(18), >98	55
2-phenyl-1,3-dithiane	$C_{10}H_{12}S_2$	*Pseudomonas putida* UV4	(18), >98	(7), >98	58
2-phenylethyl acetate	$C_{10}H_{12}O_2$	*Escherichia coli* JM109(pDTG601)	(11), >98		186

TABLE 1. MICROBIOLOGICAL OXYGENATIONS OF BENZENES (Continued)

Substrate		Microorganism, Conditions	Product(s), Yield(s) (% or g/L) and Enantiomeric Excess %		Refs.
$C_{10}H_{12}O_2$		*Pseudomonas putida* 39/D		(—), —	81
$C_{10}H_{14}$		*Pseudomonas putida* UV4	(33), >98	+ (9), >98	55
$C_{10}H_{14}$		*Pseudomonas putida* UV4	(27), >98		55
$C_{10}H_{14}$		*Pseudomonas desmolytica* S449B1 or *Pseudomonas convexa* S107B1	(—), —		195
$C_{10}H_{14}$		*Pseudomonas* strain	(—), —		196
$C_{10}H_{14}$		*Pseudomonas putida*, UV4	(15), >98		55

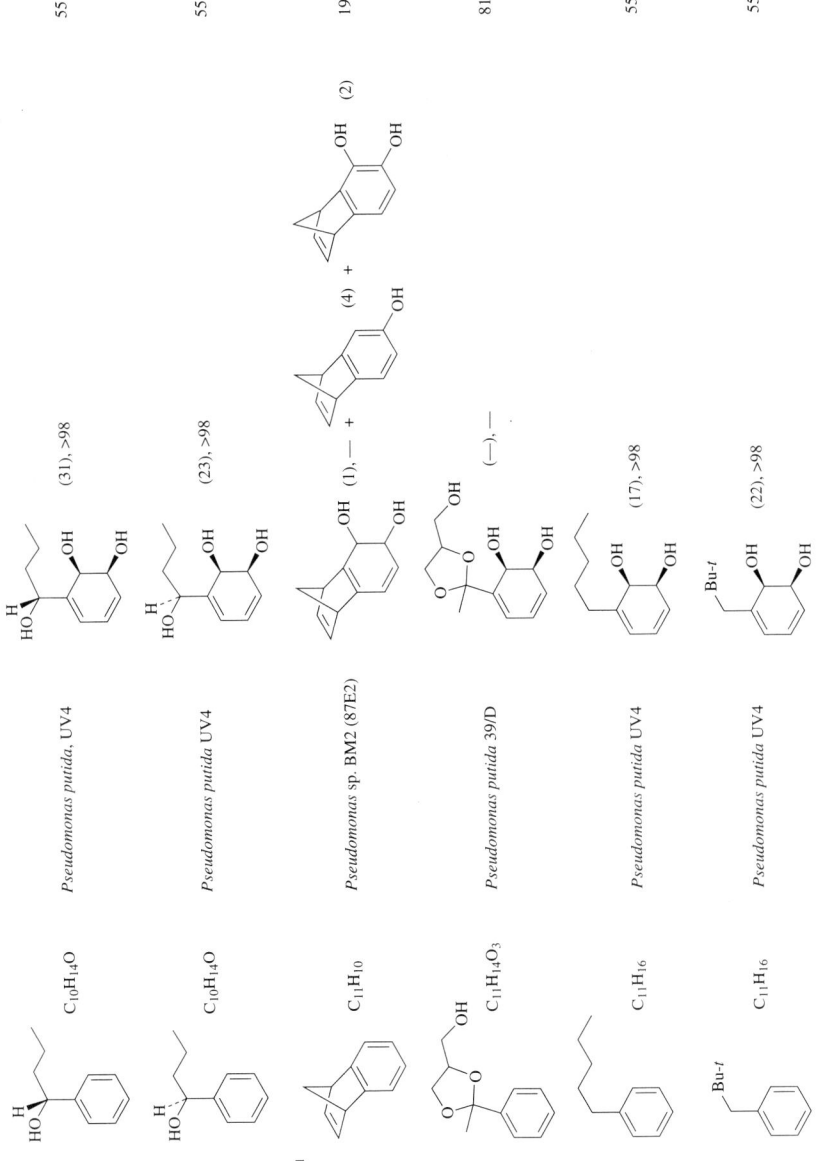

TABLE 1. MICROBIOLOGICAL OXYGENATIONS OF BENZENES (*Continued*)

Substrate	Microorganism, Conditions	Product(s), Yield(s) (% or g/L) and Enantiomeric Excess %	Refs.
$C_{11}H_{16}$ (Bu-*t*, *p*-methyl)	*Pseudomonas putida* UV4	(Bu-*t* diol), (5), >98	55
C_{12}			
$C_{12}H_{14}$ (cyclohexenyl-phenyl)	*Escherichia coli* JM109(pDTG601)	(diol), (3 g/L), —	189, 190
$C_{12}H_{14}$ (cyclohexenyl-phenyl)	*Escherichia coli* JM109(pDTG601)	(diol), (0.8 g/L), —	190
$C_{12}H_{14}O$ (2-phenylcyclohexanone)	*Escherichia coli* JM109(pDTG601)	**I** + **II** + **III** (1.0 g/L), —; **I:II:III** = 2:1:1	189, 190
$C_{12}H_{14}O$ ((S)-2-phenylcyclohexanone)	*Escherichia coli* JM109(pDTG601)	(—), —	190

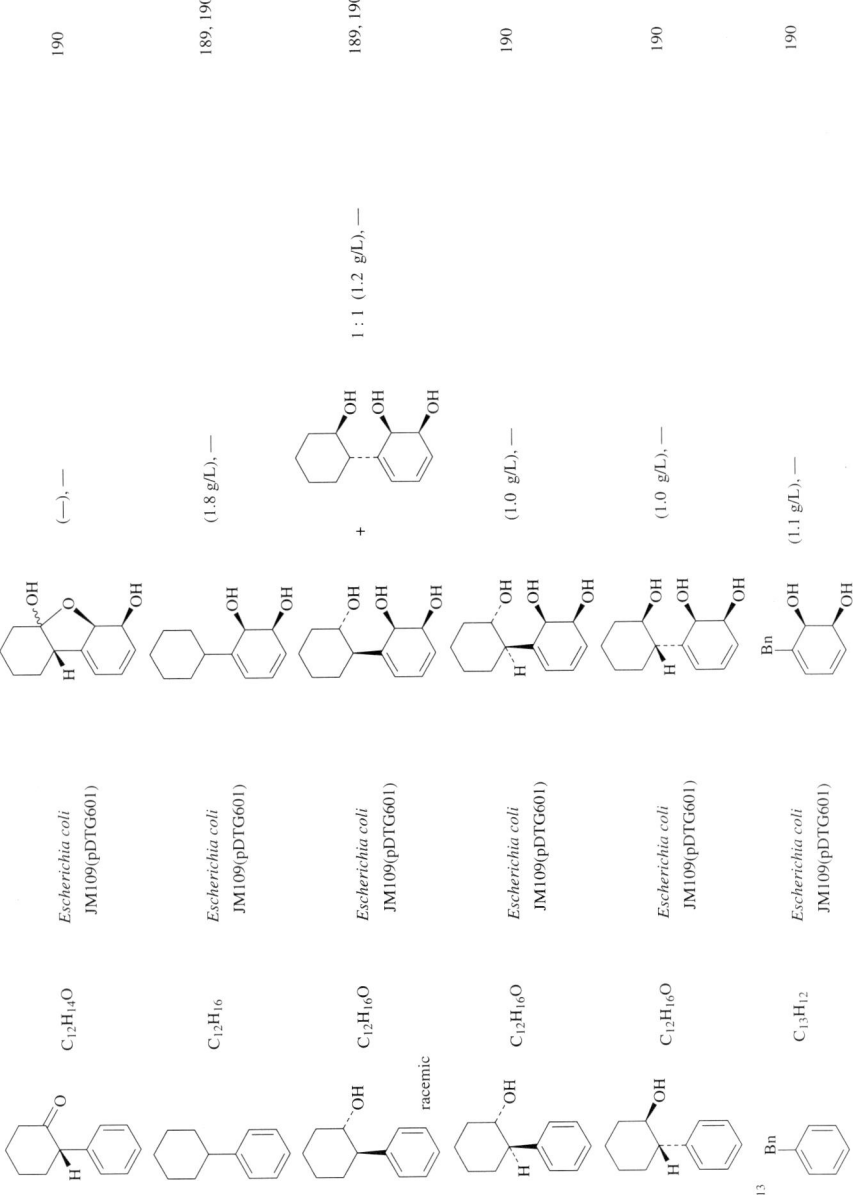

TABLE 1. MICROBIOLOGICAL OXYGENATIONS OF BENZENES (*Continued*)

Substrate	Microorganism, Conditions	Product(s), Yield(s) (% or g/L) and Enantiomeric Excess %	Refs.
C$_{13}$H$_{12}$ (diphenylmethane)	*Pseudomonas* sp. BM2	(45), >95 dihydrodiol^a + (7) catechol + (4) 2-hydroxy + (1) 3-hydroxy	197
C$_{14}$H$_9$Cl$_5$, Ar = 4-ClC$_6$H$_4$	*Alcaligenes eutrophus* A5	("near stoichiometric"),—	198
C$_{14}$H$_{12}$ (*cis*-stilbene)	*Pseudomonas fluorescens* N3 TTC1	(—),—	48
C$_{14}$H$_{14}$S (PhCH$_2$CH$_2$SPh)	*Escherichia coli* JM109(pDTG601)	(16), 98	186
C$_{20}$H$_{26}$O$_3$	*Pseudomonas abietaniphila* BKME-9; *Escherichia coli* XL1 Blue MR(pVM20)	(—),—	94

^{a} The assignment of absolute configuration for this product is tentative.

TABLE 2. MICROBIOLOGICAL OXYGENATIONS OF BENZOIC AND NAPTHOIC ACIDS AND ESTERS

Substrate		Microorganism, Conditions	Product(s), Yield(s) (% or g/L) and Enantiomeric Excess %	Refs.
C_7				
3,5-difluorobenzoic acid	$C_7H_4F_2O_2$	*Pseudomonas putida* JT103	cis-diol (—), —	75, 53
3,4-difluorobenzoic acid	$C_7H_4F_2O_2$	*Pseudomonas putida* JT103	cis-diol (—), —	53
3,5-dichlorobenzoic acid	$C_7H_4Cl_2O_2$	*Alcaligenes eutrophus* B 9	cis-diol (—), —	36, 199
2-fluorobenzoic acid	$C_7H_5FO_2$	*Alcaligenes eutrophus* B 9	cis-diol (—), —	36, 74, 200
3-fluorobenzoic acid	$C_7H_5FO_2$	*Alcaligenes eutrophus* B 9	cis-diol (—), —	35
			" (—), — + cis-triol (—), —	
4-fluorobenzoic acid	$C_7H_5FO_2$	*Alcaligenes eutrophus* B 9	cis-diol (—), —	36, 74
				35, 36, 74

183

TABLE 2. MICROBIOLOGICAL OXYGENATIONS OF BENZOIC AND NAPTHOIC ACIDS AND ESTERS (Continued)

Substrate	Microorganism, Conditions	Product(s), Yield(s) (% or g/L) and Enantiomeric Excess %	Refs.
$C_7H_5ClO_2$ 3-chlorobenzoic acid	Alcaligenes eutrophus B 9	(—), —	35
$C_7H_5ClO_2$ 4-chlorobenzoic acid	Alcaligenes eutrophus B 9	(—), — + (—), — (cis diol + cis diol with CO₂H/OH)	201, 36, 74
$C_7H_5ClO_2$	Alcaligenes eutrophus B 9	(—), —	36, 74
$C_7H_5BrO_2$ 3-bromobenzoic acid	Alcaligenes eutrophus B 9	(—), — + (—), —	36
$C_7H_5BrO_2$ 4-bromobenzoic acid	Pseudomonas putida JT 107	(5 g/L), >98	202
$C_7H_6O_2$ benzoic acid	Pseudomonas putida U103	(2-3 g/L), —	33
	Escherichia coli (pPL416, pKT570)	(—), —	203
	Alcaligenes eutrophus B 9	(ca. 50), —	35, 36
$C_7HD_5O_2$	Pseudomonas putida JT103	(—), —	204

C_8	3-(trifluoromethyl)benzoic acid $C_8H_5F_3O_2$	*Pseudomonas putida* mt-2 (ATCC 33015)	cis-diol-CF₃ carboxylic acid	(—), —	205
	"	*Alcaligenes eutrophus* B9	"	(—), —	205
	4-(trifluoromethyl)benzoic acid $C_8H_5F_3O_2$	*Pseudomonas putida* PL-pT-11/43	cis-diol-CF₃	(80), —	206
	"	*Pseudomonas putida* JT107	"	(—), —	207, 202
	terephthalic acid $C_8H_6O_4$	*Comamonas testosteroni* T-2	cis-diol dicarboxylic acid	(—), —	208
	methyl benzoate $C_8H_8O_2$	*Pseudomonas putida* UV4	cis-diol methyl ester	(42), —	61
	3-methylbenzoic acid $C_8H_8O_2$	*Alcaligenes eutrophus* B9	cis-diol-CH₃	(—), —	36, 35
	4-methylbenzoic acid $C_8H_8O_2$	*Alcaligenes eutrophus* B9	cis-diol-CH₃ + isomer	(—), —	36
	"	*Pseudomonas putida* BGXM1	"	(48), —	143

TABLE 2. MICROBIOLOGICAL OXYGENATIONS OF BENZOIC AND NAPTHOIC ACIDS AND ESTERS (*Continued*)

Substrate	Microorganism, Conditions	Product(s), Yield(s) (% or g/L) and Enantiomeric Excess %	Refs.
C_{10} $C_8H_8O_2$	*Pseudomonas* JT 107	(10.2 g/L), —	202
$C_8HD_7O_2$	*Pseudomonas* JT 107	(—), —	204
$C_{10}H_{12}O_2$	*Pseudomonas putida* PL-pT-11/43	(—), —	209
$C_{11}H_8O_2$	*Pseudomonas* sp. A3	(74), — *cis*	210, 76
$C_{12}H_{10}O_2$	*Pseudomonas* sp. C22	(—), —	210
C_{14} $C_{14}H_{10}O_3$	*Pseudomonas fluorescens* N3 TTC1	(80), —	210
	Pseudomonas sp. POB 310	(—), —	211

TABLE 3. MICROBIOLOGICAL OXYGENATIONS OF BIPHENYLS

Substrate	Microorganism, Conditions	Product(s), Yield(s) (% or g/L) and Enantiomeric Excess %	Refs.
C_{12}			
$C_{12}H_6Cl_4$	Biphenyl 2,3-dioxygenase from *Pseudomonas* sp. LB400	(—), — + (tr), —	73
	Escherichia coli FM4110(pGEM410)	" (—), —	50
$C_{12}H_7Cl_3$	Biphenyl 2,3-dioxygenase from *Pseudomonas* sp. LB400	(—), — + (—), —	73
	Burkholderia sp. LB400	" (—), —	52
$C_{12}H_7Cl_3$	*Pseudomonas pseudoalcaligenes* KF707	**I** + **II**, **I+II** (—), —; **I:II** = 97:3	212
	Pseudomonas cepacia LB400	**I+II** (—), —; **I:II** = 5:95	212
$C_{12}H_8$	*Pseudomonas putida* UV4	(66), >98	69, 70

TABLE 3. MICROBIOLOGICAL OXYGENATIONS OF BIPHENYLS (*Continued*)

Substrate		Microorganism, Conditions	Product(s), Yield(s) (% or g/L) and Enantiomeric Excess %	Refs.
$C_{12}H_8$		*Pseudomonas* sp. strain C250 cells, grown on carbazole	(—), >98	37
		Pseudomonas fluorescens N3	(—), —	48
$C_{12}H_8Cl_2$		Biphenyl 2,3-dioxygenase from *Pseudomonas* sp. LB400	(—), —	73
$C_{12}H_8Cl_2$		*Burkholderia* sp. LB400 BPDO encoded in *Escherichia coli* BL21(DE3)/pLys	(—), —	52
		Pseudomonas pseudoalcaligenes KF707	(—), —	212

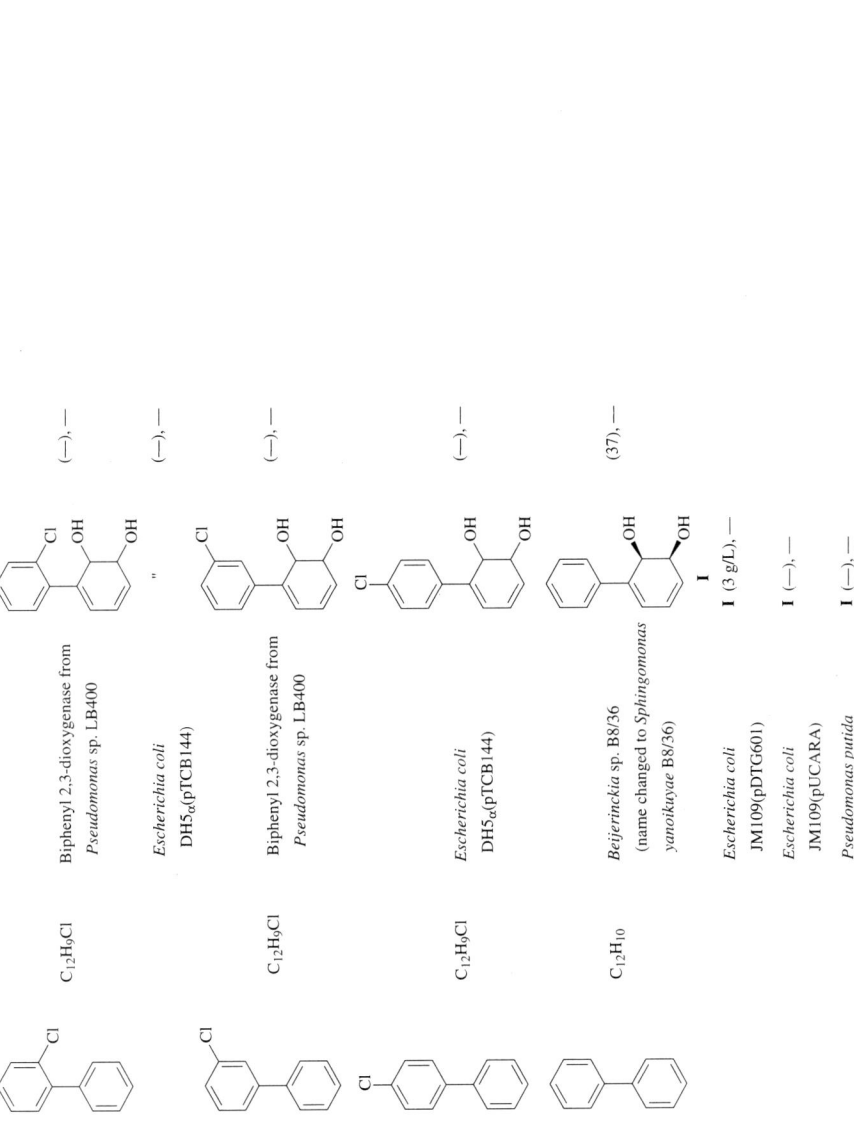

TABLE 3. MICROBIOLOGICAL OXYGENATIONS OF BIPHENYLS (*Continued*)

Substrate	Microorganism, Conditions	Product(s), Yield(s) (% or g/L) and Enantiomeric Excess %	Refs.
$C_{12}H_{10}$	Biphenyl 2,3-dioxygenase from *Pseudomonas* sp. LB400	**I** (—), —	51
	Pseudomonas fluorescens N3 TCC1	**I** (—), — + (—), —	48
	Escherichia coli JM109(DE3)(pDTG141)	**I** (—), >98	31
	Escherichia coli JM109(DE3)(pDTG141-A206I)	**I** + **II** (—); **I:II** = 95:5	91
	Escherichia coli JM109(DE3)(pDTG141-F352V)	**I** (—), >95 + (—), 75 **I:II** = 87:13	31
	Escherichia coli JM109(DE3)(pDTG141-F352V) (pDTG511)	**III** (—), >98 **I:III** = 4:96	31
	Pseudomonas pseudoalcaligenes KF748	**I** (—), —	216
	Escherichia coli DH5α(pTCB144)	**I** (—), —	77
$C_{13}H_{10}$	*Pseudomonas* sp. 9816/11	**I** (4), >95 + **II** HO (0.2)	67

		Escherichia coli JM109(DE3)(pDTG141)	**I** (57), >95 + **II** (6)	(—), —	67
		Pseudomonas fluorescens N3 TCC1	**I** (—), — +	(—), —	48
$C_{13}H_{12}O$		*Brevibacterium* DPO 1361		(—), —	217
		Pseudomonas sp. F274	"	(7), —	218, 219
		Escherichia coli JM109(pDTG601)		(2.5 g/L), —	213
$C_{14}H_{14}O_2$		*Escherichia coli* JM109(pDTG601)		(0.8 g/L), —	213

C_{14}

TABLE 4. MICROBIOLOGICAL OXYGENATIONS OF NAPHTHALENES

Substrate	Microorganism, Conditions	Product(s), Yield(s) (% or g/L) and Enantiomeric Excess %	Refs.
C₁₀			
$C_{10}H_7F$	*Pseudomonas fluorescens* N3 TTC1	(2), >95 + (2), >95	47
$C_{10}H_7Cl$	Soil bacterium	(—), —	220
	Pseudomonas fluorescens N3 TTC1	(65), >95 + (14), >95	47
$C_{10}H_7Cl$	*Pseudomonas fluorescens* N3 TTC1	(30), >95	47
$C_{10}H_7Br$	*Pseudomonas putida* NCIB 9816-11	**I** (28), >98 + **II** (14), >98 + (tr), —	221
	Pseudomonas fluorescens N3 TTC1	**I** (25), >95 + **II** (19), >95	47

Substrate	Formula	Organism	Product	Yield	Additional product	Ref
6-bromonaphthalene	$C_{10}H_7Br$	*Pseudomonas putida* NCIB 9816-11	bromo-cis-dihydrodiol (OH, OH)	(34), >98	bromo-trans-diol (tr), —	221
		Pseudomonas fluorescens N3 TTC1	"	(32), >95		47
1-nitronaphthalene	$C_{10}H_7NO_2$	*Escherichia coli* DH5α(pDTG927)	nitro-cis-dihydrodiol	(—), —		172
2-nitronaphthalene	$C_{10}H_7NO_2$	*Pseudomonas fluorescens* N3 TTC1	nitro-cis-dihydrodiol	(34), >98		47
naphthalene	$C_{10}H_8$	*Pseudomonas putida* 119	cis-dihydrodiol	(47), >98		62, 63
		Pseudomonas putida UV, immobilized in barium alginate beads	"	(1.8 mol/g dry cell weight), >95		59
		Pseudomonas fluorescens N3 TTC1	"	(80), >95		47
		Pseudomonas sp. NCIB 9816	"	(—), —		222
		Multicomponent Enzyme System from *Pseudomonas* sp. NCIB 9816	"	(97, based on NADH), —		223
		Escherichia coli DH5α(pDTG800)	"	(0.75 nmol/mg protein), 70		224
		Escherichia coli DH5α(pDTG832)	"	(0.07 nmol/mg protein), 98		224
		Escherichia coli DH5α(pDTG833)	"	(0.69 nmol/mg protein), 96		224

TABLE 4. MICROBIOLOGICAL OXYGENATIONS OF NAPHTHALENES (*Continued*)

Substrate	Microorganism, Conditions	Product(s), Yield(s) (% or g/L) and Enantiomeric Excess %	Refs.
$C_{10}H_8$	*Escherichia coli* DH5α(pDTG834)	(0.07 nmol/mg protein), 70	224
	Escherichia coli DH5α(pDTG141)	(2.14 nmol/mg protein), >99	224
	Escherichia coli JM109(DE3)(pJS48)	(0.83 nmol/mg protein), 96	224
	Escherichia coli DH5α(pDTG927)	(—), 57	172
	Escherichia coli JM109(pUCARA)	(−10), —	214
	Agmenellum quadruplicatum PR-6	(—), —	225
	Escherichia coli JM109(pUCARA)	(—), —	226
	Escherichia coli DH5α(pTCB144)	(—), —	77
	Escherichia coli JM109(DE3)(pDTG141-A206I)	(—), >98	91
	Escherichia coli JM109(DE3)(pDTG141-A206I-F352I)	(—), 40	91
	Pseudomonas putida 119/$^{18}O_2$	(—), —	63
$C_{10}H_7D$	*Pseudomonas putida* UV4	(—), —	227, 164
$C_{10}H_7D$	*Pseudomonas putida* UV4	(—), —	227, 164

Substrate	Organism	Products	Ratio	Ref.
$C_{10}D_8$	*Pseudomonas putida* 119/$^{18}O_2$	deuterated *cis*-dihydrodiol with ^{18}OH groups	(—), —	63
$C_{10}H_{10}$	*Pseudomonas putida* UV4	*cis*-1,2-dihydroxy-1,2-dihydronaphthalene + 1,2-dihydronaphthalene-1,2-diol + 2-hydroxy-1,2-dihydronaphthalene	(10), >98 + (28), >98 + (8), >98	228, 82, 229
	Sphingomonas yanoikuyae B1	naphthalene + *cis*-dihydrodiol + 1-naphthol	(tr), — + (tr), —	230
	Sphingomonas yanoikuyae B8/36	I + II; **I:II:III** = 9:10:1	(—) **I:II:III** = 5:0:1	230
	Pseudomonas putida F39/D	I + II **I**, **II**, **III**	(—) **I:II:IV:V** = 0:1.2:1.0:2.5	231
	Pseudomonas putida NCIB 9816/11	I+II+IV+V **IV**, **V**	**I:II:IV:V** = 1.9:1.0:0:0	231
	Escherichia coli JM109(pDTG601A)	I+II+IV+V	**I:II:IV:V** = 0:1.2:1.0:2.0	231
	Escherichia coli JM109(pDTG141)	I+II+IV+V	**I:II:IV:V** = 1.5:1.0:0:0	231

195

TABLE 4. MICROBIOLOGICAL OXYGENATIONS OF NAPHTHALENES (*Continued*)

Substrate	Microorganism, Conditions	Product(s), Yield(s) (% or g/L) and Enantiomeric Excess %	Refs.
C$_{10}$H$_{10}$	*Pseudomonas putida* UV4, 23 h fermentation	**I** (ca. 90), >98 + **II** (0) + **III** (0)	228, 229
C$_{10}$H$_{10}$	*Pseudomonas putida* UV4, 0.5 h fermentation	**I** (—), >98 + **II** (—), >98 + **III** (—), 3	228, 229
C$_{10}$H$_9$D	*Pseudomonas putida* UV4	(—), — + (—), —	228
C$_{10}$H$_8$D$_2$	*Pseudomonas putida* UV4	(—), —	228
C$_{10}$H$_{10}$O (racemic)	*Pseudomonas putida* UV4	(8), >98 + (7), >98	229
C$_{10}$H$_{10}$O	*Pseudomonas putida* UV4	(—), —	228

Substrate	Organism	Product	Yield, % ee	Ref.
$C_{10}H_9DO$ (1,2-dihydronaphthalen-1-ol-D)	*Pseudomonas putida* UV4	1,2-dihydroxy-8-D-naphthalene derivative	(—), —	228
$C_{10}H_{10}O$	*Pseudomonas putida* UV4	1,2-dihydrodiol	(—), —	228
$C_{10}H_9DO$	*Pseudomonas putida* UV4	dideuterated dihydrodiol	(—), —	228
$C_{11}H_8O_2$ (1-naphthoic acid)	*Pseudomonas maltophilia* CSV89	CO_2H-substituted dihydrodiol	(—), —	232
$C_{11}H_{10}$ (1-methylnaphthalene)	*Pseudomonas fluorescens* N3 TTC1	8-methyl-1,2-dihydrodiol	(13), —	47
"	*Pseudomonas putida* CSV86	1-naphthalenemethanol (CH_2OH) +	(—), —	233
			(—), —	

TABLE 4. MICROBIOLOGICAL OXYGENATIONS OF NAPHTHALENES (*Continued*)

Substrate	Microorganism, Conditions	Product(s), Yield(s) (% or g/L) and Enantiomeric Excess %	Refs.
$C_{11}H_{10}$	*Pseudomonas putida* 39D	(20 mg/L), — + (—), —	234
	Pseudomonas putida NCIB 9816	" (350 mg/L), —	234
	Pseudomonas fluorescens N3 TTC1	" (20), —	47
$C_{11}H_{10}O$	*Pseudomonas fluorescens* N3 TTC1	(25), — + (2), —	47
	Streptomyces lividans pIJ6021-phdABCD	" (—), —	78
$C_{11}H_{10}O$	*Pseudomonas putida* 39D	**I** + **II** + **III** **I+II+III** (29)^a; **I:II:III** = 12:73:15	26
	Escherichia coli JM109(pDTG601)	**I+II+III** (64)^a; **I:II:III** = 17:69:14	26
	Pseudomonas putida NCIB 9816	**I+II+III** (6)^a; **I:II:III** = 93:7:0	26

198

C₁₂			
acenaphthylene structure, C₁₂H₈		*Escherichia coli* JM109(pDTG141)	**I+II+III** (57),[a] **I:II:III** = 93:7:0	26
		Escherichia coli C534(ProR/Sac)	**I+II+III** (57),[a] **I:II:III** = 92:8:0	26
		Beijerinckia sp. B8/36	**I+II+III** (36),[a] **I:II:III** = 74:26:0	26
		Pseudomonas fluorescens N3 TTC1	**I** (64), —	47
	acenaphthene-cis-diol (3)	*Beijerinckia* sp. B8/36	(3)	235
		Pseudomonas aeruginosa PAO1(pRE695)	" (—)	236
		Pseudomonas fluorescens N3 TTC1	" (—)	48
methyl 1-naphthoate, C₁₂H₁₀O₂	lactone structure	*Pseudomonas fluorescens* N3 TTC1	(23), >95 + methyl dihydrodiol-naphthoate (3), >95	47
methyl 2-naphthoate, C₁₂H₁₀O₂	methyl dihydroxy-dihydronaphthoate	*Pseudomonas fluorescens* N3 TTC1	(21), >95	47

TABLE 4. MICROBIOLOGICAL OXYGENATIONS OF NAPHTHALENES (*Continued*)

Substrate	Microorganism, Conditions	Product(s), Yield(s) (% or g/L) and Enantiomeric Excess %	Refs.
C$_{12}$H$_{12}$ (2,3-dimethylnaphthalene)	*Pseudomonas fluorescens* N3 TTC1	*cis*-diol (21), >95	47
C$_{12}$H$_{12}$ (2,6-dimethylnaphthalene)	*Pseudomonas fluorescens* N3 TTC1	*cis*-diol (3), >95	47
C$_{12}$H$_{12}$ (1-ethylnaphthalene)	*Pseudomonas fluorescens* N3 TTC1	*cis*-diol (22), >95	47
C$_{12}$H$_{12}$ (2-ethylnaphthalene)	*Pseudomonas fluorescens* N3 TTC1	*cis*-diol (70), >95	47

[a] The yield is given as percent of crude diols.
[b] This product was isolated as the methyl ester dimethyl ketal.

TABLE 5. MICROBIOLOGICAL OXYGENATIONS OF POLYCYCLIC AROMATICS

Substrate	Microorganism, Conditions	Product(s), Yield(s) (% or g/L) and Enantiomeric Excess %	Refs.
C_{14}			
$C_{14}H_{10}$ (anthracene)	*Beijerinckia* sp. B8/36 or *Escherichia coli* JM109-(DE3)(pDTG141-F352V)	(25), >95	28, 29, 31
	Pseudomonas putida 119	" (—), —	28, 29
	Escherichia coli JM109(pUCARA)	" (—), —	214
	Pseudomonas fluorescens N3 TTC1	" (—), —	48
	Mycobacterium sp. strain PYR-1	" (—), —	237
$C_{14}H_{10}$ (phenanthrene)	*Beijerinckia* sp. B8/36	**I** (32), >95	28, 30, 31
	Pseudomonas putida 119	**I** (12), >95 + **II** (—), —	28, 30
	Escherichia coli JM109(pUCARA)	**I** (—), — + **II** (—), —	214
	Pseudomonas fluorescens N3 TTC1	**I** (—), — + **II** (—), —	48
	Mycobacterium sp. strain PYR-1	**I** (—), — + **III** (—), — **II** (8), >95 **III** (0) 1 : 5	237
	Escherichia coli JM109(DE3)(pDTG141-F352V)	**I** (—), >95 + (—), 91	31

TABLE 5. MICROBIOLOGICAL OXYGENATIONS OF POLYCYCLIC AROMATICS (Continued)

Substrate		Microorganism, Conditions	Product(s), Yield(s) (% or g/L) and Enantiomeric Excess %	Refs.
$C_{14}H_{12}$		Pseudomonas putida 9816/11	(7), >95 + (tr)	66
		Escherichia coli JM109(DE3)(pDTG141)	(0.05 g/L), >95	66
$C_{14}H_{12}$		Pseudomonas putida 9816/11	(8), >95 + (3), >98	66
		Escherichia coli JM109(DE3)(pDTG141)	(5), >95 + (2), >98	66
$C_{16}H_{10}$		Mycobacterium sp.	(tr)	238
C_{16}		Mycobacterium sp. strain PYR-1 dioxygenase genes expressed in E. coli	(—)	239
		Mycobacterium sp. Strain RJGII-135	(—)	240
		Mycobacterium sp. strain KR2	(—)	241
		Mycobacterium flavescens (ATCC 700033)	(—)	242
		Pseudomonas stutzeri strain P16	(—)	243
		Bacillus cereus strain P21	(—)	243

$C_{16}H_{10}$ (fluoranthene)	*Escherichia coli* JM109(pUCARA)	cis-OH,OH (—), —	214
$C_{18}H_{12}$ (chrysene)	*Mycobacterium* sp. strain KR20	(—), —	244
	Sphingomonas yanoikuyae B8/36	(1), >98 + (tr), —	245, 246
$C_{18}H_{12}$ (benz[a]anthracene)	*Beijerinckia* sp. B8/36	**I** (10), >95 + (1), >95	
	Mycobacterium sp. Strain RJGII-135	**I** (—) + **II** (1), >95 + **III** (1), >95 **I:II:III** = 73:15:12	247, 248
	Sphingomonas yanoikuyae B8/36	(12), >98	240
$C_{18}H_{12}$ (triphenylene)	*Pseudomonas putida* 9816/11	(ca.1), >98	249
	Pseudomonas putida 9816/11; Triton X100	(7), >98	249

203

TABLE 5. MICROBIOLOGICAL OXYGENATIONS OF POLYCYCLIC AROMATICS (*Continued*)

Substrate	Microorganism, Conditions	Product(s), Yield(s) (% or g/L) and Enantiomeric Excess %	Refs.
C$_{20}$ $C_{20}H_{12}$	*Beijerinckia* sp. B8/36	(—),— *cis* + (—),—[a]	247
$C_{20}H_{12}$	*Selenastrum capricornutum*	(—),— + (—),—	250
	Mycobacterium sp. Strain RJGII-135	" (—)	240

[a] The structure of this product is tentative.

TABLE 6. MICROBIOLOGICAL OXYGENATIONS OF HETEROCYCLES

Substrate	Microorganism, Conditions	Product(s), Yield(s) (% or g/L) and Enantiomeric Excess %	Refs.
C_4 C_4H_4S (thiophene)	*Pseudomonas putida* UV4	(diol) (34), — ⇌ 60:40 (ca. 1), 43	251
C_5 C_5H_6S (methylthiophene)	*Pseudomonas putida* UV4	(diol) ⇌ 60:40 (11), 48	251
C_6 C_6H_7NO (N-methyl-2-pyridone)	*Escherichia coli* JM109(DE3)(pDTG141)	(ca. 50), — + (3), —	252
C_8 $C_8H_6N_2$ (quinoxaline)	*Pseudomonas putida* UV4	(2), >98 + (2) + (<1)	253, 144
$C_8H_6N_2$ (quinazoline)	*Pseudomonas putida* UV4	(4), — + a	253, 144
C_8H_6O (benzofuran)	*Pseudomonas putida* UV4	(34), >98 + (32) + (12)	254, 255, 256, 251

TABLE 6. MICROBIOLOGICAL OXYGENATIONS OF HETEROCYCLES (*Continued*)

Substrate	Microorganism, Conditions	Product(s), Yield(s) (% or g/L) and Enantiomeric Excess %	Refs.
C₈H₆S (benzothiophene)	*Pseudomonas putida* UV4	**I** (9), >98 + **II** ⇌ **III** (15)	255, 251
	Pseudomonas putida RE213	**I** (12) + **III** (4)	257
	Pseudomonas putida UV4, immobilized in barium alginate beads	**I+III**; **I:III** = 1:3	59
C₈H₇N (indole)	*Pseudomonas putida* PpG7, or *Escherichia coli* HB101, or *Pseudomonas putida* 39/D, or *Beijerinckia* sp. B8/36, or *Pseudomonas Putida* UV4	[indigo via dihydroxyindoline intermediate]	258, 255
C₈H₈O (2,3-dihydrobenzofuran)	*Pseudomonas putida* UV4	(11), 73 + (tr) + (tr)	254, 256
C₈H₈O₂	*Pseudomonas putida* UV4	(5), 86 + (7), 70 + (4)	254
C₈H₈O₂ (R,S)	*Pseudomonas putida* UV4	(4), >98 + (22), >98 + (13)	254

206

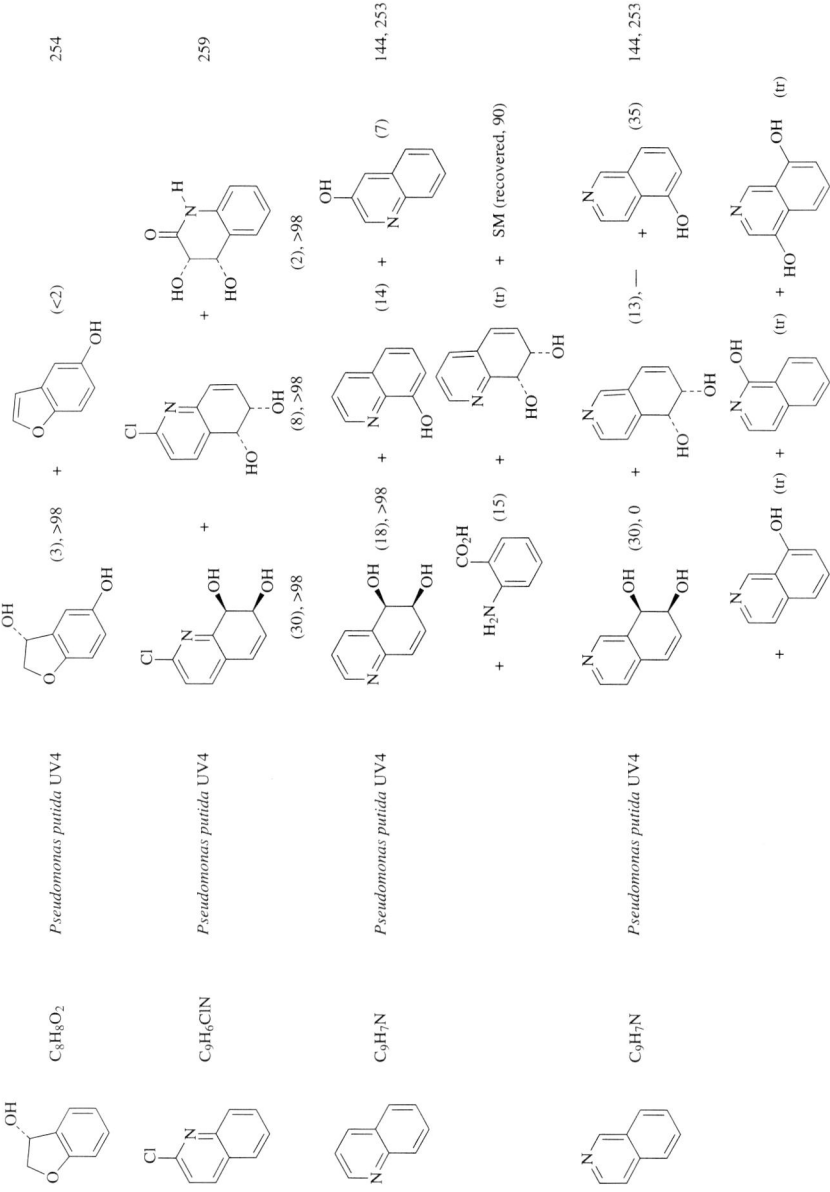

TABLE 6. MICROBIOLOGICAL OXYGENATIONS OF HETEROCYCLES (*Continued*)

Substrate	Microorganism, Conditions	Product(s), Yield(s) (% or g/L) and Enantiomeric Excess %	Refs.
C₉H₆DN	*Pseudomonas putida* UV4	(—), —	227, 164
C₉H₆DN	*Pseudomonas putida* UV4	(—), —	227, 164
C₉H₅D₂N	*Pseudomonas putida* UV4	(—), —	227, 164
C₉H₈O	*Pseudomonas putida* UV4	(8), 80	251
C₉H₈S	*Pseudomonas putida* UV4	(25), — + (3), 9 9:1 equilibrium	251
C₉H₈S	*Pseudomonas putida* UV4	78:28 equilibrium (79), >98	251

Formula	Substrate	Organism	Product(s), yield	Ref.
C_{10}	$C_{10}H_7NO_3$ (quinoline-2-carboxylic acid, 4-hydroxy)	*Pseudomonas fluorescens*	(—), —	260
	$C_{10}H_9NO_3$ (2-methoxyquinoline)	*Pseudomonas putida* UV4	(7), >98 + (2), >98 + (13), >98	259
C_{12}	$C_{12}H_8N_2$ (phenazine)	*Sphingomonas yanoikuyae* B8/36	(40), >98 + (15), >98	32
	$C_{12}H_8O$ (dibenzofuran)	*Pseudomonas* sp. NCIB 9816-11	(27), >95 + (12), >95	67
		Escherichia coli JM109(DE3)(pDTG141)	**I** (34), >95 + **II** (22), >95	67
		Beijerinckia sp. B8/36	**I** (—), — + **II** (—), —	261
		Pseudomonas fluorescens TTC1	**I** (1.6 g/L, total), >95 + **II** (—), >95; **I:II** = 3:2	262

209

TABLE 6. MICROBIOLOGICAL OXYGENATIONS OF HETEROCYCLES (*Continued*)

Substrate	Microorganism, Conditions	Product(s), Yield(s) (% or g/L) and Enantiomeric Excess %	Refs.
$C_{12}H_8O_2$	*Pseudomonas* sp. NCIB 9816-11	(4), —	46
$C_{12}H_8S$	*Beijerinckia* sp. B8/36	(1), —	263
	Beijerinckia sp. B8/36	(20), — + (7) **I** **II**	264
	Pseudomonas sp. NCIB 9816-11	**I** (15), >95 + **II** (3)	67
	Escherichia coli JM109(DE3)(pDTG141)	**I** (49), >95 + **II** (5)	67
	Pseudomonas fluorescens TTC1	**I** (0.45 g/L, total), >95 + **II** (—), >95; **I:II** = 15:1	262
$C_{12}H_8S_2$	*Pseudomonas fluorescens* TTC1	(0.22 g/L), —	262
$C_{12}H_{10}N_2O_2$	*Sphingomonas yanoikuyae* B8/36	(—), >98	32

C13	C13H9N	Sphingomonas yanoikuyae B8/36	(50), >98 + (12), >98	32
		Pseudomonas fluorescens TTC1	(0.22 g/L), —	262
	C13H11NO2	Sphingomonas yanoikuyae B8/36	(—), >98	32
C16	C16H10O	Sphingomonas yanoikuyae B8/36	(14), >98	249
		Pseudomonas putida 9816/11	(11), >98	249
	C16H10S	Sphingomonas yanoikuyae B8/36	(2), >98	249

TABLE 6. MICROBIOLOGICAL OXYGENATIONS OF HETEROCYCLES (*Continued*)

Substrate	Microorganism, Conditions	Product(s), Yield(s) (% or g/L) and Enantiomeric Excess %	Refs.
$C_{16}H_{10}S$	*Sphingomonas yanoikuyae* B8/36	(7), >98 + (3), >98	246
$C_{16}H_{14}O$	*Pseudomonas putida* 9816/11	(8), >98	249
$C_{16}H_{14}S$	*Sphingomonas yanoikuyae* B8/36	(3), >98 + (1.5), >98	249
$C_{17}H_{11}N$	*Sphingomonas yanoikuyae* B8/36	(8), >98 + (7), >98	246

[a] These two products were formed in an inseparable mixture.

TABLE 7. MICROBIOLOGICAL OXYGENATIONS OF OLEFINS

Substrate	Microorganism, Conditions	Product(s), Yield(s) (% or g/L) and Enantiomeric Excess %	Refs.
C₄			
C₄H₆ (butadiene)	*Pseudomonas putida* ML 2	(—), 25 [but-3-ene-1,2-diol]	84
C₅			
C₅H₆ (cyclopentadiene)	*Pseudomonas putida*, UV4	(ca. 30), 20 [cyclopent-2-ene-1,2-diol]	86
C₅H₈ (isoprene)	*Pseudomonas putida* ML 2; propylene glycol	(—), 34 [I] + (—), 40 [II]; I:II = 4:1	84
C₅H₈	*Pseudomonas putida* UV4; propylene glycol	I (—), 9 + II (—), 45; I:II = 2:1	84
C₅H₈	*Pseudomonas putida* 8859; propylene glycol	I (—), 12 + II (—), 16; I:II = 3:1	84
C₅H₈ (trans-piperylene)	*Pseudomonas putida* ML 2; propylene glycol	(—), 38 + (—), 33; 2:1	84
C₅H₈ (cis-piperylene)	*Pseudomonas putida* ML 2; propylene glycol	(—), 74 + (—), 70; 1:1	84
C₆			
C₆H₈ (1,3-cyclohexadiene)	*Pseudomonas putida* UV4	(ca. 30), >98 [cyclohex-2-ene-1,2-diol]	86

TABLE 7. MICROBIOLOGICAL OXYGENATIONS OF OLEFINS (*Continued*)

Substrate	Microorganism, Conditions	Product(s), Yield(s) (% or g/L) and Enantiomeric Excess %	Refs.
C₇			
C₇H₈ (cycloheptatriene)	*Pseudomonas putida* UV4	(diol) + (diol-HO) (*ca.* 30, total), >98; 2:1	86
C₇H₈ (norbornadiene)	*Pseudomonas* sp. BM2	(diol) (35) + (triol) (3)	197
C₇H₁₀ (cycloheptadiene)	*Pseudomonas putida* UV4	(diol) (*ca.* 30), >98	86
C₇H₁₂ *R,S* (3-methylcyclohexene)	*Pseudomonas putida* UV4	(diol) (—), — + (diol) (—), —	265, 153
C₈			
C₈H₈ (styrene)	Purified NDO from *Pseudomonas* sp. strain 9816-4	(PhCH(OH)CH₂OH) (—), 78	266
C₈H₁₀ (isopropylidenecyclopentadiene)	*Pseudomonas putida* RE213	(diol) (95), >98	85
C₈H₁₀	"	(*ca.* 30), >98	86
C₈H₁₂ (cyclooctadiene)	*Pseudomonas putida* UV4	(diol) (*ca.* 30), >98	86

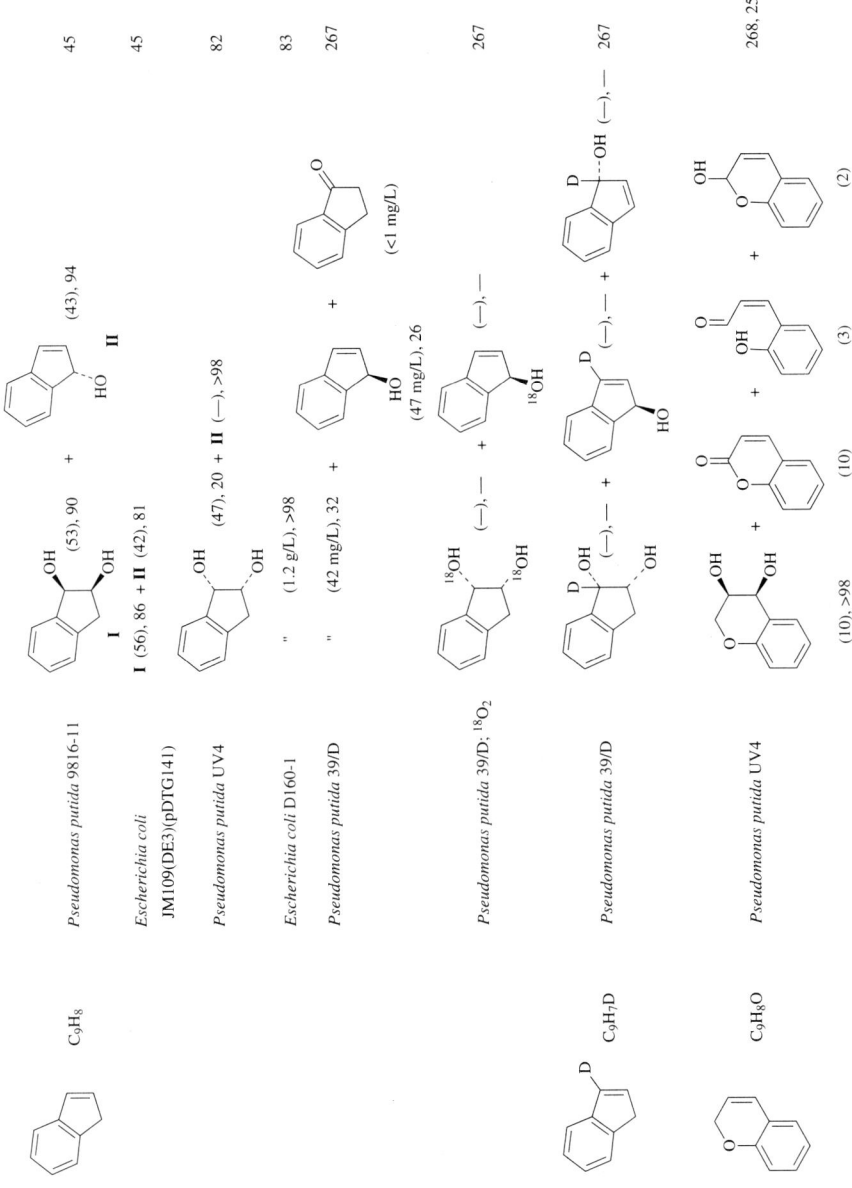

TABLE 7. MICROBIOLOGICAL OXYGENATIONS OF OLEFINS (*Continued*)

Substrate	Microorganism, Conditions	Product(s), Yield(s) (% or g/L) and Enantiomeric Excess %	Refs.
C_9H_8S	*Pseudomonas putida* UV4	(20), >98 + (—) + (—)	255
$C_{10}H_8$	*Pseudomonas putida* UV4	(*ca.* 20), >98	86
$C_{11}H_{12}$	*Pseudomonas putida* UV4	(6), >98 **I** + (0.4), >98 **II**	152, 82
	Pseudomonas putida F39/D	**I** (16), >98 + **II** (1), >98	269
	Pseudomonas sp. 9816/11	(13), >98	269
	Sphingomonas yanoikuyae B8/36	"	269
$C_{11}H_{12}O$	*Pseudomonas putida* UV4	(18), >98	268, 255

TABLE 8. MICROBIOLOGICAL BENZYLIC OXYGENATIONS

Substrate	Microorganism, Conditions	Product(s), Yield(s) (% or g/L) and Enantiomeric Excess %	Refs.
C_7			
C_7H_8	*Pseudomonas putida* UV4	(4), >98 + (ca. 60), >98	55
C_8			
C_8H_7N	*Pseudomonas putida* UV4	(20), >98 + (15), >98	55
C_8H_8	*Pseudomonas putida* UV4	(18), >98 + (13), >98 + (9), >98	69, 70
	Pseudomonas fluorescens 127-68 XVII	(1), 72 + **I** (—), 0 + **II** (—) + **I** (33), ca. 20 + **II** (2)	182
C_8H_8O	*Pseudomonas putida* UV4	(9) + (35) *cis*	69

TABLE 8. MICROBIOLOGICAL BENZYLIC OXYGENATIONS (Continued)

Substrate	Microorganism, Conditions	Product(s), Yield(s) (% or g/L) and Enantiomeric Excess %	Refs.
C_8H_8O (2,3-dihydrobenzofuran)	Pseudomonas putida UV4	(11), 73 + (tr) + (tr) + (9)	254, 256
C_8H_{10} (p-xylene)	Pseudomonas putida Biotype A strain (ATCC 39119)	(<1), —	188
C_8H_{10} (ethylbenzene)	Pseudomonas putida 39/D	(<1), — + (20), —	183
C_8H_{10}	Pseudomonas putida UV4	I (5), >98 + II (60), >98	55
C_9H_8 (indene)	Pseudomonas putida 9816-11	I (53), 90 + (43), 94 II	45
	Escherichia coli JM109 (DE3)(pDTG141)	I (56), 86 + II (42), 81	45
	Pseudomonas putida UV4	(47), 20 + II (—), >98 III	82

218

Pseudomonas putida 39/D — III (42 mg/L), 32 + (1-indanol) + (47 mg/L), 26 + (1-indanone) (<1 mg/L) — 267

Pseudomonas putida 39/D; $^{18}O_2$ — ^{18}OH-indanol + ^{18}OH-indanol (—), — + (—), — — 267 — C$_9$H$_7$D

Pseudomonas putida 39/D — (D,OH-indanol) + (D,OH-indanol) (—), — + (—), — + (D,OH-indene) (—), — 267 — C$_9$H$_8$O

Pseudomonas putida 9816/11 — (indanone-OH) (18), 62 + (indanone-OH) (2), 0 270 — C$_9$H$_8$O

Pseudomonas putida UV4 — (diol) (28), >98 + (indanol) (12) — 87

Pseudomonas putida 39/D — (hydroxy-indanone) (26), 76 + (hydroxy-indanone) (tr) I II 270

Pseudomonas putida 9816/11 — I (23), 6 + II (tr) 270

TABLE 8. MICROBIOLOGICAL BENZYLIC OXYGENATIONS (*Continued*)

Substrate	Microorganism, Conditions	Product(s), Yield(s) (% or g/L) and Enantiomeric Excess %	Refs.
C$_9$H$_9$N$_3$ (2-azidoindane)	*Pseudomonas putida* UV4	(1-OH, 2-N$_3$ indanol) (61), >98	87
C$_9$H$_9$Cl (2-chloroindane)	*Pseudomonas putida* UV4	(1-OH, 2-Cl indanol) (38), >98	87
C$_9$H$_9$Br (2-bromoindane)	*Pseudomonas putida* UV4	(1-OH, 2-Br indanol) (35), >98 + (1-OH, 2-Br, 3-OH) (26), >98	87
C$_9$H$_9$I (2-iodoindane)	*Pseudomonas putida* UV4	(1-OH, 2-I indanol) (20), >98 + (1-indenol) (13), >98 + (1-OH, 3-OH, 2-I) (0.5), >98	87
C$_9$H$_{10}$ (allylbenzene)	*Pseudomonas putida* UV4	(allyl-cis-diol) (15), >98 + (allyl-cis-diol isomer) (18), >98	55
C$_9$H$_{10}$ (indane)	*Pseudomonas putida* 9816-11	**I** (cis-1,2-indandiol) (7), 86 + **II** (1-indenol) (8), 69 + **III** (1-indanol) (64), 79	45
	Escherichia coli JM109 (DE3)(pDTG141)	**I** (18), 91 + **II** (19), 83 + **III** (54), 93	45

220

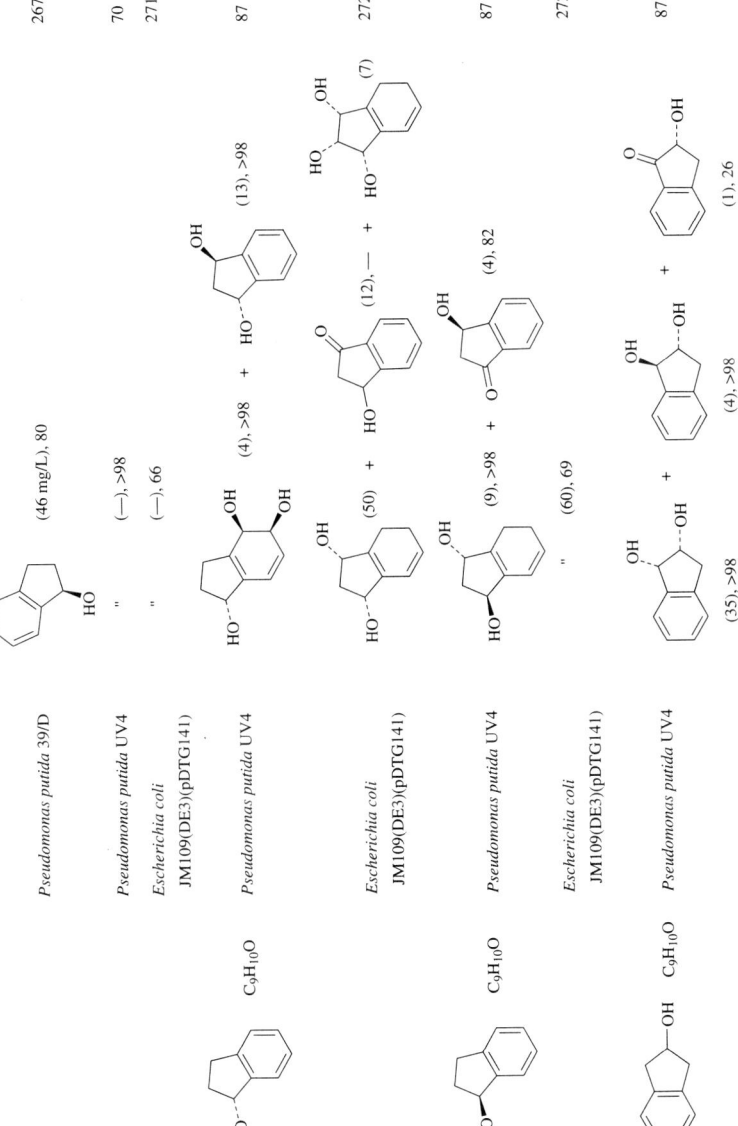

TABLE 8. MICROBIOLOGICAL BENZYLIC OXYGENATIONS (*Continued*)

Substrate	Microorganism, Conditions	Product(s), Yield(s) (% or g/L) and Enantiomeric Excess %	Refs.
C$_9$H$_{12}$ (1,2,4-trimethylbenzene)	*Pseudomonas aeruginosa* PAO1(pRE695)	2,5-dimethylbenzyl alcohol (—), — + 3,4-dimethylbenzyl alcohol (—), —	236
C$_9$H$_{12}$ (propylbenzene)	*Pseudomonas putida* UV4	*cis*-diol (26), >98 + hydroxylated product (15), >98	55
C$_{10}$H$_9$N (2-cyanoindane)	*Pseudomonas putida* UV4	*trans*-hydroxy nitrile (37), >98	87
C$_{10}$H$_{11}$NO$_2$ (2-indanyl carbamate)	*Pseudomonas putida* UV4	*trans*-hydroxy carbamate (36), >98	87
C$_{10}$H$_{12}$ (2-methylallylbenzene)	*Pseudomonas putida* UV4	*cis*-diol (18), >98 + hydroxylated diol (9), >98	55
C$_{10}$H$_{12}$ (2-methylindane)	*Pseudomonas putida* UV4	*trans*-indanol (34), >98 + *cis*-indanol (20), >98 + indanone (18), >98 + *cis*-indandiol (17), >98	87

$C_{10}H_{12}O$		*Pseudomonas putida* UV4	(9), >98 + (1), >98	87
$C_{10}H_{14}$		*Pseudomonas putida* UV4	(9), >98 + (33), >98	55
$C_{11}H_{10}$		*Sphingomonas paucimobilis* 2322	(—), — + (—), —	273
$C_{11}H_{12}$		*Pseudomonas putida* UV4	(6), >98 + (0.4), >98	152, 82
$C_{11}H_{12}O_2$		*Pseudomonas putida* F39/D	**I** (16), >98 + **II** (1), >98	269
		Pseudomonas putida UV4	(51) + (25), >98 + (3), >98 + (2), >98	87
$C_{12}H_{10}$		*Beijerinckia* sp. B8/36	(—)	235

223

TABLE 8. MICROBIOLOGICAL BENZYLIC OXYGENATIONS (Continued)

Substrate	Microorganism, Conditions	Product(s), Yield(s) (% or g/L) and Enantiomeric Excess %	Refs.
C$_{12}$H$_{10}$ (acenaphthylene)	*Pseudomonas aeruginosa* PAO1(pRE695)	acenaphthenol-OH (—), —	236
C$_{12}$H$_{12}$ (1,5-dimethylnaphthalene)	*Pseudomonas aeruginosa* PAO1(pRE695)	CH$_2$OH-substituted (—)	236
C$_{12}$H$_{12}$ (1,8-dimethylnaphthalene)	*Pseudomonas aeruginosa* PAO1(pRE695)	CH$_2$OH-substituted (—)	236
C$_{12}$H$_{12}$ (2,6-dimethylnaphthalene)	*Pseudomonas aeruginosa* PAO1(pRE695)	CH$_2$OH-substituted (—)	236
C$_{13}$H$_{10}$ (fluorene)	*Pseudomonas* sp. 9816/11	**I** (cis-diol) (4), >95 + **II** (9-fluorenol) (0.2)	67
	Escherichia coli JM109(pDTG601)	**I** (57), >95 + **II** (6)	67

C$_{13}$

TABLE 9. MICROBIOLOGICAL OXYGENATION OF THIOLS

Substrate	Microorganism, Conditions	Product(s), Yield(s) (% or g/L) and Enantiomeric Excess %	Refs.
C₄ C₄H₄S (thiophene)	*Pseudomonas putida* UV4	(11), 77 + (45), —	89
C₅ C₅H₆S (2-methylthiophene)	*Pseudomonas putida* UV4	(12), 8 + (4), —	89
C₅H₆S (3-methylthiophene)	*Pseudomonas putida* UV4	(4), 51 + (4), —	89
C₅H₆S₂ (2-(methylthio)thiophene)	*Pseudomonas putida* UV4	(41), >98	44
C₅H₆S₂	*Pseudomonas putida* NCIMB 8859	(4), 69	44
C₆ C₆H₇NS (2-(methylthio)pyridine)	*Pseudomonas putida* UV4	(20), 94	58, 44

TABLE 9. MICROBIOLOGICAL OXYGENATION OF THIOLS (*Continued*)

Substrate		Microorganism, Conditions	Product(s), Yield(s) (% or g/L) and Enantiomeric Excess %	Refs.
SMe-(2-pyridyl)	C₆H₇NS	*Pseudomonas putida* NCIMB 8859	Me-S(O)-(2-pyridyl) (18), 35	44
SMe-(4-pyridyl)	C₆H₇NS	*Pseudomonas putida* NCIMB 8859	Me-S(O)-(4-pyridyl) (5), 95	44
C₇ SMe-C₆H₄-4-F	C₇H₇FS	*Pseudomonas putida* UV4	Me-S(O)-C₆H₄-4-F (31), 78	44
		Pseudomonas putida NCIMB 8859	Me-S(O)-C₆H₄-4-F (4), 91	44
		Pseudomonas putida UV4	Me-S(O)-C₆H₄-3-F (30), 98	44
SMe-C₆H₄-3-F	C₇H₇FS	*Pseudomonas putida* NCIMB 8859	" (53), 97	44

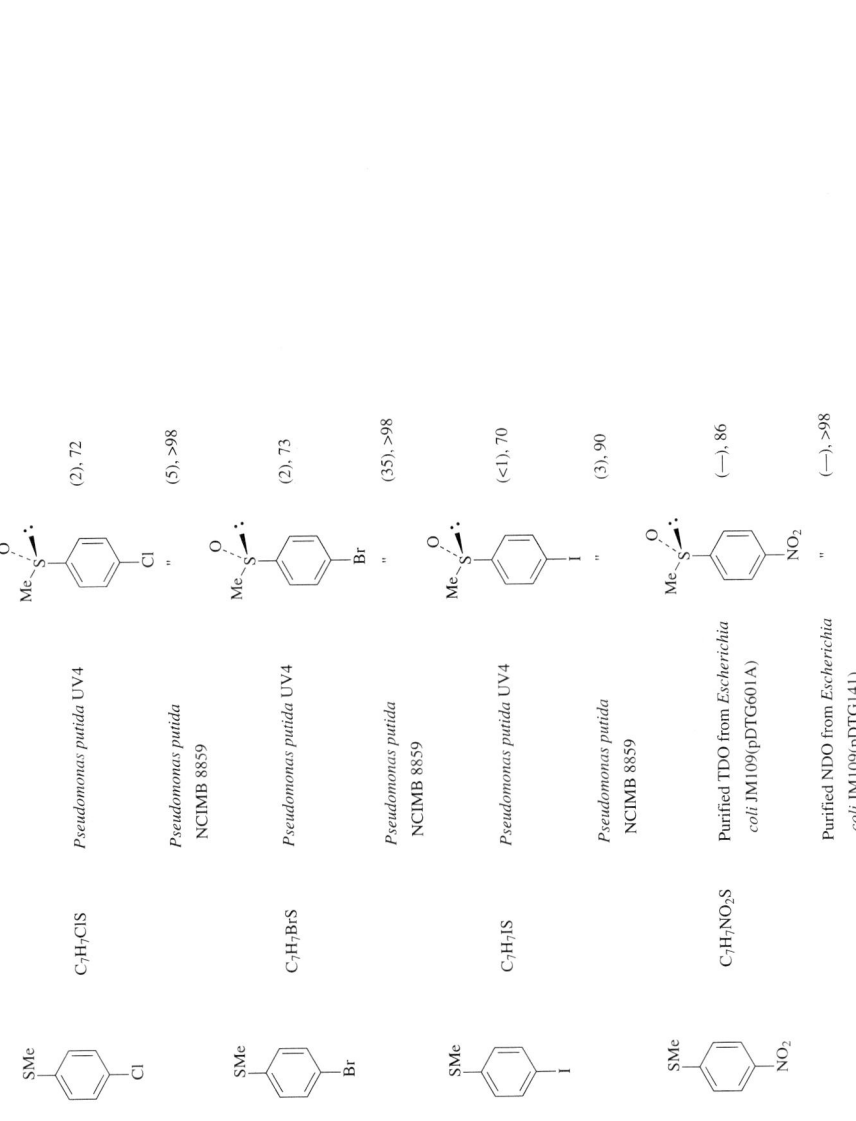

SMe-C6H4-Cl	C7H7ClS	Pseudomonas putida UV4	Me-S(O)-C6H4-Cl	(2), 72	44
		Pseudomonas putida NCIMB 8859	"	(5), >98	44
SMe-C6H4-Br	C7H7BrS	Pseudomonas putida UV4	Me-S(O)-C6H4-Br	(2), 73	44
		Pseudomonas putida NCIMB 8859	"	(35), >98	44
SMe-C6H4-I	C7H7IS	Pseudomonas putida UV4	Me-S(O)-C6H4-I	(<1), 70	44
		Pseudomonas putida NCIMB 8859	"	(3), 90	44
SMe-C6H4-NO2	C7H7NO2S	Purified TDO from *Escherichia coli* JM109(pDTG601A)	Me-S(O)-C6H4-NO2	(—), 86	88
		Purified NDO from *Escherichia coli* JM109(pDTG141)	"	(—), >98	88

TABLE 9. MICROBIOLOGICAL OXYGENATION OF THIOLS (Continued)

Substrate	Microorganism, Conditions	Product(s), Yield(s) (% or g/L) and Enantiomeric Excess %	Refs.
C₇H₈S SMe-C₆H₅	Pseudomonas putida UV4	Me-S(O)-Ph (95), >98	58, 44
	Escherichia coli (pKS T11)	" (—), >95	58
	Purified TDO from Escherichia coli JM109(pDTG601A)	" (—), >98	88
	Purified NDO from Escherichia coli JM109(pDTG141)	Me-S(O)-Ph (other enantiomer) (—), >98	88
	Pseudomonas putida NCIMB 8859	(33), 91	58, 44
C₈H₆S (benzothiophene)	Pseudomonas putida UV4	[dihydro sulfoxide] (2), 6 + [sulfoxide] (3), 3 + [sulfide] (1)	89
C₈H₇F₃S SMe-C₆H₄-CF₃	Pseudomonas putida UV4	Me-S(O)-C₆H₄-CF₃ (2), 76	44
	Pseudomonas putida NCIMB 8859	" (3), 98	44

Substrate		Organism	Product(s), % ee, % yield	Refs.
SMe-C6H4-CN	C8H7NS	*Pseudomonas putida* NCIMB 8859	Me-S(O)-C6H4-CN (<1), 73	44
PhS-CH=CH2	C8H8S	*Pseudomonas putida* UV4	Ph-S(O)-CH=CH2 (38), >98	58, 44
		Escherichia coli (pKS T11)	" (—), >95	58
		Pseudomonas putida NCIMB 8859	Ph-S(O)-CH=CH2 (—), 91	58, 44
2-methyl-benzodithiole	C8H8S2	*Pseudomonas putida* UV4	I (—), 97 + II (5), >98 I (40), >98 + II (—), 40	58
		Escherichia coli (pKS T11)	(—), 82	58
		Pseudomonas putida NCIMB 8859	benzodithiole S-oxide (—), 82	58
SMe-C6H4-OMe	C8H10OS	Purified TDO from *Escherichia coli* JM109(pDTG601A)	Me-S(O)-C6H4-OMe (—), 32 + oxide (—), 38	88

TABLE 9. MICROBIOLOGICAL OXYGENATION OF THIOLS (Continued)

Substrate	Microorganism, Conditions	Product(s), Yield(s) (% or g/L) and Enantiomeric Excess %	Refs.
C₈H₁₀OS (SMe–C₆H₄–OMe)	Purified NDO from *Escherichia coli* JM109(pDTG141)	Me-S(O)-C₆H₄-OMe (—), >98	88
C₈H₁₀OS (PhS-CH₂-OMe)	*Pseudomonas putida* UV4	MeO-CH₂-S(O)-Ph (36), >98	44
	Pseudomonas putida NCIMB 8859	" (4), 10	44
C₈H₁₀S (SMe–C₆H₄–Me)	Purified TDO from *Escherichia coli* JM109(pDTG601A)	Me-S(O)-C₆H₄-Me (—), 38 + cis SMe/OH/OH diol (—), —	88
	Purified NDO from *Escherichia coli* JM109(pDTG141)	" (—), >98	88
	Pseudomonas putida UV4	" (64), >98	58, 44
	Escherichia coli (pKS T11)	" (—), >95	58
C₈H₁₀S (SEt–C₆H₅)	Purified TDO from *Escherichia coli* JM109(pDTG601A)	Et-S(O)-Ph (—), >98	88

230

Substrate	Formula	Organism	Product	Yield (%)	Ref
PhSCH₂SMe	C₈H₁₀S₂	Purified NDO from *Escherichia coli* JM109(pDTG141)	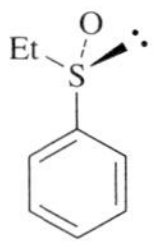 Et-S(=O)-Ph	(—), 93	88
		Pseudomonas putida NCIMB 8859	"	(27), 84	58, 44
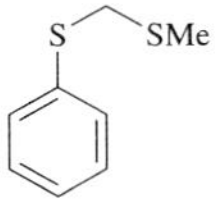		*Pseudomonas putida* UV4	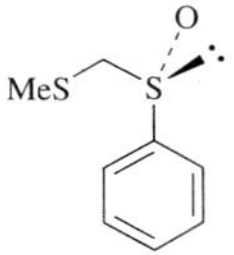 MeS-CH₂-S(=O)-Ph	(20), 97	58
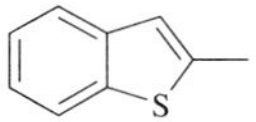 2-methylbenzothiophene	C₉H₈S	*Pseudomonas putida* UV4	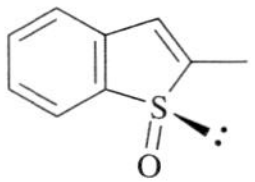 2-methylbenzothiophene S-oxide	(2), >98	89
		Pseudomonas putida 8859	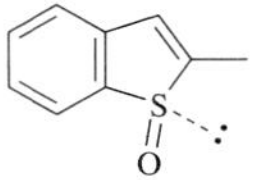 2-methylbenzothiophene S-oxide	(26), 56	89
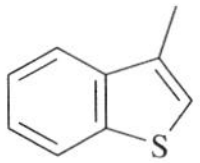 3-methylbenzothiophene	C₉H₈S	*Pseudomonas aeruginosa* PAO1(pRE695)	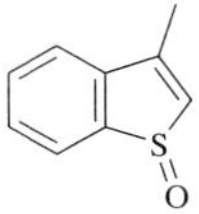 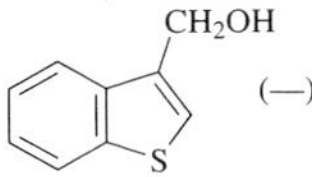 3-methylbenzothiophene S-oxide (—) + 3-(hydroxymethyl)benzothiophene (—)		236

C₉

231

TABLE 9. MICROBIOLOGICAL OXYGENATION OF THIOLS (*Continued*)

Substrate		Microorganism, Conditions	Product(s), Yield(s) (% or g/L) and Enantiomeric Excess %	Refs.
C_9H_8S		*Pseudomonas putida* 8859	(7), 41	89
$C_9H_{12}S$	SPr-*n*	*Pseudomonas putida* UV4	(5), >98	58, 44
$C_9H_{12}S$	SPr-*i*	*Pseudomonas putida* NCIMB 8859	" (33), 86	44
		Pseudomonas putida UV4	(27), 97	58
C_{10}				
$C_{10}H_{12}S_2$		*Pseudomonas putida* NCIMB 8859	(24), 76	58, 44
		Pseudomonas putida UV4	(7), >98 + (18), >98	58
$C_{10}H_{14}S$	SBu-*n*	*Pseudomonas putida* UV4	(7), 98	58

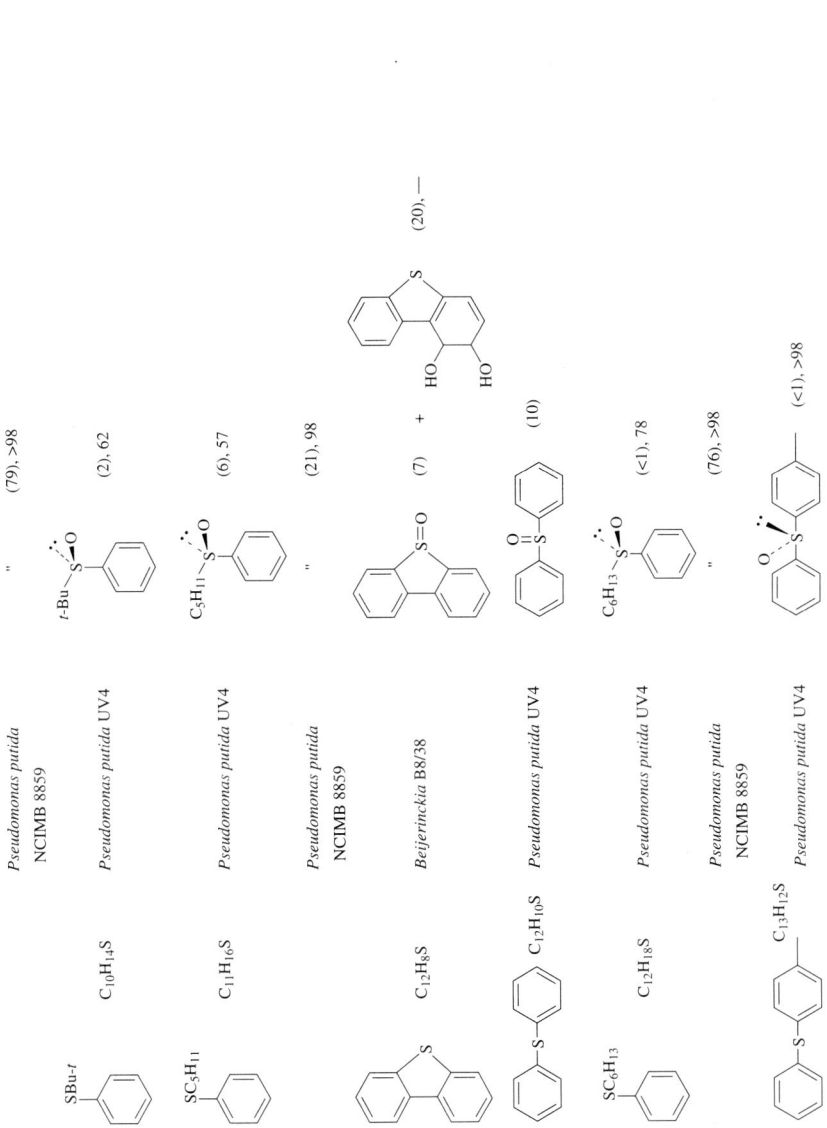

TABLE 9. MICROBIOLOGICAL OXYGENATION OF THIOLS (Continued)

Substrate		Microorganism, Conditions	Product(s), Yield(s) (% or g/L) and Enantiomeric Excess %	Refs.
Me-C₆H₄-S-C₆H₅	C₁₃H₁₂S	*Pseudomonas putida* UV4	Me-C₆H₄-S(O)-C₆H₅ (1), 86	58
SC₇H₁₅-C₆H₅	C₁₃H₂₀S	*Pseudomonas putida* UV4	C₇H₁₅-S(O)-C₆H₅ (<1), 12	44
		Pseudomonas putida NCIMB 8859	" (43), >98	44
C₁₄				
SC₈H₁₇-C₆H₅	C₁₄H₂₂S	*Pseudomonas putida* UV4	C₈H₁₇-S(O)-C₆H₅ (<1), 22	44
		Pseudomonas putida NCIMB 8859	" (25), 98	44

234

TABLE 10. DIHYDRODIOLS THAT HAVE NOT BEEN ISOLATED

Substrate		Microorganism, Conditions	Product(s), Yield(s) (% or g/L) and Enantiomeric Excess %		Refs.
C_6					
	$C_6H_2Cl_4$	Escherichia coli DHα(pTCB144)	(tetrachloro-cis-dihydrodiol)	(—)	77
	$C_6H_3Cl_3$	Escherichia coli DHα(pTCB144)	(trichloro-cis-dihydrodiol)	(—), —	77
	$C_6H_3Cl_3$	Escherichia coli DHα(pTCB144)	(trichloro-cis-dihydrodiol)	(—), —	77
	$C_6H_4Cl_2$	Pseudomonas sp.	(dichloro-cis-dihydrodiol)	(—), —	274
	$C_6H_4Cl_2$	Alcaligenes sp. OBB65	(dichloro-cis-dihydrodiol)	(—), —	275
C_7					
	$C_7H_4Cl_2O_2$	Pseudomonas putida PL-pT-11/43	(dichloro-cis-dihydrodiol carboxylic acid)	(—), —	209

TABLE 10. DIHYDRODIOLS THAT HAVE NOT BEEN ISOLATED (*Continued*)

Substrate		Microorganism, Conditions	Product(s), Yield(s) (% or g/L) and Enantiomeric Excess %	Refs.
4-Cl-C$_6$H$_4$-CO$_2$H	C$_7$H$_5$ClO$_2$	*Pseudomonas putida* PL-pT-11/43	3-Cl-cis-dihydrodiol-CO$_2$H (—), —	209
4-Br-C$_6$H$_4$-CO$_2$H	C$_7$H$_5$BrO$_2$	*Pseudomonas putida* PL-pT-11/43	3-Br-cis-dihydrodiol-CO$_2$H (—), —	209
4-I-C$_6$H$_4$-CO$_2$H	C$_7$H$_5$IO$_2$	*Pseudomonas putida* PL-pT-11/43	3-I-cis-dihydrodiol-CO$_2$H (—), —	209
C$_8$				
phthalic acid (1,2-(CO$_2$H)$_2$-C$_6$H$_4$)	C$_8$H$_6$O$_4$	*Micrococcus* sp. 12B	dihydrodiol-diacid (—), —	276
		Burkholderia cepacia DBO1	dihydrodiol-diacid (—), —	277, 278
3-F-4-Me-C$_6$H$_3$-CO$_2$H	C$_8$H$_7$FO$_2$	*Pseudomonas putida* PL-pT-11/43	F-Me-dihydrodiol-CO$_2$H (—), —	209

Substrate	Formula	Organism	Product	Yield, ee	Ref
3-chloro-4-methylbenzoic acid	$C_8H_7ClO_2$	*Pseudomonas putida* PL-pT-11/43	5-chloro-6-methyl-*cis*-dihydrodiol carboxylic acid	(—), —	209
3-bromo-4-methylbenzoic acid	$C_8H_7BrO_2$	*Pseudomonas putida* PL-pT-11/43	5-bromo-6-methyl-*cis*-dihydrodiol carboxylic acid	(—), —	209
4-(bromomethyl)benzoic acid	$C_8H_7BrO_2$	*Pseudomonas putida* PL-pT-11/43	corresponding *cis*-dihydrodiol	(—), —	209
4-methylbenzoic acid	$C_8H_8O_2$	*Pseudomonas putida* PL-pT-11/43	corresponding *cis*-dihydrodiol	(—), —	209
4-methylstyrene	C_9H_{10}	*Pseudomonas putida* PL-pT-11/43	corresponding *cis*-dihydrodiol	(—), —	209
3,4-dimethylbenzoic acid	$C_9H_{10}O_2$	*Pseudomonas putida* PL-pT-11/43	corresponding *cis*-dihydrodiol	(—), —	209

TABLE 10. DIHYDRODIOLS THAT HAVE NOT BEEN ISOLATED (*Continued*)

Substrate		Microorganism, Conditions	Product(s), Yield(s) (% or g/L) and Enantiomeric Excess %		Refs.
$C_9H_{10}O_2$	(4-ethylbenzoic acid)	*Pseudomonas putida* PL-pT-11/43	(diol structure)	(—), —	209
$C_9H_{10}O_3$	(4-ethoxybenzoic acid)	*Pseudomonas putida* PL-pT-11/43	(diol structure)	(—), —	209
$C_9H_{11}NO_2$	(phenylalanine)	A bacterium	(diol structure)	(—), —	279
C_9H_{12}	(isopropylbenzene)	*Pseudomonas desmolytica*; *Pseudomonas convexa*	(diol structure)	(—), —	195
C_{10}					
$C_{10}H_{12}$	(4-methylallylbenzene)	*Pseudomonas putida* PL-pT-11/43	(diol structure)	(—), —	209

238

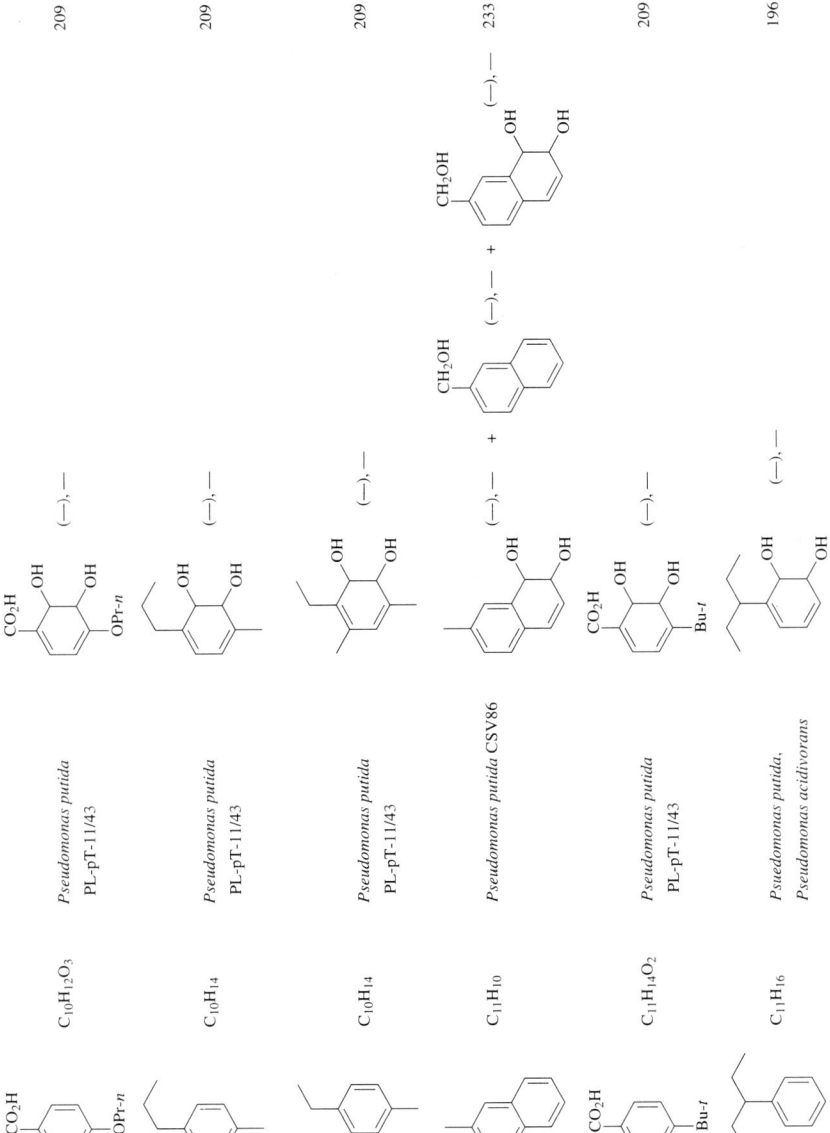

TABLE 10. DIHYDRODIOLS THAT HAVE NOT BEEN ISOLATED (*Continued*)

Substrate	Microorganism, Conditions	Product(s), Yield(s) (% or g/L) and Enantiomeric Excess %	Refs.
C_{12}			
$C_{12}H_5Cl_5$	*Burkholderia* sp. LB400	(—), —	52
$C_{12}H_5Cl_5$	*Escherichia coli* JM105(S4.43)	(—), —	280
$C_{12}H_6Cl_4$	*Burkholderia* sp. LB400	(—), —	52
$C_{12}H_6Cl_4$	*Escherichia coli* JM105(S4.43)	(—), —	280

C₁₂H₆Cl₄	*Escherichia coli* JM105(S4.43)	(—), —	280
C₁₂H₆Cl₄	*Escherichia coli* JM105(S4.43)	(—), —	280
C₁₂H₆Cl₄	*Escherichia coli* JM105(S4.43)	(—), —	280
C₁₂H₆Cl₄	*Escherichia coli* JM105(S4.43)	(—), —	280
	Burkholderia sp. LB400	(—), —	281

TABLE 10. DIHYDRODIOLS THAT HAVE NOT BEEN ISOLATED (*Continued*)

Substrate		Microorganism, Conditions	Product(s), Yield(s) (% or g/L) and Enantiomeric Excess %	Refs.
$C_{12}H_6Cl_4$		*Burkholderia* sp. LB400	(—), —	281
$C_{12}H_6Cl_4$		*Burkholderia* sp. LB400	(—), —	281
$C_{12}H_7Cl_3$		*Burkholderia* sp. LB400	(—), —	281
$C_{12}H_7Cl_3$		*Burkholderia* sp. LB400	(—), —	281

Substrate	Formula	Organism	Product	Yield, ee	Ref
2,4,4'-trichlorobiphenyl	$C_{12}H_7Cl_3$	*Burkholderia* sp. LB400	dihydrodiol	(—), —	52
2,3,4'-trichlorobiphenyl	$C_{12}H_7Cl_3$	*Burkholderia* sp. LB400	dihydrodiol	(—), —	52
2,4,3'-trichlorobiphenyl	$C_{12}H_7Cl_3$	*Escherichia coli* JM105(S4.43)	dihydrodiol	(—), —	280
2,5,3'-trichlorobiphenyl	$C_{12}H_7Cl_3$	*Burkholderia* sp. LB400	dihydrodiol	(—), —	52, 281
		Biphenyl 2,3-dioxygenase from *Pseudomonas* sp. LB400		(—), —	73

TABLE 10. DIHYDRODIOLS THAT HAVE NOT BEEN ISOLATED (*Continued*)

Substrate	Microorganism, Conditions	Product(s), Yield(s) (% or g/L) and Enantiomeric Excess %	Refs.
$C_{12}H_7Cl_3$ (2,4,3'-trichlorobiphenyl)	*Burkholderia* sp. LB400	(—), — + (—), —	52
$C_{12}H_7Cl_3$ (2,3,3'-trichlorobiphenyl)	*Burkholderia* sp. LB400	(—), —	52
$C_{12}H_7Cl_3$ (3,5,2'-trichlorobiphenyl)	*Burkholderia* sp. LB400	(—), —	52
$C_{12}H_7Cl_3$ (2,3,2'-trichlorobiphenyl)	*Burkholderia* sp. LB400	(—), —	52

244

Substrate	Formula	Organism	Product	Yield, ee	Ref
3,4,2'-trichlorobiphenyl	$C_{12}H_7Cl_3$	*Burkholderia* sp. LB400	3,4-dichloro-2'-(dihydrodiol)biphenyl	(—), —	52
2,4,2'-trichlorobiphenyl	$C_{12}H_7Cl_3$	*Burkholderia* sp. LB400	2,4-dichloro-2'-(dihydrodiol)biphenyl	(—), —	52
3-chlorodibenzofuran	$C_{12}H_7ClO$	*Escherichia coli* DH5α(pTCB144)	chlorodibenzofuran dihydrodiol	(—), —	77
1-chlorodibenzo-*p*-dioxin	$C_{12}H_7ClO_2$	*Beijerinckia* sp. B8/36	Dihydrodiol isolated, regiochemistry not determined		263
2-chlorodibenzo-*p*-dioxin	$C_{12}H_7ClO_2$	*Beijerinckia* sp. B8/36	Dihydrodiol isolated, regiochemistry not determined		263

TABLE 10. DIHYDRODIOLS THAT HAVE NOT BEEN ISOLATED (*Continued*)

Substrate	Microorganism, Conditions	Product(s), Yield(s) (% or g/L) and Enantiomeric Excess %	Refs.
$C_{12}H_8Cl_2$ (3,4'-dichlorobiphenyl)	*Burkholderia* sp. LB400	(—), — + (—), —	52
$C_{12}H_8Cl_2$ (3,3'-dichlorobiphenyl)	Biphenyl 2,3-dioxygenase from *Pseudomonas* sp. LB400	(—), — + (—), —	73
$C_{12}H_8Cl_2$	*Burkholderia* sp. LB400	"	52
$C_{12}H_8Cl_2$ (4,2'-dichlorobiphenyl)	*Burkholderia* sp. LB400	(—), —	52
$C_{12}H_8Cl_2$ (2,2'-dichlorobiphenyl)	*Burkholderia* sp. LB400	(—), — + (—), —	52
	Biphenyl 2,3-dioxygenase from *Pseudomonas* sp. LB400	"	73

$C_{12}H_8Cl_2$	*Burkholderia* sp. LB400	(—), —	**I** + **I** (—), — (—), —	52
	Biphenyl 2,3-dioxygenase from *Pseudomonas* sp. LB400	(—), —		73
$C_{12}H_8Cl_2$	*Burkholderia* sp. LB400	(—), —		281
$C_{12}H_8O$	*Pseudomonas* sp. HH69	(—), —		282
	Sphingomonas sp. RW1	(—), —		283
	Escherichia coli DH5α(pTCB144)	(—), —		77

TABLE 10. DIHYDRODIOLS THAT HAVE NOT BEEN ISOLATED (*Continued*)

Substrate	Microorganism, Conditions	Product(s), Yield(s) (% or g/L) and Enantiomeric Excess %	Refs.
$C_{12}H_8O_2$	*Pseudomonas* sp. HH69	(—), —	284
$C_{12}H_9Cl$	*Escherichia coli* DH5α(pTCB144)	(—), —	77
$C_{12}H_9Cl$	*Burkholderia* sp. LB400	(—), —	281
	Purified NDO, nap dox$_{G7}$, from *Pseudomonas putida* G7	(—), —	285
$C_{12}H_9Cl$	*Burkholderia* sp. LB400	(—), —	281
	Purified NDO, nap dox$_{G7}$, from *Pseudomonas putida* G7	(—), —	285
	Pseudomonas pseudoalcaligenes KF707B1	(—), —	286
$C_{12}H_9Cl$	Purified NDO, nap dox$_{G7}$, from *Pseudomonas putida* G7	(—), —	285

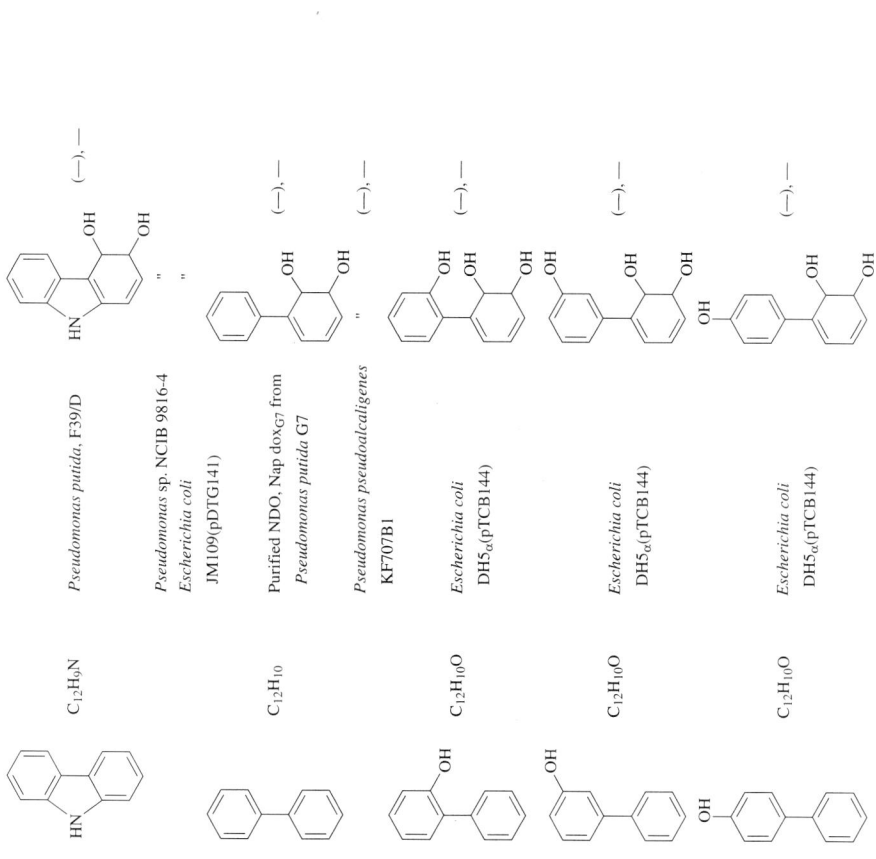

TABLE 10. DIHYDRODIOLS THAT HAVE NOT BEEN ISOLATED (*Continued*)

Substrate		Microorganism, Conditions	Product(s), Yield(s) (% or g/L) and Enantiomeric Excess %	Refs.
C$_{12}$H$_{10}$O		*Escherichia coli* DH5α(pTCB144)	(—), —	77
C$_{12}$H$_{11}$N		*Escherichia coli* DH5α(pTCB144)	(—), —	77
C$_{13}$				
C$_{13}$H$_7$FO		*Pseudomonas* sp. F274	(—), —	288
C$_{13}$H$_9$Br		*Pseudomonas* sp. F274	(—), —	288

![CN-biphenyl]	$C_{13}H_9N$	![CN-cyclohexadiene-diol] *Escherichia coli* DH5α(pTCB144)	(—), —	77
![biphenyl-CO2H]	$C_{13}H_{10}O_2$	![CO2H-cyclohexadiene-diol] *Pseudomonas putida* PL-pT-11/43	(—), —	209
![diphenylmethane]	$C_{13}H_{12}$	![benzyl-cyclohexadiene-diol] *Pseudomonas pseudoalcaligenes* KF707B1	(—), —	286
![benzhydrol]	$C_{13}H_{12}O$	![phenyl-hydroxymethyl-cyclohexadiene-diol] *Escherichia coli* DH5α(pTCB144) *Pseudomonas pseudoalcaligenes* KF707B1	(—), — (—), —	77 286
![bibenzyl] C_{14}	$C_{14}H_{14}$	![phenethyl-catechol] *Pseudomonas pseudoalcaligenes* KF707B1	(—), —	286

TABLE 10. DIHYDRODIOLS THAT HAVE NOT BEEN ISOLATED (Continued)

Substrate		Microorganism, Conditions	Product(s), Yield(s) (% or g/L) and Enantiomeric Excess %	Refs.
C_{14}				
	$C_{14}H_8O_3$	Pseudomonas sp. F274	(—), —	288
	$C_{14}H_{12}$	Pseudomonas sp. F274	(—), —	288
	$C_{14}H_{12}O_2$	Pseudomonas putida PL-pT-11/43	(—), —	209
C_{15}				
	$C_{15}H_{10}O_2$	Mycobacterium sp. KR2	(—), —	241

TABLE 11. CHEMICAL TRANSFORMATIONS OF DIHYDRODIOLS

Substrate		Conditions	Product(s), Yield(s) (% or g/L) and Enantiomeric Excess %		Refs.
C_6					
(iodo-fluoro dihydrodiol)	$C_6H_6FIO_2$	H_2, Pd/C, MeOH	(fluoro diol)	(75), >98	27
(iodo-fluoro dihydrodiol)	$C_6H_6FIO_2$	H_2, Pd/C, MeOH	(fluoro diol)	(75), >98	27
(iodo-fluoro dihydrodiol)	$C_6H_6FIO_2$	H_2, Pd/C, MeOH	(fluoro diol)	(70), 88	27
		1. H_2, Pd/C 2. *P. putida* NCIMB 8859	"	(—), >98	108
(iodo-chloro dihydrodiol)	$C_6H_6ClIO_2$	H_2, Pd/C, MeOH	(chloro diol)	(67), >98	27
(iodo-chloro dihydrodiol)	$C_6H_6ClIO_2$	1. H_2, Pd/C 2. *P. putida* NCIMB 8859	(chloro diol)	(—), >98	108
(iodo-bromo dihydrodiol)	$C_6H_6BrIO_2$	H_2, Pd/C, MeOH	(bromo diol)	(65), >98	27

TABLE 11. CHEMICAL TRANSFORMATIONS OF DIHYDRODIOLS (*Continued*)

Substrate	Conditions	Product(s), Yield(s) (% or g/L) and Enantiomeric Excess %	Refs.
C$_6$H$_6$BrIO$_2$	H$_2$, Pd/C, MeOH	(80), >98	27
C$_6$H$_6$BrIO$_2$	1. H$_2$, Pd/C 2. *P. putida* NCIMB 8859	(—), >98	108
C$_6$H$_7$ClO$_2$	CH$_3$C(OCH$_3$)$_2$CH$_3$, acetone, *p*-TsOH	(—), >98	289
C$_6$H$_7$BrO$_2$	CH$_3$C(OCH$_3$)$_2$CH$_3$, CH$_2$Cl$_2$, *p*-TsOH	(100), >98	290
C$_6$H$_7$BrO$_2$	HC≡CSiMe$_3$, Pd(PPh$_3$)$_4$, CuI, *n*-BuNH$_2$	(78), >98	105

TABLE 11. CHEMICAL TRANSFORMATIONS OF DIHYDRODIOLS (*Continued*)

Substrate	Conditions	Product(s), Yield(s) (% or g/L) and Enantiomeric Excess %	Refs.
$C_6H_7IO_2$	$Bu_3SnC≡CH$, $Pd(OAc)_2$, PPh_3, THF	(35), >98	25
	$HC≡CSiMe_3$, $Pd(OAc)_2$, Ph_3P, Et_3N	(39), >98	105
	Bu_3SnOMe, $Pd(PPh_3)_4$, THF	(11), >98	105, 25
	$Bu_3SnCH_2CH=CH_2$, $Pd(PPh_3)_4$, THF	(31), >98	105, 25
	Bu_3SnSCH_3, $Pd(PPh_3)_4$, THF	(55), >98	25
	Bu_3SnSEt, $Pd(PPh_3)_4$, THF	(61), >98	25

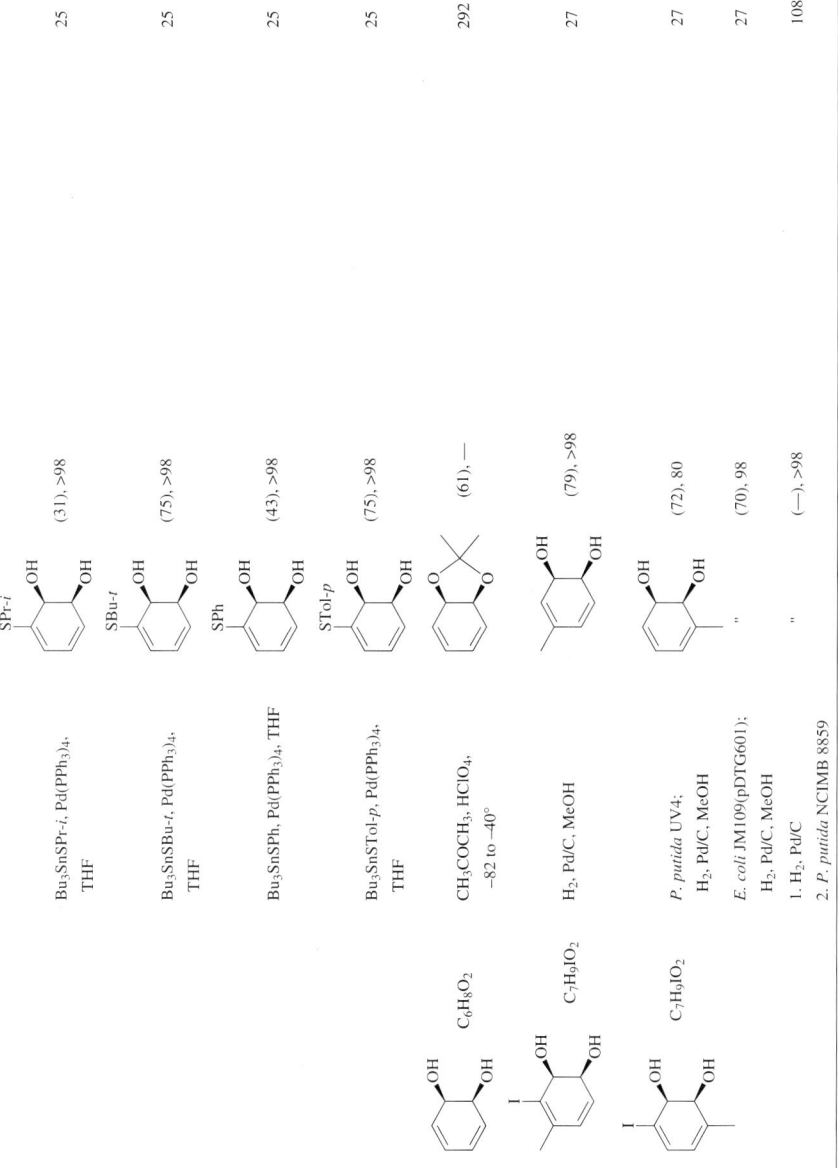

REFERENCES

[1] Walker, N.; Wiltshire, G. H. *J. Gen. Microbiol.* **1953**, *8*, 273.
[2] Gibson, D. T.; Koch, J. R.; Kallio, R. E. *Biochemistry* **1968**, *7*, 2653.
[3] Gibson, D. T. *CRC Crit. Rev. Microbiol.* **1971**, *1*, 199.
[4] Gibson, D. T.; Cardini, G. E.; Maseles, F. C.; Kallio, R. E. *Biochemistry* **1970**, *9*, 1631.
[5] Gibson, D. T.; Subramanian, V. In *Microbial Degradation of Organic Compounds*; Gibson, D. T., Ed.; Marcel Dekker: New York, 1984; pp 181–252.
[6] Reineke, W. In *Microbial Degradation of Organic Compounds*; Gibson, D. T., Ed.; Marcel Dekker: New York, 1984; pp 319–360.
[7] Safe, S. H. In *Microbial Degradation of Organic Compounds*, Gibson, D. T., Ed.; Marcel Dekker: New York, 1984; pp 361–369.
[8] Ribbons, D. W.; Keyser, P.; Eaton, R. W.; Anderson, B. N.; Kunz, D. A.; Taylor, B. F. In *Microbial Degradation of Organic Compounds*; Gibson, D. T., Ed.; Marcel Dekker: New York, 1984; pp 371–397.
[9] Sutherland, J. B.; Rafii, F.; Khan, A. A.; Cerniglia, C. E. In *Microbial Transformation and Degradation of Toxic Organic Chemicals*; Young, L. Y.; Cerniglia, C. E., Eds.; Wiley-Liss: New York, 1995; pp 269–306.
[10] Resnick, S. M.; Lee. K.; Gibson, D. T. *J. Indust. Microbiol.* **1996**, *17*, 438.
[11] Gibson, D.T. *J. Indust. Microbiol. Biotech.* **1999**, *23*, 284.
[12] Butler, C. S.; Mason, J. R. *Adv. Microbial Physiol.* **1997**, *38*, 47.
[13] Solomon, E. I.; Brunold, T. C.; Davis, M. I.; Kemsley, J. N.; Lee, S.-K.; Lehnert, N.; Neese, F.; Skulan, A. J.; Yang, Y.-S.; Zhou, J. *Chem. Rev.* **2000**, *100*, 235.
[14] Sheldrake, G. N. In *Chirality in Industry*; Collins, A. N., Sheldrake, G. N., Crosby, J., Eds.; Wiley: New York, 1992; pp 127–166.
[15] Hudlicky, T.; Gonzalez, D.; Gibson, D. T. *Aldrichim. Acta* **1999**, *32*, 35.
[16] Boyd, D. R.; Sharma, N. D.; Allen, C. C. R. *Curr. Opin. Biotech.* **2001**, *12*, 564.
[17] Gibson, D. T. *J. Indust. Microbiol. Biotech.* **1997**, *19*, 312.
[18] Kauppi, B.; Lee, K.; Carredano, E.; Parales, R. E.; Gibson, D. T.; Eklund, H.; Ramaswamy, S. *Structure* **1998**, *6*, 571.
[19] Karlsson, A.; Parales, J. V.; Parales, R. E.; Gibson, D. T.; Eklund, H.; Ramaswamy, S. *Science* **2003**, *299*, 1039.
[20] Lee, K.; Kauppi, B.; Parales, R. E.; Gibson, D. T.; Ramaswamy, S. *Biochem. Biophys. Res. Commun.* **1997**, *241*, 553.
[21] Carredano, E.; Karlsson, A.; Kauppi, B.; Choudhury, D.; Parales, R. E.; Parales, J. V.; Lee, K.; Gibson, D. T.; Eklund, H.; Ramaswamy, S. *J. Mol. Biol.* **2000**, *296*, 701.
[22] Berry, A.; Dodge, T. C.; Pepsin, M.; Weyler, W. *J. Indust. Microbiol. Biotech.* **2002**, *28*, 127.
[23] Boyd, D. R.; Sharma, N. D.; Hand, M. V.; Groocock, M. R.; Kerley, N. A.; Dalton, H.; Chima, J.; Sheldrake, G. N. *J. Chem. Soc., Chem. Commun.* **1993**, 974.
[24] Boyd, D. R.; Dorrity, M. R. J.; Hand, M. V.; Malone, J. F.; Sharma, N. D.; Dalton, H.; Gray, D. J.; Sheldrake, G. N. *J. Am. Chem. Soc.* **1991**, *113*, 666.
[25] Boyd, D. R.; Sharma, N. D.; Byrne, B.; Hand, M. V.; Malone, J. F.; Sheldrake, G. N.; Blacker, J.; Dalton, H. *J. Chem. Soc., Perkin Trans. 1* **1998**, 1935.
[26] Whited, G. M.; Downie, J. C.; Hudlicky, T.; Fearnley, S. P.; Dudding, T. C.; Olivo, H. F.; Parker, D. *Bioorg. Med. Chem.* **1994**, *2*, 727.
[27] Boyd, D. R.; Sharma, N. D.; Barr, S. A.; Dalton, H.; Chima, J.; Whited, G.; Seemayer, R. *J. Am. Chem. Soc.* **1994**, *116*, 1147.
[28] Jerina, D. M.; Selander, H.; Yagi, H.; Wells, M. C.; Davey, J. F.; Mahadevan, V.; Gibson, D. T. *J. Am. Chem. Soc.* **1976**, *98*, 5988.
[29] Akhtar, M. N.; Boyd, D. R.; Thompson, N. J.; Koreeda, M.; Gibson, D. T.; Mahadevan, V.; Jerina, D. M. *J. Chem. Soc., Perkin Trans. 1* **1975**, 2506.
[30] Koreeda, M.; Akhtar, M. N.; Boyd, D. R.; Neill, J. D.; Gibson, D. T.; Jerina, D. M. *J. Org. Chem.* **1978**, *43*, 1023.
[31] Parales, R. E.; Resnick, S. M.; Yu, C.-L.; Boyd, D. R.; Sharma, N. D.; Gibson, D. T. *J. Bacteriol.* **2000**, *182*, 5495.
[32] Boyd, D. R.; Sharma, N. D.; Carroll, J. G.; Allen, C. C. R.; Clarke, D. A.; Gibson, D. T. *Chem. Commun.* **1999**, 1201.

[33] Jenkins, G. N.; Ribbons, D. W.; Widdowson, D. A.; Slawin, A. M. Z.; Williams, D. J. *J. Chem. Soc., Perkin Trans. 1* **1995**, 2647.
[34] Myers, A. G.; Siegel, D. R.; Buzard, D. J.; Charest, M. G. *Org. Lett.* **2001**, *3*, 2923. See supporting information for extensive details of experimental conditions.
[35] Reiner, A. M.; Hegeman, G. D. *Biochemistry* **1971**, *10*, 2530.
[36] Reineke, W.; Otting, W.; Knackmuss, H.-J. *Tetrahedron* **1978**, *34*, 1707.
[37] Resnick, S. M.; Torok, D. S.; Gibson, D. T. *J. Org. Chem.* **1995**, *60*, 3546.
[38] Stephenson, G. R.; Howard, P. W. *J. Chem. Soc., Perkin Trans. 1* **1994**, 2873.
[39] Gibson, D. T.; Hensley, M.; Yoshioka, H.; Mabry, T. J. *Biochemistry* **1970**, *9*, 1626.
[40] Hudlicky, T.; Stabile, M. R.; Gibson, D. T.; Whited, G. M. *Org. Synth.* **1999**, *76*, 77.
[41] Zylstra, G. J.; Gibson, D. T. *J. Biol. Chem.* **1989**, *264*, 14940.
[42] Ballard, D. G. H.; Courtis, A.; Shirley, I. M.; Taylor, S. C. *Macromolecules* **1988**, *21*, 294.
[43] Heald, S. C.; Jenkins, R. O. *Appl. Microbiol. Biotechnol.* **1996**, *45*, 56.
[44] Boyd, D. R.; Sharma, N. D.; Haughey, S. A.; Kennedy, M. A.; McMurray, B. T.; Sheldrake, G. N.; Allen, C. C. R.; Dalton, H.; Sproule, K. *J. Chem. Soc., Perkin Trans. 1* **1998**, 1929.
[45] Gibson, D. T.; Resnick, S. M.; Lee, K.; Brand, J. M.; Torok, D. S.; Wackett, L. P.; Schocken, M. J.; Haigler, B. E. *J. Bacteriol.* **1995**, *177*, 2615.
[46] Klěcka, G. M.; Gibson, D. T. *Biochem. J.* **1979**, *180*, 639.
[47] Bestetti, G.; Bianchi, D.; Bosetti, A.; Di Gennaro, P.; Galli, E.; Leoni, B.; Pelizzoni, F.; Sello, G. *Appl. Microbiol. Biotechnol.* **1995**, *44*, 306.
[48] Di Gennaro, P.; Sello, G.; Bianchi, D.; D'Amico, P. *J. Biol. Chem.* **1997**, *272*, 30254.
[49] Gibson, D. T.; Roberts, R. L.; Wells, M. C.; Kobal, V. M. *Biochem. Biophys. Res. Commun.* **1973**, *50*, 211.
[50] Mondello, F. J. *J. Bacteriol.* **1989**, *171*, 1725.
[51] Haddock, J. D.; Nadim, L. M.; Gibson, D. T. *J. Bacteriol.* **1993**, *175*, 395.
[52] Seeger, M.; Zielinski, M.; Timmis, K. N.; Hofer, B. *Appl. Environ. Microbiol.* **1999**, *65*, 3614.
[53] Rossiter, J. T.; Williams, S. R.; Cass, A. E. G.; Ribbons, D. W. *Tetrahedron Lett.* **1987**, *28*, 5173.
[54] Gibson, D. T.; Koch, J. R.; Schuld, C. L.; Kallio, R. E. *Biochemistry* **1968**, *7*, 3795.
[55] Boyd, D. R.; Sharma, N. D.; Bowers, N. I.; Duffy, J.; Harrison, J. S.; Dalton, H. *J. Chem. Soc., Perkin Trans. 1* **2000**, 1345.
[56] Ballard, D. G. H.; Blacker, A. J.; Woodley, J. M.; Taylor, S. C. In *Plastics from Microbes*; Mobley, D. P., Ed.; Hanser: Munich, 1994; p 139.
[57] D. R. Boyd, School of Chemistry, Queen's University of Belfast, Belfast BT9 5AG, U.K. personal communication.
[58] Allen, C. C. R.; Boyd, D. R.; Dalton, H.; Sharma, N. D.; Haughey, S. A.; McMordie, R. A. S.; McMurray, B.T.; Sheldrake, G. N.; Sproule, K. *J. Chem. Soc., Chem. Commun.* **1995**, 119.
[59] Quintana, M. G.; Dalton, H. *Enzyme Microb. Technol.* **1999**, *24*, 232.
[60] Carragher, J. M.; McClean, W. S.; Woodley, J. M.; Hack, C. J. *Enzyme Microb. Technol.* **2001**, *28*, 183.
[61] Blacker, A. J.; Booth, R. J.; Davies, G. M.; Sutherland, J. K. *J. Chem. Soc., Perkin Trans. 1* **1995**, 2861.
[62] Jerina, D. M.; Daly, J. W.; Jeffrey, A. M.; Gibson, D. T. *Arch. Biochem. Biophys.* **1971**, *142*, 394.
[63] Jeffrey, A. M.; Yeh, H. J. C.; Jerina, D. M.; Patel, T. R.; Davey, J. F.; Gibson, D. T. *Biochemistry* **1975**, *14*, 575.
[64] Suen, W.-C.; Gibson, D. T. *Gene* **1994**, *143*, 67.
[65] Suen, W.-C. Ph.D. Dissertation, University of Iowa, Iowa City, 1991; Univ. Microfilms Int., Order No. 9137003.
[66] Resnick, S. M.; Gibson, D. T. *Appl. Environ. Microbiol.* **1996**, *62*, 3355.
[67] Resnick, S. M.; Gibson, D. T. *Appl. Environ. Microbiol.* **1996**, *62*, 4073.
[68] Khan, A. A.; Wang, R.-F.; Cao, W.-W.; Franklin, W.; Cerniglia, C. E. *Int. J. Syst. Bacteriol.* **1996**, *46*, 466.
[69] Boyd, D. R.; Sharma, N. D.; Evans, T. A.; Groocock, M.; Malone, J. F.; Stevenson, P. J.; Dalton, H. *J. Chem. Soc., Perkin Trans. 1* **1997**, 1879.
[70] Boyd, D. R.; Sharma, N. D.; Stevenson, P. J.; Chima, J.; Gray, D. J.; Dalton, H. *Tetrahedron Lett.* **1991**, *32*, 3887.
[71] Bopp, L. H. *J. Ind. Microbiol.* **1986**, *1*, 23.
[72] Viallard, V.; Poirier, I.; Cournoyer, B.; Haurat, J.; Wiebkin, S.; Ophel-Keller, K.; Balandreau, J. *Int. J. Syst. Bacteriol.* **1998**, *48*, 549.
[73] Haddock, J. D.; Horton, J. R.; Gibson, D. T. *J. Bacteriol.* **1995**, *177*, 20.
[74] Reineke, W.; Knackmuss, H.-J. *Biochim. Biophys. Acta* **1978**, *542*, 412.

[75] Cass, A. E. G.; Ribbons, D. W.; Rossiter, J. T.; Williams, S. R. *FEBS Lett.* **1987**, *220*, 353.
[76] Knackmuss, H.-J.; Beckmann, W.; Otting, W. *Angew. Chem., Int. Ed. Engl.* **1976**, *15*, 549.
[77] Raschke, H.; Meier, M.; Burken, J. G.; Hany, R.; Müller, M. D.; van der Meer, J. R.; Kohler, H.-P. E. *Appl. Environ. Microbiol.* **2001**, *67*, 3333.
[78] Chun, H.-K.; Ohnishi, Y.; Misawa, N.; Shindo, K.; Hayashi, M.; Harayama, S.; Horinouchi, S. *Biosci. Biotechnol. Biochem.* **2001**, *65*, 1774.
[79] Hudlicky, T.; Boros, E. E.; Boros, C. H. *Tetrahedron: Asymmetry* **1993**, *4*, 1365.
[80] Hudlicky, T.; Luna, H.; Barbieri, G.; Kwart, L. D. *J. Am. Chem. Soc.* **1988**, *110*, 4735.
[81] Williams, M. G.; Olson, P. E.; Tautvydas, K. J.; Bitner, R. M.; Mader, R. A.; Wackett, L. P. *Appl. Microbiol. Biotechnol.* **1990**, *34*, 316.
[82] Boyd, D. R.; McMordie, R. A. S.; Sharma, N. D.; Dalton, H.; Williams, P.; Jenkins, R. O. *J. Chem. Soc., Chem. Commun.* **1989**, 339.
[83] Reddy, J.; Lee, C.; Neeper, M.; Greasham, R.; Zhang, J. *Appl. Microbiol. Biotechnol.* **1999**, *51*, 614.
[84] Boyd, D. R.; Clarke, D.; Cleij, M. C.; Hamilton, J. T. G.; Sheldrake, G. N. *Monatsh. Chem.* **2000**, *131*, 673.
[85] Eaton, R. W.; Selifonov, S. A. *Appl. Environ. Microbiol.* **1996**, *62*, 756.
[86] Bowers, N. I.; Boyd, D. R.; Sharma, N. D.; Kennedy, M. A.; Sheldrake, G. N.; Dalton, H. *Tetrahedron: Asymmetry* **1998**, *9*, 1831.
[87] Bowers, N. I.; Boyd, D. R.; Sharma, N. D.; Goodrich, P. A.; Groocock, M. R.; Blacker, A. J.; Goode, P.; Dalton, H. *J. Chem. Soc., Perkin Trans. 1* **1999**, 1453.
[88] Lee, K.; Brand, J. M.; Gibson, D.T. *Biochem. Biophys. Res. Commun.* **1995**, *212*, 9.
[89] Boyd, D. R.; Sharma, N. D.; Haughey, S. A.; Malone, J. F.; McMurray, B. T.; Sheldrake, G. N.; Allen, C. C. R.; Dalton, H. *Chem. Commun.* **1996**, 2363.
[90] Parales, R. E.; Lee, K.; Resnick, S. M.; Jiang, H.; Lessner, D. J.; Gibson, D. T. *J. Bacteriol.* **2000**, *182*, 1641.
[91] Yu, C.-L.; Parales, R. E.; Gibson, D. T. *J. Indust. Microbiol. Biotech.* **2001**, *27*, 94.
[92] Suenaga, H.; Mitsuoka, M.; Ura, Y.; Watanabe, T.; Furukawa, K. *J. Bacteriol.* **2001**, *183*, 5441.
[93] Sakamoto, T.; Joern, J. M.; Arisawa, A.; Arnold, F. H. *Appl. Environ. Microbiol.* **2001**, *67*, 3882.
[94] Martin, V. J. J.; Mohn, W. W. *J. Bacteriol.* **1999**, *181*, 2675.
[95] Senanayake, C. H.; DiMichele, L. M.; Liu, J.; Fredenburgh, L. E.; Ryan, K. M.; Roberts, F. E.; Larsen, R. D.; Verhoeven, T. R.; Reider, P. J. *Tetrahedron Lett.* **1995**, *36*, 7615.
[96] Carless, H. A. J. *Tetrahedron: Asymmetry* **1992**, *3*, 795.
[97] Boyd, D. R.; Sheldrake, G. N. *Nat. Prod. Rep.* **1998**, 309.
[98] Brown, S. M.; Hudlicky, T. In *Organic Synthesis: Theory and Applications*; Hudlicky, T., Ed.; JAI: Greenwich, CT, 1993; Vol. 2, pp 113–177.
[99] Hudlicky, T. In *Green Chemistry. Designing Chemistry for the Environment*; Anastas, P. T., Williamson, T. C., Eds.; ACS Symposium Series 626, American Chemical Society: Washington, DC, 1996; pp 180-197.
[100] Hudlicky, T. In *Green Chemistry. Frontiers in Benign Chemical Syntheses and Processes*; Anastas, P. T., Williamson, T. C., Eds.; Oxford University Press: Oxford, U.K., 1998; pp 166–177.
[101] Hudlicky, T.; Abboud, K. A.; Entwistle, D. A.; Fan, R.; Maurya, R.; Thorpe, A. J.; Bolonick, J.; Myers, B. *Synthesis* **1996**, 897.
[102] Hudlicky, T.; Seoane, G.; Price, J. D.; Gadamasetti, K. G. *Synlett* **1990**, 433.
[103] Hudlicky, T. *Chem. Rev.* **1996**, *96*, 3.
[104] Hudlicky, T.; Thorpe, A. J. *Chem. Commun.* **1996**, 1993.
[105] Boyd, D. R.; Hand, M. V.; Sharma, N. D.; Chima, J.; Dalton, H.; Sheldrake, G. N. *J. Chem. Soc., Chem. Commun.* **1991**, 1630.
[106] Hudlicky, T.; Boros, E. E. *Tetrahedron: Asymmetry* **1992**, *3*, 217.
[107] Akgün, H.; Hudlicky, T. *Tetrahedron Lett.* **1999**, *40*, 3081.
[108] Allen, C. C. R.; Boyd, D. R.; Dalton, H.; Sharma, N. D.; Brannigan, I.; Kerley, N. A.; Sheldrake, G. N.; Taylor, S. C. *J. Chem. Soc., Chem. Commun.* **1995**, 117.
[109] Ballard, D. G. H.; Courtis, A.; Shirley, I. M.; Taylor, S. C. *J. Chem. Soc., Chem. Commun.* **1983**, 954.
[110] Gin, D. L.; Conticello, V. P.; Grubbs, R. H. *J. Am. Chem. Soc.* **1994**, *116*, 10507.
[111] Wagaman, M. W.; Grubbs, R. H. *Macromolecules* **1997**, *30*, 3978.
[112] Ley, S. V.; Sternfeld, F.; Taylor, S. *Tetrahedron Lett.* **1987**, *28*, 225.
[113] Banwell, M. G.; Corbett, M.; Mackay, M. F.; Richards, S. L. *J. Chem. Soc., Perkin Trans. 1* **1992**, 1329.
[114] Ley, S. V.; Parra, M.; Redgrave, A. J.; Sternfeld, F. *Tetrahedron* **1990**, *46*, 4995.

[115] Carless, H. A. J.; Oak, O. Z. *Tetrahedron Lett.* **1989**, *30*, 1719.
[116] Noh, T.; Gan, H.; Halfon, S.; Hrnjez, B. J.; Yang, N. C. *J. Am. Chem. Soc.* **1997**, *119*, 7470.
[117] Pu, L.; Grubbs, R. H. *J. Org. Chem.* **1994**, *59*, 1351.
[118] Boyd, D. R.; Blacker, J.; Byrne, B.; Dalton, H.; Hand, M. V.; Kelly, S. C.; More O'Ferrall, R. A.; Rao, S. N.; Sharma, N. D.; Sheldrake, G. N. *J. Chem. Soc., Chem. Commun.* **1994**, 313.
[119] Carless, H. A. J.; Malik, S. S. *J. Chem. Soc., Chem. Commun.* **1995**, 2447.
[120] Banwell, M. G.; Dupuche, J. R. *Chem. Commun.* **1996**, 869.
[120a] Banwell, M. G.; Darmos, P.; McLeod, M. D.; Hockless, D. C. R. *Synlett* **1998**, 897.
[121] Banwell, M.; Blakey, S.; Harfoot, G.; Longmore, R. *J. Chem. Soc., Perkin Trans. 1* **1998**, 3141.
[122] Oppolzer, W.; Spivey, A. C.; Bochet, C. G. *J. Am. Chem. Soc.* **1994**, *116*, 3139.
[123] Tian, X.; Hudlicky, T.; Königsberger, K. *J. Am. Chem. Soc.* **1995**, *117*, 3643.
[124] Tian, X.; Maurya, R.; Königsberger, K.; Hudlicky, T. *Synlett* **1995**, 1125.
[125] Banwell, M. G., Forman, G. S. *J. Chem. Soc., Perkin Trans. 1* **1996**, 2565.
[126] Butora, G.; Hudlicky, T.; Fearnley, S. P.; Gum, A. G.; Stabile, M. R.; Abboud, K. *Tetrahedron Lett.* **1996**, *37*, 8155.
[127] Butora, G.; Hudlicky, T.; Fearnley, S. P.; Stabile, M. R.; Gum, A. G.; Gonzalez, D. *Synthesis* **1998**, 665.
[128] Butora, G.; Gum, A. G.; Hudlicky, T.; Abboud, K. A. *Synthesis* **1998**, 275.
[129] Endoma, M. A.; Butora, G.; Claeboe, C. D.; Hudlicky, T.; Abboud, K. A. *Tetrahedron Lett.* **1997**, *38*, 8833.
[130] Gonzalez, D.; Schapiro, V.; Seoane, G.; Hudlicky, T.; Abboud, K. *J. Org. Chem.* **1997**, *62*, 1194.
[131] Gonzalez, D.; Martinot, T.; Hudlicky, T. *Tetrahedron Lett.* **1999**, *40*, 3077.
[132] Hudlicky, T.; Seoane, G.; Pettus, T. *J. Org. Chem.* **1989**, *54*, 4239.
[133] Desjardins, M.; Lallemand, M.-C.; Hudlicky, T.; Abboud, K. A. *Synlett* **1997**, 728.
[134] Wendeborn, S.; De Mesmaeker, A.; Brill, W. K.-D. *Synlett* **1998**, 865.
[135] Hoffmann, R. *Angew. Chem., Int. Ed. Engl.* **2001**, *40*, 3337.
[136] Berteina, S.; De Mesmaeker, A.; Wendeborn, S. *Synlett* **1999**, 1121.
[137] Nakajima, M.; Tomida, I.; Takei, S. *Chem. Ber.* **1959**, *92*, 163.
[138] Yang, N. C.; Chen, M.-J.; Chen, P.; Mak, K. T. *J. Am. Chem. Soc.* **1982**, *104*, 853.
[139] Motherwell, W. B.; Williams, A. S. *Angew. Chem., Int. Ed. Engl.* **1995**, *34*, 2031.
[140] Whited, G. M., Genencor International, Palo Alto, CA 94304, personal communication.
[141] Gibson, D. T.; Mahadevan, V.; Davey, J. F. *J. Bacteriol.* **1974**, *119*, 930.
[142] Resnick, S. M.; Gibson, D. T. *Biodegradation* **1993**, *4*, 195.
[143] Whited, G. M.; McCombie, W. R.; Kwart, L. D.; Gibson, D. T. *J. Bacteriol.* **1986**, *166*, 1028.
[144] Boyd, D. R.; Sharma, N. D.; Dorrity, M. R. J.; Hand, M. V.; McMordie, R. A. S.; Malone, J. F.; Porter, H. P.; Dalton, H.; Chima, J.; Sheldrake, G. N. *J. Chem. Soc., Perkin Trans. 1* **1993**, 1065.
[145] Fonken, G. S.; Johnson, R. A. *Chemical Oxidations with Microorganisms*; Marcel Dekker: New York, 1972; pp 243–255.
[146] Stanier, R. Y.; Palleroni, N. J.; Doudoroff, M. *J. Gen. Microbiol.* **1966**, *43*, 159.
[147] Cohen-Bazire, G.; Sistrom, W. R.; Stanier, R. Y. *J. Cell. Comp. Physiol.* **1957**, *49*, 25.
[148] Vishniac, W.; Santer, M. *Bacteriol. Rev.* **1957**, *21*, 195.
[149] Ribbons, D. W. In *Biocatalysts for Fine Chemical Synthesis*; Roberts, S. M., Ed.; Wiley: Chichester, U.K., 1999; pp 3:5.1–3:5.15.
[150] Jenkins, R. O.; Dalton, H. *FEMS Microbiol. Lett.* **1985**, *30*, 227.
[151] Sambrook, J.; Maniatis, T.; Fritch, E. F. In *Molecular Cloning: A Laboratory Manual*, 2nd ed.; Cold Spring Harbor Laboratory: Cold Spring Harbor, NY, 1989; Vol. 3, p A.3.
[152] Boyd, D. R.; Dorrity, M. R. J.; Malone, J. F.; McMordie, R. A. S.; Sharma, N. D.; Dalton, H.; Williams, P. *J. Chem. Soc., Perkin Trans. 1* **1990**, 489.
[153] Ziffer, H.; Kabuto, K.; Gibson, D. T.; Kobal, V. M.; Jerina, D. M. *Tetrahedron* **1977**, *33*, 2491.
[154] Hudlicky, T.; Gonzalez, D.; Stabile, M.; Endoma, M. A. A.; Deluca, M.; Parker, D.; Gibson, D. T.; Resnick, S. M.; Whited, G. M. *J. Fluorine Chem.* **1998**, *89*, 23.
[155] Sander, P.; Wittich, R.-M.; Fortnagel, P.; Wilkes, H.; Francke, W. *Appl. Environ. Microbiol.* **1991**, *57*, 1430.
[156] Spiess, E.; Sommer, C.; Görisch, H. *Appl. Environ. Microbiol.* **1995**, *61*, 3884.
[157] Spain, J. C.; Gibson, D. T. *Appl. Environ. Microbiol.* **1988**, *54*, 1399.
[158] Spain, J. C.; Nishino, S. F. *Appl. Environ. Microbiol.* **1987**, *53*, 1010.
[159] Haigler, B. E.; Spain, J. C. *Appl. Environ. Microbiol.* **1991**, *57*, 3156.
[160] Jung, K.-H.; Lee, J.-Y.; Kim, H.-S. *Biotechnol. Bioeng.* **1995**, *48*, 625.

[161] Schofield, J. A.; Betteridge, P. R.; Ryback, G.; Geary, P. J. U.S. Patent 4,876,200 (1989); *Chem. Abstr.* **1988**, *108*, 203317e.
[162] Reineke, W.; Knackmuss, H.-J. *Appl. Environ. Microbiol.* **1984**, *47*, 395.
[163] Taylor, S. C. U.S. Patent 4,508,822 (1985); *Chem. Abstr.* **1983**, *99*, 54282f.
[164] Barr, S. A.; Bowers, N.; Boyd, D. R.; Sharma, N. D.; Hamilton, L.; Austin, R.; McMordie, S.; Dalton, H. *J. Chem. Soc., Perkin Trans. 1* **1998**, 3443.
[165] Högn, T.; Jaenicke, L. *Eur. J. Biochem.* **1972**, *30*, 369.
[166] Schofield, J. A. U.S. Patent 4,863,861 (1989); *Chem. Abstr.* **1988**, *108*, 203317e.
[167] Taylor, S. C. U.S. Patent 4,740,638 (1988); *Chem. Abstr.* **1988**, *108*, 148916x.
[168] Pollmann, K.; Beil, S.; Pieper, D. H. *Appl. Environ. Microbiol.* **2001**, *67*, 4057.
[169] Haigler, B. E.; Spain, J. C. *Appl. Environ. Microbiol.* **1989**, *55*, 372.
[170] Ziffer, H.; Jerina, D. M.; Gibson, D. T.; Kobal, V. M. *J. Am. Chem. Soc.* **1973**, *95*, 4048.
[171] Kobal, V. M.; Gibson, D. T.; Davis, R. E.; Garza, A. *J. Am. Chem. Soc.* **1973**, *95*, 4420.
[172] Lessner, D. J.; Johnson, G. R.; Parales, R. E.; Spain, J. C.; Gibson, D. T. *Appl. Environ. Microbiol.* **2002**, *68*, 634.
[173] Jenkins, R. O.; Stephens, G. M.; Dalton, H. *Biotechnol. Bioeng.* **1987**, *29*, 873.
[174] Wahbi, L. P.; Phumathon, P.; Brown, A.; Minter, S.; Stephens, G. M. *Biotechnol. Lett.* **1997**, *19*, 961.
[175] Tsai, J. T.; Wahbi, L. P.; Dervakos, G. A.; Stephens, G. M. *Biotechnol. Lett.* **1996**, *18*, 241.
[176] Vanderberg, L. A.; Krieger-Grumbine, R.; Taylor, M. N. *Appl. Microbiol. Biotechnol.* **2000**, *53*, 447.
[177] Hudlicky, T.; Boros, E. E.; Boros, C. H. *Synlett* **1992**, 391.
[178] Königsberger, K.; Hudlicky, T. *Tetrahedron: Asymmetry* **1993**, *4*, 2469.
[179] Novak, B. H.; Hudlicky, T. *Tetrahedron: Asymmetry* **1999**, *10*, 2067.
[180] Warhurst, A. M.; Clarke, K. F.; Hill, R. A.; Holt, R. A.; Fewson, C. A. *Appl. Environ. Microbiol.* **1994**, *60*, 1137.
[181] Bestetti, G.; Galli, E.; Benigni, C.; Orsini, F.; Pelizzoni, F. *Appl. Microbiol. Biotechnol.* **1989**, *30*, 252.
[182] Swanson, P. E. *Appl. Envrion. Microbiol.* **1992**, *58*, 3404.
[183] Gibson, D. T.; Gschwendt, B.; Yeh, W. K.; Kobal, V. M. *Biochemistry* **1973**, *12*, 1520.
[184] Howard, P. W.; Stephenson, G. R.; Taylor, S. C. *J. Chem. Soc., Chem. Commun.* **1990**, 1182.
[185] Stabile, M. R.; Hudlicky, T.; Meisels, M. L.; Butora, G.; Gum, A. G.; Fearnley, S. P.; Thorpe, A. J.; Ellis, M. R. *Chirality* **1995**, *7*, 556.
[186] Hudlicky, T.; Endoma, M. A. A.; Butora, G. *J. Chem. Soc., Perkin Trans. 1* **1996**, 2187.
[187] Stabile, M. R.; Hudlicky, T.; Meisels, M. L. *Tetrahedron: Asymmetry* **1995**, *6*, 537.
[188] Hagedorn, S. U.S. Patent 4,532,209 (1985); *Chem. Abstr.* **1984**, *110*, 110524.
[189] Bui, V.; Hansen, T. V.; Stenstrøm, Y.; Ribbons, D. W.; Hudlicky, T. *J. Chem. Soc., Perkin Trans. 1* **2000**, 1669.
[190] Bui, V. P.; Hansen, T. V.; Stenstrøm, Y.; Hudlicky, T.; Ribbons, D. W. *New J. Chem.* **2001**, *25*, 116.
[191] Astley, S. T.; Meyer, M.; Stephenson, G. R. *Tetrahedron Lett.* **1993**, *34*, 2035.
[192] Omori, T.; Jigami, Y.; Minoda, Y. *Agr. Biol. Chem.* **1974**, *38*, 409.
[193] Taylor, S. C. U.S. Patent 5,073,640 (1991); *Chem. Abstr.* **1991**, *114*, 41053k.
[194] de Frenne, E.; Eberspächer, J.; Lingens, F. *Eur. J. Biochem.* **1973**, *33*, 357.
[195] Jigami, Y.; Omori, T.; Minoda, Y. *Agr. Biol. Chem.* **1975**, *39*, 1781.
[196] Baggi, G.; Catelani, D.; Galli, E.; Teccani, V. *Biochem. J.* **1972**, *126*, 1091.
[197] Geary, P. J.; Pryce, R. J.; Roberts, S. M.; Ryback, G.; Winders, J. A. *J. Chem. Soc., Chem. Commun.* **1990**, 204.
[198] Nadeau, L. J.; Sayler, G. S.; Spain, J. C. *Arch. Microbiol.* **1998**, *171*, 44.
[199] Reineke, W.; Knackmuss, H.-J. *J. Bacteriol.* **1980**, *142*, 467.
[200] Engesser, K. H.; Schmidt, E; Knackmuss, H.-J. *Appl. Environ. Microbiol.* **1980**, *39*, 68.
[201] Dorn, E.; Hellweg, M.; Reineke, W.; Knackmuss, H.-J. *Arch. Microbiol.* **1974**, *99*, 61.
[202] Taylor, S. J. C.; Ribbons, D. W.; Slawin, A. M. Z.; Widdowson, D. A.; Williams, D. J. *Tetrahedron Lett.* **1987**, *28*, 6391.
[203] Zeyer, J.; Lehrbach, P. R.; Timmis, K. N. *Appl. Environ. Microbiol.* **1985**, *50*, 1409.
[204] Morawski, B.; Casy, G.; Illaszewicz, C.; Griengl, H.; Ribbons, D. W. *J. Bacteriol.* **1997**, *179*, 4023.
[205] Engesser, K. H.; Cain, R. B.; Knackmuss, H. J. *Arch. Microbiol.* **1988**, *149*, 188.
[206] DeFrank, J. J.; Ribbons, D. W. *Biochem. Biophys. Res. Commun.* **1976**, *70*, 1129.
[207] Engesser, K. H.; Rubio, M. A.; Ribbons, D. W. *Arch. Microbiol.* **1988**, *149*, 198.
[208] Schläfli, H. R.; Weiss, M. A.; Leisinger, T.; Cook, A. M. *J. Bacteriol.* **1994**, *176*, 6644.

209 DeFrank, J. J.; Ribbons, D. W. *J. Bacteriol.* **1977**, *129*, 1356.
210 Brilon, C.; Beckmann, W.; Knackmuss, H.-J. *Appl. Environ. Microbiol.* **1981**, *42*, 44.
211 Engesser, K. H.; Fietz, W.; Fischer, P.; Schulte, P.; Knackmuss, H.-J. *FEMS Microbiol. Lett.* **1990**, *69*, 317.
212 Kimura, N.; Nishi, A.; Goto, M.; Furukawa, K. *J. Bacteriol.* **1997**, *179*, 3936.
213 Gonzalez, D.; Schapiro, V.; Seoane, G.; Hudlicky, T. *Tetrahedron: Asymmetry* **1997**, *8*, 975.
214 Nojiri, H.; Nam, J.-W.; Kosaka, M.; Morii, K.-I.; Takemura, T.; Furihata, K.; Yamane, H.; Omori, T. *J. Bacteriol.* **1999**, *181*, 3105.
215 Catelani, D.; Sorlini, C.; Treccani, V. *Experientia* **1971**, *27*, 1173.
216 Fujikawa, K.; Sakai, M.; Furukawa, K. *Kyushu Sangyo Daigaku Kogakubu Kenkyu Hokoku* **1998**, *35*, 147; *Chem. Abstr.* **1999**, *131*, 57840b.
217 Engesser, K. H.; Strubel, V.; Christoglou, K.; Fischer, P.; Rast, H. G. *FEMS Microbiol. Lett.* **1989**, *65*, 205.
218 Grifoll, M.; Selifonov, S. A.; Chapman, P. J. *Appl. Environ. Microbiol.* **1994**, *60*, 2438.
219 Selifonov, S. A.; Grifoll, M.; Gurst, J. E.; Chapman, P. J. *Biochem. Biophys. Res. Commun.* **1993**, *193*, 67.
220 Walker, N.; Wiltshire, G. H. *J. Gen. Microbiol.* **1955**, *12*, 478.
221 Hudlicky, T.; Endoma, M. A. A.; Butora, G. *Tetrahedron: Asymmetry* **1996**, *7*, 61.
222 Catterall, F. A.; Murray, K.; Williams, P. A. *Biochim. Biophys. Acta* **1971**, *237*, 361.
223 Ensley, B. D.; Gibson, D. T.; Laborde, A. L. *J. Bacteriol.* **1982**, *149*, 948.
224 Parales, J. V.; Parales, R. E.; Resnick, S. M.; Gibson, D. T. *J. Bacteriol.* **1998**, *180*, 1194.
225 Cerniglia, C. E.; Gibson, D. T.; Van Baalen, C. *Biochem. Biophys. Res. Commun.* **1979**, *88*, 50.
226 Cerniglia, C. E.; Van Baalen, C.; Gibson, D. T. *J. Gen. Microbiol.* **1980**, *116*, 485.
227 Barr, S. A.; Boyd, D. R.; Sharma, N. D.; Hamilton, L.; McMordie, R. A. S.; Dalton, H. *J. Chem. Soc., Chem. Commun.* **1994**, 1921.
228 Boyd, D. R.; Sharma, N. D.; Kerley, N. A.; McMordie, R. A. S.; Sheldrake, G. N.; Williams, P.; Dalton, H. *J. Chem. Soc., Perkin Trans. 1* **1996**, 67.
229 Agarwal, R.; Boyd, D. R.; McMordie, R. A. S.; O'Kane, G. A.; Porter, P.; Sharma, N. D.; Dalton, H.; Gray, D. J. *J. Chem. Soc., Chem. Commun.* **1990**, 1711.
230 Eaton, S. L.; Resnick, S. M.; Gibson, D. T. *Appl. Environ. Microbiol.* **1996**, *62*, 4388.
231 Torok, D. S.; Resnick, S. M.; Brand, J. M.; Cruden, D. L.; Gibson, D. T. *J. Bacteriol.* **1995**, *177*, 5799.
232 Phale, P. S.; Mahajan, M. C.; Vaidyanathan, C. S. *Arch. Microbiol.* **1995**, *163*, 42.
233 Mahajan, M. C.; Phale, P. S.; Vaidyanathan, C. S. *Arch. Microbiol.* **1994**, *161*, 425.
234 Deluca, M. E.; Hudlicky, T. *Tetrahedron Lett.* **1990**, *31*, 13.
235 Schocken, M. J.; Gibson, D. T. *Appl. Environ. Microbiol.* **1984**, *48*, 10.
236 Selifonov, S. A.; Grifoll, M.; Eaton, R. W.; Chapman, P J. *Appl. Environ. Microbiol.* **1996**, *62*, 507.
237 Moody, J. D.; Freeman, J. P.; Doerge, D. R.; Cerniglia, C. E. *Appl. Environ. Microbiol.* **2001**, *67*, 1476.
238 Heitkamp, M. A.; Freeman, J. P.; Miller, D. W.; Cerniglia, C. E. *Appl. Environ. Microbiol.* **1988**, *54*, 2556.
239 Khan, A. A.; Wang, R.-F.; Cao, W.-W.; Doerge, D. R.; Wennerstrom, D.; Cerniglia, C. E. *Appl. Environ. Microbiol.* **2001**, *67*, 3577.
240 Schneider, J.; Grosser, R.; Jayasimhulu, K.; Xue, W.; Warshawsky, D. *Appl. Environ. Microbiol.* **1996**, *62*, 13.
241 Rehmann, K.; Noll, H. P.; Steinberg, C. E. W.; Kettrup, A. A. *Chemosphere* **1998**, *36*, 2977.
242 Dean-Ross, D.; Cerniglia, C. E. *Appl. Microbiol. Biotechnol.* **1996**, *46*, 307.
243 Kazunga, C.; Aitken, M.D. *Appl. Environ. Microbiol.* **2000**, *66*, 1917.
244 Rehmann, K.; Hertkorn, N.; Kettrup, A. A. *Microbiol.* **2001**, *147*, 2783.
245 Boyd, D. R.; Sharma, N. D.; Agarwal, R.; Resnick, S. M.; Schocken, M. J.; Gibson, D. T.; Sayer, J. M.; Yagi, H.; Jerina, D. M. *J. Chem. Soc., Perkin Trans. 1* **1997**, 1715.
246 Boyd, D. R.; Sharma, N. D.; Hempenstall, F.; Kennedy, M. A.; Malone, J. F.; Allen, C. C. R.; Resnick, S. M.; Gibson, D. T. *J. Org. Chem.* **1999**, *64*, 4005.
247 Gibson, D. T.; Mahadevan, V.; Jerina, D. M.; Yagi, H.; Yeh, H. J. C. *Science* **1975**, *189*, 295.
248 Jerina, D. M.; van Bladeren, P. J.; Yagi, H.; Gibson, D. T.; Mahadevan, V.; Neese, A. S.; Koreeda, M.; Sharma, N. D.; Boyd, D. R. *J. Org. Chem.* **1984**, *49*, 3621.
249 Boyd, D. R.; Sharma, N. D.; Harrison, J.; Kennedy, M.; Allen, C.C.R.; Gibson, D.T. *J. Chem. Soc., Perkin Trans. 1* **2001**, 1264.
250 Lindquist, B.; Warshawsky, D. *Experientia* **1985**, *41*, 767.
251 Boyd, D. R.; Sharma, N. D.; Brannigan, I. N.; Haughey, S. A.; Malone, J. F.; Clarke, D. A.; Dalton, H. *Chem. Commun.* **1996**, 2361.

[252] Modyanova, L.; Azerad, R. *Tetrahedron Lett.* **2000**, *41*, 3865.
[253] Boyd, D. R.; McMordie, R. A. S.; Porter, H. P.; Dalton, H.; Jenkins, R. O.; Howarth, O. W. *J. Chem. Soc., Chem. Commun.* **1987**, 1722.
[254] Boyd, D. R.; Sharma, N. D.; Boyle, R.; Malone, J. F.; Chima, J.; Dalton, H. *Tetrahedron: Asymmetry* **1993**, *4*, 1307.
[255] Boyd, D. R.; Sharma, N. D.; Boyle, R.; McMurray, B. T.; Evans, T. A.; Malone, J. F.; Dalton, H.; Chima, J.; Sheldrake, G. N. *J. Chem. Soc., Chem. Commun.* **1993**, 49.
[256] Boyd, D. R.; Sharma, N. D.; Boyle, R.; McMordie, R. A. S.; Chima, J.; Dalton, H. *Tetrahedron Lett.* **1992**, *33*, 1241.
[257] Eaton, R. W.; Nitterauer, J. D. *J. Bacteriol.* **1994**, *176*, 3992.
[258] Ensley, B. D.; Ratzkin, B. J.; Osslund, T. D.; Simon, M. J.; Wackett, L. P.; Gibson, D. T. *Science* **1983**, *222*, 167.
[259] Boyd, D. R.; Sharma, N. D.; Carroll, J. G.; Malone, J. F.; Mackerracher, D. G.; Allen, C. C. R. *Chem. Commun.* **1998**, 683.
[260] Taniuchi, H.; Hayaishi, O. *J. Biol. Chem.* **1963**, *238*, 283.
[261] Cerniglia, C. E.; Morgan, J. C.; Gibson, D. T. *Biochem. J.* **1979**, *180*, 175.
[262] Bianchi, D.; Bosetti, A.; Cidaria, D.; Bernardi, A.; Gagliardi, I.; D'Amico, P. *Appl. Microbiol. Biotechnol.* **1997**, *47*, 596.
[263] Klěcka, G. M.; Gibson, D. T. *Appl. Environ. Microbiol.* **1980**, *39*, 288.
[264] Laborde, A. L.; Gibson, D. T. *Appl. Environ. Microbiol.* **1977**, *34*, 783.
[265] Ziffer, H.; Gibson, D. T. *Tetrahedron Lett.* **1975**, 2137.
[266] Lee, K.; Gibson, D. T. *J. Bacteriol.* **1996**, *178*, 3353.
[267] Wackett, L. P.; Kwart, L. D.; Gibson, D. T. *Biochemistry* **1988**, *27*, 1360.
[268] Boyd, D. R.; Sharma, N. D.; Boyle, R.; Evans, T. A.; Malone, J. F.; McCombe, K. M.; Dalton, H.; Chima, J. *J. Chem. Soc., Perkin Trans. 1* **1996**, 1757.
[269] Resnick, S. M.; Gibson, D. T. *Appl. Environ. Microbiol.* **1996**, *62*, 1364.
[270] Resnick, S. M.; Torok, D. S.; Lee, K.; Brand, J. M.; Gibson, D. T. *Appl. Environ. Microbiol.* **1994**, *60*, 3323.
[271] Brand, J. M.; Cruden, D. L.; Zylstra, G. J.; Gibson, D.T. *Appl. Environ. Microbiol.* **1992**, *58*, 3407.
[272] Lee, K.; Resnick, S. M.; Gibson, D. T. *Appl. Environ. Microbiol.* **1997**, *63*, 2067.
[273] Dutta, T. K.; Selifonov, S. A.; Gunsalus, I. C. *Appl. Environ. Microbiol.* **1998**, *64*, 1884.
[274] Haigler, B. E.; Nishino, S. F.; Spain, J. C. *Appl. Environ. Microbiol.* **1988**, *54*, 294.
[275] de Bont, J. A. M.; Vorage, M. J. A. W.; Hartmans, S.; van den Tweel, W. J. J.; *Appl. Environ. Microbiol.* **1986**, *52*, 677.
[276] Eaton, R. W.; Ribbons, D. W. *J. Bacteriol.* **1982**, *151*, 48.
[277] Chang, H.-K.; Zylstra, G. J. *J. Bacteriol.* **1998**, *180*, 6529.
[278] Chang, H.-K.; Zylstra, G. J. *J. Bacteriol.* **1999**, *181*, 3069.
[279] Buck, R.; Eberspächer, J.; Lingens, F. *Hoppe-Seyler's Z. Physiol. Chem.* **1979**, *360*, 957.
[280] Brühlmann, F.; Chen, W. *FEMS Microbiol. Lett.* **1999**, *179*, 203.
[281] Arnett, C. M.; Parales, J. V.; Haddock, J. D. *Appl. Environ. Microbiol.* **2000**, *66*, 2928.
[282] Fortnagel, P.; Harms, H.; Wittich, R.-M.; Krohn, S.; Meyer, H.; Sinnwell, V.; Wilkes, H.; Francke, W. *Appl. Environ. Microbiol.* **1990**, *56*, 1148.
[283] Bünz, P. V.; Cook, A. M. *J. Bacteriol.* **1993**, *175*, 6467.
[284] Harms, H.; Wittich, R.-M.; Sinnwell, V.; Meyer, H.; Fortnagel, P.; Francke, W. *Appl. Environ. Microbiol.* **1990**, *56*, 1157.
[285] Barriault, D.; Sylvestre, M. *Appl. Microbiol. Biotechnol.* **1999**, *51*, 592.
[286] Kimura, N.; Kato, M.; Nishi, A.; Furukawa, K. *Biosci. Biotech. Biochem.* **1996**, *60*, 220.
[287] Resnick, S. M.; Torok, D. S.; Gibson, D. T. *FEMS Microbiol. Lett.* **1993**, *113*, 297.
[288] Grifoll, M.; Selifonov, S. A.; Chapman, P. J. *Appl. Environ. Microbiol.* **1995**, *61*, 3490.
[289] Hudlicky, T.; Luna, H.; Olivo, H. F.; Andersen, C.; Nugent, T.; Price, J. D. *J. Chem. Soc., Perkin Trans. 1* **1991**, 2907.
[290] Ley, S. V.; Redgrave, A. J.; Taylor, S. C.; Ahmed, S.; Ribbons, D. W. *Synlett* **1991**, 741.
[291] Entwistle, D. A.; Hudlicky, T. *Tetrahedron Lett.* **1995**, *36*, 2591.
[292] Banwell, M. G. *Org. Prep. Proced. Int.* **1989**, *21*, 255.

CHAPTER 3

Cu, Ni, AND Pd MEDIATED HOMOCOUPLING REACTIONS IN BIARYL SYNTHESES: THE ULLMANN REACTION

TODD D. NELSON

Department of Process Research, Merck Research Laboratories, Merck & Co., Wayne, Pennsylvania 19087

R. DAVID CROUCH

Department of Chemistry, Dickinson College, Carlisle, Pennsylvania 17013

CONTENTS

	PAGE
ACKNOWLEDGMENTS	267
INTRODUCTION	267
COPPER-MEDIATED HOMOCOUPLINGS	268
Cu(0) Preparation/Activation	268
Scope and Limitations	270
Mechanism	273
Coupling via Cuprates	274
NICKEL-MEDIATED HOMOCOUPLINGS	275
Ni(0) Preparation/Activation	275
Scope and Limitations	277
Mechanism	280
PALLADIUM-MEDIATED HOMOCOUPLINGS	283
Scope and Limitations	283
Mechanism	286
AXIALLY CHIRAL BIARYLS	287
POLYMERIZATIONS	291
COMPARISON WITH OTHER BIARYL SYNTHESES	294
EXPERIMENTAL PROCEDURES	298
5,5′-Dimethyl-3,3′-dinitro-2,2′-bipyridyl (Preparation Using Copper Powder)	298

---

todd_nelson@merck.com
crouch@dickinson.edu

This chapter is dedicated to Professor Albert I. Meyers on the occasion of his retirement.

Organic Reactions, Vol. 63, Edited by Larry E. Overman et al.
ISBN 0-471-44532-0 © 2004 Organic Reactions, Inc. Published by John Wiley & Sons, Inc.

1,1'-Dinaphthalene-2,2'-dicarbonitrile (Preparation Using Copper Bronze) 299
3,3''',5,5'''-Tetraiodo-4'',6'-dimethoxy-2,2'''-dimethyl-4,4'''-bis(1-methylethoxy)-2',2''-dinitro-
1,1':3',1'':3'',1''''-quaterphenyl (Preparation Using Copper and Copper(II) Triflate) . . 299
3,3'Dihydroxy-6,6'-dimethyl-2,2'-bipyridine (Preparation Using Copper(I)
 thiophenecarboxylate) 300
6,7-Dihydro-3,9-dimethoxy-5*H*-dibenzo[*a,c*]cycloheptene (Intramolecular Coupling
 Using Ni(PPh$_3$)$_4$) 301
5,5'-Bis(methoxycarbonyl)-3,3'-bipyridyl (Using Catalytic Ni(PPh$_3$)$_2$Br$_2$) 301
4,4'-Dimethoxybiphenyl (Preparation Using Rieke Ni) 302
2,2'-Diamino-3,3'-Bipyridyl (Preparation Using Nickel Complex Reducing Agents (NiCRA)) 302
4,4'-Diphenyl-2,2'-bithienyl (Preparation Using NiCl$_2$) 303
6,6'-Dicarbomethoxy-2,2'-binaphthyl from 6-Carbomethoxy-2-naphthyl Mesylate
 (Preparation Using Ni(PPh$_3$)$_2$Cl$_2$) 303
4,4'-Dinitrobiphenyl (Preparation using Pd(OAc)$_2$) 304
Biphenyl (Preparation using Pd/C) 304
TABULAR SURVEY 304
 INTERMOLECULAR NON-HETEROCYCLE HOMOCOUPLING
  TABLE 1. UNSUBSTITUTED BIPHENYL 306
  TABLE 2. 2,2'-DISUBSTITUTED BIARYLS 313
  TABLE 3. 3,3'-DISUBSTITUTED BIARYLS 330
  TABLE 4. 4,4'-DISUBSTITUTED BIARYLS 338
  TABLE 5. 2,2',3,3'-TETRASUBSTITUTED BIARYLS 365
  TABLE 6. 2,2',4,4'-TETRASUBSTITUTED BIARYLS 366
  TABLE 7. 2,2',5,5'-TETRASUBSTITUTED BIARYLS 372
  TABLE 8. 2,2',6,6'-TETRASUBSTITUTED BIARYLS 378
  TABLE 9. 3,3',4,4'-TETRASUBSTITUTED BIARYLS 382
  TABLE 10. 3,3',5,5'-TETRASUBSTITUTED BIARYLS 387
  TABLE 11. 2,2',3,3',4,4'-HEXASUBSTITUTED BIARYLS 388
  TABLE 12. 2,2',3,3',5,5'-HEXASUBSTITUTED BIARYLS 391
  TABLE 13. 2,2',3,3',6,6'-HEXASUBSTITUTED BIARYLS 393
  TABLE 14. 2,2',4,4',5,5'-HEXASUBSTITUTED BIARYLS 395
  TABLE 15. 2,2',4,4',6,6'-HEXASUBSTITUTED BIARYLS 400
  TABLE 16. 3,3',4,4',5,5'-HEXASUBSTITUTED BIARYLS 409
  TABLE 17. 2,2',3,3',4,4',5,5'-OCTASUBSTITUTED BIARYLS 410
  TABLE 18. 2,2',3,3',4,4',6,6'-OCTASUBSTITUTED BIARYLS 412
  TABLE 19. 2,2',3,3',5,5',6,6'-OCTASUBSTITUTED BIARYLS 418
  TABLE 20. 2,2',3,3',4,4',5,5',6,6'-DECASUBSTITUTED BIARYLS 420
  TABLE 21. 1,1'-BINAPHTHYLS 423
  TABLE 22. 2,2'-BINAPHTHYLS 433
  TABLE 23. BISTETRAHYDRONAPHTHYLS 436
  TABLE 24. MISCELLANEOUS BIARYLS 437
  TABLE 25. 1,1'-BIANTHRAQUINONES 444
  TABLE 26. 2,2'-BIANTHRAQUINONES 447
  TABLE 27. BICOUMARINS, BICHROMONES, AND BIFLAVONES 448
  TABLE 28. BIPHENYLENES VIA INTERMOLECULAR COUPLING 451
  TABLE 29. TRIARYLENES 453
  TABLE 30. TETRAARYLENES 454
 INTRAMOLECULAR NON-HETEROCYCLIC HOMOCOUPLING
  TABLE 31. INTRAMOLECULAR COUPLINGS FORMING SYMMETRIC BIPHENYLS AND BINAPHTHYLS . 455
  TABLE 32. INTRAMOLECULAR COUPLINGS FORMING UNSYMMETRIC BIPHENYLS AND BINAPHTHYLS 466
  TABLE 33. BIPHENYLENES VIA INTRAMOLECULAR COUPLING 474
 INTERMOLECULAR HETEROCYCLE HOMOCOUPLING
  TABLE 34. BIPYRROLES 478
  TABLE 35. UNSUBSTITUTED-2,2'-BIPYRIDYL 480

TABLE 36. 3,3′-Disubstituted-2,2′-Bipyridyls	482
TABLE 37. 4,4′-Disubstituted-2,2′-Bipyridyls	483
TABLE 38. 5,5′-Disubstituted-2,2′-Bipyridyls	484
TABLE 39. 6,6′-Disubstituted-2,2′-Bipyridyls	485
TABLE 40. Polysubstituted-2,2′-Bipyridyls	489
TABLE 41. 3,3′-Bipyridyls	492
TABLE 42. 4,4′-Bipyridyls	494
TABLE 43. 2,2′-Bipyrimidyls	495
TABLE 44. 4,4′-Bipyrimidyls	496
TABLE 45. 5,5′-Bipyrimidyls	497
TABLE 46. 2,2′-Biquinolines	498
TABLE 47. Other Biquinolines	500
TABLE 48. Bi-isoquinolines	501
TABLE 49. 8,8′-Biquinolyls	503
TABLE 50. Bifurans	504
TABLE 51. Bis-Dibenzofurans and Bis-Dibenzodioxanes	505
TABLE 52. Unsubstituted-2,2′-Bithienyl	506
TABLE 53. Substituted-2,2′-Bithienyls	507
TABLE 54. 3,3′-Bithienyls	512
TABLE 55. Bibenzothiophenes	513
TABLE 56. Biselenyls	514
TABLE 57. Bimetallocenes and Polymetallocenylenes	515
TABLE 58. Miscellaneous Bi(heterocycles)	516
Intramolecular Heterocycle Homocoupling	
TABLE 59. Intramolecular Couplings Forming Symmetric Heterocyclic Biaryls	518
TABLE 60. Intramolecular Couplings Forming Unsymmetric Heterocyclic Biaryls	521
Polymerization	
TABLE 61. Poly(paraphenylenes)	526
TABLE 62. Poly(metaphenylenes) and Poly(biphenylenes)	530
TABLE 63. Poly(thiophenes)	531
TABLE 64. Miscellaneous Poly(arenes)	532
References	535

ACKNOWLEDGMENTS

The authors gratefully acknowledge the patience and guidance of Lou Hegedus (Colorado State University) and the numerous editorial and scientific comments from Michael Holden (Dickinson College), Doug Frantz (Merck), Carl LeBlond (Merck), and Jeff Marcoux (Merck). We are greatly indebted to Ulf Dolling (Merck) for translating a score of papers from the German literature; likewise, we thank Nobu Yasuda (Merck) for his translating assistance. We also acknowledge the help of Dr. Linda Press and Dr. Susan Curran during the preparation of this chapter.

INTRODUCTION

The traditional Ullmann reaction is the homocoupling of aromatic halides mediated by copper at elevated temperatures (Eq. 1). Ullmann first reported this reaction in 1901.[1,2] Although the coupling conditions that were first reported are still widely used, a host of modifications have been made to the reaction. Some of these modifications include the use of activated and alternative metals, often resulting in much lower coupling temperatures. Nickel and palladium are the most utilized source of alternative metals to effect this transformation. Periodic comprehensive reviews have been published, the most recent of which appeared more than 25 years ago.[3-6] In addition, brief overviews of the Ullmann reaction have occurred in review articles on biaryl construction.[7-12] This review covers the literature on Cu-, Ni-, and Pd-mediated homocoupling reactions in biaryl synthesis from 1901 through 2000. The focus of this review is on the scope and limitation of these processes, preparation of the activated metal, and mechanism of the homocoupling process. The application and utility of biaryl compounds is beyond the scope of this review.

$$\text{Ar-X} \xrightarrow{\text{Cu, Ni, or Pd}} \text{Ar-Ar} \qquad \text{(Eq. 1)}$$

COPPER-MEDIATED HOMOCOUPLINGS

Cu(0) Preparation/Activation

Traditionally, copper bronze is used to dimerize aryl halides. Although there are many reports regarding the capricious nature of the Ullmann homocoupling, the reports are much less frequent when activated copper is used. Therefore, as a matter of course, copper bronze is activated immediately prior to use. The activity of the copper powder may even result in a different product distribution (Eq. 2).[13]

$$\text{(Eq. 2)}$$

X		A	B
I	Unactivated Cu, neat, 205°	(66%)	(—)
I	Activated Cu, neat, 205°	(—)	(81%)
Cu	200°	(—)	(75%)

The primary surface impurities on copper powder are copper oxides. In fact, the manufacturing process for J. T. Baker purified copper utilizes a method in which molten copper is sprayed into an "oxidizing atmosphere" at 1400°.[14] These copper oxides are relatively insoluble in water and organic solvents, but are much more soluble in acids and bases. One method for cleaning the copper surface is by washing with an acetone solution of iodine followed by a HCl/acetone solution.[15,16] The activated copper is thoroughly rinsed with acetone or benzene and dried under vacuum prior to use. An alternative procedure is to clean the copper with nitrogen-containing complexing agents (e.g., EDTA, ethylenediamine, biquinolyl, NH_3), which frees the copper surface of cuprous ions.[17] In contrast, ethanolamine can inhibit dimerization by blocking sites on the surface of copper contained on an alumina catalyst.[18]

In addition to cleaning the copper surface, there are alternative methods for the preparation of highly activated copper. Copper powder can also be prepared from the reduction of copper (II) sulfate with zinc metal[19] or with CrO_2Cl_2.[20]

A number of ways are used to prepare highly reactive zero-valent copper.[21–23] Potassium naphthalide can be used to reduce CuI to an activated zero-valent copper slurry, which allows the Ullmann coupling to proceed at lower temperatures.[24] A drawback to this protocol is that the activated copper slurry is prone to sintering upon excessive stirring, which results in decreased reactivity. The reduction of CuI with potassium naphthalide requires an age time of 8 hours; however, a much faster (30 minutes) reduction occurs when CuCl is reduced by lithium naphthalide.[25] The decreased age time necessary for complete copper salt reduction also decreases the amount of slurry sintering. An even more reactive copper slurry can be prepared from the reduction of soluble copper salt complexes [CuI-PEt_3 or CuCl-SMe_2] with lithium naphthalide.[26]

Sonication is also successful at cleaning the metal surface of copper. The surface morphology of copper dramatically changes upon exposure to ultrasound. For example, a batch of copper powder that contains a coat of 1.2 μm copper oxide (Cu_2O) layer can be activated upon sonication. After 1 hour of sonication, the thickness of the Cu_2O layer is virtually unchanged; however, continued ultrasonic irradiation for an additional 3 hours completely removes the copper oxide layer. After sonication, thin films of carbon and nitrogen are detected on the surface of copper.[27] Sonication of the copper metal prior to and after addition of an aryl halide can increase the rate of biaryl coupling by a factor of more than 50.[28] A four-fold excess of pre-sonicated copper is necessary to obtain such a rate enhancement. When this amount of pre-sonicated copper and 2-iodonitrobenzene are sonicated at 64° for 2 hours, 70% of the biaryl is obtained. A reduced rate of reaction is observed when the amount of pre-sonicated copper is limited to a two-fold excess. Increasing the amount of copper from a four-fold excess to an eight-fold excess gives the same yield of biaryl product (ca. 71%) after a two hour period. In addition to cleaning the metal surface, sonication also reduces the particle size of the copper metal. The direct result of this phenomenon is that the number of active sites is increased, thereby increasing the rate of reaction. The average particle size of the copper powder reaches a constant minimum (25 μm from 87 μm) after 45 minutes of sonication in DMF.[29] This reduction in particle size, however, is dependent on the type of copper being used.[27] Although the use of ultrasound to facilitate the Ullmann reaction is not widely practiced, it seems that it is a convenient method that avoids the high temperatures that are utilized in the traditional Cu(0)-promoted dimerization of aryl halides. In fact, sonication of either a nitromethane or a nitrobenzene solution of picryl bromide and copper forms 2,2',4,4',6,6'-hexanitrobiphenyl in a much higher yield than in the absence of ultrasound (Eq. 3).[30] When the nitrated solvents are replaced by xylene, only 16% (^{1}H NMR) of the dimer is observed.[30]

$$\text{2,4-(O}_2\text{N)}_2\text{C}_6\text{H}_3\text{-Br(NO}_2\text{)} \xrightarrow[\text{PhNO}_2]{\text{Cu, ultrasound}} \text{dimer} \quad (77\%) \quad \text{(Eq. 3)}$$

Copper supported on alumina is also effective in this type of homocoupling.[31] The copper is both highly dispersed as well as having a smaller particle size. Copper supported on SiO_2, ZrO_2, TiO_2, and Fe_2O_3 is ineffective.

Scope and Limitations

Temperatures of >100° are usually necessary to initiate coupling in the traditional Ullmann reaction with copper powder. The Ullmann coupling proceeds most rapidly with aryl halides that are substituted in the ortho position with groups that contain lone pairs of electrons, regardless of whether the groups are electron-donating or withdrawing.[32] Examples of this are illustrated in Eq. 4[33] and Eq. 5.[34]

$$\text{2,6-diiodo-CO}_2\text{Me-C}_6\text{H}_3 \xrightarrow{\text{Cu}, 110\text{-}115°} \text{biaryl} \quad (61\%) \quad \text{(Eq. 4)}$$

$$\text{2,5-dibromo-4-NHCOMe-C}_6\text{H}_3 \xrightarrow[60°]{\text{Cu, DMSO}} \text{biaryl} \quad (84\%) \quad \text{(Eq. 5)}$$

Both 2-iodonitrobenzene and iodoferrocene give near quantitative yields when treated with copper powder at 60°.[35] In contrast, both iodobenzene and 1-iodo-3-nitrobenzene are unreactive under these conditions. Additional electron-withdrawing groups render the aryl halides even more reactive. The admixture of picryl chloride and copper at 135° results in an explosion.[36]

The order of reactivity for halides is I > Br >> Cl, with aromatic fluorides being inert. Aryl sulfonate dimerization does not occur with copper; however, in nickel couplings, aryl bromides are more reactive than aryl sulfonates.[37] In some reactions, aryl mesylates, tosylates, triflates, stannanes, silanes, and sulfonyl halides are also used as participants in biaryl formation via aryl homocoupling with copper, nickel, or palladium. Unless the aromatic ring is sufficiently reactive (e.g., by the attachment of electron-withdrawing groups), aromatic chlorides do not participate in this dimerization under traditional coupling conditions (e.g., copper/bronze). However, these types of couplings smoothly occur when alternative sources of metal are used. Ester, aldehyde, and nitro groups are compatible with the standard reaction conditions. In general, the aryl halide is activated when these groups are ortho to the halide. On occasion, elevated temperatures (>200°)[38] or excess copper [39] can cause reduction of nitro groups. Carboxylic acids, however, are usually protected in some fashion, so as to preclude decarboxylation.[40,41] Free amino and free hydroxy groups are likewise avoided. In these instances, competition from diarylamine and diaryl ether formation arises, and often predominates. (Although these later types of couplings are also referred to as Ullmann couplings or condensations, and are synthetically useful, these reactions will not be covered by this review since a new carbon-carbon aryl bond is not created.)[42–45]

Although the presence of primary and secondary unprotected amine functional groups in the aryl halide results in poor dimerization under traditional copper-mediated conditions,[46] slightly improved yields are obtained when palladium[47,48] or nickel[49–52] is used. Similarly, phenolic aryl halides afford the dimerized bisphenols when either nickel[53,54] or palladium is employed[55] instead of copper as the coupling agent.

The Ullmann coupling between unlike aryl halides occurs when there are sufficient electronic and/or steric differences between the two arene partners. In general, owing to steric hindrance, bulky groups ortho to the aryl halide retard the reaction. However, biaryls that contain four flanking groups about the biaryl axis can often be formed. Although in some cases steric hindrance does inhibit Ullmann couplings, this effect is more pronounced in other biaryl couplings such as the Suzuki reaction.[56] Lower coupling yields are observed as the ortho substituents become prohibitively large.

The most common solvent for the Ullmann reaction is dimethylformamide.[57] Nitrobenzene is also used frequently; when higher temperatures are needed, p-nitrotoluene can be used.[58] Other solvents and diluents that can be used include decalin, quinoline, biphenyl, p-cymene, tetramethylurea, naphthalene, pyridine, collidine, bitolyl, pseudocumene, xylene, diphenyl ether, tetralin, chlorobenzene, benzene, decane, toluene, anthracene, ethylene dichloride, 2,4-dimethylsulfolane, α-methylnaphthalene, dimethyl sulfoxide, and 1,3-dimethyl-2-imidazolidinone. A wide array of other substituted arenes can also be used as diluents.[46] Of these, 1,2,4-trichlorobenzene and substituted benzaldehydes function well. Phenols, primary amines, and aromatic aldehydes function poorly as diluents.

In a few reactions, improved yields are obtained when acetic acid is added to the reaction mixture[59] or when the copper is contaminated with fatty acids.[60] Benzoic acid promotes the reductive dimerization of iodobenzene.[46] In general, nitrobenzene is used at <200° in order to avoid unwanted nitro reduction. On occasion, the use of DMF promotes dehalogenation.[61,62] Utilization of m-dinitroaryls as solvents results in cross-coupling reactions between the substrate and the solvent.[63,64] An aqueous solution of copper sulfate is used as the solvent system in the reductive coupling of halogenated aromatic sulfonates (Ar-O scission). In the older literature, sand and salt are often used in the absence of a solvent, especially in large-scale work. In addition to serving as a heat exchanger, these additives aid in the break up of the reaction mixture, which in turn assists in the solubilization of the coupled product upon extraction.

Copper (I) oxide can be used to dimerize aryl halides; however, the yields are typically lower than when copper powder is used.[65] Copper (I) sulfide [Cu$_2$S] is more reactive than Cu$_2$O, but less reactive than copper powder.[66] Increased reduction of aryl halides results when these copper salts mediate the reaction. Copper halides are not useful in promoting the homocoupling of haloarenes.[66] Cu(I)-induced homogeneous Ullmann couplings of aryl halides exist.[67–69] Typical conditions for these reactions are Copper(I) triflate (CuOTf) in aqueous NH$_3$/CH$_3$CN. Solvated cuprous ions most likely initiate the coupling process. Couplings under these conditions seem to require the presence of an ortho electron-withdrawing group. Excellent selectivity is observed upon coupling triiodobiphenyls (Eq. 6).[70]

Mild Ullmann couplings of aryl bromides and iodides are also accomplished with Cu(I) thiophene-2-carboxylate (Eq. 7).[71] Low yields of products from Ullmann homocoupling are realized with 2-iodobenzoate and CuCl or CuBr, but not with CuI in N-methylpyrrolidinone at room temperature.[71] An ortho ligating group is generally required, but these conditions are tolerated by many functional groups.

Symmetric biaryls are formed by the homocoupling of aryl chloro- and fluorodimethylsilanes (Eq. 8).[72] This copper-catalyzed dimerization affords good yields and occurs rapidly at room temperature. In addition to zero-valent copper, nickel, and palladium (*vide infra*), homocouplings of aryl halides occur by using metallic titanium[73] or indium.[74]

Mechanism

The actual mechanism of the copper-mediated homocoupling of aryl halides has not been entirely elucidated. The two most likely processes by which coupling may occur involve either formation of aryl radicals or the formation of discrete copper-aryl species. Although neither mechanism may be entirely discounted, and to some extent both may be operative, considerable evidence exists for the intermediacy of discrete aryl copper species.

To a lesser extent, however, there is evidence for a homolytic coupling pathway. An initial step is complexation of solubilized copper atoms with the haloarene. Subsequent outer-sphere single-electron transfer from copper to the aryl halide can then produce an aryl radical.[75] Direct aryl radical dimerization can result in the termination of this sequence. However, an ArCu(II)X species can potentially be generated via this radical intermediate by a net oxidative addition, which is present in the alternative oxidative addition/reductive elimination mechanistic iteration (later in the text).

$$ArX + Cu(0) \longrightarrow [ArX]^{-\bullet} + [Cu(I)]^{+}$$
$$[ArX]^{-\bullet} + [Cu(I)]^{+} \longrightarrow Ar^{\bullet} + Cu(I)X$$
$$Ar^{\bullet} + {}^{\bullet}Ar \longrightarrow Ar-Ar$$

$$Ar^{\bullet} + Cu(I)X \xrightarrow{\text{oxidative addition}} Ar-Cu(II)X$$

An aryl radical can possibly account for the facile coupling of 2,6-disubstituted aryl halides to form a biaryl with four flanking groups about the biaryl axis.[76,77] Decreased biaryl formation from the copper powder mediated coupling of 2-iodonitrobenzene, 4-iodonitrobenzene, and 2-bromonitrobenzene occurs upon the addition of radical traps to the reaction mixture.[78] Similar results are obtained in the coupling of aryl halides with copper(I) oxide.[66] Treatment of iodobenzene vapor with copper in the presence of ethyl benzoate results in the partial formation of biphenyl-2-carboxylic acid and biphenyl-4-carboxylic acid (after hydrolysis), which the authors regard as evidence for hydrogen atom abstraction from ethyl benzoate by an aryl radical intermediate.[76]

Cu(111) surface analysis techniques can be used to probe the mechanism of homocoupling on the copper surface, although these ultra-high vacuum conditions are very different from the actual reaction conditions.[79] Two different mechanisms are postulated to be operative on the Cu(111) crystal. A radical mechanism is supported by the fact that the temperature at which carbon-iodide scission occurs is the same as that of the coupling of molecular iodobenzene and phenyl groups adsorbed to Cu(111).[79] Complementary Cu(111) surface work includes variable heating rate temperature-programmed reactions studies.[80] Meta and para electron-withdrawing groups on the phenyl radical, derived from the corresponding iodobenzene, lower the activation barrier to dimerization. Electron-releasing groups raise the activation barrier.

The most widely accepted mechanism for the Ullmann homocoupling involves the formation of an aryl copper intermediate. A direct co-condensation reaction of copper vapors with aryl halides results in the conclusion that an initial oxidative addition step is followed by a disproportionation and reductive elimination sequence.[81] Reaction initiation results from copper atoms on copper clusters or crystallites. Nickel vapors are more reactive than copper vapors. Under ultra-high vacuum conditions, π-complexation between the aromatic nucleus and copper occurs.[82] The resulting species is postulated to exist either as an aryl copper species or as an aryl anion. Regardless of the exact species of the absorbed phenyl group, the aryl π system is held parallel to the surface plane of copper.[82]

$$ArX + Cu(0) \longrightarrow ArCu(II)X$$
$$ArCu(II)X + Cu(0) \longrightarrow ArCu(I) + Cu(I)X$$
$$ArCu(I) + ArX \longrightarrow ArCu(III)XAr$$
$$ArCu(III)XAr \longrightarrow Ar\text{-}Ar + Cu(I)X$$

The mechanism for the Cu(I)-induced homogeneous Ullmann coupling invokes the possibility of a Cu(III) intermediate.[68,83] Aryl radicals are not operative in the Cu(I)-mediated homocouplings as shown by the fact that the dimerization of o-iodo-N,N-dimethylbenzamide occurs without a 1,5-hydrogen atom transfer.[68] In analogous couplings initiated by Cu(I), iodomaleate and iodofumarate esters couple stereospecifically, thus discounting the intermediacy of vinyl radicals.[84]

Highly reactive zero-valent copper, which is prepared by the reduction of soluble copper(I) phosphine complexes with lithium naphthalide, reacts with aryl halides to form organocopper species.[22,23,26] As in traditional Ullmann couplings, iodoarenes are more active than bromoarenes. Likewise, an activating group in the ortho position enhances the reactivity.[85] Other ligating additives can be used to stabilize such aryl copper intermediates.[86] Many of these ArCu compounds are stable at room temperature and can be isolated. Moderate heating of these ArCu species results in biaryl formation;[87,88] however, some of the intermediates show a high degree of stability. For example, 4-MeOC$_6$H$_4$Cu is stable in refluxing tetrahydrofuran.[25] Likewise, pentafluorophenylcopper is formed in a nearly quantitative yield from C$_6$F$_5$I and "Rieke copper" as an isolable tan solid.[24] Refluxing mesitylene (151°) as solvent is required to effect homocoupling.[25] Heating the solid C$_6$F$_5$Cu under argon affords the corresponding biaryl along with copper metal. These observations support the premise that an oxidative addition manifold is operative.[25]

In addition to heat, homocoupling of aryl copper intermediates may be caused by the introduction of molecular oxygen.[25,26] The stable ArCu species can also be quenched with either H$_2$O or D$_2$O giving the reduced arenes ArH or ArD, respectively.[25] Electrophiles such as alkyl and acyl halides also can react with the ArCu to provide either the alkylated or acylated aromatic systems.[85]

Coupling via Cuprates

The oxidation of homocuprates (Ar$_2$CuLi) results in the coupling of the two organic ligands to form dimers.[89,90] Although oxidation of diaryl cuprates leads to the formation of biaryls,[91] a detailed consideration of such couplings is not included in this chapter because the aryl copper species are derived directly from an aryllithium, not an aryl halide. However, the growing significance of this approach to biaryl formation warrants mention.

Although the oxidation of homocuprates to form dimers was initially observed as a side reaction, the intentional introduction of an oxidant into the reaction can be synthetically useful.[92–96] More recently, higher order cyanocuprates have been dimerized under oxidative conditions.[97–99] The addition of aryllithiums to a solution of a lower order cyanocuprate produces an unsymmetrical diarylcuprate which in turn yields the unsymmetrical biaryl upon treatment with O_2 (Eq. 9).[97]

$$\text{ArLi} \xrightarrow{\text{CuCN}} \text{ArCu(CN)Li} \xrightarrow{\text{Ar'Li}} \text{(Ar)(Ar')Cu(CN)Li}_2 \xrightarrow[-125°]{O_2} \text{Ar-Ar'} \quad (78\text{-}90\%) \quad \text{(Eq. 9)}$$

Formation of the higher order cuprate and the subsequent oxidation are very sensitive to temperature with significantly less cross-coupled product being formed at higher temperatures due to increased dimerization.[99] Ortho-substituted aryllithium compounds participate in this reaction even when both aryl rings bear such substituents.[97] Although coupling to form symmetrical biaryls is possible, the reaction is primarily used for cross-coupling and the products include unsymmetrical biphenyls,[97–99] unsymmetrical binaphthyls,[98,100] unsymmetrical bithienyls,[101] and thienylnaphthalenes.[98] Intramolecular coupling reactions[101] allow the use of chiral templates to direct biaryl formation, resulting in asymmetric binaphthyls.[102,103] This asymmetric coupling has been applied toward the synthesis of optically pure biflavones,[104] (+)-o-permethyltellimagrandin II,[105] and (+)-kotanin.[106]

NICKEL-MEDIATED HOMOCOUPLINGS

The introduction of Ni as an agent in the coupling of aryl halides represents a major advance in the field.[9,107] A 1971 report by Semmelhack and co-workers described the use of $Ni(cod)_2$ in DMF as an alternative to copper in the reductive homocoupling of aryl halides (Eq. 10).[50] A major advantage of this new reagent is that the coupling reaction can be accomplished at or near room temperature.

$$\text{PhBr} \xrightarrow[\text{DMF, 52°}]{\text{Ni(cod)}_2} \text{Ph-Ph} \quad (82\%) \quad \text{(Eq. 10)}$$

Ni(0) Preparation/Activation

The earliest application of nickel reagents to reductive coupling reactions of aryl halides required their activation prior to use.[23] $Ni(acac)_2$ is reduced to Ni(0) by triethylaluminum in the presence of 1,5-cyclooctadiene to form $Ni(cod)_2$, an isolable, though air-sensitive, yellow solid.[108,109]

Nickel-Complex Reducing Agents (NiCRA's) are prepared by the reduction of Ni(OAc)$_2$ with alkali metal hydrides and alcohol in THF or THF-benzene.[54,110] Excess NaH is required for the reduction of Ni(II) to Ni(0) with alkoxides and bipyridine serving to stabilize the nickel complex that forms. Catalytic quantities of NiCRA reagents fail to promote dimerization of aryl halides. However, another report describing the use of LiH and Ni(OAc)$_2$ allows for catalytic quantities of the latter in the coupling of aryl chlorides and bromides.[110]

Rieke and co-workers described a method for the preparation of Ni(0) via the reduction of nickel(II) halides with lithium metal/naphthalene in glyme.[21,111,112] After the activated nickel forms, the excess of naphthalene remains in the reaction mixture or, alternatively, it is removed by replacing the supernatant. Nickel-aluminum bimetallic clusters can also be prepared by treatment of a mixture of Ni(acac)$_2$, Al(acac)$_3$, and bipyridine with NaH;[113] they facilitate the coupling of aryl chlorides and bromides.[113] Although not yet applied to reductive biaryl couplings, Ni(0) on carbon is another means of introducing highly dispersed Ni(0) into a reaction.[114–116] More recently, Ni(0) on charcoal has been prepared in situ for the coupling of aryl chlorides with organozinc reagents[117–119] and other cross-coupling reactions.[116]

Activated Ni(0) can also be prepared via electrolysis of aqueous NiSO$_4$ with a mercury cathode.[120] Mercury is removed from the resulting nickel amalgam by distillation under reduced pressure to yield activated Ni(0), which is used to couple aryl halides. Electrochemically activated Ni powder is vastly superior to commercially available Ni(0) in biaryl couplings. Ni(0) can also be prepared by electrolysis of a DMF solution of Bu$_4$NBF$_4$ using a platinum cathode and a nickel anode, yielding highly activated Ni(0).[121]

The development of in situ preparations of Ni(PPh$_3$)$_4$ greatly simplifies the use of these reagents,[122] leading to catalytic methods.[123] Stoichiometric quantities of Ni(PPh$_3$)$_2$Cl$_2$, PPh$_3$, and Zn dust in O$_2$-free DMF react at 50° to form Ni(PPh$_3$)$_4$, which efficiently promotes the reductive homocoupling of aryl halides.[122] Although oxygen must be rigorously excluded, no special techniques or equipment are required.

In a catalytic process, Ni(0) that is consumed in reductive biaryl coupling reactions can be regenerated by using a stoichiometric quantity of Zn dust.[123] Such catalytic processes often produce higher isolated yields than the stoichiometric version of this reaction.[123] Dimerization of slower reacting aryl bromides is accelerated by the addition of an equimolar amount of potassium iodide to the reaction. To form the Ni(0) complex, triphenylphosphine is required in large excess relative to Ni(PPh$_3$)$_2$Cl$_2$, because complexation to zinc is also observed.

An alternative method to generate Ni(0) in situ is electrolysis of NiBr$_2$ or NiCl$_2$ in THF/HMPA,[124] leading to biaryl products.[125] Catalytic amounts of Ni(0) are also prepared in situ from Ni(II)-bpy complexes using electrochemical methods, leading to efficient biaryl coupling reactions.[126–128]

A detailed study of the in situ formation of Ni(0) reagents shows that the ease of reduction of Ni(II) salts by metal reducing agents in N,N-dimethylacetamide (DMAc) is NiI$_2$ > NiBr$_2$ > NiCl$_2$.[51] On the basis of this observation, NiI$_2$ would appear to be the choice as the Ni source in these reactions. However, substantial amounts of non-coupled, reduced arene product are obtained when NiI$_2$ is used to generate Ni(0). Fortunately, the rate of reduction of NiCl$_2$ can be accelerated by the addition of halide salts in the sequence $I^- > Br^- > Cl^- >$ no salt. Substantial amounts of reduced, non-coupled aryl halide remain when Ni(OAc)$_2$ or Ni(acac)$_2$ is used as the nickel source. Other nickel sources are ineffective in generating an active Ni species and only starting aryl halide remains upon attempted reductive homocoupling reactions.[51] In addition, Zn, Mn, and Mg are useful as reducing agents for the conversion of Ni(II) into Ni(0), with Zn being the most effective. Fe, Al, Na, and Ca are ineffective. Another report, however, describes a lack of catalytic activity when Cr or Mn powder or Na(Hg) are used in place of Zn in HMPA/DMF.[129] *tert*-Butylmagnesium bromide is also useful in place of these metal reducing agents.[130] The addition of 2,2′-bipyridine also greatly accelerates the reduction of NiCl$_2$ to Ni(0).[51]

Scope and Limitations

The reactivity of aryl halides in Ni-mediated coupling reactions depends to a large extent on the nature of Ni(0) used and reaction conditions. As with the more traditional copper agents, when Ni(cod)$_2$ or Ni(PPh$_3$)$_4$ are used, the reactivity of aryl halides in homocoupling reactions is $I > Br > Cl$.[49] When Rieke Nickel is used, the order of reactivity is also $I > Br >> Cl$.[111] When aryl iodides are coupled using Rieke Nickel, no difference is observed when the source of Ni(0) is NiCl$_2$, NiBr$_2$, or NiI$_2$. However, when aryl bromides are coupled, the best results are obtained when Rieke Ni is generated from NiI$_2$, implying that halogen-halogen exchange facilitates the reaction.[111] Aryl chlorides undergo efficient coupling using Ni(0) generated from NiCl$_2$, PPh$_3$, and excess reducing metal.[51] Similarly, the reactivity of aryl halides using NiCRA is $Cl > Br > I$, with substantial reduced, non-coupled arene being recovered for aryl iodides.[131]

Homocoupling of aryl triflates using Ni(cod)$_2$ under photolytic conditions has been described,[132] whereas tosylates are generally unreactive toward Ni(cod)$_2$ or Ni(PPh$_3$)$_4$.[49] However, aryl sulfonates dimerize in the presence of Ni(0) generated in situ from Ni(II) halides, triphenylphosphine, zinc, and iodide salts.[133–136] The homocoupling of aryl sulfonates using NiCl$_2$, Zn, PPh$_3$, and NaI in DMF has also been shown to occur under sonication.[135,136] Although not necessary,[137] sonication effectively doubles the rate of reaction.[136] The highest yields and fastest reactions are achieved with aryl triflates. Aryl tosylates have also been shown to undergo Ni-mediated homocouplings in a variety of solvents including DMF, DMAc, THF, 1,4-dioxane, and acetonitrile.[137] Aryl mesylates also can be effectively coupled using Ni(PPh$_3$)$_2$Cl$_2$ and Zn in NaI with DMF (Eq. 11).[133,134]

$$\text{MeO}_2\text{C-C}_6\text{H}_4\text{-X} \xrightarrow{\text{NiCl}_2(\text{PPh}_3)_2,\ \text{Zn},\ \text{Et}_4\text{NI}}_{\text{THF, 67°}} \text{MeO}_2\text{C-C}_6\text{H}_4\text{-C}_6\text{H}_4\text{-CO}_2\text{Me} \quad (83\text{-}99\%) \quad \text{(Eq. 11)}$$

X = tosylate, triflate, mesylate

Intramolecular coupling with Ni(cod)$_2$ proves to be a low-yielding reaction. Instead, much higher yields are obtained by the use of Ni(PPh$_3$)$_4$ in DMF as the coupling agent (Eq. 12).[138] It is noteworthy that several intramolecular coupling reactions that are resistant to traditional Ullmann coupling methods can be achieved using Ni(PPh$_3$)$_4$ in DMF.[49]

$$\text{MeO-Ar-CH}_2\text{CH}_2\text{-Ar-OMe (diiodide)} \longrightarrow \text{phenanthrene-type product} \quad \text{(Eq. 12)}$$

Ni(cod)$_2$, DMF, reflux   (0-15%)
Ni(PPh$_3$)$_4$, DMF, 55° (81%)
Cu-Bronze, heat   (70%)

Although ortho substituents typically inhibit nickel-mediated biaryl coupling reactions,[49,111,120,122,123] increased catalytic quantities of Ni(0) in the presence of stoichiometric amounts of Et$_4$NI produce good yields of biaryls from ortho-substituted aryl bromides.[139] NiCRA-mediated reactions also allow for easy coupling of a number of ortho-substituted aryl halides.[54]

Generally, electron-withdrawing substituents favor coupling[51] whereas electron-donating groups decrease the reactivity of aryl halides,[111] leading to side reactions that can be minimized by the use of appropriately chosen ligands.[51] Many functional groups have proven to be compatible with Ni-mediated reactions including ketones, aldehydes, esters, and nitriles.[49,111,139] Acetals and ketals are stable to NiCRA conditions.[54] Other functional groups are stable under some conditions but not others. Amines, for example, are compatible with Ni(cod)$_2$[49] and NiCRA[54] but not with Ni(0) generated in situ from Ni(II) salts.[51] Similarly, hydroxy groups may prevent coupling using Ni(cod)$_2$,[49] lead to arene reduction as a major by-product when Ni(0) is generated in situ from Ni(II) salts,[51] or couple without complication under NiCRA conditions.[54] However, 2-bromo-3-hydroxypyridine couples to form bipyridyl using Ni(0) generated in situ from NiCl$_2$.[140] Nitro[111] and carboxylic acid groups lead to undesired by-products with no evidence of biaryl formation.[49,51,54] The ability of nitro-containing substrates to destroy Ni(0) catalysts is also observed in the coupling of arylzinc halides with aryl halides.[141]

In situ generation of stoichiometric quantities of Ni(0) has been successfully applied to the reductive homocoupling of halopyridines, haloquinolines, and halopyrimidines to form bipyridyls,[142] biquinolines,[142] and bipyrimidines.[143] Although the catalytic method is less successful in a number of cases,[142] Ni(0) reagents are vastly superior to Cu- and Pd-mediated methods in coupling halopyridines.[144] Catalytic Ni(CO)$_2$(PPh$_3$)$_2$ is useful in forming biphenyls and bipyridyls from the corresponding aryl bromide.[145]

Generally, DMF is the solvent of choice, but other solvents also allow efficient coupling. Use of dimethylacetamide results in successful coupling of ketone- and aldehyde-containing aryl halides, a reaction that is unsuccessful in DMF.[146] Co-solvents can drastically improve the usefulness of some solvents. In reactions employing Ni(PPh$_3$)$_4$, toluene or THF are effective only if DMF or PPh$_3$ is added.[49] Reactions utilizing NiCRAs allow the use of nonpolar solvents such as benzene, xylene, hexane, and cyclohexane in combination with THF.[54] DMF is not useful as a solvent in NiCRA-mediated reactions.

Additives are also important in improving yields in nickel-mediated coupling reactions (Eq. 13). Aryl bromides[120,127] and iodides[120] dimerize in comparable yields using activated Ni(0) in DMF; however, aryl chlorides are unreactive unless KI is added.[120] The addition of Et$_4$NI improves the reaction yields, allows THF to be used in place of DMF, and removes the necessity for an excess of PPh$_3$ in reductive homocoupling of aryl halides.[139] Other iodide salts improve the yield when aryl bromides are used as substrates.[129] Addition of KI to reactions using NiCRA results in higher yields of biaryl product and shorter reaction times.[54] Other promoter salts include alkali, alkali earth, Zn, Mn, and Al halides, with iodides being especially effective.[147] Addition of such promoters accelerates reactions by 20 fold.[147] With such modifications, suitable solvents include DMF, DMAc, DMSO, and other polar aprotic solvents.[147,148]

$$\text{PhBr} \longrightarrow \text{Ph-Ph} \qquad \text{(Eq. 13)}$$

Conditions	Yield
Ni(PPh$_3$)$_2$Cl$_2$ / Zn / PPh$_3$ (1:1:2), DMF, 50°	(73%)[122]
Ni(PPh$_3$)$_2$Cl$_2$ / Zn / PPh$_3$ (0.05:1:0.4), DMF, 50°	(89%)[123]
Ni(PPh$_3$)$_2$Cl$_2$ / Zn / PPh$_3$, KI (0.05:1:0.4:1), DMF, rt	(81%)[123]
Ni(PPh$_3$)$_2$Cl$_2$ / Zn / PPh$_3$, KI (0.1:1.5:1), THF, 50°	(92%)[139]

Ligands are also required in reactions employing in situ generation of the Ni(0). Monodentate triaryl phosphines better stabilize Ni(0) than either trialkyl or bidentate aryl phosphines.[51] Triaryl phosphites are effective ligands in biaryl couplings and may be used in lower concentrations than the corresponding phosphines.[149] However, although Ni(PPh$_3$)$_2$Cl$_2$ is a stable reagent, the Ni(II)/phosphite complex is not isolable and must be formed in situ. The addition of 2,2′-bipyridine minimizes the side reactions that occur through ligand exchange when the aryl chloride bears an electron-donating group. The 2,2′-bipyridine is prepared in situ from 2-chloropyridine.[147,148] 1,10-Phenanthroline has been used in place of bipyridine[146] and other bidentate ligands are also suitable.

Electrochemical reductive homocoupling of aryl halides using catalytic Ni(0) reagents has been reported.[53,128,150–152] Using an iron or duralumin anode, the reactions are conducted using a catalytic amount of NiBr$_2$(bpy) in DMF/EtOH or EtOH/MeOH containing an electrolyte, indicating that polar protic solvents such as alcohols can be used in reductive homocoupling reactions.[53] Additionally, NaBr is used as a less expensive alternative to Bu$_4$NF, and hydroxy-containing aryl halides are efficiently coupled via this protocol. Aryl chlorides are not reactive under these conditions. The use of 2,2′-dipyridylamine (dpa) as ligand extends electrocatalytic methods to allow EtOH to serve as the sole solvent.[153] Electrocatalyzed couplings require relatively large amounts of nickel halide because of catalyst consumption during each cycle. Turnover is greatly improved through using dibromo[1,2-bis(di-2-propylphosphino)benzene]Ni(II).[154] However, the resulting Ni(0) species is only effective in coupling aryl chlorides. The numerous sources of Ni(0) and the many methods for its generation combined with the variety of reaction conditions allow a wide range of aryl halides and sulfonates to be coupled and permit a multitude of biaryl compounds to be accessible via nickel-mediated methods.

Mechanism

A number of mechanistic hypotheses have been advanced to explain Ni-mediated reductive biaryl formation.[155] Although the details may differ, two key steps are commonly accepted by the various theories: in an early step, oxidative addition of aryl halide to Ni(0) occurs, forming an ArNi(II) species, and in a final step, reductive elimination of a diaryl nickel species yields the biaryl product. The controversy lies in the intervening steps and in the nature of the diaryl nickel species. Arriving at a single mechanism to explain all Ni-mediated reactions has been complicated by the considerable variability in effective reaction conditions as well as the difficulty in obtaining kinetic data at metal surfaces in heterogeneous reactions.[155]

The earliest proposed mechanisms were based largely on semi-quantitative observations of the coupling reactions. Tsou and Kochi recognized the crucial role played by ArNi(II)X, the product of the oxidative addition of ArX to Ni(0), and an investigation of its fate in these processes provided considerable mechanistic insight.[156] ArNi(II)X can be prepared and isolated and in this study, its reactivity in subsequent reactions is followed.[156]

The possibility that a second molecule of ArX might add to ArNi(II)X to form a $Ar_2Ni(IV)X_2$ species which might undergo reductive elimination[49,50] was considered.[156] Such a reaction would be reversible, meaning that if ArNi(II)X were combined with a different aryl halide, scrambling of aryl groups in the Ni(II) species would be expected. Since aryl scrambling of this nature was not observed when Ar-X was added to ArNi(II)X, it is unlikely that this step is a part of the coupling process. It is also important to note that the existence of the key $Ar_2Ni(IV)X_2$ species has never been demonstrated.[139]

$$ArNi(II)X + Ar'-X \rightleftharpoons \underset{Ar'}{\overset{Ar}{Ni(IV)X_2}} \rightleftharpoons Ar'Ni(II)X + ArX$$

Another possibility that was considered is a metathesis reaction of two ArNi(II)X molecules to form $Ar_2Ni(II)$.[156] For this mechanism to be valid, pre-formed ArNi(II)X should decompose to form biaryl which is also not observed. Thus, it is unlikely that the diarylnickel species is $Ar_2Ni(II)$ under the conditions used in this study.[156] Interestingly, another report described a different outcome for a similar experiment.[157] When PhNi(II)Br was stirred in DMF at room temperature, biphenyl formed in 80% isolated yield. The role of solvent appears critical, as Tsou and Kochi performed their experiments in benzene.[156] A metathesis mechanism has been invoked to explain the formation of biaryl from a stable $Ar_2Ni(II)(bpy)$ complex in the presence of O_2, much like from diarylcuprates.[158]

Under the conditions employed by Tsou and Kochi, aryl radicals were rigorously excluded, rendering their involvement unlikely.[156] However, selective radical inhibition was observed by the addition of electron acceptors such as quinones, suggesting that Ni(I) and/or Ni(III) might be intermediates. ArNi(II)X remains as a key intermediate, indicating that one-electron transfer to ArNi(II)X is required, as demonstrated with electrochemical methods,[159] or two competing reactions are occurring, one producing ArNi(II)X and the other producing ArNi(III)X$_2$ from oxidative addition of ArX to NiX.[160]

The possibility that more than one mechanism may be needed has long been recognized.[51,161] Arylnickel(II) complexes produce biaryls in high yield in polar solvents without consuming aryl halide. In nonpolar solvents, the same complexes form little biaryl unless additional aryl halide is added. The current theory holds that it is unlikely that one mechanism is in operation for Ni-mediated biaryl couplings given the considerable variation in ligands,[49,51] the type of aryl halide used,[161] source and quantity of Ni,[51] and reducing agents used to form Ni(0).[51] It has been proposed that a shift in mechanism might occur in the homocoupling of aryl sulfonates as substrate is consumed while the amount of arylnickel species remains constant.[133]

Two mechanisms have emerged that represent opposite ends of a continuum in Ni-mediated biaryl couplings.[125] Their "relative importance is a function of the exact experimental conditions."[125] One mechanism, shown below, is most appropriate when a relatively small amount of Ni is used with a large excess of reductant such as zinc powder.[51] The catalytic cycle begins with oxidative addition of ArX to Ni(0) followed by the one-electron reduction of ArNi(II)X, the key step in the cycle. Oxidative addition of a second molecule of ArX to the resultant ArNi(I)X yields Ar$_2$Ni(III)X, which undergoes reductive elimination to release biaryl. Ni(I)XL$_n$ can then proceed along either of two paths, undergoing oxidative addition of ArX or a one-electron reduction to form Ni(0).[125] A similar mechanism that does not include the competition for Ni(I) has been proposed for reactions using triaryl phosphates in place of triarylphosphines.[149]

Data on electrocatalytic biaryl couplings are consistent with this mechanism.[162] When aryl sulfonates (X = OTs, OMs, OTf) are used, the Ni(II) species may have an ionic structure.[133] When iodide salts are employed as additives, an additional step includes the substitution of iodide for X$^-$ in ArNi(II)X.[123,129] Iodide has been implicated in bridging between Ni and Zn to aid in electron transfer.[51,139] Zinc and other reducing metals[51] provide the electron source for reducing steps and can also be substituted by electrochemical techniques that provide a large reductive driving force.[125] In all cases, however, an excess of reductant must be present for this mechanism to be valid.

Electrochemical methods have been especially useful in gaining a mechanistic insight into the chemistry near the metal surfaces.[155] Based on electrochemically driven reactions, a similar mechanism has been developed which does not include a second competing pathway for Ni(I).[125] In essence, the mechanisms are identical with oxidative addition of ArX to Ni(I) not included in the more recent version. In general, however, the electrochemical data support the mechanism proposed for the coupling of aryl halides in aprotic, polar solvents with an excess of reducing metal.

$$
\begin{array}{c}
X^- \quad\quad ArX \\
\nearrow\quad Ni(0)L_n \quad\searrow \\
+1e^- \\
Ni(I)XL_n \quad\quad ArNi(II)XL_n \\
\quad\quad\quad\quad +1e^- \\
Ar\text{-}Ar \\
Ar_2Ni(III)XL_n \;\longleftarrow\; ArNi(I)L_n
\end{array}
$$

The other catalytic cycle that has emerged is similar to the mechanism proposed by Tsou and Kochi[160] in which ArNi(II)X and ArNi(III)X_2 react to form Ar_2Ni(III)X. In the catalytic version, a double chain sequence operates with two interdependent but cooperative catalytic cycles.[125] Unlike the previous mechanism in which the two cycles compete for intermediates, these cycles feed one another and operate cooperatively.

$$
\begin{array}{c}
ArX \quad ArNi(II)X \quad ArNi(III)X_2 \quad ArX \\
\downarrow\quad\quad\quad\quad\quad\quad\quad\quad\quad\quad\quad\downarrow \\
Ni(0) \quad\quad\quad\quad\quad\quad\quad\quad Ni(I)X \\
\uparrow\quad\quad\quad\quad\quad\quad\quad\quad\quad\quad\uparrow \\
2e^- \quad Ni(II)X_2 \quad\quad Ar_2Ni(III)X \\
\quad\quad\quad\quad\quad\quad\quad\quad\quad\quad\quad Ar\text{-}Ar
\end{array}
$$

A key aspect of this mechanism is the required interaction of ArNi(II)X and ArNi(III)X_2. Under the reducing environment provided when excess Zn is present, neither of these intermediates would exist in more than trace quantities and their bimolecular reaction would be so slow as to prevent this mechanism from functioning.[125] ArNi(II)X and ArNi(III)X_2 are distinguished by the presence of a second halide atom in the latter.[156] The paramagnetic ArNi(III)X_2 would also be expected to be labile and capable of transferring a ligand.[156] Thus, two competing reactions can be envisioned: aryl and halide transfers.[156] A bridging mechanism explains these transfers and is supported by the observation that aryl iodides are more reactive than aryl bromides which mirrors the bridging capabilities of the two ligands.[156]

$$ArNi(II)X + Ar'Ni(III)X_2 \rightleftharpoons ArNi(III)X_2 + Ar'Ni(II)X$$

$$ArNi(II)X + Ar'Ni(III)X_2 \rightleftharpoons \underset{Ar}{\overset{Ar'}{\diagdown}}Ni(III)X + Ni(II)X_2$$

Only the second possible ligand transfer leads to the diaryl nickel(III) species required for biaryl coupling. The formation of biaryl is an intramolecular process.[156]

PALLADIUM-MEDIATED HOMOCOUPLINGS

Scope and Limitations

The first reported palladium-mediated dimerizations of aryl halides occurred in the presence of hydrazine.[48] Typically, $PdCO_3$, N_2H_2, NaOH, aqueous MeOH, and bromo, or chloroarene substrates are used to form the biaryl in moderate yields (Eq. 14).[163] Palladium-mercury amalgam can be used to homocouple aryl halides under these conditions (Eq. 15).[164]

Arylhydrazines may be used in place of hydrazine.[165] The addition of $HgCl_2$ promotes these reactions. A competing side reaction is the heterocoupling between the aryl halide and the aryl hydrazine. The desired products are also contaminated with products resulting from the reductive dimerization of the aryl hydrazine.

Symmetric biaryls can be formed from the palladium-mediated homocoupling of aryl sulfinates.[166] When an aqueous solution of an arenesulfinic acid and 100 mol% of Na_2PdCl_4 is heated an overall reductive desulfinylation occurs, with extrusion of sulfur dioxide, to afford the corresponding biaryl. Although these conditions result in only poor yields of the biaryl, conducting the reaction in the presence of a catalytic amount of $HgCl_2$ increases the product yield two-fold. However, $HgCl_2$ is not necessary to obtain useful yields of biaryls from the desulfonylative homocoupling of arenesulfonic acids. Slightly modified conditions use an aqueous solution of 100 mol% of $PdCl_2$ as the palladium source.[167] This reaction is sensitive to substitution at the ortho position of the aromatic ring. Dechlorination and decarboxylation of the substituted arenesulfonic acid do not occur.

Arylsulfonyl chlorides are dimerized in the presence of Ti(OPr-i)$_4$ with a palladium catalyst loading of 2.5 mol %.[168] Both PdCl$_2$(PhCN)$_2$ and Pd(OAc)$_2$ perform well as catalysts in this methodology whereas PdCl$_2$(PPh$_3$)$_2$ and Pd black function poorly. Decreased yields are obtained when Ti(OPr-i)$_4$ is replaced with Ti(OBu-n)$_4$, while other metal alkoxides [Ti(OBu-t)$_4$, B(OPr-i)$_3$, and Al(OPr-i)$_3$] are completely ineffective.[168] Good dimerization yields are obtained when halogens (Br, Cl, F) are present meta or para to the sulfonyl chloride without competitive dimerization at the halide-bearing carbon. It can be assumed, based on other Pd-mediated desulfonylative couplings, that an ortho substituent hinders biaryl formation. Substituted naphthalenes are poor coupling substrates under these conditions.

Aromatic iodides, when treated with a catalytic amount of Pd(OAc)$_2$ in Et$_3$N at 100°, undergo homocoupling.[169] Homocoupling occurs in the presence of phenylhydrazine upon the in situ generation of Pd(0) from PdCl$_2$ in methanol at reflux.[165] Increased amounts of side products are formed when Bu$_3$N is used in place of Et$_3$N, and the presence of water increases the level of dehalogenated arene that is formed.[169] Steric constraints are much more pronounced than in Ni and Cu homocouplings. The yield of biphenyl product is decreased when ortho substituents are introduced in the aromatic iodide. Homocoupling is not observed when the aryl iodide is 2,6-disubstituted.[169] Slightly modified conditions [Pd(OAc)$_2$, PPh$_3$] have been used to couple 5-iodopyrimidines.[170]

When triphenylarsine is used as the ligand, the Pd(OAc)$_2$-mediated aryl halide dimerization can be extended to the homocoupling of bromobenzenes.[171,172] Triphenylarsine is preferable to PPh$_3$, NPh$_3$, SbPh$_3$, and BiPh$_3$. Substituted arylbromides form the symmetric biaryl in modest yields. The main side reaction is the reduction of the aryl bromide to the parent arene. The yield of this side reaction is typically 20–30%. The ligand is not necessary in order to couple iodoarenes.[169] Ester, Cl, and F substituents are unaffected under the reaction conditions; however, bromoarenes that contain hydroxy, cyano, carboxy, or amino substituents do not couple.[172] Interestingly, 4-iodobenzoic acid forms the corresponding dimer in 70% yield (in the absence of AsPh$_3$) whereas the 4-bromobenzoic acid does not form the desired product. Stoichiometric amounts of amines can be used as hydrogen donors to regenerate the catalyst in these coupling reactions.

Similar conditions [5 mol% Pd(OAc)$_2$, n-Bu$_4$NBr, K$_2$CO$_3$, DMF, 100°] can be used to convert iodobenzene into biphenyl in 75% yield.[173] Biphenyl formation from bromobenzene under these conditions occurs in 25% yield.[173] Other substituted aryl bromides readily undergo homocoupling under similar conditions (Eq. 16).[174,175] Dimerization also occurs in the absence of n-Bu$_4$NBr by using Pd(OAc)$_2$ and diisopropylethylamine in DMF.[176]

A mixture of a catalytic amount of Pd(OAc)$_2$ and 200 mol% of N-benzyl-1,4-dihydronicotinamide (BNAH) can be used to dimerize 4-iodonitrotoluene.[177] The BNAH may function as an electron-transfer agent. The yield can be substantially improved by adding Et$_3$N to the reaction mixture. It is not known whether this Pd coupling occurs with Et$_3$N and in the absence of BNAH, i.e., conditions that homocouple aryl iodides.[169]

Another interesting Pd(OAc)$_2$-mediated homocoupling of aryl halides uses 50 mol% hydroquinone, Cs$_2$CO$_3$, and 2–4 mol% As(o-tolyl)$_3$.[178] Aryl triflates undergo hydrolysis under the reaction conditions. The couplings proceed more rapidly with As(o-tolyl)$_3$ than with P(o-tolyl)$_3$. Hydroquinone is a requisite reagent; replacement with benzoquinone or catechol results in only unreacted starting material.

Aryl chlorides, bromides, and iodides are coupled upon treatment with Pd/C, sodium formate, NaOH, and a surfactant.[179] The choice of surfactant has only modest effects on the coupling yield; however, biaryl formation is dramatically decreased when the surfactant is eliminated. A more recent investigation of this methodology reports that PEG-400[180] and CTAB[181] are the preferred phase-transfer catalysts. Polyethers also effectively enhance homocoupling yields under these conditions.[182] This methodology can be applied to the coupling of heterocycles (Eq. 17).[183] As is typically the case, ortho substitution hinders homocoupling.[179] Nitro groups are reduced and the resulting anilines are converted into benzidines. Acetates are inert to the reductive coupling conditions.

$$\text{6-methyl-2-bromopyridine} \xrightarrow[\text{TEBAC, NaOH, H}_2\text{O, reflux}]{\text{Pd/C, NaOH, HCO}_2\text{Na}} \text{6,6'-dimethyl-2,2'-bipyridine} \quad (67\%) \quad \text{(Eq. 17)}$$

Alternative reducing agents such as carbon monoxide, glycerin, ethylene glycol, alcohols,[184] and formic hydrazide[55] can be used to promote aryl dimerization under biphasic conditions. Water can function as a hydrogen source in these types of dimerizations when zinc is used as a reducing agent.[180,185] Hydrogen gas can also be used to regenerate the active palladium catalyst.[186,187] The solid Pd/C catalyst retains >99% of its activity upon recycling. Although the dimerization pathway predominates under the reaction conditions of Pd/C and PEG-400 in water at 90–120°, competitive aryl halide reduction occurs to the extent of 20–40%. In fact, similar conditions can be used for the simple dehalogenation of aryl halides.[188] A heterogeneous trimetallic catalyst (4% Pd, 1% Pt, and 5% Bi on carbon) can be used under similar conditions to dimerize aryl halides.[187] An active carbon-supported palladium iron catalyst can be used with sodium hydroxide and glycerin to achieve aryl halide homocoupling.[189]

Aryl iodides and bromides and pyridyl bromides undergo Pd(0)-catalyzed electro-reductive coupling to form biaryls in excellent yields.[190] Aryl chlorides slowly form arenes under the reaction conditions. This reduction is a competitive process with naphthyl iodides and is the sole pathway for ortho-substituted aryl bromides. Aryl triflates also undergo Pd(0)-catalyzed electro-reductive coupling to form biaryls.[191,192] Palladium complexes of monodentate ligands are preferable to those of bidentate ligands in the dimerization of aryl and naphthyl triflates.[192] When these reactions are performed in the absence of Pd, only the corresponding phenol is formed.[191] Zinc may also be used as the reducing agent in this reaction.[193]

Palladacycle catalysts with either a tertiary phosphine[194] or K_2CO_3/Bu_4NBr[195] also dimerize aryl halides. Homocoupled compounds can also be formed as the predominant products from low-yielding cross-coupling reactions between aryl halides and siloxanes with either catalytic systems of $Pd(OAc)_2$, $P(o$-tolyl$)_3$, KF, and DMF[196] or $Pd(dba)_2$, tetrabutylammonium fluoride (TBAF), and DMF.[197] TBAF is essential in the $[PdCl(\pi$-$C_3H_5)]_2$-induced homocoupling of aryl halides in DMSO.[198]

Mechanism

The mechanism of palladium-catalyzed systems has been studied in great detail.[192,199–204] In the most widely accepted postulated mechanism, only Pd(0) and Pd(II) intermediates are involved in the catalytic cycle. This mechanism is shown below (X = leaving group, L = ligand). The initial step in this mechanism is a two-electron reduction of the palladium(II) source to either an anionic or neutral Pd(0) species, which is followed by an oxidative addition of this ligated Pd(0) intermediate to an aryl halide. This results in either the formation of an anionic $Ar[Pd(II)X_2L_2]^-$ intermediate or the corresponding neutral $ArPd(II)XL_2$ intermediate. The latter species can also be arrived at by dissociation of X^- from the former Pd(II) species. Further reductive dissociation provides a ligated anionic Pd(0) complex that undergoes a second oxidative addition to the aryl halide. The palladium reduction is typically accomplished by electrochemical means, Zn, hydrogen gas, or other miscellaneous agents (see above). Once again, halide dissociation occurs to arrive at the penultimate intermediate $Ar_2Pd(II)L_2$. A final reductive elimination step provides the desired biaryl and regenerates the catalytic Pd(0) species to propagate the aryl halide dimerization process.

Slightly modified mechanisms have been proposed for palladacycle catalysis[194] and for the $Pd(OAc)_2$/hydroquinone redox system.[178] In addition, a single-electron mechanism can be invoked, in which the aryl halide coordinates to the surface of a Pd cluster.[181,186] Single-electron transfer from the catalyst to the aryl halide occurs, which results in a radical anion. This species decomposes to aryl radicals. In all likelihood, the aryl radicals then dimerize on the palladium surface. Chlorobiphenyls have not been observed during the coupling of chlorobenzene,[181,186] thus negating the direct attack of phenyl radicals on chlorobenzene. Once coupled, the newly formed biphenyl dissociates from the metal surface.

AXIALLY CHIRAL BIARYLS

A requisite for successful diastereoselective Ullmann couplings is that rotation about the chiral axis of the newly formed carbon-carbon bond be restricted. This can be accomplished by incorporation of sufficient steric hindrance about the biaryl axis (generally tri-or tetra-substitution in the ortho positions). To date, chiral catalysts have not been reported to promote these types of asymmetric coupling reactions. In successful intramolecular couplings, pre-existing stereochemical elements within the molecule induce asymmetry.

Enantiomerically pure 2,2′-binaphthol has been used in this protocol as a template for the synthesis of chiral biaryl and binaphthyls.[205] 2-Haloaroyl chlorides were converted into the corresponding diesters. These halogenated aromatic rings were then juxtaposed so that only a single diastereomer could be formed upon reductive coupling to form the biaryl (Eq. 18).[206] Hydrolysis of the ester releases the starting chiral binaphthol as well as the newly created chiral biaryl. Nonsymmetric chiral biaryls can also be formed by this method. This is done by monoesterification of 2,2′-binaphthol with one equivalent of an appropriate benzoyl donor. Subsequent esterification with a different benzoyl unit then completes the template construction[207]

(Eq. 18)

(47%)

A variety of other chiral scaffolds (**1,2,3**) have been used to juxtapose aryl halides prior to reductive dimerization.[206] Carbohydrate scaffolds **4–9** have also been used to tether two iodinated galloyl esters and subsequent reductive homocoupling occurred with varying levels of diastereoselectivity.[208] These systems were somewhat flexible, which allowed for a higher degree of conformational freedom. This in turn resulted in lower levels of asymmetric induction.

In another scaffold, the element of chirality was introduced by an asymmetric dihydroxylation of a stilbene precursor. This conformationally flexible system was then locked into a more restricted state (**10**) by carbonate formation. The corresponding axially chiral biaryl was then formed with a high degree of diastereoselectivity via a nickel-mediated reductive homocoupling in 33% yield.[209] No mention was made to account for the remaining 67% of material. This unactivated system did not undergo a copper-mediated reductive homocoupling. Studies on the conformation of the dibromide indicate that a change in twist (from M to P) occurs during the coupling reaction.[210,211]

Other intramolecular dimerizations have been accomplished in which the resulting biaryl's chirality has been relayed from an enantiomerically pure scaffold such as **11**[104] or **12**.[102,106] Coupling in both cases was effected by the initial generation of the higher order cyanocuprate, which was followed by the oxidative release of the desired biaryl.[97–99,101] It was not mentioned how the intramolecular coupling proceeded under more traditional conditions [e.g., Cu(0) or Ni(0)]. Other successful couplings with the 1,2-diol linker **1** have been reported.[105,212] These types of asymmetric, intramolecular Ullmann couplings have also utilized optically active 1,3-diols **13** and **14** as scaffolds.[103]

Chiral biaryls can also be formed from intermolecular Ullmann couplings. Initial attempts to induce asymmetry were unsuccessful. The highest level of asymmetric induction was only 13% de when chiral esters such as **15** were used.[213] Likewise, carbohydrate esters such as **16** have been poor transmitters of chiral information during the formation of the biaryl axis.[208]

15
R* = *l*-menthyl
cholesteryl
(+)-1-phenylethyl
(−)-1-phenylethyl
(−)-2-octyl

16
R* = 1,2:5,6-di-*O*-isopropylidene-α-D-glucosyl
1,2:5,6-di-*O*-isopropylidene-α-D-galactosyl

However, successful stereoselective intermolecular Ullmann reactions have been reported in which an oxazoline is employed as the chiral controller element.[214–218] The rigidity of the oxazoline and the presence of nitrogen atoms are the key structural features that allow this approach to be successful. It is not necessary to have high levels of asymmetric induction during the carbon-carbon bond formation as long as rotation about the chiral biaryl axis can occur under the reaction conditions. In one case, the Ullmann coupling of a bromooxazoline resulted in only a modest 62:38 diastereomeric ratio of atropisomers ($S:R$) after 1 hour. Although the coupling reaction was rapid, a thermodynamically controlled resolution is operative under the reaction conditions (Eq. 19).[214,215] In this manner, diastereomerically enriched biaryls were formed. After equilibration of the reaction for 40 hours, a steady state diastereomeric mixture of 93:7 ($S:R$) remained.[214,215] Interconversion of atropisomers occurred thermally in the absence of metal. This was the first report of a deracemization of chiral biaryls.[219] This phenomenon was later applied to the in situ deracemization of C_2-symmetric bisoxazolines by Cu(I).[220,221] Subsequently, other covalent methods have been utilized to deracemize biaryls.[222,223] Depending on the nature of the aryl component, increasingly higher levels of diastereoselectivity have been observed as the size of the chiral oxazoline substituent is decreased.[224]

(Eq. 19)

(59%) de = 93:7

Naturally occurring chiral biaryls have been synthesized by the diastereoselective intermolecular copper-mediated homocoupling of aromatic halides that contain a chiral oxazoline. These couplings usually proceed with high levels of diastereoselectivity; however, a case of poor selectivity has been reported.[225,226] This methodology has resulted in the synthesis of chiral biaryl-containing natural products such as (S)-gossypol,[227–229] isokotanin A,[230] (−)-herbetenediol,[224] (−)-mastigophorene A,[224] (−)-mastigophorene B,[224] (+)-kotanin,[106] and an ellagitannin.[218]

An extension of this methodology has resulted in the synthesis of C_2-symmetric chiral binaphthyls (Eq. 20).[216] Since atropisomerism of the binaphthyl system is precluded under the reaction conditions, the major biaryl stereoisomer was determined on the basis of kinetic preference in the transition state.

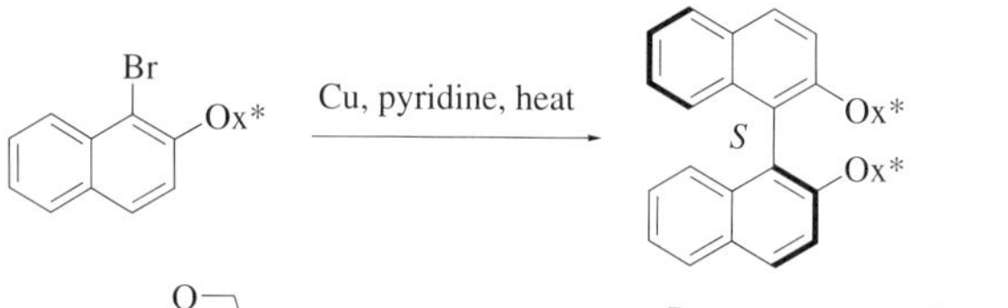

R		S:R
Ph	(75%)	2:1
i-Pr	(60%)	4:1
t-Bu	(79%)	32:1

(Eq. 20)

Chiral 1,1′,8,8′-binaphthyls have also been synthesized in this fashion.[220,231] Interestingly, the subtle change of solvent (pyridine to DMF) completely changed the predominant atropisomer from S to R about the biaryl axis (Eq. 21).[220]

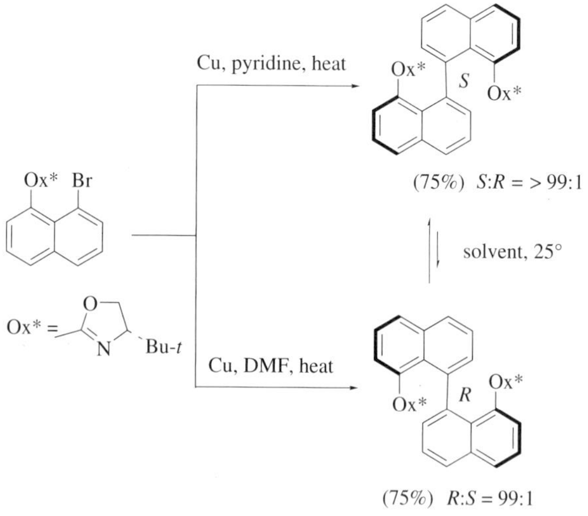

(75%) S:R = >99:1

(75%) R:S = 99:1

(Eq. 21)

POLYMERIZATIONS

Reductive homocoupling reactions provide the capability of coupling aromatic rings into polymers of arenes and, as such, have received considerable attention from polymer chemists. A logical target for such an application of Ullmann and similar reactions would be poly(p-phenylene)[232] and related polymers which exhibit desirable characteristics such as thermal stability and electrical conductivity.[233]

Classic copper-mediated Ullmann couplings of 1,4-dihaloarenes lead to the formation of smaller "oligo(p-phenylenes)"[232] such as biphenyls and terphenyls.[234] Often, however, attempts to prepare higher molecular weight poly(p-phenylene) derivatives result in the formation of unwanted by-products which decrease the yield of desired product and make purification difficult. The molecular weights of the polymers produced in metal-catalyzed reactions are also frequently limited by the solubility of the polymeric product in the reaction medium.[235]

Some successes with copper-mediated Ullmann methods have been reported. Heating 4,4'-diiodo-3,3'-dimethylbiphenyl with copper powder in α-methylnaphthalene yields a polymeric product with molecular weights ranging from 1000 to 300,000.[236,237] Biphenyl is also useful as a solvent in the preparation of polymers via copper-mediated reductive homocoupling. Upon heating with activated copper and mercury in biphenyl, poly(3,3'-biphenylylenarylmethane) forms. In both instances, however, side reactions lead to the formation of non-linear products.[234]

As with simple reductive homocoupling reactions to form biaryls, the use of DMF as a solvent allows milder conditions for polymer formation with fewer side reactions. Treatment of 4,4'-diiodo-3,3'-dinitrobiphenyl with Cu powder in DMF yields a mostly linear, substituted poly(p-phenylene) with an average length of 52 monomer units (Eq. 22).[238]

$$I-\underset{O_2N}{\bigcirc}-\underset{NO_2}{\bigcirc}-I \xrightarrow[DMF, 120-140°]{Cu} I-[\underset{O_2N}{\bigcirc}-\underset{NO_2}{\bigcirc}]_n-I \quad (97-99\%) \quad \text{(Eq. 22)}$$

Copper-mediated reductive homocoupling reactions are also useful in the preparation of other poly(arenes),[239,240] oligo(pyrrole-2,5-diyls),[241] polythiophenes,[242] and pyrrole-derived zwitterionic polymers.[243]

Both traditional copper-mediated Ullmann conditions and Ni(cod)$_2$ are useful in the reductive coupling of substituted p-dihaloarenes to form substituted poly(p-phenylenes).[244] Similarly, copper and nickel-mediated systems have been compared in the preparation of poly(alkyl thiophene-3-carboxylates).[242] Although the polymeric products are of comparable average molecular weights, the polymer obtained from the copper-mediated reaction was of higher quality with narrower polydispersity.[242,245] A comparison of Rieke Nickel with in situ generated Ni(0) produces polymers of similar molecular weights, but different polydispersity, with Rieke Ni yielding a free flowing powder as opposed to the sticky solid produced when NiBr$_2$, Zn, and PPh$_3$ are used.[245]

The lower reaction temperatures that are required for Ni(0)-mediated reactions have made nickel the preferred catalyst for the polymerization of dihaloarenes. In situ generated Ni(cod)$_2$ has been used in the synthesis of poly(*p*-phenylenes),[246,247] poly(9,10-dihydrophenanthrenes),[248] and poly(thiophenes).[246,249] These reactions produce polymers with regular structure in high yield (Eq. 23).[246,247]

$$Br-\text{C}_6H_4-\text{C}_6H_4-Br \xrightarrow{\text{Ni(cod)}_2,\ \text{bpy, DMF}} Br-[\text{C}_6H_4-\text{C}_6H_4]_n-Br \quad (99\%) \quad \text{(Eq. 23)}$$

Dibromothiophenes may also undergo polymerization in reductive homocoupling reactions as demonstrated by the formation of oligo(3-formyl-2,5-thienyl) in 81% yield from 3-formyl-2,5-dibromothiophene and in situ generated Ni(0).[250] In situ electrochemical generation of Ni(0) is also used to prepare poly(*p*-phenylenes).[251]

Modifications to the in situ generation of catalytic Ni(0) allow aryl chlorides to be reductively coupled under mild conditions.[51] The use of an excess of reducing metals such as zinc and catalytic NiCl$_2$ has been applied with great success to the preparation of polymers from dichloroarenes.[252] The first description of this methodology is in the synthesis of polyaryl ether sulfones via nickel-mediated reductive homocoupling of the corresponding dichloroarenes.[253,254] The novelty and versatility of this method is also demonstrated in the polymerization of 2,5-dichlorothiophene (Eq. 24).[253] Poly(thiophene) had previously been prepared from 2,5-dibromothiophene. Although no yield was provided, this represents the first synthesis from 2,5-dichlorothiophene.

Several reaction parameters are critical to the preparation of high polymer. Moisture must be rigorously excluded as aryl halides are readily reduced in the presence of Ni(0) and water. Oxygen must also be excluded because of its capacity to deactivate the catalyst. The reducing agent, zinc, should have a high surface area but low oxide content. Indeed, the most common cause of failure in these coupling reactions is the use of poor quality zinc metal. Triaryl phosphites can also be used in place of triarylphosphines with good results.[255]

$$Cl-\text{(thiophene)}-Cl \xrightarrow[\text{DMAc, 70°}]{\text{NiCl}_2,\ \text{PPh}_3,\ \text{Zn, bpy}} Cl-[\text{thiophene}]_n-Cl \quad \text{(Eq. 24)}$$

This general method has been applied to the synthesis of substituted poly(*p*-phenylenes). Poly(3-methoxycarbonylphenylene-1,4-diyl) is prepared from the corresponding dichloroarene with a degree of polymerization of approximately 100. (Eq. 25).[233,256] The polymer contains a mixture of "head-to-head" and "head-to-tail" units. Substituted poly(*m*-phenylene) can also be prepared using nickel-catalyzed reductive homocoupling.[257]

$$\text{Cl}\underset{\text{Cl}}{\overset{\text{CO}_2\text{Me}}{\bigcirc}} \xrightarrow[\text{DMF, 80°}]{\text{NiBr}_2,\ \text{Zn, PPh}_3} \left[\text{Cl}\underset{}{\overset{\text{CO}_2\text{Me}}{\bigcirc}}\text{Cl}\right]_n \quad (85\%) \qquad \text{(Eq. 25)}$$

The cooordinating ligands of nickel have dramatic effects on the polymerization of 2-benzoyl-1,4-dichlorobenzene.[255] The addition of bipyridyl (bpy) results in shorter reaction times and, apparently, different microstructures of the polymer product.

The wide range of conditions under which high polymer can be prepared has also been noted.[258] Polar aprotic solvents such as DMF, NMP, and HMPA as well as benzene have proven useful. Also, ratios of monomer to NiX_2 can vary from about 10 to nearly 5000 and ratios of PPh_3, and NaI to Ni catalyst may vary from 1.0 to about 10. The zinc to monomer ratio must be at least 1.0. Although the preferred reaction temperature is 50°, reactions have been successful from 25° to about 100°. In studies toward the polymerization of 2,5-bis(4-chloro-1-naphthyl)biphenyl, differences in the conditions for generating the nickel catalyst have been shown to have a profound impact on polymerization reactions.[259] When the catalyst is prepared from $NiCl_2$, PPh_3, Zn, and bpy in DMF or DMAc at 90°, yields range from 89–91%. However, when the catalyst is generated using an alternative method,[139] the yields are considerably lower (Eq. 26). It is not clear whether this difference arises from temperature or solvent effects. However, other work that incorporates similar temperatures produces comparable polymer yields, although unreacted monomer is identified in the reaction using THF as solvent.[260]

(Eq. 26)

$NiCl_2$, PPh_3, Zn, bpy / DMF, 90°	(89%)
$NiCl_2$, PPh_3, Zn, bpy / DMAc, 90°	(91%)
$Ni(PPh_3)_2Cl_2$, Zn, Et_4NI / THF, 67°	(2%)

The ability to couple aryl sulfonates using Ni-mediated reductive homocoupling techniques has been widely applied to the preparation of polymers.[135,136,261,262] Bistriflates of substituted 1,4-arenediols are converted into substituted polyphenylene polymers with unusual solubility in organic solvents.[261,262] This represents a major advance in the synthesis of poly(p-phenylenes) as readily available bisphenols and hydroquinones can enter into polymerization reactions via their sulfonates.[252] The expense of preparing triflates is such that a less expensive alternative is desirable. As a result, aryl tosylates and mesylates have been shown to undergo reductive homocoupling to yield biphenyls, although at slower reaction rates. Polymerization of mesylates allows biphenyls to be prepared in high yield[134] and application of this methodology to mesylates of bisphenols leads to poly(*p*-phenylenes) in good yield.[263]

COMPARISON WITH OTHER BIARYL SYNTHESES

Many other methods have been developed to prepare biaryls via the coupling of aromatic rings[8,9,264] but detailed discussions of them are beyond the scope of this review. Broadly speaking, however, these methods can be divided into reductive and oxidative couplings with subcategories of stoichiometric and catalytic quantities of reagent. These methods are most often applied to cross-coupling reactions but many examples of homocoupling reactions are known. Reduction of aryldiazonium salts allows coupling with another arene to form a biaryl. The intermolecular version of the reaction is known as the Gomberg-Bachmann-Hey reaction.[265] When the reaction is carried out intramolecularly, the reaction is known as the Pschorr reaction.[266] Typically, the Pschorr reaction is carried out under acid conditions in the presence of copper powder although variations are known.[8] This reaction is often plagued by low yields and is infrequently used. However, modified conditions allow its successful use in phenanthrene synthesis[9] and diazonium salts of anthranilic acid are coupled using Cu(I) to give diphenic acid in up to 90% yield.[267] The Gomberg-Bachmann-Hey reaction is typically carried out in aqueous base.[265] When phase-transfer conditions and nonaqueous solvents are used, yields are improved considerably.[268] These reactions are most successful on electron-poor substrates.[9] Reductive coupling of hypervalent iodides, which was useful in the formation of biphenylenes,[269] is also useful in forming biaryls.[270–272]

Photochemical coupling reactions to form biaryls have also been described.[8] Photolysis of aryl iodides with aromatic solvents generates unsymmetrical biaryls.[273] Arylthallium bis(trifluoroacetate) couples with an aromatic solvent under photolytic conditions in high yield.[274]

Aryl halides reductively couple using silver powder [275–278] in place of copper but the yields tend to be somewhat lower.[279] Other reductive coupling reactions include the intramolecular coupling of aryl iodides using Al_2O_3,[280,281] simple heating of naphthyl halides to yield binaphthyls,[282,283] and the use of $Hg(OAc)_2$ and Ce(IV) to dimerize halobenzenes to biphenyls.[284] 2,6-Diiodophenols dimerize in a buffered two-phase system.[285]

Conversion of aryl halides into aryl metals is another method to form biaryls. Treatment of aryl halides with lithium metal in a Wurtz-type coupling generates biaryls.[286,287] Preformed pyridyllithium compounds undergo cross-coupling with pyridyl sulfoxides[288] and pyridyl sulfoxides undergo homocoupling upon treatment with CH_3MgBr.[289] Lithiated dibenzofuran couples when treated with Fe(acac)[290] and aryllithium compounds form symmetrical biaryls upon treatment with Co(II) salts[291–293] or Ni(II) salts.[294] Copper (II) salts mediate the coupling of aryllithium compounds.[295,296]

Treatment of aryllithium with CuI forms arylcopper species that couple with aryl iodides, forming biaryls.[297–299] The aryllithium is formed from an aryl bromide or iodide bearing a ligand at the ortho position, which is required to stabilize the arylcopper intermediate. Although the requirement of this ligand limits the generality of this ambient temperature method, this methodology allows access to biaryls containing two ortho substituents[297] and has been used in the synthesis of Steganacin[298] and oxygenated phenanthrenes.[300]

Aryl halides are converted into arylmagnesium halides that participate in cross-coupling reactions upon treatment with metal catalysts (Eq. 27).[264]

$$\text{ArMgX} + \text{Ar'X} \xrightarrow{\text{Ni(0)}} \text{Ar-Ar'} \quad \text{(Eq. 27)}$$

Ni(II) salts serve as catalysts in preparing heterobiaryls[301,302] and poly(arenes)[303,304] via this method. Ni(II) catalysts bearing chiral ligands are useful in the preparation of enantiomerically pure biaryls.[305-308] Cu(II) salts may also be used in the Kharasch method.[309,310] Other metal catalysts[264] include Pd(II), Fe(II), Fe(III), Co(II),[292] Ni(0), and Cr(III).[303] Aryl Grignard reagents dimerize upon treatment with triflic anhydride[311] or with 1,4-dichloro-2-butyne.[312] Arylmagnesium bromides form symmetric biaryls upon treatment with $TiCl_4$.[313]

The polar nature of the Grignard reagent required in the Kharasch method precludes the use of potentially reactive groups on the partner aryl halide. However, the method provides some versatility by allowing the replacement of aryl halides with aryl sulfonates,[314] thioethers, carbamates, and ethers.[264] Reactivity differences among these groups also allow for selective coupling at one site of a doubly substituted aromatic compound.

A related method of biaryl synthesis is the Negishi reaction in which arylzinc halides couple with aryl halides in the presence of Ni(0) or Pd(0) (Eq. 28).[264,315,316]

$$\text{ArZnX} + \text{Ar'X} \xrightarrow{\text{Ni(0) or Pd(0)}} \text{Ar-Ar'} \quad \text{(Eq. 28)}$$

Arylzinc halides, prepared from the aryllithium intermediates, cross-couple with aryl bromides and iodides in the presence of Pd(II) and DIBAL-H or $Ni(PPh_3)_4$.[317] Catalytic amounts of $Pd[P(t\text{-}Bu)_3]_2$ couple arylzinc halides with aryl chlorides.[318] Arylzinc halides are also formed using a sacrificial anode process and Co(II) catalyst.[319] This method can be used to couple a pyrimidine and an indole[320] and allow for sulfonates to be used in place of aryl halides as the coupling partner with the arylzinc halide.[264] (Phenylzinc chloride)chromium complexes couple with aryl halides in the presence of a Pd catalyst.[321] Like the reactions involving Grignard reagents, however, the required use of aryllithium compounds limits the nature of substituents that the reaction can tolerate. Highly reactive Rieke zinc allows the conversion of aryl halides into arylzinc halides and subsequent cross-coupling with aryl halides in the presence of Pd(0) to produce biaryls in high yield.[322] Rieke zinc also allows for the homocoupling of dibromothiophenes to form poly(thiophenes) or 1,4-diiodobenzene to form poly(p-phenylenes).[316] The regioregularity of the substituted polymer depends on the catalyst used. Aryl halides may also be converted into arylzinc halides using electrochemical methods, avoiding the troublesome aryllithium intermediate.[323]

Other aryl metals also undergo coupling to form biaryls.[324] Diarylgold(III) complexes produce biaryls upon treatment with $NaClO_4$ and triphenylphosphine.[325] Arylmercuric chloride and diphenylmercury dimerize in the presence of $[RhCl(CO)_2]_2$ to yield symmetrical biaryls.[326] Diarylmercury species undergo homocoupling when heated with Ni(0), Pd(0), Cu(0), Pt(0), Ag(0),[327] or $Cu/PdCl_2$.[328] Aryllead triacetates form biaryls when treated with CuI or $Pd_2(dba)_3\text{-}CHCl_3$.[272] In the presence of $Ni(dppe)Cl_2$ catalyst, 2-(phosphininyl)halogenozirconocene forms 2,2'-biphosphinine.[329] Triarylgallium(III) undergoes homocoupling when reacted with H_2O in toluene.[330]

In situ generated hypervalent siloxanes undergo cross-coupling with aryl iodides and electron-deficient aryl bromides in the presence of $Pd(dba)_2$(Eq. 29).[197] Treatment of phenyl trimethoxysilane with tetrabutylammonium fluoride generates the hypervalent fluorosilicate ion which transmetallates with the arylpalladium halide complex and subsequently couples with aryl iodide. Pd catalysts and fluoride salts[331] or NaOH[332] are also used to cross-couple arylchlorosilanes and aryl halides.

$$ArSi(OMe)_3 \xrightarrow{F^-} \begin{bmatrix} MeO & OMe \\ Ar-Si-OMe \\ F \end{bmatrix}^- \xrightarrow[Pd(dba)_2]{Ar'X} Ar\text{-}Ar' \qquad \text{(Eq. 29)}$$

In the late 1970's, arylstannanes were shown to couple with aryl halides and triflates in the presence of Pd(0) catalysts in what is now known as the Stille reaction (Eq. 30).[333] A major advantage of the Stille reaction is the compatibility of the reaction conditions with substituents that are not tolerated under conditions involving more reactive main-group metals.[264] Typically, arylstannanes containing tributyltin or trimethyltin are used[264] but the reaction has been expanded to allow the use of arylstannoates in aqueous solution.[334,335] In general, arylstannanes couple best with aryl bromides and iodides, with aryl bromides requiring more rigorous conditions.[333] Aryl triflates also undergo coupling with arylstannanes and have reactivities comparable to the aryl bromides.[336] Diazonium salts[337] and hypervalent iodides also serve as coupling partners with the stannanes. Pd(0) is the catalytic species in these reactions, but Pd(II) salts are also useful because they can be reduced to Pd(0) by the organostannane.[336] Although Pd-catalysts are preferred, yields can be improved by the addition of copper salts.[264] Pd(0) coordinated by a bidentate iminophosphine ligand shows remarkable catalytic activity.[338] Steric effects are important in Stille couplings with ortho substituents significantly slowing the reaction rate.[333] The Stille reaction is useful in the preparation of heterobiaryls in which either aryl moiety can be derived from the arylstannane.[264] Poly(arylenes) have also been prepared via the Stille reaction[339] and solid-phase biaryl synthesis using microwave irradiation has been reported.[340]

The Stille reaction can also be used to form symmetrical biaryls.[338] Treatment of iodopyrimidines with hexamethylditin and Pd(0) yields bipyrimidyl.[341] Cu(II) or Mn(II) salts are used to form symmetrical biaryls from arylstannanes in the presence of I_2.[342,343]

$$\text{ArSnR}_3 + \text{Ar'X} \xrightarrow{\text{Pd(0)}} \text{Ar-Ar'} \qquad \text{(Eq. 30)}$$
$$X = \text{I, Br, OTf}$$

Suzuki couplings in which an arylboronic acid or ester couples with an aryl halide in the presence of Pd(0) and a base have become extensively used (Eq. 31).[264,344–346] Pd(0) in the form of Pd(PPh$_3$)$_4$ is most commonly used as a catalyst but the more easily handled PdCl$_2$(PPh$_3$)$_2$ and Pd(OAc)$_2$ with triphenylphosphine are readily converted into Pd(0), allowing good yields to be obtained.[344] Phosphine-free Pd(0) is actually more reactive in the Suzuki coupling but is less stable to heat. In a related reaction, Ni(PPh$_3$)$_2$Cl$_2$ mediates the cross-coupling of chloronaphthalene with an intramolecularly stabilized arylaluminum reagent.[347] Triflates add to arylboronic acids with the relative reactivity being I > Br > OTf >> Cl.[344] In general, aryl chlorides are inert unless newly developed catalytic systems are employed.[348–355] Aryl triflates give better results when coupled with arylboronate esters.[336] Aryl mesylates and aryl arenesulfonates also participate in the Suzuki reaction using Ni(0) catalysts.[356] The sluggish nature of this reaction can be improved by treating the arylboronate ester with BuLi, forming a more reactive borate that couples with aryl mesylates to form biaryls.[357] The use of a base is required in Suzuki couplings; Na$_2$CO$_3$, NaHCO$_3$, Et$_3$N, and Cs$_2$CO$_3$ have proven useful.[344] A variety of fluoride salts have been successfully used as well.[358] For more sterically demanding coupling reactions, strong bases such as NaOH or Ba(OH)$_2$ markedly accelerate the reaction.[344] Strong bases, however, are incompatible with triflates.[345] In all cases, homogenous conditions result in shorter reaction times.[344]

$$\text{Ar-B}\begin{smallmatrix}\text{OH}\\\text{OH}\end{smallmatrix} + \text{Ar'X} \xrightarrow[\text{base}]{\text{Pd(0)}} \text{Ar-Ar'} \qquad \text{(Eq. 31)}$$

Although more widely used for cross-coupling reactions, the Suzuki reaction has also been applied to the preparation of symmetrical biaryls. Homocoupling was first observed as a competing reaction when cross-coupling was slow.[359] It was subsequently demonstrated that homocoupling can occur as a synthetically useful reaction. In situ formation of arylboronate esters allows for symmetrical biaryl formation in moderate to excellent yield.[360]

Widely used in polymer and natural product synthesis, the Suzuki reaction is applicable to the synthesis of axially chiral biaryls.[264,344,345] Planar chiral (haloarene)chromium complexes couple with arylboronic acids in the presence of Pd(0) and base to yield axially chiral biaryls.[321,361] Chiral ligands also induce asymmetry in binaphthyl products of Suzuki couplings.[362]

Oxidative coupling of phenols dates back to before 1900 (Eq. 32).[363,364] The variety of conditions available to effect coupling of phenols is so vast that no single mechanism can account for all of the reactions. However, radical processes appear to be at work. A number of metal salts are useful in mediating phenolic coupling including Fe(ClO$_4$)$_3$,[365] FeCl$_3$,[366] AlCl$_3$,[367] Cu(II),[368] Ru(IV),[369] molten SbCl$_3$,[370] Mn(III),[371] and Tl(III).[372] Catalytic quantities of the CuCl(OH)-TMEDA complex in the presence of O_2 mediate the intramolecular coupling of naphthols.[373] Naphthols undergo dimerization when treated with Cu(II) on alumina.[374] VOF_3[375] and $VOCl_3$[376] are also used in coupling phenolic systems, and electrochemical methods have proven useful in the preparation of poly(2,6-dihydroxynaphthalene).[377] Other metals have also been employed to couple phenols.[8]

$$2 \; \text{PhOH} \xrightarrow{\text{metal salts}} \text{2,2'-biphenol} \quad \text{(Eq. 32)}$$

Chiral biaryls can also be prepared by oxidative coupling of naphthols using Cu(II)-chiral amine complexes.[378–380] Chiral biphenanthrols are also prepared using Cu(II)-chiral amine complexes.[381] High enantiomeric excess levels can be achieved using CuCl and chiral 1,5-diaza-cis-decalins.[382] Coupling of chiral tetrahydronaphthols using $K_3Fe(CN)_6$ also leads to chiral biaryls.[383] Electrolysis of 2-naphthol, 2-methoxynaphthalene, and 10-hydroxyphenanthrene on a TEMPO-modified graphite electrode in the presence of (−)-sparteine produces enantiomerically enriched dimers.[384]

Non-phenolic systems also undergo oxidative coupling.[8,363] Pd(II) is useful in forming symmetrical biaryls by promoting the oxidative homocoupling of toluene or bitolylmercury in acetic acid,[385] monosubstituted benzenes in trifluoroacetic acid,[386] substituted benzenes in acetylacetone and O_2,[387] and substituted benzenes with thallium tris(trifluoroacetate).[388] Pd(II) is also used in a Heck-type reaction to couple iodothiophenes to form poly(thiophenes).[389] Arylmagnesium halides, treated with TlBr, dimerize to form symmetrical biaryls.[390] Arylthallium bis(trifluoroacetate) cross-couples with benzene under photolytic conditions.[274] Thiophenes undergo homocoupling when treated with $Tl(O_2CCF_3)_3$ in trifluoroacetic acid.[391] A mixture of $AlCl_3$ and $CuCl_2$ can be used to couple biphenyl with mesitylene.[392] Raney Nickel dimerizes pyridine but in low yields.[393] Aluminum chloride induces the cross-coupling of chlorothiophenes and substituted benzenes or naphthalenes.[394] NO^+ ions, generated by treatment of sodium nitrite with triflic acid, produce aromatic radical cations which couple to convert benzenes and naphthalenes into biaryls.[395] A number of other systems can also be employed to form biaryls.[8]

EXPERIMENTAL PROCEDURES

$$2 \; \text{(5-methyl-3-nitro-2-chloropyridine)} \xrightarrow[\text{DMF, 100–105°}]{\text{Cu}} \text{5,5'-dimethyl-3,3'-dinitro-2,2'-bipyridyl}$$

5,5′-Dimethyl-3,3′-dinitro-2,2′-bipyridyl (Preparation Using Copper Powder).[396] To a stirred mixture of 8.6 g (0.050 mol) of 2-chloro-5-methyl-3-nitropyridine in 50 mL of DMF was added 10 g (0.157 mol) of copper powder. The mixture was heated to 100–105° and maintained at this temperature for 4 hours. After cooling to room temperature, the mixture was filtered and the filter cake was washed with boiling DMF (2 × 20 mL). The collected filtrate was diluted with 200 mL of water and 50 mL of 25% aqueous ammonium hydroxide. The resulting precipitate was collected and washed with water to give 4.3 g (63%) of 5,5'-dimethyl-3,3'-dinitro-2,2'-bipyridyl as yellow crystals, mp 198–200°; ^{1}H NMR (CDCl$_3$) δ 9.0 (s, 4H), 2.8 (s, 6H); Anal. Calcd for $C_{12}H_{10}N_4O_4$: C, 52.55; H, 3.68; N, 20.43. Found: C, 52.68; H, 3.66; N, 20.53.

1,1'-Dinaphthalene-2,2'-dicarbonitrile (Preparation Using Copper Bronze).[397] To 2 g (0.008 mol) of I$_2$ in 100 mL of acetone was added 10 g (0.157 mol) of copper bronze. After the mixture was stirred for 10 minutes, the residue was filtered and then stirred for 2 minutes with 200 mL of a 1:1 mixture of acetone and concentrated HCl. After filtration by suction, a bright red copper powder was obtained, which was dried in vacuo. 1-Bromo-2-naphthalenecarbonitrile (7.7 g, 0.0332 mol) and 30 mL of DMF were added to the dried metal powder. The mixture was heated to reflux under an argon atmosphere for 36 hours. After cooling to room temperature, the mixture was poured into a beaker containing 300 mL of water and 300 mL of CH$_2$Cl$_2$. The organic layer was separated and the aqueous layer was extracted with CH$_2$Cl$_2$. The combined organic phase was washed with water (3 × 100 mL) and dried with Na$_2$SO$_4$. After filtration, the solvent was removed in vacuo and the residue was filtered through silica gel using 1:1 CH$_2$Cl$_2$/hexane as the eluent to yield 3.2 g (63%) of 1,1'-binaphthalene-2,2'-dicarbonitrile as white needles, mp 232°; ^{1}H NMR (CDCl$_3$) δ 8.10 (br d, J = 8.4 Hz, 2H), 8.00 (br d, J = 8.1 Hz, 2H), 7.82 (d, J = 8.7 Hz, 2H), 7.63 (ddd, J = 1.4, 7.0, 8.2 Hz, 2H), 7.41 (ddd, J = 1.4, 7.0, 8.4 Hz, 2H), 7.16 (d, J = 8.4 Hz, 2H); ^{13}C NMR (CDCl$_3$) d 140.5, 134.8, 131.8, 130.4, 129.2, 128.7, 128.5, 126.7, 126.3, 117.5, 111.5; Anal. Calcd for $C_{22}H_{12}N_2$: C, 86.82; H, 3.97; N, 9.20. Found: C, 86.56; H, 3.93; N, 9.14.

3,3''',5,5'''-Tetraiodo-4'',6'-dimethoxy-2,2'''-dimethyl-4,4'''-bis(1-methylethoxy)-2',2''-dinitro-1,1':3',1'':3'',1'''-quaterphenyl (Preparation Using Copper and Copper(II) Triflate).[70] A mixture of 9.2 g (25.4 mmol) of hydrated copper (II) trifluoromethanesulfonate, 1.2 g (19 mmol) of Cu powder, 185 mL of acetone, and 11 mL of acetonitrile was heated at reflux under a nitrogen atmosphere for 2.5 hours. The mixture was cooled in an ice bath and a solution of 8.6 g (12.7 mmol) of 3,3′,5′-triiodo-4′-isopropoxy-6-methoxy-2′-methyl-2-nitrobiphenyl in 35 mL of DMSO was added dropwise, followed by 42 mL of ammonium hydroxide. The mixture was stirred at 0° for 8 hours, then at 5° for 12 hours. Water (250 mL) and CHCl$_3$ (250 mL) were added, and the layers were separated. The organic layer was washed with 5% HCl and water, dried over MgSO$_4$, passed through a pad of alumina, and evaporated. The residue was crystallized from MeOH to yield 6.3 g (90%) of a pale yellow solid, mp 282–283°; NMR (CDCl$_3$) δ 1.43 (d, J = 6 Hz, 12 H), 2.28 (br s, 6H), 3.82 (s, 6H), 4.86 (m, 2H), 7.06 (d, J = 9 Hz, 2H), 7.39 (d, J = 9 Hz, 2H); Anal. Calcd for C$_{34}$H$_{32}$I$_2$N$_2$O$_8$: C, 37.0; H, 2.9; N, 2.5. Found: C, 36.7; H, 3.0; N, 2.5.

3,3′-Dihydroxy-6,6′-dimethyl-2,2′-bipyridine (Preparation Using Copper(I) thiophenecarboxylate).[71] *Copper(I) Thiophenecarboxylate:*[71] A 500-mL round-bottomed flask was charged with 100 g (0.78 mol) of thiophene-2-carboxylic acid, 28 g (0.196 mol) of CuO, and 300 mL of toluene. The flask was equipped with a Dean-Stark trap, condenser, and magnetic stirring bar, and the mixture was brought to reflux under N$_2$ and stirred overnight with azeotropic removal of water. The yellow-brown suspension was cooled to 60° and the product was collected on a medium porosity sintered-glass filter funnel. Under a stream of N$_2$, the filter cake was washed with 300 mL of de-oxygenated MeOH and then with Et$_2$O until the eluant was colorless. The filter cake was washed with small amounts of hexanes. The product was dried on the filter under a flow of N$_2$ for 20 minutes, then transferred to a flask and dried in vacuo. The product was obtained as a tan, air-stable powder that was ground with a mortar and pestle to produce 69.8 g (94%) of a finely divided powder. Elemental analysis was consistent with about 5 mol % of unreacted CuO, which does not affect use. Anal. Calcd for C$_5$H$_3$CuSO$_2$(CuO)$_{0.05}$: C, 30.40; H, 1.50; S, 16.20. Found: C, 30.35; H, 1.55; S, 16.11.

To a stirred solution of 0.470 g (2.00 mmol) of 6-iodo-2-picolin-5-ol in 8 mL of *N*-methylpyrrolidinone at room temperature was added 1.14 g (6.00 mmol) of copper(I) thiophenecarboxylate under nitrogen. After 4 hours at room temperature, the reaction mixture was diluted with 15 mL of EtOAc and the resulting slurry was filtered through a plug of silica gel using 120 mL of EtOAc as the eluant. The solution was washed with 100 mL of dilute aqueous NH$_4$OH and the aqueous layer was back extracted with EtOAc (4 × 40 mL). The combined organic layers were washed with 50 mL of water and the resulting aqueous layer was back-extracted with 30 mL of EtOAc. The combined organic layers were dried with MgSO$_4$, filtered, and concentrated at reduced pressure. The crude product was purified by flash chromatography to give 0.15 g (69%) of 3,3′-dihydroxy-6,6′-dimethyl-2,2′-bipyridyl as a green-yellow solid, mp 186.5–188°; ^{1}H NMR (CDCl$_3$) δ 7.29 (d, J = 8.4 Hz, 2H), 7.08 (d, J = 8.4 Hz, 2H), 6.03 (br s, 2H), 2.50 (s, 6H); ^{13}C NMR (CDCl$_3$) δ 153.9, 144.6, 138.5, 126.3, 124.5, 22.7; Anal. Calcd for C$_{12}$H$_{12}$N$_2$O$_2$: C, 66.65; H, 5.59; N, 12.95. Found: C, 66.74; H, 5.65; N, 12.98.

6,7-Dihydro-3,9-dimethoxy-5H-dibenzo[a,c]cycloheptene (Intramolecular Coupling Using Ni(PPh$_3$)$_4$).[49]

To a suspension of 332 mg (0.3 mmol) of Ni(PPh$_3$)$_4$ in 14 mL of DMF at −78° under argon was added, in one portion, 127 mg (0.25 mmol) of α,ω-bis(iodomethoxyphenyl)propane. The reaction mixture was warmed to 60° for 40 hours, during which time the gold-brown starting mixture gradually became green and a black solid appeared. The mixture was cooled to 25°, poured into 50 mL of Et$_2$O, and washed sequentially with 20 mL of 1 M aqueous HCl and 40 mL of brine. After the ether layer was dried, concentration in vacuo afforded a yellow solid which was purified by preparative TLC (25% CH$_2$Cl$_2$ in hexane) to yield 188 mg (74%) of cyclic product, mp 99.5–101°; NMR (CDCl$_3$) δ 7.2 (dd, J = 7, 3 Hz, 2H), 6.9–6.6 (m, 4H), 3.78 (s, 6H), 2.65–1.9 (m, 6H); Anal. Calcd for C$_{17}$H$_{18}$O$_2$: C, 79.97; H, 6.71. Found: C, 79.94; H, 6.63.

5,5′-Bis(methoxycarbonyl)-3,3′-bipyridyl (Using Catalytic Ni(PPh$_3$)$_2$Br$_2$).[139]

To a dry flask containing a magnetic stirring bar, filled with argon, and stoppered with a rubber septum was added 1.12 g (1.5 mmol) of Ni(PPh$_3$)$_2$Br$_2$, 0.49 g (7.5 mmol) of Zn, and 1.29 g (1.5 mmol) of Et$_4$NI. After the system was purged of air and refilled with argon, 15 mL of THF was added via syringe. After about 30 minutes, a solution of 1.08 g (5 mmol) of methyl 5-bromonicotinate was added, and the mixture was stirred at 50° for 2 hours. The mixture was cooled to room temperature and poured into 30 mL of 2 M aqueous NH$_4$OH. Chloroform (100 mL) was added and the precipitates were removed by filtration. The layers were separated and the aqueous layer was extracted twice with 50 mL of CHCl$_3$. The combined organic layers were washed with water and brine, dried with MgSO$_4$, and concentrated in vacuo. The residual solid was triturated with CH$_2$Cl$_2$ to yield 459 mg (67%) of 5,5′-bis(methoxycarbonyl)-3,3′-bipyridyl, mp 226–226.5°; ^{1}H NMR (CDCl$_3$) δ 9.29 (br s, 2H), 9.05 (br s, 2H), 8.55 (m, 2H), 4.02 (s, 6H); ^{13}C NMR (CDCl$_3$) δ 165.3,

151.6, 150.7, 135.5, 132.5, 126.4, 52.7; Anal. Calcd for $C_{14}H_{12}N_2O_4$: C, 61.54; H, 4.39; N, 10.18. Found: C, 61.76; H, 4.44; N, 10.29.

$$2 \ CH_3O\text{-}C_6H_4\text{-}I \xrightarrow{\text{Rieke Ni}} CH_3O\text{-}C_6H_4\text{-}C_6H_4\text{-}OCH_3$$

4,4'-Dimethoxybiphenyl (Preparation Using Rieke Ni).[111] *Rieke Nickel Powder:*[111] A 50-mL, two-necked flask was equipped with a magnetic stirring bar, a rubber septum, and a condenser topped with an argon inlet. The flask was charged with 3.82 g (12.22 mmol) of NiI_2, 0.195 g (28.1 mmol) of freshly cut lithium wire, 0.16 g (1.25 mmol) of naphthalene, and 30 mL of glyme. The mixture was stirred vigorously at room temperature for 12 hours. The nickel powder precipitated as a black slurry in a colorless, clear solution after the stirring stopped. After the supernatant glyme was removed by syringe, 20 mL of freshly distilled glyme was added and the mixture was stirred for 5 minutes. This procedure was repeated two times to remove the naphthalene. Glyme may be replaced with other solvents such as DMF or DMSO after evaporation of glyme under reduced pressure.

To 12.22 mmol of the activated Ni prepared above in glyme was added 1.88 g (8.02 mmol) of 4-iodomethoxybenzene, and the mixture was stirred at 85° for 2 hours. The reaction mixture changed to a reddish brown color and most of the Ni powder was consumed. The reaction mixture was cooled, poured into 100 mL of Et_2O, and filtered. The filtrate was washed with 100 mL of water and dried over $MgSO_4$. The solution was concentrated to 10 mL and 10 mL of EtOH was added to precipitate the product. Crystallization from Et_2O/EtOH (1:1) gave 0.58 g (68%) of 4,4'-dimethoxybiphenyl as colorless flakes, mp 176–177°; NMR ($CDCl_3$) δ 7.50 (d, $J = 9$ Hz, 4H), 6.95 (d, $J = 9$ Hz, 4H), 3.83 (s, 6H).

$$2 \ \text{(3-Cl-2-aminopyridine)} \xrightarrow[\text{DME, 65°}]{t\text{-BuONa, Ni(OAc)}_2, \text{PPh}_3} \text{(2,2'-diamino-3,3'-bipyridyl)}$$

2,2'-Diamino-3,3'-Bipyridyl (Preparation Using Nickel Complex Reducing Agents (NiCRA)).[398] To a refluxing suspension of 1.44 g (60 mmol) of NaH, 1.77 g (10 mmol) of nickel(II) acetate and 10.48 g (40 mmol) of triphenylphosphine in 30 mL of DME was added 1.48 g (20 mmol) of *t*-BuOH in 10 mL of DME. After stirring at 65° for 2 hours, 1.29 g (10 mmol) of 2-amino-3-chloropyridine in 20 mL of DME was added and the temperature was maintained at 65° for 17 hours, at which time GC analysis showed the reaction to be complete. The flask was cooled to 25° and excess hydride was carefully destroyed by addition of EtOH until H_2 evolution ceased. Water was added, and the organic phase was extracted with Et_2O, and the extracts were dried over $MgSO_4$. After removal of solvents under reduced pressure and column chromatography using EtOAc/hexane eluents, 0.744 g (40%) of 2,2'-

diamino-3,3'-bipyridyl was isolated, mp 132°; ^{1}H NMR (CDCl$_3$) δ 8.7 (t, 1H), 6.9 (d, 2H), 3.6 (s, 2H); ^{13}C NMR (CDCl$_3$) δ 142.6, 139.6, 137.2, 123.6, 121.3.

$$2 \text{ [4-phenyl-2-chlorothiophene]} \xrightarrow[\text{DMF, 50°}]{\text{NiCl}_2, \text{PPh}_3, \text{Zn}} \text{4,4'-diphenyl-2,2'-bithienyl}$$

4,4'-Diphenyl-2,2'-bithienyl (Preparation Using NiCl$_2$).[399] To a solution of 0.65 g (5 mmol) of NiCl$_2$ and 5.2 g (20 mmol) of PPh$_3$ in 25 mL of DMF was added 0.32 g (5 mmol) of Zn powder under N$_2$. The mixture was stirred for 1 hour at 50° and 0.973 g (5 mmol) of 4-phenyl-2-chlorothiophene was added to the resultant red brown solution. After stirring for 30 minutes, the mixture was poured into 50 mL of H$_2$O. The resulting precipitate was collected by filtration and washed with H$_2$O. The dried precipitate was washed with CHCl$_3$ (2 × 20 mL) then crystallized from CHCl$_3$ to yield 1.0 g (63%) of 4,4'-diphenyl-2,2'-bithienyl, mp 224–225°; NMR (CDCl$_3$) δ 7.2–7.7 (m).

$$2 \text{ [MeO}_2\text{C-naphthyl-OSO}_2\text{CH}_3\text{]} \xrightarrow[\text{THF, 67°}]{\text{Ni(PPh}_3)_2\text{Cl}_2, \text{Zn, Et}_4\text{NI}} \text{[6,6'-dicarbomethoxy-2,2'-binaphthyl]}$$

6,6'-Dicarbomethoxy-2,2'-binaphthyl from 6-Carbomethoxy-2-naphthyl Mesylate (Preparation Using Ni(PPh$_3$)$_2$Cl$_2$).[133] A 125-mL Schlenk tube was charged with 65.2 mg (0.10 mmol) of Ni(PPh$_3$)$_2$Cl$_2$, 111 mg (1.7 mmol) of zinc powder, 385.5 mg (1.5 mmol) of tetraethylammonium iodide, and a magnetic stirring bar. After the tube was sealed with a rubber septum, the contents were dried at 22° at reduced pressure (10^{-3} torr) for 10 hours. The contents of the tube were then placed under an argon atmosphere by filling with Ar followed by three evacuation-filling cycles. Freshly distilled THF (0.50 mL) was added via syringe and the mixture was stirred at room temperature for 5 minutes during which time the color of the mixture became deep red-brown. A solution of 6-carbomethoxy-2-naphthyl mesylate (280 mg, 1.0 mmol) in 0.50 mL of THF was added via syringe. The reaction mixture was heated to reflux and stirred for 10 hours. The mixture was then cooled, filtered, diluted with water, and extracted with CHCl$_3$. The organic phase was dried with MgSO$_4$ and concentrated in vacuo. Purification using column chromatography and crystallization from CHCl$_3$/hexanes yielded 337 mg (91%) of 6,6'-dicarbomethoxy-2,2'-binaphthyl as white crystals, mp 275°; NMR (CDCl$_3$)

δ 8.65 (s, 2H), 8.17–8.09 (m, 4H), 4.03 (s, 6H); Anal. Calcd for $C_{24}H_{18}O_4$: C, 77.82; H, 4.90. Found: C, 77.16; H, 4.81.

$$2 \; O_2N\text{–}C_6H_4\text{–}I \xrightarrow[\text{DMA, 75°}]{\text{Pd(OAc)}_2, \text{(o-tolyl)}_3\text{As,} \\ \text{hydroquinone, Cs}_2\text{CO}_3} O_2N\text{–}C_6H_4\text{–}C_6H_4\text{–}NO_2$$

4,4′-Dinitrobiphenyl (Preparation using Pd(OAc)$_2$).[178] To a mixture of 299 mg (1.201 mmol) of 4-iodonitrobenzene, 69.8 mg (0.634 mmol) of hydroquinone, and 409.4 mg (1.257 mmol) of Cs$_2$CO$_3$ was added a pre-stirred solution of 5.7 mg (0.021 mmol) of Pd(OAc)$_2$ and 9.0 mg (0.026 mmol) of tri-o-tolylarsine in 3.0 mL of DMA. The mixture was immediately degassed and the flask was refilled with N$_2$. The reaction mixture was stirred for 1 hour at 75°, then cooled to room temperature. The reaction was quenched with 20 mL of 2 N HCl and the mixture was extracted with EtOAc (3 × 30 mL). The combined organic layers were washed with 10% aqueous NaOH solution (4 × 40 mL) and brine (40 mL), dried over MgSO$_4$, filtered, and concentrated under reduced pressure. Column chromatography on silica gel (eluant: 14% Et$_2$O in hexane) afforded 125.6 mg (86%) of 4,4′-dinitrobiphenyl as a tan crystalline solid, mp 235°; ^{1}H NMR (DMSO-d$_6$) δ 8.37 (d, J = 8.8 Hz, 4H), 8.08 (d, J = 8.8 Hz, 4H); ^{13}C NMR (DMSO-d$_6$) δ 147.6, 144.1, 128.7, 124.3.

$$2 \; C_6H_5\text{–}Cl \xrightarrow[\text{H}_2\text{O, 100°}]{\text{Pd/C, Zn, NaOH, PEG-400}} C_6H_5\text{–}C_6H_5$$

Biphenyl (Preparation using Pd/C).[180] In a 300-mL stainless steel autoclave equipped with a six-blade impeller and an external heating mantle, 5.0 g (44 mmol) of chlorobenzene, 3.3 g (50 mmol) of Zn powder, 5.0 g (125 mmol) of NaOH, 1.5 g (8.4 mol%) of PEG-400, and 1.0 g of 5% w/w Pd/C (1.0 mol% Pd) were combined and diluted to a total volume of 50 mL with H$_2$O. The autoclave was heated to 100° and the mixture was stirred at 950 rpm for 2 hours. After cooling, the mixture was extracted with 40 mL of CH$_2$Cl$_2$, the organic layer was dried, and the solvent was evaporated. Crystallization from EtOH afforded 2.35 g (68%) of biphenyl, mp 69°; NMR (CDCl$_3$) δ 7.59 (dq, 4H), 7.46 (dt, 4H), 7.39 (tt, 2H); Anal. Calcd for $C_{12}H_{10}$: C, 93.46; H, 6.54. Found: C, 93.26; H, 6.74.

TABULAR SURVEY

The tables have been prepared by categorizing biaryls by the nature of the aromatic rings being coupled. Tables 1 through 20 include biphenyls with different substitution patterns while Tables 21 through 33 include examples of biaryls of other non-heterocyclic arenes. Tables 34 through 60 include examples of biheteroarenes. Polymers have been gathered into separate tables. Within each table, entries are listed by increasing number of carbon atoms.

Yields in parentheses are based on isolated products, unless noted otherwise. Examples with unspecified yields have been excluded. For additional pertinent data on polyarenes (beyond chemical yields) for entries in Tables 61–64 (polyarenes), readers are directed to the original publications. The literature has been reviewed through 2000.

The following abbreviations are used in the tables:

acac	acetylacetonate
AcOH	acetic acid
t-AmOH	*tert*-amyl alcohol
BNAH	*N*-benzyl-1,4-dihydronicotinamide
bpy	2,2'-bipyridine
Bs	benzenesulfonyl
cod	1,5-cyclooctadiene
CRA	complex reducing agent
CTA	cetyltrimethylammonium
CTAB	cetyltrimethylammonium bromide
CTBPB	cetyltributylphosphonium bromide
dba	dibenzylideneacetone
DIPEA	diisopropylethylamine
diphos	1,2-bis(diphenylphosphino)ethane
DMAc	*N,N*-dimethylacetamide
DME	1,2-dimethoxyethane
DMF	*N,N*-dimethylformamide
dmpb	1,4-bis(dimethylphosphino)butane
DMS	dimethylsulfide
DMSO	dimethyl sulfoxide
dpa	2,2'-dipyridylamine
dppe	1,2-bis(diphenylphosphino)ethane
dppf	1,3-bis(diphenylphosphino)ferrocene
dppp	1,3-bis(diphenylphosphino)propane
EDTA	ethylenediamine tetraacetic acid
ETOXCA	ethylene oxide/cetyl alcohol condensate
Fs	*p*-fluorobenzenesulfonyl
HMPA	hexamethylphosphoramide
HMPT	hexamethylphosphorous triamide
IPA	isopropyl alcohol
NMP	*N*-methylpyrrolidinone
Ms	methanesulfonyl
PEG	polyethyleneglycol
SDPNS	sodium diisopropylnaphthalene
Tf	trifluoromethanesulfonyl
Ts	*p*-toluenesulfonyl
TBAF	tetrabutylammonium fluoride
TEBAC	triethylbenzylammonium chloride
TMEDA	*N,N,N',N'*-tetramethylethylenediamine
TMS	trimethylsilyl

TABLE 1. UNSUBSTITUTED BIPHENYL

Substrate	Conditions	Product(s) and Yield(s) (%)	Refs.
C_6 — PhCl	Ni(cod)$_2$, DMF, 50°	(14)[a]	49, 50
	Ni(CO)$_2$(PPh$_3$)$_2$, DMSO, 70°	(35)[a]	145
	Ni(OAc)$_2$, NaH, t-AmONa, bpy, THF, 63°	(86), (90)[a]	131
	Ni(OAc)$_2$, NaH, t-AmONa, PPh$_3$, THF, 63°	(90)	131
	NiBr$_2$, Zn, KI, HMPA, 50°	(0)	129
	NiCl$_2$, Zn, NaBr, DMF, bpy, 60-80°	(81)	146
	NiBr$_2$, NMP, 2e$^-$, bpy	(90)	128
	Ni(OAc)$_2$, t-BuOLi, LiH, bpy, THF, 63°	(89), (94)[a]	110
	Ni(OAc)$_2$, t-AmONa, NaH, bpy, THF, KI, 63°	(92)	54
	Ni powder, DMF, 130°	(0)	120
	Ni powder, DMF, KI, 130°	(90)[a]	120
	Ni(OAc)$_2$, t-BuONa, LiH, bpy, THF, reflux	(93)[a]	400
	NiCl$_2$(PEt$_3$)$_2$, Zn, HMPA, 80°	(25)[a]	401
	NiCl$_2$(PBu$_3$)$_2$, Zn, HMPA, 80°	(18)[a]	401
	[Ni(P^P)Br$_2$ complex], DMSO, 2e$^-$, 65°, R = Pr-i	(80)[a]	154
	NiBr$_2$, Zn, PPh$_3$, DMAc, 80°	(99)[a]	51
	NiI$_2$·6H$_2$O, Zn, PPh$_3$, DMAc, 80°	(58)[a]	51
	Ni(OAc)$_2$·4H$_2$O, Zn, PPh$_3$, DMAc, 80°	(66)[a]	51
	Ni(acac)$_2$·2H$_2$O, Zn, PPh$_3$, DMAc, 80°	(82)[a]	51
	Ni(NO$_3$)$_2$·6H$_2$O, Zn, PPh$_3$, DMAc, 80°	(0)[a]	51
	NiO, Zn, PPh$_3$, DMAc, 80°	(66)[a]	51

	Conditions	Yield (%)	Refs.
	NiF$_2$, Zn, PPh$_3$, DMAc, 80°	(82)^a	51
	NiCl$_2$, Zn, PPh$_3$, DMAc, 80°	(99)^a	51,52,148
	NiCl$_2$, Mg, PPh$_3$, DMAc, 80°	(99)^a	51,52,148
	NiCl$_2$, Mn, PPh$_3$, DMAc, 80°	(99)^a	52,148
	NiCl$_2$, Zn, P(C$_6$H$_4$OMe-p)$_3$, DMAc, 80°	(89)^a	52,148
	NiCl$_2$, NaI, PPh$_3$, Zn, bpy, DMAc, 80°	(78)^a	51,52,148
	NiCl$_2$, NaI, PPh$_3$, Zn, DMAc, 80°	(99)^a	51,52,148
	NiBr$_2$, NaI, PPh$_3$, Zn, DMAc, 80°	(91)^a	51,52,148
	NiI$_2$, NaI, PPh$_3$, Zn, DMAc, 80°	(100)^a	51,52,148
	Ni(acac)$_2$, NaI, PPh$_3$, Zn, DMAc, 80°	(94)^a	51,52,148
	Ni(OAc)$_2$, NaI, PPh$_3$, Zn, DMAc, 80°	(99)^a	51,52,148
	Pd/C, Zn, PEG-400, H$_2$O, NaOH, heat	(68)	180
	Pd/C, H$_2$, PEG-400, H$_2$O, NaOH, 110°	(71), (76)^a	186
	Pd/C, HCO$_2$Na, NaOH, CTAB, H$_2$O, 110°	(83), (87)^a	181
	Pd/C, HCO$_2$K, NaOH, CTAB, H$_2$O, 110°	(83), (88)^a	181
	PdCl$_2$, N$_2$H$_4$·H$_2$O, MeOH, reflux	(trace)	47
	Pd(OAc)$_2$, K$_2$CO$_3$, n-Bu$_4$NBr, DMF, 100°	(0)	173
	Pd/C, NaOH, H$_2$O, formic hydrazide, 90°	(54)^a	55
	Pd/C, HCO$_2$Na, NaOH, H$_2$O, 95°, CTAB	(48)	179
	PdCl$_2$, phenylhydrazine, NaOH, MeOH, reflux	(12)	165
	PdCl$_2$, HgCl$_2$, phenylhydrazine, NaOH, MeOH, reflux	(12)	165

Ph–Br

	Conditions	Yield (%)	Refs.
	Ni(cod)$_2$, DMF, 55°	(82)^{a,b}	49,50
	Ni(cod)$_2$, PPh$_3$, bpy, THF	(81)	402
	Ni(cod)$_2$, PPh$_3$, bpy, C$_6$H$_6$	(83)	402
	Ni(OAc)$_2$, NaH, t-AmONa, PPh$_3$, THF, 63°	(70)	131
	Ni(OAc)$_2$, NaH, t-AmONa, bpy, THF, 63°	(70), (76-84)^a	131
	NiCl$_2$(PEt$_3$)$_2$, Zn, HMPA, 30°	(95)^a	40

TABLE1. UNSUBSTITUTED BIPHENYL (Continued)

Substrate	Conditions	Product(s) and Yield(s) (%)	Refs.
Ph-Br	NiCl$_2$(PBu$_3$)$_2$, Zn, NMP, 30°	(81)	401
	NiCl$_2$(P(c-C$_6$H$_{11}$)$_3$)$_2$, Zn, NMP, 80°	(75)^a	401
	NiBr$_2$, Zn, KI, DMF, HMPA	(98)^a	403
	NiBr$_2$, Zn, KI, PBu$_3$, DMF, 150°	(88)^a	129
	NiBr$_2$, Zn, KI, PBu$_3$, NMP, 150°	(89)^a	129
	NiBr$_2$, Zn, KI, HMPA, 50°	(98)	129
	NiCl$_2$, Zn, NaBr, DMF, bpy, 60-80°	(78)	146
	NiBrPh(PBu$_3$)$_2$, PPh$_3$, CH$_3$CN, 2e$^-$	(82)	159
	NiBr$_2$, NMP, bpy, 2e$^-$	(75)	128
	Ni powder, DMF, 140°	(71)	120
	Ni powder, DMSO, 140°	(99)^a	120
	Ni powder, pyridine, 140°	(71)^a	120
	Ni powder, mesitylene, 140°	(24)^a	120
	Ni powder, ethylene glycol, 140°	(0)	120
	Ni powder, neat, 140°	(23)^a	120
	Ni(OAc)$_2$, t-BuOH, LiH, PPh$_3$, THF, reflux	(95)^a	400
	Ni(OAc)$_2$, t-BuOH, LiH, bpy, THF, reflux	(98)^a	400
	NiBr$_2$(PPh$_3$)$_2$, Zn, 50°, THF	(75)	139
	NiCl$_2$(PPh$_3$)$_2$, Zn, Et$_4$NI, 50°, THF	(92)	139,404
	NiI$_2$(PPh$_3$)$_2$, Zn, Et$_4$NI, 50°, THF	(94-99)	139,404
	NiI$_2$(PPh$_3$)$_2$, Zn, Et$_4$NI, 50°, THF	(94)	139,404
	NiBr$_2$(PPh$_3$)$_2$, Zn, Et$_4$NI, 50°, DMF	(84)	139,404
	NiBr$_2$(PPh$_3$)$_2$, Zn, Et$_4$NI, 50°, CH$_3$CN	(88)	139,404
	NiBr$_2$(PPh$_3$)$_2$, Zn, Et$_4$NI, 50°, acetone	(80)	139,404
	Ni(OAc)$_2$, t-BuOLi, LiH, bpy, THF, 63°	(91), (100)^a	110

308

Conditions	Yield (%)	Ref.
NiCl$_2$(PPh$_3$)$_2$, PPh$_3$, DMF, n-Bu$_4$NBr, 2e$^-$	(80)	127
Ni(OAc)$_2$, NaH, t-AmONa, bpy, THF, KI, 63°	(82)	54
NiCl$_2$(PPh$_3$)$_2$, Zn, PPh$_3$, DMF, 50°	(73)	122
NiCl$_2$(PPh$_3$)$_2$, Zn, PPh$_3$, DMF, 50°	(89)	123
Ni(OAc)$_2$, bpy, Et$_4$NBr, CH$_3$CN, 2e$^-$	(92)	405
Ni(CO)$_2$, (PPh$_3$)$_2$, DMSO, 70°	(75)^a	145
Ni(CO)$_2$, (PPh$_3$)$_2$, toluene, 70°	(65)^a	145
Ni(CO)$_2$, (PPh$_3$)$_2$, hexane, 70°	(25)^a	145
NiBr$_2$, EtOH, dpa, H$_2$O, 2e$^-$	(95)	153
NiBr$_2$, EtOH, MeOH, bpy, 2e$^-$	(84)	153
NiBr$_2$, bpy, EtOH, DMF, Bu$_4$NBF$_4$, 2e$^-$	(80)	53
NiBr$_2$, bpy, EtOH, MeOH, NaBr, 2e$^-$	(84)	53
NiBr$_2$, bpy, NMP, Bu$_4$NBF$_4$, 2e$^-$, 45°	(75)	53
NiBr$_2$, bpy, DMF, Bu$_4$NBF$_4$, 2e$^-$, 45°	(85)	53
NiBr$_2$, bpy, MeOH, Bu$_4$NBF$_4$, 2e$^-$	(75)	53
NiBr$_2$, bpy, MeOH, EtOH, Bu$_4$NBF$_4$, 2e$^-$	(80-84)	53
NiBr$_2$, bpy, MeOH, EtOH, H$_2$O, Bu$_4$NBF$_4$, 2e$^-$	(40)	53
NiBr$_2$, bpy, EtOH, H$_2$O, Bu$_4$NBF$_4$, 2e$^-$	(36)	53
NiCl$_2$, CrCl$_2$, Mn, THF, RT	(88)	406
Co powder, DMF, heat	(71)	120
PdCl$_2$, N$_2$H$_4$•H$_2$O, MeOH, reflux	(30)	47
PdCl$_2$, HgCl$_2$, N$_2$H$_4$•H$_2$O, MeOH, reflux	(4)	47
Pd(OAc)$_2$, N$_2$H$_4$•H$_2$O, K$_2$CO$_3$, n-Bu$_4$NBr, DMF, 100°	(25)	173
Pd/CaCO$_3$, MeOH, KOH, heat	(77)	48
Pd/C, HCO$_2$Na, NaOH, H$_2$O, 95°	(30)	179
Pd/C, HCO$_2$Na, NaOH, H$_2$O, 95°, CTAB	(65)	179
Pd/C, HCO$_2$Na, NaOH, H$_2$O, 95°, CTBPB	(60)	179

TABLE 1. UNSUBSTITUTED BIPHENYL (Continued)

Substrate	Conditions	Product(s) and Yield(s) (%)	Refs.
Ph–Br	Pd/C, HCO$_2$Na, NaOH, H$_2$O, 95°, SDPNS	(65)	179
	Pd/C, HCO$_2$Na, NaOH, H$_2$O, 95°, ETOXCA	(51)	179
	Pd/C, Zn, air, H$_2$O, acetone, 25°	(92)[a]	185
	Pd/C, H$_2$, PEG-400, H$_2$O, NaOH, 110°	(73), (79)[a]	186
	Pd(OAc)$_2$, As(o-tol)$_3$, hydroquinone, Cs$_2$CO$_3$, 100°	(56)	178
	Pd(OAc)$_2$, AsPh$_3$, Bu$_3$N, DMF, 140°	(68-78)	171,172
	PdCl$_2$, HgCl$_2$, p-tolylhydrazine hydrochloride, NaOH, MeOH, reflux	(41)	165
	PdCl$_2$, HgCl$_2$, phenylhydrazine, NaOH, MeOH, reflux	(97)	165
	PdCl$_2$, phenylhydrazine, NaOH, MeOH, reflux	(62)	165
Ph–I	Cu, DMF, reflux	(80)	407
	Cu, neat, 190°	(30-78)	32
	Cu, neat, 230°	(82)	2
	Cu, neat, 60°	(0)	35
	Cu, neat, heat	(78-80)	46,63
	Cu, neat, 60°	(0)	35
	Cu, diluents, heat	(0-75)	46
	CuI-PEt$_3$, lithium naphthalide, DME, 85°	(66)[a]	25,26
	Ni(OAc)$_2$, NaH, t-AmONa, bpy, THF, 63°	(37), (40-50)[a]	131
	Ni(cod)$_2$, DMF, 40°	(71)[a]	49,50
	Ni (electrogenerated), DMF, 100°	(58)	121
	NiBr$_2$, Zn, HMPA, 50°	(0)[a]	129
	NiBr$_2$, Zn, KI, HMPA, 50°	(94), (98)[a]	129,403

310

Conditions	Yield	Ref
NiCl$_2$(PEt$_3$)$_2$, Zn, HMPA, 25°	(93)[a]	401
NiCl$_2$(PEt$_3$)$_2$, Zn, NMP, 25°	(88)[a]	401
NiBr$_2$, NMP, 2e$^-$, bpy	(70)	128
NiBr$_2$, EtOH, MeOH, NaBr, 2e$^-$, bpy	(85)	53
NiBr$_2$, EtOH, dpa, H$_2$O, 2e$^-$	(85)	153
NiBr$_2$, EtOH, MeOH, bpy, 2e$^-$	(85)	153
Ni powder, DMF, 120°	(73)	120
NiI$_2$, lithium naphthalide, DME, 80°	(83)[a]	111
NiCl$_2$, CrCl$_2$, Mn, THF, rt	(98)	406
Pd(OAc)$_2$, Et$_3$N, 100°	(54)	169
Pd(OAc)$_2$, Bu$_3$N, 100°	(38)	169
Pd(OAc)$_2$, Bu$_3$N, DMF, 140°	(54)	172
Pd(OAc)$_2$, K$_2$CO$_3$, n-Bu$_4$NBr, DMF, 100°	(75)	173
Pd(OAc)$_2$, PPh$_3$, K$_2$CO$_3$, 115°	(48)	408
Pd(OAc)$_2$, DIPEA, DMF, 80°	(96)	176
[PdCl(π-C$_3$H$_5$)]$_2$, TBAF, DMSO, 120°	(82)	198
Pd(PPh$_3$)$_4$, Et$_4$NOTs, DMF, 2e$^-$	(94)	190
PdCl$_2$, N$_2$H$_4$·H$_2$O, MeOH, reflux	(49)	47
PdCl$_2$, HgCl$_2$, N$_2$H$_4$·H$_2$O, MeOH, reflux	(90)	47
Pd(OAc)$_2$, As(o-tol)$_3$, hydroquinone, Cs$_2$CO$_3$, 25°	(39)	178
Pd(OAc)$_2$, As(o-tol)$_3$, hydroquinone, Cs$_2$CO$_3$, 75°	(96)	178
Pd(OAc)$_2$, P(o-tol)$_3$, hydroquinone, Cs$_2$CO$_3$, 75°	(37)[b]	178
Pd/C, Zn, air, H$_2$O, acetone, 25°	(94)[a]	185
PdCl$_2$, p-tolylhydrazine hydrochloride, NaOH, MeOH, reflux	(55)	165
PdCl$_2$, phenylhydrazine, NaOH, MeOH, reflux	(63)	165
PdCl$_2$, phenylhydrazine, NaOH, MeOH, reflux	(quantitative)	165
![Pd complex with OAc, R = o-tolyl], EtN(i-Pr)$_2$, DMF, 110°	(85)	194

TABLE 1. UNSUBSTITUTED BIPHENYL (*Continued*)

Substrate	Conditions	Product(s) and Yield(s) (%)	Refs.
X = OTf	PdCl$_2$(PPh$_3$)$_2$, DMF, *n*-Bu$_4$NBF$_4$, 2e$^-$, 90°	(76)	192
OTf	PdCl$_2$(PPh$_3$)$_2$, DMF, Zn, 90°	(30)[b]	192,193
OTf	NiCl$_2$(dppe), KI, DMF, THF, Zn, 67°	(99)	192,193
OTf	Ni(cod)$_2$, *hv*, toluene, KI, *N*-methylimidazole	(90)	132
OMs	NiCl$_2$(PPh$_3$)$_2$, Zn, Et$_4$NI, THF, 67°	(91)[a]	133
OBs	NiCl$_2$(PPh$_3$)$_2$, PPh$_3$, NaBr, Zn, DMF, 100°	(80)	137
OFs	NiCl$_2$(PPh$_3$)$_2$, Zn, Et$_4$NI, THF, 67°	(99)[a]	133
SO$_2$Cl	PdCl$_2$(PhCN)$_2$, Ti(OPr-*i*)$_4$, *m*-xylene, 140°	(51)	168
SiMe$_2$Cl	CuI, TBAF, CH$_3$CN	(73)	72
SiMe$_2$F	CuI, TBAF, CH$_3$CN	(76)	72

[a] The yield was determined by gas chromatography.
[b] The yield was determined by NMR spectroscopy.

TABLE 2. 2,2'-DISUBSTITUTED BIARYLS

Substrate	Conditions	Product(s) and Yield(s) (%)	Refs.
C_6 (2-R-C₆H₄-X)		2,2'-R₂-biphenyl	
R X			
F Cl	Ni(OAc)₂, NaH, t-AmONa, bpy, THF, C₆H₆, KI, 63°	(60)	54
F Br	NiBr₂, bpy, EtOH, MeOH, 2e⁻	(trace)	153
F Br	NiBr₂, EtOH, 2e⁻, dpa, H₂O	(68)	153
F Br	Ni(OAc)₂, NaH, t-AmONa, bpy, THF, C₆H₆, KI, 63°	(57)	54
F Br	Pd(OAc)₂, Bu₃N, AsPh₃, DMF, 140°	(44)	172
F I	Cu, neat, heat	(65)	409
Cl I	Cu, neat, 230°	(53)	410
Cl I	Cu, neat, 190°	(35)	32
Cl I	Cu, neat, 190-260°	(39)	60
Cl OTf	PdCl₂(PPh₃)₂, DMF, 2e⁻, n-Bu₄NBF₄, 90°	(34)	191,192
Cl OTf	NiBr₂(PPh₃)₂, THF, Zn, Et₄NI, 50°	(56)	139
Cl OTf	NiCl₂, Zn, KI, HMPA, 20-50°	(25-84)[a]	129,403
Br I	Cu, neat, 220°	(7)	411
2-Cl-C₆H₄-NO₂	Cu, neat, heat	(32) after reduction to diamine	412
	Cu, neat, 215-225°	(61-66) 2,2'-(NO₂)₂-biphenyl	291,413

TABLE 2. 2,2'-DISUBSTITUTED BIARYLS (Continued)

Substrate	Conditions	Product(s) and Yield(s) (%)	Refs.
2-Cl-NO₂	Cu, neat, 240-250°	2,2'-dinitrobiphenyl (71)	414
	Cu, sand, 215-225°	(52-61)	15,415
	Cu, sand, 260-265°	(52-61)	416
	Cu, sand, 200-240°	(43)	417
	Cu, neat, 200-220°	(40)	418
	Cu, sand, 200-245°	(60)	1
	Cu, sand, 240-245°	(41)	419
	Cu, nitrobenzene, reflux	(0)	418
	Cu, nitrobenzene, 210°	(77)	58
	Cu, DMF, reflux	(80)	57
2-Br-NO₂	Cu, neat, 210-220°	(76)	1
	Cu, neat, 190°	(75)	32
	Cu, neat, 200-220°	(64)	418
	Cu, nitrobenzene, reflux	(45)	418
	Cu, methyl benzoate, 190°	(71)	46
	Cu, 1,2,4-trichlorobenzene, 190°	(71)	46
	Ni(cod)₂, DMF, 36°	(0)	49,50
	Cu(I) thiophene-2-carboxylate, NMP, 70°	(86)	71
	CuCl·DMS, dioxane, lithium naphthalide, 101°	(87)[a]	25
	Cu(OTf)₂, Cu, CH₃CN, acetone, aq. NH₃	(79),(90)[a]	68,83
2-I-NO₂	Cu, neat, heat	(23) after reduction to diamine	413
	Cu, neat, 200-220°	(65)	418
	Cu, neat, 190-240°	(96)	19
	Cu, neat, 60°	(99)	35

R	X			
OH	Br	Cu, nitrobenzene, reflux	(43)	418
		Cu, xylene, 120-140°	(93-97)	238
		CuCl·DMS, dioxane, lithium naphthalide, 101°	(87)	25,26
		CuI-PEt$_3$, lithium naphthalide, DME, 85°	(87)[a]	26
		CuI(I) thiophene-2-carboxylate, NMP, 23°	(92)	71
		Cu$_2$O, pyridine, 115°	(81)	66
		Cu(OTf), CH$_3$CN	(0)	67
		Cu(OTf), CH$_3$CN, sulfolane	(0)	67
		Cu(OTf), CH$_3$CN, IPA	(0)	67
		Cu(OTf), CH$_3$CN, acetone	(0)	67
		Cu(OTf), CH$_3$CN, acetone, aq. NH$_3$	(92)[a]	67
		Cu, acetone, CH$_3$CN, aq. NH$_3$	(82)	420
		Cu, CuSO$_4$, acetone, CH$_3$CN, aq. NH$_3$	(83)	420
		Ni$_2$, DME, lithium naphthalide, 80°	(<1)[a]	111
PO(OH)$_2$	Br	Ni(OAc)$_2$, NaH, t-AmONa, bpy, THF, C$_6$H$_6$, KI, 63°	(63)	54
		1. Pd/CaCO$_3$, NaOH, MeOH, reflux 2. HCl	(0)	420
NH$_2$	I	Pd/CaCO$_3$, N$_2$H$_4$·H$_2$O, KOH, MeOH, 135-140°	(20)	48
		1. Cu, CuSO$_4$, H$_2$O, reflux 2. PCl$_5$	(12)	421
		1. Cu, CuSO$_4$, H$_2$O, reflux 2. PCl$_5$	(75)	422

TABLE 2. 2,2'-DISUBSTITUTED BIPHENYL (Continued)

Substrate	Conditions	Product(s) and Yield(s) (%)	Refs.
C₇			
Cl	Ni(OAc)₂, NaH, t-AmONa, THF, 63°	(82), (84)[a]	131
Cl	Ni(OAc)₂, NaH, t-AmONa, bpy, THF, KI, 63°	(90)	54
Cl	Ni(OAc)₂, NaH, t-AmONa, bpy, THF, C₆H₆, KI, 63°	(80)	54
Cl	NiBr₂, bpy, PPh₃, THF, t-BuMgCl, reflux	(85)[a]	130
Cl	Pd/C, H₂, PEG-400, H₂O, NaOH, 100-110°	(33)[a]	186
Cl	Pd/C, Zn, PEG-400, H₂O, NaOH, heat	(27)	180
Br	Ni powder, DMF, 140°	(27)[a]	120
Br	NiBr₂(PPh₃)₂, Zn, THF, Et₄NI, 50°	(83)	139
Br	Ni(OAc)₂, NaH, t-AmONa, bpy, THF, KI, 63°	(75)	54
Br	Ni(OAc)₂, NaH, t-AmONa, bpy, THF, C₆H₆, KI, 63°	(80)	54
Br	NiCl₂, t-BuOLi, LiH, bpy, THF, 63°	(61), (76)[a]	110
Br	NiCl₂(PPh₃)₂, Zn, PPh₃, DMF, 40°	(12)[a]	123
Br	NiBr₂, EtOH, 2e⁻, dpa, H₂O	(43)	153
Br	NiBr₂, bpy, EtOH, MeOH, 2e⁻	(trace)	153
Br	Ni(cod)₂, DMF, 34°	(41)	49, 50
Br	Pd(OAc)₂, Et₃N, DMF, 115°	(60)	175
Br	Pd(OAc)₂, Bu₃N, DMF, AsPh₃, 140°	(0)	172
Br	Pd/C, HCO₂Na, SDPNS, NaOH, H₂O, 95°	(33)	179
Br	PdCl₂, o-tolylhydrazine hydrochloride, NaOH, MeOH, reflux	(99)	165
I	Cu, neat, sealed tube, 230°	(63)	2
I	Cu, neat, 190°	(25)	32
I	Cu, neat, 260°	(65)	423
I	Cu, neat, sealed tube, 230-240°	(75)	424
I	NiBr₂, Zn, KI, HMPA, 50°	(83)	129, 403

X	Conditions	Product	Yield (%)	Refs.
I	PdCl$_2$, N$_2$H$_4$·H$_2$O, MeOH, reflux		(17-54)	47
I	PdCl$_2$, N$_2$H$_4$·H$_2$O, MeOH, reflux, HgCl$_2$		(67-77)	47
I	Pd(OAc)$_2$, Bu$_3$N, DMF, 140°		(0)	172
I	Pd(OAc)$_2$, Et$_3$N, 100°		(10)	169
I	Pd(OAc)$_2$, n-Bu$_3$N, 100°		(1)	169
I	Pd/C, Zn, air, H$_2$O, acetone, 25°		(84)^a	185
I	[Pd(OAc)(P(o-tolyl)$_2$-benzyl)]$_2$ dimer, EtN(i-Pr)$_2$, DMF, 110°		(74)	194
OTf	NiCl$_2$, Zn, PPh$_3$, NaI, DMF, sonication, 60°		(82)^a	135
OTf	PdCl$_2$(PPh$_3$)$_2$, Zn, DMF, 90°		(0)^b	192
OFs	NiCl$_2$(PPh$_3$)$_2$, Zn, THF, Et$_4$NI, 67°		(72)^a	133

R	X	Conditions	Yield (%)	Refs.
CH$_2$OH	I	Cu(I) thiophene-2-carboxylate, NMP, 70°	(48) based on 62% conversion	71
CH$_2$OH	I	In, DMF, reflux	(78)	74
CF$_3$	Cl	Ni(OAc)$_2$, NaH, t-AmONa, bpy, THF, 63°	(71)	54
CF$_3$	Cl	Ni(OAc)$_2$, NaH, t-AmONa, bpy, THF, C$_6$H$_6$, KI, 63°	(70)	54
CF$_3$	Br	Ni(OAc)$_2$, NaH, t-AmONa, bpy, THF, 63°	(42)	54
CF$_3$	Br	Ni(OAc)$_2$, NaH, t-AmONa, bpy, THF, C$_6$H$_6$, KI, 63°	(62)	54

TABLE 2. 2,2′-DISUBSTITUTED BIARYLS (Continued)

Substrate	Conditions	Product(s) and Yield(s) (%)	Refs.
R = CF₃, X = I	Cu, neat, reflux	biaryl (R,R) (75)	425
R = CF₃, X = I	NiBr₂, Zn, KI, HMPA, 35°	(75)[a]	129
R = CN, X = Br	Cu, neat, 190°	(3)	32
R = CN, X = Br	Pd(OAc)₂, EtN(i-Pr)₂, DMF, 115°	(86), (89)[a]	175
X = Br (R = CHO)	Cu, neat, 190°	I (0)	32
X = Br	Pd(OAc)₂, EtN(Pr-i)₂, DMF, 115°	(62), (69)[a] I:II 84:16	175
X = Br	Ni(cod)₂, bpy, DMF, 60°	(96 overall) I:II 8:92	426
X = Br	1.2 eq Ni(cod)₂	(100 overall)	426
X = I	Cu, neat, 220°	(70)	427
X = I	Cu, DMF, reflux	(65)	428
R = H, X = I (CO₂R)	Indium, DMF, reflux	(75)	74
R = K, X = Br	1. Cu, H₂O, reflux; 2. acid	(76)	429

[a]

Products: 2,2′-disubstituted biaryl (I) and 9,10-dihydroxy-9,10-dihydrophenanthrene (II)

Substrate	Conditions	Product	Yield (%)	Ref
X=Cl	Ni(OAc)₂, NaH, t-AmONa, bpy, THF, 63°	2,2'-(OMe)₂-biphenyl	(91)	54
Cl	Ni(OAc)₂, NaH, t-AmONa, bpy, THF, C₆H₆, KI, 63°		(74)	54
Br	Ni(OAc)₂, NaH, t-AmONa, bpy, THF, 63°		(66-77), (69-79)ᵃ	131
Br	NiBr₂, EtOH, MeOH, NaBr, bpy, 2e⁻		(28)	53
Br	NiBr₂(PPh₃)₂, Zn, THF, Et₄NI, 50°		(81)	139
Br	Cu, neat, reflux		(0)	32
Br	Pd(OAc)₂, Bu₃N, DMF, AsPh₃, 140°		(0)	172
Br	Ni(OAc)₂, NaH, t-AmONa, bpy, THF, 63°		(74)	54
Br	Ni(OAc)₂, NaH, t-AmONa, bpy, THF, C₆H₆, KI, 63°		(68)	54
I	Cu, neat, 210-260°		(88)	2
I	Cu, neat, 190°		(70)	32
I	Indium, DMF, reflux		(88)	74
I	Pd(OAc)₂, hydroquinone, As(o-tol)₃, Cs₂CO₃, 100°		(82)	178
I	Pd/C, Zn, air, H₂O, acetone, 25°		(73)ᵃ	185
I	NiI₂, lithium naphthalide, DME, 80°		(<1)ᵃ	111
2-I-SMe-C₆H₄	Cu, neat, 200-250°	2,2'-(SMe)₂-biphenyl	(81)	430

TABLE 2. 2,2'-DISUBSTITUTED BIARYLS (Continued)

Substrate		Conditions	Product(s) and Yield(s) (%)	Refs.
C_8 (2-R-C$_6$H$_4$-X)			(2,2'-R,R-biphenyl)	
R	X			
Et	I	Cu, neat, 240°	(60)	431
COMe	Br	[PdCl(π-C$_3$H$_5$)]$_2$, Bu$_4$NF, DMSO, 120°	(0)	198
COMe	I	Cu, neat, heat	(45)	432
COMe	I	Cu, DMF, reflux	(59)	432
COMe	I	NiBr$_2$, KI, Zn, HMPA, 20°	(68), (96)[a]	129
(2-CO$_2$Me-C$_6$H$_4$-X)			(2-CO$_2$Me, 2'-CO$_2$Me biphenyl)	
X				
Cl		Cu, neat, 190°	(0)	32
Br		Cu, neat, 190°	(80)	32
Br		NiBr$_2$(PPh$_3$)$_2$, Zn, THF, 67°	(68)	139
Br		NiBr$_2$(PPh$_3$)$_2$, Zn, Et$_4$NI, THF, 67°	(79–90)	139
Br		Ni(cod)$_2$, DMF, 41-54°	(81)	49
Br		NiCl$_2$(PPh$_3$)$_2$, Zn, PPh$_3$, DMF, 40°	(33)[a]	123
C_{8-9} (2-R-C$_6$H$_4$-X)				
R	X			
NHCOCF$_3$	I	Cu(I) thiophene-2-carboxylate, NMP, 23°	(79) based on 89% conversion	71

	R	Conditions	Product	Yield (%)	Ref
C₉					
Me-N(H)-C(=O)-Me	I	Cu(I) thiophene-2-carboxylate, NMP, 23°	biphenyl with R at 2,2'	(90)	71
CH₂NHMe	Br	Cu(I) thiophene-2-carboxylate, NMP, 23°		(99)	71
CH₂NMe₂		Cu(I) thiophene-2-carboxylate, NMP, 23°		(97)	71
Me-N(Me)-C(=O)-Me		Cu(I) thiophene-2-carboxylate, NMP, 70°		(83)	71
H₂C-NHMe (C(=O))		Cu(I) thiophene-2-carboxylate, NMP, 23°		(51) based on 82% conversion	71
S=C(NMe₂)		Cu(I) thiophene-2-carboxylate, NMP, 23°		(94)	71
O-C(=O)-NMe₂		Cu(I) thiophene-2-carboxylate, NMP, 23°		(41) based on 50% conversion	71
2-(1,3-dioxolan-2-yl)phenyl-X	X = Cl	1. Ni(OAc)₂, NaH, t-AmONa, bpy, THF, C₆H₆, KI, 63° 2. acid	2,2'-diformylbiphenyl	(64)	54
	X = Br	1. Ni(OAc)₂, NaH, t-AmONa, bpy, THF, C₆H₆, KI, 63° 2. acid		(57)	54

TABLE 2. 2,2'-DISUBSTITUTED BIARYLS (Continued)

Substrate	Conditions	Product(s) and Yield(s) (%)	Refs.
C₁₀ 2-iodo-phenyl OTMS	1. Cu, quinoline, 240° 2. HCl	2,2'-dihydroxybiphenyl (25)	433
2-bromo-(t-Bu)phenyl	PdCl₂(PPh₃)₂, DMF, Et₄NOTs, 2e⁻	2,2'-di-t-Bu-biphenyl (0)	190
2-iodo-phenyl C(O)CH(CH₃)₂	Cu, DMF, reflux	2,2'-bis(isobutyryl)biphenyl (66)	432
2-iodo-phenyl CH₂CH(OMe)₂	Cu, neat, 235–260°	2,2'-bis(CH₂CH(OMe)₂)biphenyl (82)	434
2-X-phenyl-(2-methyl-1,3-dioxolane); X = Cl	1. Ni(OAc)₂, NaH, t-AmOH, bpy, THF, C₆H₆, KI, 63° 2. acid	2,2'-bis(COMe)biphenyl (65)	54
X = Br	1. Ni(OAc)₂, NaH, t-AmOH, bpy, THF, C₆H₆, KI, 63° 2. acid	(62)	54

Substrate	Conditions	Product	(Yield)	Ref.
2-I-C6H4-CO2SiMe3	1. Cu, quinoline, 200° 2. NaOH/HCl	2,2'-biphenyl-dicarboxylic acid	(70)	433
2-I-C6H4-CH2OSiMe3	1. Cu, quinoline, 240° 2. HCl	2,2'-bis(hydroxymethyl)biphenyl	(35)	433
2-I-C6H4-C6H4-R (C12)	Cu, neat, heat (R=H)	4,4''-R2-o-terphenyl	(69)	435
	Cu, neat, 260° (R=H)		(72)	436
	Cu, neat, 300° (R=H)		(52)	437
	Cu, neat, 220-225° (R=NO2)		(50)[c]	39
	Cu, neat, 225-235° (R=NO2)		(51)	438
2-I-C6H4-OPh	Cu, neat, 210-270°	2,2'-diphenoxybiphenyl	(75)	410

323

TABLE 2. 2,2'-DISUBSTITUTED BIARYLS (Continued)

Substrate	Conditions	Product(s) and Yield(s) (%)	Refs.
2-iodo-phenyl phenylsulfonate (SO₃Ph)	Cu, neat, 195°	2,2'-bis(SO₃Ph)biphenyl (78)	439
	Cu, neat, 180–210°	(81)	421
C₁₃ 2-iodo-phenyl tosylate (OTs)	Cu, neat, 260°	2,2'-bis(OTs)biphenyl (50)	2

R	X	Conditions	Product yield	Refs.
H	Cl	NiCl₂, NaI, triphenyl phosphite, NMP, Zn, 70°	(76)[a]	149
H	Cl	NiCl₂, NaI, tris(2-methylphenyl) phosphite, NMP, ZnCl₂, 60°	(100)[a]	149
H	Cl	NiCl₂, NaI, tris(phenylphenyl) phosphite, NMP, Zn, 60°	(99)[a]	149
H	Cl	NiCl₂, NaI, tris(2-methoxyphenyl) phosphite, NMP, Zn, 60°	(97)[a]	149
H	I	Cu, DMF, reflux	(76)	432
H	I	NiCl₂(PEt₃)₂, Zn, NMP, 30°	(67)	432
F	Cl	NiBr₂, Zn, DMAc, PPh₃, 80–90°	(82)	401
F	Br	Cu, neat, 200°	(30)	440
Cl	Br	Cu, neat, 200°	(0)	441
				441

Substrate	Conditions	Product(s)	Refs.
2-iodo-2'-methylbiphenyl	Cu, neat, 265–285°	2,2''-dimethyl-o-terphenyl (70)	442
C₁₄: 2-iodo-2'-(di-t-butylphosphinoyl)biphenyl	Cu, DMF, heat	bis[2-(di-t-butylphosphinoyl)biphenyl] (96)	443
2-bromo-3'-methylbenzophenone	Cu, neat, 250°	I + II (38)	441
2-bromo-4'-methylbenzophenone	Ni(cod)₂ (2 eq), bpy, DMF, 60°	I:II 76:24 (96 overall)	426
	Ni(cod)₂ (2 eq), bpy, DMF, 60°, H⁺	I:II 0:100 (99 overall)	426

Ar = C₆H₄Me-4

TABLE 2. 2,2'-DISUBSTITUTED BIARYLS (Continued)

Substrate	Conditions	Product(s) and Yield(s) (%)	Refs.
C₁₄ 2-bromophenyl-(N-R-imidazol-2-yl), R = OCH₂CH₂OSiMe₃	Ni(cod)₂, hv, toluene	2,2'-bis(N-R-imidazol-2-yl)biphenyl (66)	132
C₁₆ 2-bromo-4'-ethoxybenzophenone	Cu, neat, 250°	2,2'-bis(4-ethoxybenzoyl)biphenyl (57)	441
2-iodophenylferrocene	Cu, neat, 147°	2,2'-diferrocenylbiphenyl (11)	444
2-bromo-2',4',6'-trimethylbenzophenone	Cu, neat, 210°	2,2'-bis(2,4,6-trimethylbenzoyl)biphenyl (80)	445

C₁₇

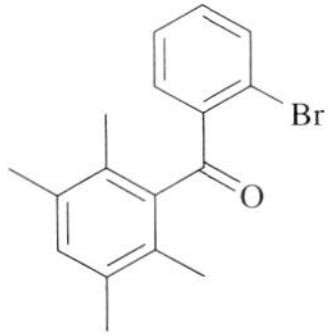

Cu, sand, 220-250° (67) 446

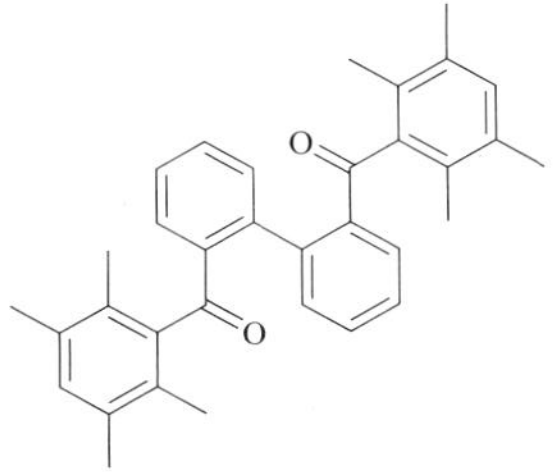

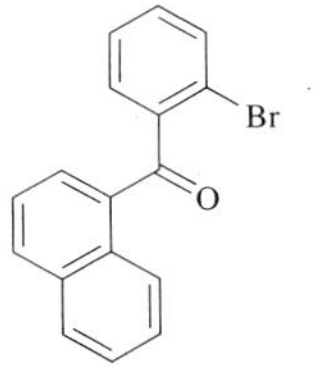

C₁₈

Cu, neat, 200° (66) 441

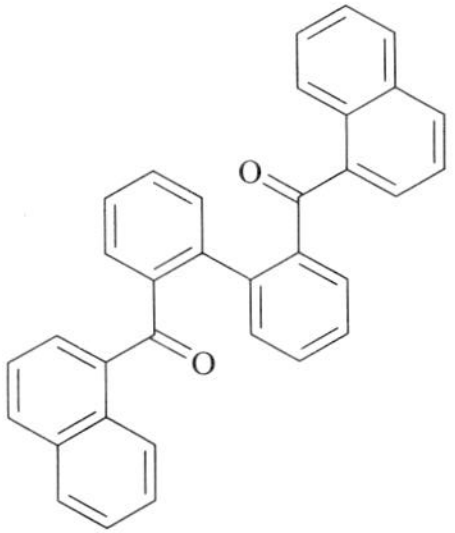

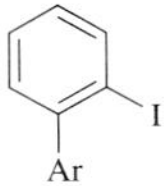

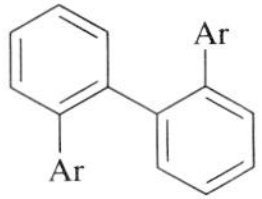

Ar		
biphenyl-2-yl	Cu, neat, 275-280°	(21)	447
biphenyl-3-yl	Cu, neat, 255-260°	(40)	447
biphenyl-4-yl	Cu, neat, 230-290°	(50)	291
biphenyl-4-yl	Cu, neat, 255-260°	(49)	447

TABLE 2. 2,2'-DISUBSTITUTED BIARYLS (Continued)

Substrate	Conditions	Product(s) and Yield(s) (%)	Refs.
C19, 2-iodo-Ph$_2$P(=O)-phenyl	Cu, DMF, reflux	2,2'-bis(Ph$_2$P=O)biphenyl (78)	448
2-bromo-ArC(=O)-phenyl; Ar = biphenyl-4-yl	Cu, neat, 200°	2,2'-bis(ArC=O)biphenyl (22)	441
C24, 2-iodo-2'-Ar-biphenyl		2',2'''-diAr-1,1':2',1'':2'',1'''-quaterphenyl	
Ar = biphenyl-2-yl	Cu, neat, 255-265°	(3)	449
Ar = biphenyl-3-yl	Cu, neat, 275-285°	(23)	449
Ar = biphenyl-4-yl	Cu, neat, 225-235°	(12)	449

328

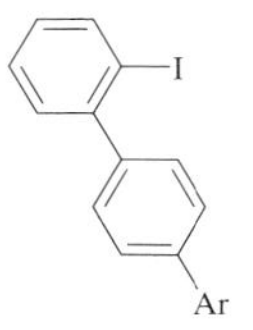

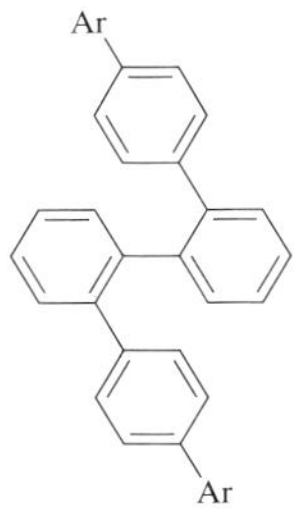

Ar		
biphenyl-2-yl	Cu, neat, 265-275°	(67)
biphenyl-3-yl	Cu, neat, 265-275°	(63)
biphenyl-4-yl	Cu, neat, 225-235°	(53)

438
438
438

C$_{25}$

	I:II	
Ni(cod)$_2$ (2 eq), bpy, DMF, 60°.	(92 overall) I:II	78:22	426
Ni(cod)$_2$ (2 eq), bpy, DMF, 60°, H$^+$	(97 overall) I:II	0:100	426

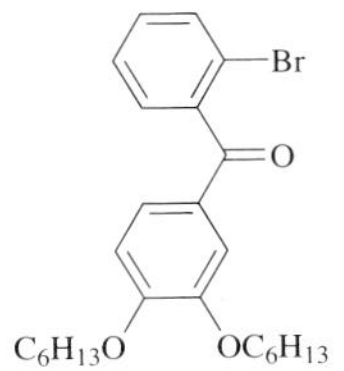

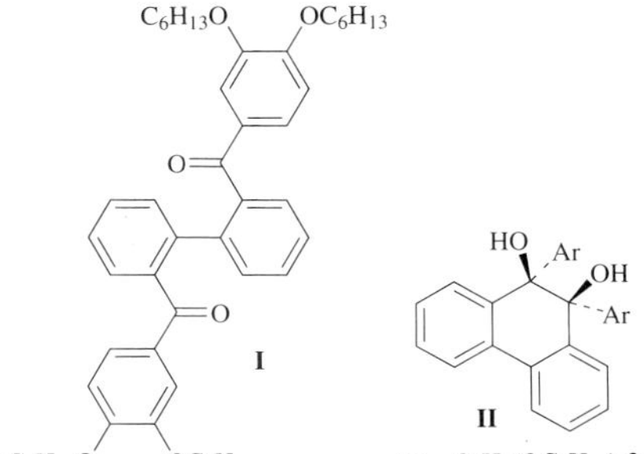

Ar = C$_6$H$_3$(OC$_6$H$_{13}$)-3,4

[a] The yield was determined by gas chromatography.
[b] The yield was determined by NMR spectroscopy.
[c] Excess Cu powder (>1.4 eq) reduced nitro to amine.

TABLE 3. 3,3'-DISUBSTITUTED BIARYLS

Substrate		Conditions	Product(s) and Yield(s) (%)	Refs.
C_6				
3-F-C$_6$H$_4$-X	X		3,3'-F$_2$-biphenyl	
	Cl	Ni(OAc)$_2$, NaH, t-AmONa, bpy, THF, C$_6$H$_6$, KI, 63°	(84)	54
	Br	Ni(OAc)$_2$, NaH, t-AmONa, bpy, THF, C$_6$H$_6$, KI, 63°	(74)	54
	Br	Ni(OAc)$_2$, t-BuOLi, bpy, THF, 63°	(87), (100)^a	110
	Br	Pd(OAc)$_2$, n-Bu$_4$NBr, H$_2$O, IPA, K$_2$CO$_3$, DMF, 115°	(86)	174
	Br	NiBr$_2$, bpy, 2e$^-$, EtOH, DMF, Bu$_4$NBF$_4$	(53)	53
	Br	NiBr$_2$, bpy, 2e$^-$, EtOH, MeOH, NaBr	(46)	53
	Br	Pd(OAc)$_2$, H$_2$O, IPA, K$_2$CO$_3$, DMF, 115°	(86), (97)^a	175
	I	Cu, neat, heat	(23)	409
	I	[PdCl(π-C$_3$H$_5$)]$_2$, TBAF, DMSO, 120°	(68)	198
3-Cl-C$_6$H$_4$-X	X		3,3'-Cl$_2$-biphenyl	
	Br	Pd(OAc)$_2$, n-Bu$_4$NBr, H$_2$O, K$_2$CO$_3$, DMF, 115°	(42), (83)^a	174, 175
	I	Cu, neat, 180-260°	(55)	59
	I	Cu, neat, 230-250°	(64)	410
	I	Cu, neat, 250°	(96)	450
	I	Cu, neat, 250°	(67)	2
	I	Cu, neat, 190°	(40)	32
	I	Pd(OAc)$_2$, K$_2$CO$_3$, DMF, H$_2$O, IPA, 115°	(82), (94)^a	175
	I	Cu, neat, 190-260°	(55)	59
	SO$_2$Cl	PdCl$_2$(PhCN)$_2$, Ti(OPr-i)$_4$, m-xylene, 140°	(67)	168

Starting material	X	Conditions	Product	Yield	Ref
3-X-C6H4-NH2	Cl	Pd/CaCO3, N2H4·H2O, MeOH, KOH, 135-140°	3,3'-diaminobiphenyl	(47)	48
	Br	Pd/CaCO3, N2H4·H2O, MeOH, KOH, 135-140°		(54)	48

X	Conditions	Yield	Ref
Cl	Cu, neat, 200-220°	(0)	418
Cl	Cu, nitrobenzene, reflux	(0)	418
Br	Cu, neat, 190°	(3)	32
Br	Cu, neat, 200-220°	(15)	418
Br	Cu, nitrobenzene, reflux	(0)	418
I	Cu, neat, 190°	(30)	32
I	Cu, neat, heat	(26)	417
I	Cu, nitrobenzene, 210°	(48)	58
I	Cu, neat, 200-220°	(0)	418
I	Cu, nitrobenzene, reflux	(36)	418
I	Cu, neat, 60°	(0)	35
I	Cu, DMF, 210°	(82)	451
I	Pd(OAc)2, As(tol-o)3, hydroquinone, Cs2CO3, 75°	(86)	178
I	Cu, neat, 200-225°	(52)	1

3,3'-dinitrobiphenyl product (O2N–C6H4–C6H4–NO2)

X	R	Conditions	Yield	Ref
I	SO2F	Cu, neat, 200-230°	(54)	452
Br	PO(OH)2	1. Pd/CaCO3, NaOH, MeOH, reflux; 2. HCl	(23)	453
Cl	OH	Ni(OAc)2, NaH, t-AmOH, bpy, THF, C6H6, KI, 63°	(70)	54

TABLE 3. 3,3'-DISUBSTITUTED BIARYLS (*Continued*)

Substrate	Conditions	Product(s) and Yield(s) (%)	Refs.
C_7			
X = Cl	NiBr$_2$, bpy, 2e$^-$, NMP	(80)	128
Cl	Ni powder, KI, DMF, 140°	(72)	120
Cl	Pd/C, HCO$_2$Na, NaOH, H$_2$O, CTAB, 95°	(36)	179
Cl	Ni(OAc)$_2$, NaH, *t*-AmONa, THF, bpy, 63°	(75), (80)^a	131
Cl	Ni(OAc)$_2$, NaH, *t*-AmONa, bpy, THF, KI, 63°	(85)	54
Cl	Ni(OAc)$_2$, NaH, *t*-AmONa, bpy, THF, C$_6$H$_6$, KI, 63°	(77)	54
Br	Ni(OAc)$_2$, NaH, *t*-AmONa, bpy, THF, 63°	(88)	54
Br	Ni(OAc)$_2$, NaH, *t*-AmONa, bpy, THF, C$_6$H$_6$, KI, 63°	(83)	54
Br	Pd/CaCO$_3$, N$_2$H$_4$•H$_2$O, MeOH, KOH, 135-140°	(52)	48
Br	Ni powder, DMF, 140°	(68)	120
Br	Ni(OAc)$_2$, LiH, *t*-BuOLi, bpy, THF, 63°	(92), (100)^a	110
Br	PdCl$_2$, *m*-tolylhydrazine hydrochloride, NaOH, MeOH, HgCl$_2$, reflux	(quantitative)	165
I	Cu, neat, 205-240°	(35)	2
I	Cu, neat, 190°	(50)	32
I	Cu, DMF, reflux	(55)	57
I	PdCl$_2$, N$_2$H$_4$•H$_2$O, MeOH, reflux	(28)	47
I	Pd/C, Zn, air, H$_2$O, acetone, 25°	(89)^a	185
OTf	NiCl$_2$, Zn, PPh$_3$, NaI, DMF, sonication, 60°	(82)^a	135

Substrate	X	Conditions	Product	Yield (%)	Ref
3-CF₃-C₆H₄-X	Cl	Ni(OAc)₂, NaH, t-AmONa, bpy, KI, 63°	3,3'-(CF₃)₂-biphenyl	(93)	54
	Cl	Ni(OAc)₂, NaH, t-AmONa, bpy, THF, C₆H₆, KI, 63°		(77)	54
	Br	Ni(OAc)₂, NaH, t-AmONa, bpy, THF, 63°		(87)	54
	Br	Ni(OAc)₂, NaH, t-AmONa, bpy, THF, C₆H₆, KI, 63°		(80)	54
	Br	Pd(OAc)₂, Bu₃N, DMF, AsPh₃, 140°		(0)	172
	I	Cu, neat, reflux		(72)	425
	I	NiCl₂(PPh₃)₂, PPh₃, Zn, NaBr, DMF, 100°		(69)	137
3-NC-C₆H₄-Br		Pd(OAc)₂, n-Bu₄NBr, DMF, IPA, Et₃N, 115°	3,3'-(CN)₂-biphenyl	(57), (64)ᵃ	174
		Pd(OAc)₂, DMF, IPA, Et₃N, 115°		(57), (64)ᵃ	175
C₇₋₁₄ 3-OHC-C₆H₄-X	Cl	Pd(OAc)₂, n-Bu₄NBr, DMF, Et₃N, 115°	3,3'-(CHO)₂-biphenyl	(60), (67)ᵃ	174
	Br	Pd(OAc)₂, DMF, Et₃N, 115°		(53), (67)ᵃ	175
	OTs	NiCl₂(PPh₃)₂, PPh₃, Zn, NaBr, DMF, 100°		(62)	137
3-HO₂C-C₆H₄-Br		1. Pd/CaCO₃, N₂H₄·H₂O, MeOH, KOH, 135-140°; 2. acid	3,3'-(CO₂H)₂-biphenyl	(57)	48
		1. Pd/C, NaOH, H₂O, formic hydrazide, 85°; 2. acid		(88)ᵇ	55

TABLE 3. 3,3'-DISUBSTITUTED BIARYLS (Continued)

Substrate	Conditions	Product(s) and Yield(s) (%)	Refs.
C7 (MeO-C6H4-X)		3,3'-(MeO)2-biphenyl	
X = Cl	Ni(OAc)2, NaH, t-AmONa, bpy, THF, 63°	(82)	54
Cl	Ni(OAc)2, NaH, t-AmONa, bpy, THF, C6H6, KI, 63°	(73)	54
Cl	Ni(OAc)2, LiH, t-BuOLi, bpy, THF, 63°	(70), (79)[a]	110
Br	Ni(OAc)2, NaH, t-AmONa, bpy, THF, 63°	(82)	54
Br	Ni(OAc)2, NaH, t-AmONa, bpy, THF, C6H6, KI, 63°	(82)	54
Br	Ni(OAc)2, LiH, t-BuOLi, bpy, THF, 63°	(93), (100)[a]	110
Br	Pd(OAc)2, Bu3N, DMF, AsPh3, 140°	(57)	172
Br	Ni(OAc)2, NaH, t-AmONa, THF, bpy, 63°	(71), (75)[a]	131
Br	NiCl2(PEt3)2, Zn, NMP, 30°	(91)	401
I	Cu, neat, 190°	(65)	32
I	Pd(OAc)2, As(o-tol)3, hydroquinone, Cs2CO3, 75°	(96)	178
I	Pd/C, Zn, air, H2O, acetone	(79)[a]	185
C7 (3-Br-styrene)	Pd(PPh3)4, DME, n-Bu3N, reflux	3,3'-divinylbiphenyl (23)	454
C8 (3-X-acetophenone)		3,3'-diacetylbiphenyl	
X = Br	NiCl2(PPh3)2, Zn, Et4NI, THF, 50°	(75)	455
I	Cu, neat, 240-260°	(36)	456

$C_{8,9}$

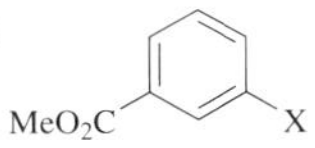

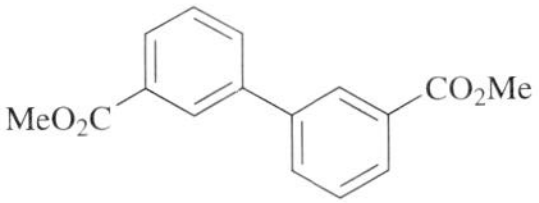

X		
Cl	NiBr$_2$(PPh$_3$)$_2$, Zn, Et$_4$NI, THF, 50°	(81)	139
Br	NiBr$_2$(PPh$_3$)$_2$, Zn, Et$_4$NI, THF, 50°	(85)	139
Br	NiBr$_2$(PPh$_3$)$_2$, Zn, PPh$_3$, DMF, 50°	(85)	122
I	NiBr$_2$, Zn, KI, HMPA, 50°	(96)	129, 403
I	NiCl$_2$(PPh$_3$)$_2$, PPh$_3$, Zn, DMF, 54°	(65)	457
I	Cu, neat, 260°	(58)	2
I	Cu, neat, 240°	(32)	458
I	Cu, neat, 190°	(25)	32
I	Cu, DMF, reflux	(89)	57
OMs	NiBr$_2$(PPh$_3$)$_2$, Zn, Et$_4$NI, THF, 67°	(93)^a	133

$C_{9,10}$

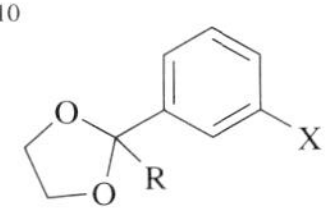

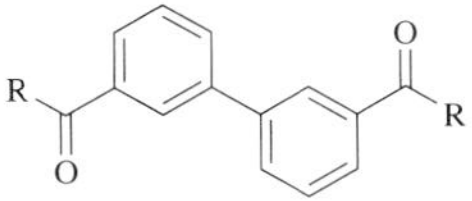

R	X		
H	Cl	Ni(OAc)$_2$, NaH, t-AmONa, bpy, THF, C$_6$H$_6$, KI, 63°	(75)	54
H	Br	Ni(OAc)$_2$, NaH, t-AmONa, bpy, THF, C$_6$H$_6$, KI, 63°	(71)	54
Me	Cl	Ni(OAc)$_2$, NaH, t-AmONa, bpy, THF, C$_6$H$_6$, KI, 63°	(73)	54
Me	Br	Ni(OAc)$_2$, NaH, t-AmONa, bpy, THF, C$_6$H$_6$, KI, 63°	(68)	54

C_{10}

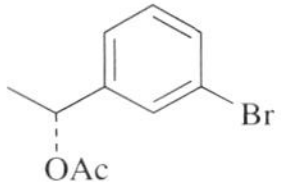

NiCl$_2$(PPh$_3$)$_2$, Zn, pyridine   (65)   459

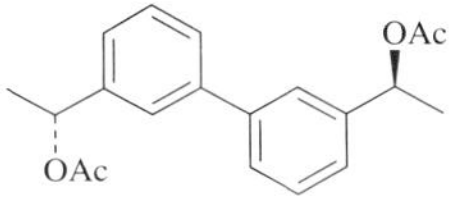

TABLE 3. 3,3'-DISUBSTITUTED BIARYLS (Continued)

Substrate	Conditions	Product(s) and Yield(s) (%)	Refs.
C₁₂ 3-iodobiphenyl	Cu, neat, 225-312° Cu, sand, 250-260°	(16) (83) crude	437 460
3-iodo-3'-nitrobiphenyl	Cu, neat, 225-235°	(62)	449
3-iodophenyl phenyl ether	Cu, neat, 210-270°	(68)	410
C₁₃ 3-chlorobenzophenone	NiCl₂, tris(2-phenylphenyl) phosphite, NMP, Zn, NaI, 60°	(97)^a	149
3-iodophenyl benzoate	Cu, neat, 215-260°	(41)	2

C$_{18}$			
[3-iodo-biphenyl structure, Ar at 3-position]	Cu, neat, 260–265°	[m,m-biphenyl Ar-Ar] (60)	447
Ar = biphenyl-2-yl			
C$_{24}$			
[3,3'-diiodobiphenyl with Ar substituents]			
	Ar		
	biphenyl-2-yl		
	biphenyl-3-yl		
	Cu, neat, 260–265°	(60)	449
	Cu, neat, 225–235°	(54)	449
[3'-iodo-4-Ar-biphenyl structure]			
	Ar		
	biphenyl-2-yl		
	biphenyl-3-yl		
	Cu, neat, 225–235°	(64)	438
	Cu, neat, 225–235°	(50)	438
[3,3'-diiodobiphenyl with Ar substituents]	Cu, neat, 265–270°	(63)	449
Ar = biphenyl-4-yl			

[a] The yield was determined by gas chromatography.
[b] The yield was determined by HPLC.

TABLE 4. 4,4′-DISUBSTITUTED BIARYLS

Substrate		Conditions	Product(s) and Yield(s) (%)		Refs.
C_6 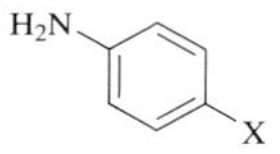					
H_2N–C$_6H_4$–X			H_2N–C$_6H_4$–C$_6H_4$–NH_2		
	X				
	I	$PdCl_2$, $HgCl_2$, $N_2H_4 \cdot H_2O$, MeOH, reflux	(26)		47
	Br	$Pd(OAc)_2$, Bu_3N, $AsPh_3$, DMF, 140°	(0)		172
	Br	$Pd/CaCO_3$, $N_2H_4 \cdot H_2O$, KOH, MeOH, 135–140°	(60) derivative		48
	Br	$Ni(cod)_2$, DMF, 35–45°	(30), (54)[a]		49, 50
	Cl	$Pd/CaCO_3$, $N_2H_4 \cdot H_2O$, KOH, MeOH, 135–140°	(47)		48
	Cl	$NiCl_2$, PPh_3, Zn, DMAc, 80°	(31)[b]		51, 52
O_2N–C$_6H_4$–I			O_2N–C$_6H_4$–C$_6H_4$–NO_2		
		Cu, neat, 220–235°	(52)		1
		Cu, neat, 190°	(25)		32
		Cu, neat, 200–220°	(54)		418
		Cu, neat, 140°	(<8)		238
		Cu, nitrobenzene, 210°	(61)		58
		Cu, nitrobenzene, reflux	(25)		418
		Cu, xylene, 140°	(92)		238
		Cu, biphenyl, 200°	(58)		238
		CuI, potassium naphthalide, DME, 85°	(3)		24
		Cu, DMF, reflux	(75)		407
		$NiBr_2 \cdot KI$, Zn, HMPA, 50°	(0)		129
		$NiCl_2$, $CrCl_2$,	(0)		406
		Mn, THF, rt			

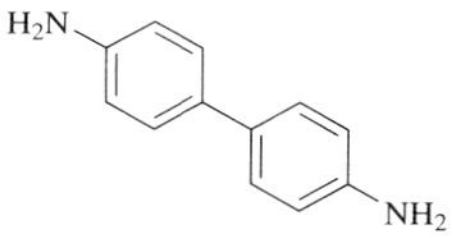

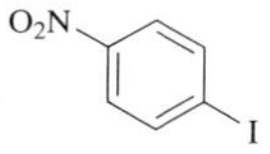

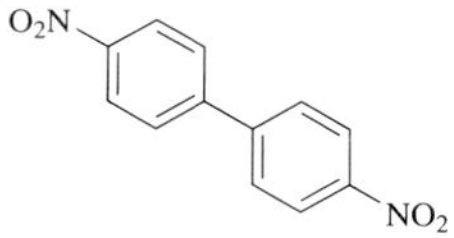

[Structure shown: a bicyclic pinene-fused quinoline-type pyridine product associated with the $NiCl_2, CrCl_2$ entry]

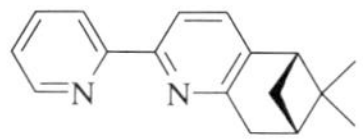

R⁴	X	Conditions	Yield	Ref.
NO_2	Br	$Pd(OAc)_2$, Et_3N, 100°	(54)	169
NO_2	Br	$Pd(OAc)_2$, n-Bu_3N, 100°	(0)	169
NO_2	Br	$Pd(OAc)_2$, Et_3N, CH_3CN, BNAH, 100°	(77)	177
NO_2	Br	$Pd(OAc)_2$, n-Bu_3N, DMF, 140°	(47)	172
NO_2	Br	$Pd(OAc)_2$, PPh_3, NaOAc, DMF, 110-115°	(46)	461
NO_2	Br	$[PdCl(p-C_3H_5)]_2$, TBAF, DMSO, 120°	(77)	198
NO_2	Br	$Pd(OAc)_2$, $As(o\text{-}tol)_3$, hydroquinone, Cs_2CO_3, 75°	(86)	178
NO_2	Br	$Pd(OAc)_2$, $P(o\text{-}tol)_3$, hydroquinone, Cs_2CO_3, 75°	(89)	178
NO_2	Br	[Pd catalyst with OAc, R = o-tolyl], $EtN(i\text{-}Pr)_2$, DMF, 110°	(80)	194
NO_2	Br	Cu, neat, 190°	(3)	32
NO_2	Br	Cu, neat, 200-220°	(36)	418
NO_2	Br	Cu, nitrobenzene, reflux	(15)	418
NO_2	Br	Ni powder, DMF, 140°	(0)	120
NO_2	Br	$Ni(CO_2(PPh_3)_2$, DMSO, 70°	(0)[b]	145
NO_2	Br	$Ni(cod)_2$, DMF, 40°	(0)	49, 50
NO_2	Br	$Pd(OAc)_2$, n-Bu_3N, DMF, $AsPh_3$, 140°	(30)	172
NO_2	Br	$Pd(OAc)_2$, DMF, K_2CO_3, H_2O, IPA, 115°	(31), (74)[b]	175
NO_2	Br	$Pd(OAc)_2$, DMF, K_2CO_3, H_2O, IPA, n-Bu_4NBr, 115°	(85)	174
NO_2	Br	$Pd(OAc)_2$, $As(o\text{-}tol)_3$, hydroquinone, Cs_2CO_3, 100°	(88)	178
NO_2	Cl	Cu, neat, 200-220°	(0)	418
NO_2	Cl	Cu, nitrobenzene, reflux	(0)	418
NO_2	Cl	$NiCl_2$, PPh_3, Zn, DMAc, 80°	(0)[b]	51
NO_2	Cl	$NiBr_2$, bpy, NMP, 2e⁻	(0)	128
NO_2	Cl	$Pd(OAc)_2$, Et_3N, DMF, 115°	(17),(20)[b]	175

TABLE 4. 4,4'-DISUBSTITUTED BIARYLS (Continued)

Substrate		Conditions	Product(s) and Yield(s) (%)	Refs.
R⁴	X		R⁴–⟨⟩–⟨⟩–R⁴	
NO₂	OMs	NiCl₂(PPh₃)₂, Zn, Et₄NI, THF, 67°	(0)	133
OH	Br	NiBr₂, bpy, 2e⁻, EtOH, MeOH, NaBr	(86)	53
OH	Br	NiBr₂, bpy, 2e⁻, EtOH, DMF, Bu₄NBF₄	(84)	53
OH	Br	Ni(OAc)₂, NaH, t-AmONa, bpy, THF, 63°	(68)	54
OH	Br	Ni(OAc)₂, NaH, t-AmONa, bpy, KI, C₆H₆, THF, 63°	(76)	54
OH	Br	NiBr₂, bpy, NMP, 2e⁻	(0)	128
OH	Br	Ni(cod)₂, DMF, 40°	(0)	49, 50
O⁻Na⁺	Br	Ni(cod)₂, DMF, 37-60°	(3)	49, 50
O⁻Na	Br	Pd/CaCO₃, N₂H₄·H₂O, MeOH, KOH, 135-140°	(13)	48
O⁻Na	Br	Pd(OAc)₂, n-Bu₃N, DMF, AsPh₃, 140°	(0)	172
O⁻Na⁺	Br	Pd/BaSO₄, NaOH, H₂O, formic hydrazide, 80°	(79)ᵇ	55
OH	Cl	NiCl₂, PPh₃, Zn, DMAc, 80°	(25)ᵇ	51
F	Br	NiBr₂, bpy, 2e⁻, EtOH, MeOH, NaBr	(82)	53
F	Br	NiBr₂, bpy, 2e⁻, EtOH, DMF, Bu₄NBF₄	(63)	53
F	Br	NiCl₂, bpy, DMF, NaBr, Zn, 60-80°	(78)	146
F	Br	NiBr₂, dpa, H₂O, 2e⁻, EtOH	(90)	153
F	Br	NiBr₂, bpy, 2e⁻, EtOH, MeOH	(82)	153
F	Br	Ni(OAc)₂, NaH, t-AmONa, bpy, THF, 63°	(85)	54
F	Br	Ni(OAc)₂, NaH, t-AmONa, bpy, THF, KI, C₆H₆, 63°	(64)	54
F	Br	Pd/C, NaOH, diethylene glycol dimethyl ether, polyethylene glycol dimethyl ether, ethylene glycol, 100°	(74-89)ᵇ	462
F	Cl	Ni(OAc)₂, NaH, t-AmONa, bpy, THF, KI, C₆H₆, 63°	(75)	54
F	Cl	Pd/C, NaOH, diethylene glycol dimethyl ether, polyethylene glycol dimethyl ether, ethylene glycol, 100°	(66)ᵇ	462
F	I	Cu, neat, heat	(60)	409
F	I	Pd/C, Zn, air, H₂O, acetone, 25°	(96)ᵇ	185

F	OMs	NiCl$_2$(PPh$_3$)$_2$, Zn, Et$_4$NI, THF, 67°	(54)	133
F	OTf	NiCl$_2$(dppe), Zn, KI, DMF, THF, 67°	(85)	192
F	SO$_2$Cl	PdCl$_2$(PhCN)$_2$, Ti(OPr-i)$_4$, m-xylene, N$_2$, 140°	(74)	168
PO(OH)$_2$	Br	1. Pd/CaCO$_3$, MeOH, NaOH, reflux 2. HCl	(28)	453
Cl	Br	Pd(OAc)$_2$, n-Bu$_3$N, DMF, AsPh$_3$, 140°	(48)	172
Cl	Br	Ni(OAc)$_2$, NaH, t-AmONa, bpy, THF, 63°	(60)	54
Cl	Br	NiBr$_2$, lithium naphthalide, DME, 80°	(36)^b	111
Cl	Br	NiI$_2$, lithium naphthalide, DME, 80°	(61)^b	111
Cl	Br	NiBr$_2$, lithium naphthalide, DMSO, 80°	(41)^b	111
Cl	Br	NiBr$_2$, lithium naphthalide, DMF, 80°	(<1)^b	111
Cl	Br	NiCl$_2$, CrCl$_2$, Mn, THF, rt	(25)	406
Cl	I	Cu, neat, 180-260°	(63)	60
Cl	I	Cu, neat, 230-250°	(48)	410
Cl	I	Cu, neat, 200-250°	(82)	2
Cl	I	Cu, neat, 190°	(40)	32
Cl	I	Cu, methyl benzoate, 190°	(15)	46
Cl	I	Cu, 1,2,4-trichlorobenzene, 190°	(30)	46
Cl	I	Pd(OAc)$_2$, Et$_3$N, 100°	(57)	169
Cl	I	Pd(OAc)$_2$, n-Bu$_3$N, 100°	(67)	169
Cl	I	Pd/C, Zn, air, H$_2$O, acetone, 25°	(94)^b	185
Cl	I	NiI$_2$, lithium naphthalide, DME, 80°	(75)^b	111
Cl	I	NiBr$_2$, lithium naphthalide, DME, 80°	(77)^b	111
Cl	I	NiCl$_2$, lithium naphthalide, DME, 80°	(74)^b	111
Cl	OTf	PdCl$_2$(PPh$_3$)$_2$, DMF, 2e$^-$, n-Bu$_4$NBF$_4$ 90°	(57)	191, 192
Cl	OTf	PdCl$_2$(PPh$_3$)$_2$, DMF, Zn, 90°	(17)^a	191, 193
Cl	OTf	NiCl$_2$(dppe), KI, DMF, THF, Zn, 67°	(0)	191, 193
Cl	SO$_2$Cl	PdCl$_2$(PhCN)$_2$, Ti(OPr-i)$_4$, m-xylene, N$_2$, 140°	(74), (76)^b	168
Cl	SO$_2$Cl	Pd(OAc)$_2$, Ti(OPr-i)$_4$, m-xylene, N$_2$, 140°	(71)^b	168
Cl	SO$_2$Na	PdCl$_2$, H$_2$O, 90°	(73)	167

TABLE 4. 4,4'-DISUBSTITUTED BIARYLS (Continued)

Substrate	Conditions	Product(s) and Yield(s) (%)	Refs.
C₆ R⁴–C₆H₄–X		R⁴–C₆H₄–C₆H₄–R⁴	
Br, X = I	Pd(OAc)₂, n-Bu₃N, DMF, 140°	(41)	172
Br, X = SO₂Cl	PdCl₂(PhCN)₂, Ti(OPr-i)₄, m-xylene, N₂, 140°	(56)	168
C₇ 4-Cl-C₆H₄-CH₃		4-CH₃-C₆H₄-C₆H₄-4-CH₃	
	NiBr₂(PPh₃)₂, Zn, Et₄NI, THF, 50°	(81)	139
	NiCl₂(quinoline)₂, Zn, pyridine, 80°	(4)[b]	463
	NiCl₂, PPh₃, DMAc, Zn, 80°	(90)[b]	51
	Ni(OAc)₂, NaH, t-AmONa, THF, 63°, PPh₃	(63)	131
	Ni(OAc)₂, NaH, t-AmONa, THF, 63°, bpy	(75), (80)[b]	131
	NiCl₂, NaI, Zn, NMP, 60-80°, tris(2-methylphenyl) phosphite	(97)[b]	149
	NiCl₂, NaI, Zn, NMP, 60°, tris(2-phenylphenyl) phosphite	(99)[b]	149
	NiCl₂, NaI, Zn, NMP, 60-80°, tris(2-methoxyphenyl) phosphite	(99)[b]	149
	NiCl₂, NaI, Zn, NMP, 60-65°, tris(2-t-butylphenyl) phosphite	(99)[b]	149
	Ni(OAc)₂, t-BuOLi, LiH, bpy, THF, 63°	(91), (94)[b]	110
	Ni(OAc)₂, NaH, t-AmONa, bpy, THF, 63°	(90)	54
	Ni(OAc)₂, NaH, t-AmONa, bpy, KI, C₆H₆, THF, 63°	(74)	54
	Ni powder, DMF, 140°	(<2)[b]	120
	Ni powder, DMF, KI, 140°	(83)	120
	Pd/C, HCO₂Na, NaOH, H₂O, CTAB, 95°	(55)	179
	Pd/C, HCO₂Na, NaOH, H₂O, CTAB, 110°	(60), (67)[b]	181
	Pd/C, H₂, PEG-400, H₂O, NaOH, 100-110°	(51), (56)[b]	186
	Pd/C, NaOH, H₂O, formic hydrazide, 80°	(59)[b]	55
	Pd/C, Zn, PEG-400, H₂O, NaOH, heat	(51)	180

	DMSO, 2e⁻, 65°	(91)[b]	154
R = Pr-i	THF, 2e⁻, 65°	(96)[b]	154

4-Br-C₆H₄-CH₃ → 4,4'-dimethylbiphenyl

Ni powder, DMF, 140°	(71)	120
Ni(CO)₂(PPh₃)₂, DMSO, 70°	(70)[b]	145
NiCl₂, bpy, NaBr, DMF, Zn, 60-80°	(80)	146
NiBr₂, bpy, 2e⁻, NaBr, EtOH, MeOH	(90)	53
NiBr₂(PPh₃)₂, Zn, Et₄NI, THF, 50°	(89)	139
Ni(OAc)₂, NaH, bpy, t-AmONa, THF, 63°	(70-80)	131
NiCl₂(PPh₃)₂, Zn, PPh₃, DMF, 50°	(73)[b]	123
NiCl₂(PPh₃)₂, Zn, PPh₃, DMF, 50°	(60)	122
NiCl₂(PPh₃)₂, PPh₃, DMF, n-Bu₄NBr, 2e⁻	(75)	127
Ni(OAc)₂, t-BuOLi, LiH, bpy, THF, 63°	(74), (93)[b]	110
Ni(OAc)₂, NaH, t-AmONa, bpy, THF, KI, 63°	(84)	54
Ni(OAc)₂, NaH, t-AmONa, bpy, KI, C₆H₆, THF, 63°	(82)	54
NiCl₂, CrCl₂, Mn, THF, rt	(90)	406
Pd/C, NaOH, H₂O, formic hydrazide, 95°	(97)[b]	55
Pd(OAc)₂, n-Bu₃N, DMF, AsPh₃, 140°	(58-60)	171, 172
Pd/CaCO₃, N₂H₄·H₂O, NaOH, MeOH, 135-140°	(66)	48
PdCl₂, HgCl₂, p-tolylhydrazine hydrochloride, NaOH, MeOH, reflux	(quantitative)	165
PdCl₂, HgCl₂, phenylhydrazine, NaOH, MeOH, reflux	(76)	165
PdCl₂, phenylhydrazine, NaOH, MeOH, reflux	(27)	165

TABLE 4. 4,4'-DISUBSTITUTED BIARYLS (Continued)

Substrate	Conditions	Product(s) and Yield(s) (%)	Refs.
4-I-C6H4-CH3	Cu, neat, 210-260°	4,4'-bitolyl (54)	2
	Cu, neat, 260°	(60)	464
	Cu, neat, 190°	(45)	32
	Cu, methyl benzoate, 190°	(4)	46
	Cu, 1,2,4-trichlorobenzene, 190°	(14)	46
	Cu, DMF, reflux	(68)	57
	Ni(cod)$_2$, DMF, 40°	(63)	49, 50
	NiCl$_2$, KI, Zn, HMPA, 50°	(92)	129, 403
	NiCl$_2$(PEt$_3$)$_2$, Zn, NMP, 30°	(87)	401
	NiCl$_2$(quinoline)$_2$, Zn, pyridine, 80°	(4)	463
	Pd(OAc)$_2$, n-Bu$_3$N, 100°	(34)	169
	Pd(OAc)$_2$, Et$_3$N, 100°	(50)	169
	Pd(OAc)$_2$, n-Bu$_3$N, DMF, 140°	(54)	172
	Pd(dba)$_2$, TBAF, DMF, PhSi(OMe)$_3$	(40)	197
	PdCl$_2$, HgCl$_2$, phenylhydrazine, NaOH, MeOH, reflux	(77)	165
	PdCl$_2$, phenylhydrazine, NaOH, MeOH, reflux	(44)	165
	PdCl$_2$, N$_2$H$_4$•H$_2$O, MeOH, reflux	(48)	47
	PdCl$_2$, HgCl$_2$, N$_2$H$_4$•H$_2$O, MeOH, reflux	(quantitative)	47
	Pd/C, Zn, air, H$_2$O, acetone, 25°	(92)[b]	185
	[Pd(OAc)(P(R)$_2$-o-C$_6$H$_4$)]$_2$, EtN(Pr-i)$_2$, DMF, 110° R = o-tolyl	(87)	194

[Structure: 4,4'-dimethylbiphenyl]

X			
OTf	NiCl$_2$(dppe), KI, Zn, DMF, 90°	(94)	192, 193
OTf	NiCl$_2$(dppe), KI, Zn, THF, DMF, 67°	(59)[a]	193
OTf	NiCl$_2$, Zn, PPh$_3$, NaI, DMF, sonication, 60°	(85)[b]	135
OTf	PdCl$_2$(PPh$_3$)$_2$, Zn, DMF, 90°	(7)[a]	192, 193
OMs	NiCl$_2$(PPh$_3$)$_2$, Zn, Et$_4$NI, THF, 67°	(84)[b]	133
OBs	NiCl$_2$(PPh$_3$)$_2$, Zn, Et$_4$NI, THF, 67°	(90)[b]	133
OFs	NiCl$_2$(PPh$_3$)$_2$, Zn, Et$_4$NI, THF, 67°	(80), (93)[b]	133
SO$_2$Cl	PdCl$_2$(PhCN)$_2$, Ti(OPr-i)$_4$, N$_2$, m-xylene, 140°	(40)	168
SO$_2$Na	PdCl$_2$, H$_2$O, 90°	(71)	167
SO$_2$Na	Na$_2$PdCl$_4$, HgCl$_2$, H$_2$O, reflux	(63)	166

[Structure: 4,4'-bis(trifluoromethyl)biphenyl]

X			
Cl	NiBr$_2$, bpy, 2e⁻, Bu$_4$NBF$_4$, EtOH, DMF	(80)	53
Cl	NiCl$_2$(quinoline)$_2$, Zn, pyridine, 80°	(65), (68)[b]	463
Cl	Ni(OAc)$_2$, t-BuOLi, LiH, bpy, THF, 63°	(84), (86)[b]	110
Cl	Ni(OAc)$_2$, NaH, t-AmONa, bpy, THF, KI, 63°	(88)	54
Cl	Ni(OAc)$_2$, NaH, t-AmONa, bpy, C$_6$H$_6$, KI, THF, 63°	(77)	54
Cl	Pd/C, HCO$_2$Na, NaOH, H$_2$O, CTAB, 100°	(75), (82)[b]	181
Cl	Pd/C, H$_2$, PEG-400, H$_2$O, NaOH, 110°	(69), (77)[b]	186
Cl	Pd/C, Zn, PEG-400, H$_2$O, NaOH, heat	(69)	180
Br	Ni(OAc)$_2$, NaH, t-AmONa, bpy, THF, 63°	(93)	54
Br	Ni(OAc)$_2$, NaH, t-AmONa, bpy, C$_6$H$_6$, KI, THF, 63°	(87)	54
Br	NiBr$_2$, EtOH, 2e⁻, dpa, H$_2$O	(90)	153
Br	NiBr$_2$, EtOH, MeOH, 2e⁻, bpy	(36)	153

TABLE 4. 4,4'-DISUBSTITUTED BIARYLS (Continued)

Substrate	Conditions	Product(s) and Yield(s) (%)	Refs.
C_{7-8} F_3C-C_6H_4-X, X = I	Cu, neat, reflux	F_3C-C_6H_4-C_6H_4-CF_3 (41)	425
OTf	NiCl_2(dppe), KI, Zn, THF, DMF, 67°	(93)	192, 193
OTf	PdCl_2(PPh_3)_2, Zn, DMF, 90°	(76)	192, 193
OTf	PdCl_2(PPh_3)_2, 2e−, DMF, n-Bu_4NBF_4, 90°	(45), (68)[a]	191, 192
C_7 OHC-C_6H_4-X, X = Cl	NiBr_2(PPh_3)_2, Zn, Et_4NI, THF, 50°	OHC-C_6H_4-C_6H_4-CHO (70)	139
Cl	NiCl_2, Zn, PPh_3, DMAc, 80°	(62), (94)[b]	51
Cl	NiCl_2, Zn, pyridine, 80°	(84-85)[b]	463
Cl	NiCl_2(quinoline)_2, Zn, pyridine, 80°	(84), (90)[b]	463
Cl	NiCl_2(pyridine)_2, Zn, pyridine, 80°	(88)[b]	463
Br	NiCl_2(PPh_3)_2, Zn, PPh_3, DMF, 50°	(70)	122
Br	NiBr_2(PPh_3)_2, Zn, Et_4NI, THF, 50°	(75)	139
Br	NiCl_2, Zn, bpy, NaBr, DMAc, 60-80°	(80)	146
Br	Ni(CO)_2(PPh_3)_2, DMSO, 70°	(75)[b]	145
Br	Ni(CO)_2(PPh_3)_2, toluene, 70°	(65)[b]	145
Br	Ni(CO)_2(PPh_3)_2, hexane, 70°	(40)[b]	145
Br	Ni(cod)_2, DMF, 35°	(79)	49, 50
OHC-C_6H_4-I	1. Cu, neat, 260°; 2. PhNHNH_2	PhNHN=CH-C_6H_4-C_6H_4-CH=NNHPh (59)	2

C_{7,8}

X	Conditions	Product (yield)	Refs.
Cl	NiCl$_2$, Zn, PPh$_3$, DMAc, 80°	(98)[b]	51
Cl	NiCl$_2$(quinoline)$_2$, Zn, pyridine, 80°	(80), (85)[b]	463
Cl	Ni(OAc)$_2$, NaH, t-AmONa, bpy, THF, C$_6$H$_6$, KI, 63°	(80)	54
Cl	Pd/C, NaOH, H$_2$O, formic hydrazide, 85°	(86)[b]	55
Br	Ni (electrogenerated), DMF, 130°	(77)	121
Br	Ni(cod)$_2$, DMF, 36°	(81)	49, 50
Br	NiI$_2$, naphthalene, Li, DME, 80°	(71)	111
Br	Ni(OAc)$_2$, NaH, t-AmONa, bpy, THF, C$_6$H$_6$, KI, 63°	(72)	54
Br	NiBr$_2$, lithium naphthalide, DME, 80°	(55)	111
Br	Pd(OAc)$_2$, n-Bu$_3$N, DMF, AsPh$_3$, 140°	(0)	172
I	Ni (electrogenerated), DMF, 100°	(68)	121
I	NiI$_2$, lithium naphthalide, DME, 80°	(85)	111
I	NiI$_2$, lithium naphthalide, 20°	(85)[b]	112
OMs	NiCl$_2$(PPh$_3$)$_2$, Zn, Et$_4$NI, THF, 67°	(94)[b]	133
OTf	PdCl$_2$(PPh$_3$)$_2$, 2e$^-$, DMF, n-Bu$_4$NBF$_4$, 90°	(55), (70)	191, 192
OTf	PdCl$_2$(PPh$_3$)$_2$, Zn, DMF, 90°	(85)[a]	192, 193
OTf	NiCl$_2$(dppe), KI, Zn, DMF, THF, 67°	(45), (70)[a]	192, 193

HO$_2$C–C$_6$H$_4$–X

X	Conditions	Product (yield)	Refs.
I	Pd(OAc)$_2$, n-Bu$_3$N, DMF, 140°	(70)	172
Br	Pd(OAc)$_2$, n-Bu$_3$N, DMF, AsPh$_3$, 140°	(0)	172
Br	Ni(cod)$_2$, DMF, 40-60°	(0)	49, 50
Br	Pd/CaCO$_3$, N$_2$H$_4$·H$_2$O, KOH, MeOH, 135-140°	(40)	48
Br	Pd/C, NaOH, H$_2$O, formic hydrazide, 85°	(64)[c]	55

TABLE 4. 4,4'-DISUBSTITUTED BIARYLS (Continued)

Substrate	Conditions	Product(s) and Yield(s) (%)	Refs.
NaO₂C—⟨⟩—X X = Br	1. Ni(cod)₂, DMF, 30-65° 2. acid	HO₂C—⟨⟩—⟨⟩—CO₂H (0)	49,50
X = SO₂Na	1. PdCl₂, H₂O, 90° 2. acid	(70)	167
MeO—⟨⟩—Cl	NiCl₂, Zn, PPh₃, DMAc, 80°	MeO—⟨⟩—⟨⟩—OMe (69)[b]	51
	NiCl₂, Zn, PPh₃, DMAc, bpy, 80°	(96)	51
	Ni(OAc)₂, NaH, t-AmONa, bpy, THF, 63°	(73), (79)[b]	131
	NiBr₂(PPh₃)₂, Zn, Et₄NI, THF, 50°	(67)	139
	Ni(OAc)₂, t-BuOLi, LiH, bpy, THF, 63°	(63), (68)[b]	110
	Ni(OAc)₂, NaH, t-AmONa, bpy, THF, 63°	(87)	54
	Ni(OAc)₂, NaH, t-AmONa, bpy, C₆H₆, KI, THF, 63°	(71)	54
	NiBr₂(PPh₃)₂, Zn, Et₄NI, THF, 50°	(67)	139
	Ni(OAc)₂, t-BuOLi, LiH, bpy, THF, 63°	(63), (68)[b]	110
	NiCl₂, Zn, PPh₃, NaI, bpy, DMAc, 80°	(80)[b]	52
	NiCl₂, Zn, PPh₃, bpy, DMAc, 70°	(96)[b]	52
	NiCl₂, Zn, PPh₃, NaBr, DMF, 2-methylaminopyridine, 70°	(63)[b]	52
	NiCl₂, Zn, PPh₃, NaBr, DMAc, 2-chloropyridine, 70°	(92)[b]	52
	Pd/C, HCO₂Na, NaOH, H₂O, CTAB, 110°	(20), (21)[b]	181
	Pd/C, H₂, PEG-400, H₂O, NaOH, 110°	(36)[b]	186

348

MeO-C6H4-Br → MeO-C6H4-C6H4-OMe		

Conditions	Yield	Ref
Pd(OAc)₂, DMF, H₂O, K₂CO₃, 115°	(48), (95)[b]	175
Pd/C, HCO₂Na, NaOH, H₂O, CTAB, 95°	(49)	179
Pd(OAc)₂, n-Bu₃N, DMF, AsPh₃, 140°	(50-57)	171, 172
Pd/CaCO₃, N₂H₄·H₂O, MeOH, KOH, 135-140°	(35)	48
Pd(OAc)₂, As(o-tol)₃, hydroquinone, Cs₂CO₃, 100°	(54)	178
Pd(dba)₂, TBAF, DMF, PhSi(OMe)₃	(34)[b]	197
Pd/C, NaOH, H₂O, formic hydrazide, 95°	(87)[b]	55
Cu, neat, 190°	(0)	32
Ni(cod)₂, DMF, 40°	(83)	49, 50
NiCl₂(PPh₃)₂, Zn, PPh₃, DMF, 50°	(42)	122
NiCl₂(PPh₃)₂, PPh₃, DMF, n-Bu₄NBr, 2e⁻	(36)	127
NiCl₂, Zn, bpy, NaBr, DMF, 60-80°	(65)	146
Ni(OAc)₂, NaH, THF, t-AmONa, PPh₃, 63°	(60)	131
Ni(OAc)₂, NaH, THF, t-AmONa, bpy, 63°	(75), (78)[b]	131
NiBr₂, bpy, 2e⁻, Bu₄NBF₄, EtOH, DMF	(80)	53
NiBr₂, bpy, 2e⁻, NaBr, EtOH, MeOH	(46)	53
NiBr₂, 2e⁻, dpa, H₂O, EtOH	(32)	153
NiBr₂, 2e⁻, bpy, MeOH, EtOH	(46)	153
NiCl₂(PPh₃)₂, Zn, PPh₃, DMF, 40°	(73)[b]	123
NiBr₂, lithium naphthalide, DME, 80°	(0)[b]	111
NiBr₂(PPh₃)₂, Zn, THF, Et₄NI, 50°	(66-72)	139
NiBr₂(PPh₃)₂, Zn, THF, 50°	(61)	139
Ni powder, DMF, 140°	(68)	120
Ni(OAc)₂, NaH, t-AmONa, bpy, THF, 63°	(70)	54
Ni(OAc)₂, NaH, t-AmONa, bpy, C₆H₆, KI, THF, 63°	(70)	54
Ni(OAc)₂, t-BuOLi, LiH, bpy, THF, 63°	(39), (47)[b]	110

TABLE 4. 4,4'-DISUBSTITUTED BIARYLS (Continued)

Substrate	Conditions	Product(s) and Yield(s) (%)	Refs.
MeO–C6H4–I		MeO–C6H4–C6H4–OMe	
	Cu, neat, 230-240°	(85)	2
	Cu, neat, 190°	(70)	32
	Cu, methyl benzoate, 190°	(8)	46
	Cu, 1,2,4-trichlorobenzene, 190°	(45)	46
	Ni (electrogenerated), DMF, 80°	(75)	121
	NiI_2, lithium naphthalide, DME, 80°	(68), (85)[b]	111
	$NiBr_2$, bpy, 2e−, Bu_4NBF_4, EtOH, DMF	(58)	53
	$NiBr_2$, KI, Zn, HMPA, 50°	(87)	129, 403
	$NiCl_2$, $CrCl_2$,	(96)	406
	Mn, THF, rt (2-pyridyl pinene ligand)		
	$Pd(PPh_3)_4$, 2e−, DMF, Et_4NOTs	(87)	190
	$Pd(OAc)_2$, Et_3N, 100°	(39)	169
	$Pd(OAc)_2$, n-Bu_3N, 100°	(25)	169
	$Pd(OAc)_2$, n-Bu_3N, DMF, 140°	(58)	172
	$Pd(OAc)_2$, n-Bu_4NBr, DMF, H_2O, IPA, K_2CO_3, 115°	(81)	174
	$Pd(OAc)_2$, DMF, H_2O, IPA, K_2CO_3, 115°	(80)	175
	$Pd(OAc)_2$, As(o-tol)$_3$, hydroquinone, Cs_2CO_3, 75°	(95)	178
	$Pd(OAc)_2$, P(o-tol)$_3$, hydroquinone, Cs_2CO_3, 50°	(94)	178
	$Pd(OAc)_2$, P(o-tol)$_3$, Cs_2CO_3, 50°	(0)	178
MeO–C6H4–I		MeO–C6H4–C6H4–OMe	
	Pd/C, n-Bu_4NBr, DMF, K_2CO_3, 135°	(60)	195
	$Pd(OAc)_2$, n-Bu_4NBr, DMF, K_2CO_3, 135°	(60)	195
	Pd/C, Zn, air, H_2O, acetone, 25°	(92)[b]	185

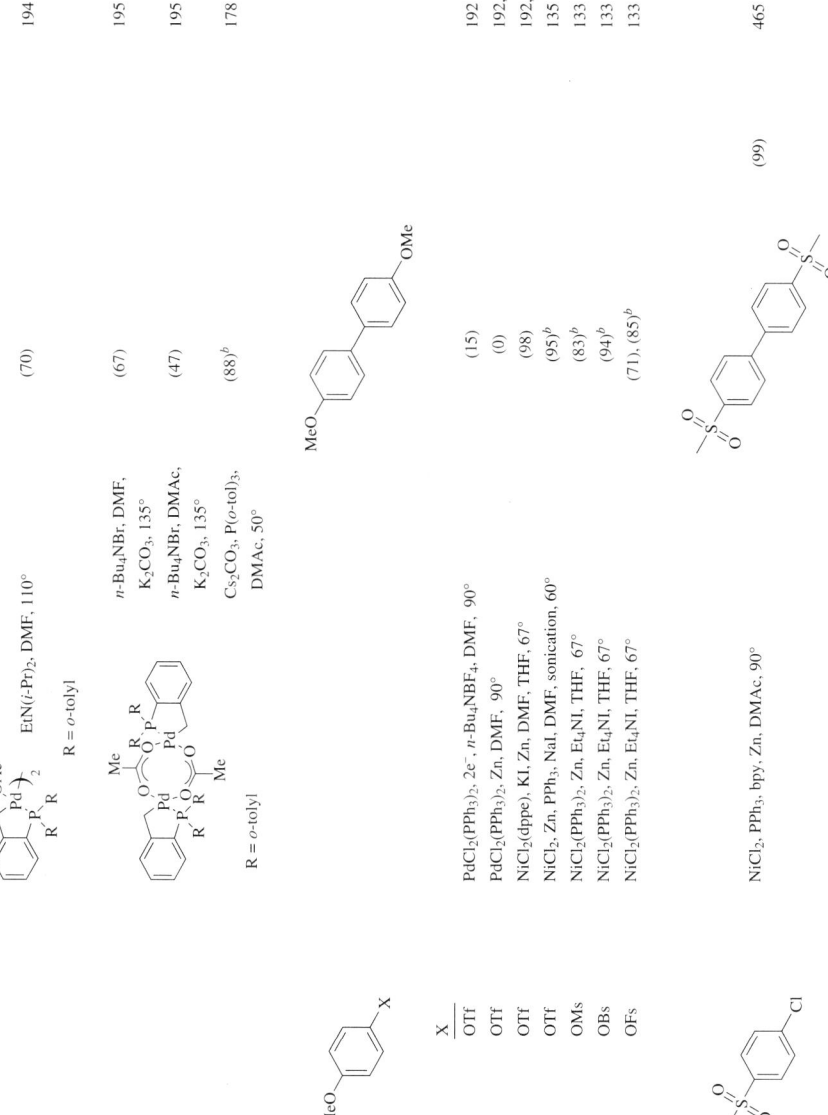

TABLE 4. 4,4'-DISUBSTITUTED BIARYLS (Continued)

Substrate	Conditions	Product(s) and Yield(s) (%)	Refs.
MeS—C₆H₄—Br	Ni(OAc)₂, t-BuOLi, LiH, bpy, THF, 63°	MeS—C₆H₄—C₆H₄—SMe (10)[b]	110
C₈ Me₂N—C₆H₄—X		Me₂N—C₆H₄—C₆H₄—NMe₂	
X = Br	NiBr₂, bpy, EtOH, MeOH	(trace)	153
Br	NiBr₂, EtOH, 2e⁻, dpa, H₂O	(30)	153
Br	Ni(OAc)₂, NaH, t-AmONa, bpy, THF, C₆H₆, KI, 63°	(81)	54
Br	Pd/CaCO₃, N₂H₄·H₂O, MeOH, KOH, 135-140°	(60)	48
Br	Pd(OAc)₂, n-Bu₃N, DMF, AsPh₃, 140°	(45)	172
I	Pd(PPh₃)₄, 2e⁻, Et₄NOTs, DMF	(93)	190
AcHN—C₆H₄—Br	NiCl₂(PPh₃)₂, Zn, PPh₃, DMF, 40°	AcHN—C₆H₄—C₆H₄—NHAc (37)	123
	Ni(cod)₂, DMF, 50-60°	(0)	49

R	X			
Et	I	Cu, neat, heat	(53)	466
Et	I	PdCl$_2$, N$_2$H$_4$·H$_2$O, MeOH, HgCl$_2$, reflux	(82)	47
CH=CH$_2$	Br	NiCl$_2$, CrCl$_2$, Mn, THF, rt	(84)	406
CH$_2$CN	Br	Ni(cod)$_2$, DMF, 33°	(79)	49, 50
OEt	I	Cu, neat, 230-240°	(75)	2
OEt	I	Cu, DMF, reflux	(77)	57
OAc	Cl	NiCl$_2$, PPh$_3$, DMAc, Zn, 80°	(66)[b]	51
OAc	Cl	NiCl$_2$, PPh$_3$, DMAc, Zn, bpy, 80°	(75), (90)[b]	51
OAc	Cl	NiCl$_2$, PPh$_3$, DMAc, Zn, NaBr, 80°	(66)[b]	147

X			
Cl	NiCl$_2$, PPh$_3$, DMAc, Zn, 80°	(100)[b]	51
Cl	Ni powder, DMF, KI, 140°	(85)	120
Cl	NiBr$_2$(PPh$_3$)$_2$, Zn, Et$_4$NI, THF, 50°	(73)	139
Cl	NiCl$_2$, bpy, NaBr, DMAc, Zn, 60-80°	(75)	146
Cl	NiCl$_2$(quinoline)$_2$, Zn, pyridine, 80°	(80), (84)[b]	463
Br	NiCl$_2$(PPh$_3$)$_2$, PPh$_3$, DMF, n-Bu$_4$NBr, 2e$^-$	(47)	127
Br	NiBr$_2$, lithium naphthalide, DME, 80°	(57)[b]	111
Br	NiI$_2$, lithium naphthalide, DME, 80°	(46)	111
Br	NiCl$_2$(PPh$_3$)$_2$, PPh$_3$, Zn, DMF, 50°	(68)	122

TABLE 4. 4,4'-DISUBSTITUTED BIARYLS (Continued)

Substrate	Conditions	Product(s) and Yield(s) (%)	Refs.
X — ⟨Ar⟩ — C(O)CH₃		CH₃C(O) — ⟨Ar⟩ — ⟨Ar⟩ — C(O)CH₃	
X			
Br	NiBr₂, EtOH, 2e⁻, dpa, H₂O	(63)	153
Br	Ni powder, DMF, 140°	(77)	120
Br	NiI₂, lithium naphthalide, DME, 20°	(46)	112
Br	NiBr₂(PPh₃)₂, Zn, Et₄NI, THF, 50°	(71)	139
Br	NiCl₂(PPh₃)₂, Zn, PPh₃, DMF, 50°	(58)	123
Br	NiBr₂, bpy, EtOH, MeOH, 2e⁻	(trace)	153
Br	Ni(cod)₂, DMF, 45°	(93)	49, 50
Br	[PdCl(π-C₃H₅)]₂, TBAF, DMSO, 120°	(75)	198
Br	Pd/C, HCO₂Na, NaOH, H₂O, SDPNS, 95°	(41)	179
I	Cu, neat, 235-260°	(26)	467
I	NiBr₂, Zn, KI, HMPA, 50°	(98)	129, 403
OTs	NiCl₂(PPh₃)₂, Zn, NaBr, DMF, PPh₃, 100°	(67)	137
OMs	NiCl₂(PPh₃)₂, Zn, Et₄NI, THF, 67°	(73)[b]	133
OMs	NiCl₂(PPh₃)₂, Zn, Et₄NI, DMAc, 100°	(85)[b]	133
OMs	NiCl₂(PPh₃)₂, Zn, KI, DMAc, 67°	(93)[b]	133
OMs	NiCl₂(PPh₃)₂, Zn, NaI, DMAc, 100°	(91)[b]	133
OMs	NiCl₂(PPh₃)₂, Zn, NaBr, DMAc, 100°	(96)[b]	133
OMs	NiCl₂(PPh₃)₂, Zn, NaBr, NMP, 100°	(trace)[b]	133
OMs	NiCl₂(PPh₃)₂, Zn, NaBr, DMF, PPh₃, 100°	(83)	137
OFs	NiCl₂(PPh₃)₂, Zn, Et₄NI, THF, 67°	(98)[b]	133

X	Conditions	Product (%)	Ref.
Cl	NiBr₂(PPh₃)₂, Zn, Et₄NI, THF, 50°	(85)	139
Cl	NiCl₂, PPh₃, Zn, pyridine, heat	(92)	468
Cl	NiCl₂, PPh₃, Zn, 4-methylpyridine, heat	(60)	468
Cl	NiCl₂, PPh₃, Zn, DMAc, heat	(82)	468
Cl	NiCl₂, Zn, pyridine, 85°	(78)[b]	463
Cl	NiCl₂, Zn, 4-methylpyridine, 85°	(83)[b]	463
Cl	NiCl₂(pyridine)₂, Zn, pyridine, 85°	(81)[b]	463
Cl	NiCl₂(quinoline)₂, Zn, pyridine, 85°	(88), (89)[b]	463
Cl	NiCl₂(PPh₃)₂, PPh₃, DMF, n-Bu₄NBr, 2e⁻	(51)	127
Br	Ni (electrogenerated), DMF, 100°	(5)	121
Br	Ni (electrogenerated), DMF, 130°	(37)	121
Br	NiCl₂(PPh₃)₂, Zn, PPh₃, DMF, 50°	(83)	123
Br	NiCl₂(PPh₃)₂, Zn, PPh₃, DMF, 50°	(76)	122
I	Cu, neat, 190°	(15)	32
I	Cu, DMF, reflux	(74)	57
I	Cu, neat, 220–260°	(70)	2
I	Pd(OAc)₂, As(o-tol)₃, hydroquinone, Cs₂CO₃, 75°	(99)	178
I	Pd(OAc)₂, P(o-tol)₃, hydroquinone, Cs₂CO₃, 75°	(95)	178
I	NiBr₂(PPh₃)₂, Zn, Et₄NI, THF, 50°	(86–90)	139
I	NiBr₂(PPh₃)₂, Zn, THF, 50°	(85)	139
I	Ni, DMF, 100°	(0)	121
I	Ni (electrogenerated), DMF, 25°	(20)	121
I	Ni (electrogenerated), DMF, 50°	(60)	121
I	Ni (electrogenerated), DMF, 100°	(90)	121
I	[Pd(OAc)(PR₂)]₂ cyclopalladated dimer, R = o-tolyl, EtN(i-Pr)₂, DMF, 110°	(76)	194

TABLE 4. 4,4'-DISUBSTITUTED BIARYLS (Continued)

Substrate	Conditions	Product(s) and Yield(s) (%)	Refs.
C_{8-14}			
MeO$_2$C–C$_6$H$_4$–X		MeO$_2$C–C$_6$H$_4$–C$_6$H$_4$–CO$_2$Me	
X = OMs	NiCl$_2$(PPh$_3$)$_2$, Zn, Et$_4$NI, THF, reflux	(99)^b	133
OMs	NiCl$_2$(PPh$_3$)$_2$, Zn, Et$_4$NI, DMAc	(73)^b	133
OMs	NiCl$_2$(PPh$_3$)$_2$, Zn, KI, THF	(93)^b	133
OMs	NiCl$_2$(PPh$_3$)$_2$, Zn, KBr, THF	(86)^b	133
OMs	NiCl$_2$(PPh$_3$)$_2$, Zn, NaBr, DMAc	(78)^b	133
OMs	NiCl$_2$(PPh$_3$)$_2$, Zn, KBr, DMAc	(79)^b	133
OTf	NiCl$_2$(PPh$_3$)$_2$, Zn, Et$_4$NI, THF, reflux	(99)^b	133
4-YC$_6$H$_4$SO$_2$–			
Y = H	NiCl$_2$(PPh$_3$)$_2$, Zn, Et$_4$NI, THF, reflux	(83), (97)^b	133
Me	NiCl$_2$(PPh$_3$)$_2$, Zn, Et$_4$NI, THF, reflux	(99)^b	133
F	NiCl$_2$(PPh$_3$)$_2$, Zn, Et$_4$NI, THF, reflux	(85), (99)	133
Cl	NiCl$_2$(PPh$_3$)$_2$, Zn, Et$_4$NI, THF, reflux	(79)^b	133
C_9			
R–C$_6$H$_4$–X		R–C$_6$H$_4$–C$_6$H$_4$–R	
R = i-Pr, X = I	Cu, neat, 260–270°	(79)	431
CO$_2$Et, Br	Ni(cod)$_2$, DMF, 40°	(81)	49, 50
CO$_2$Et, OTf	PdCl$_2$(PPh$_3$)$_2$, Zn, DMF, 90°	(69)^a	192, 193
CO$_2$Et, OTf	NiCl$_2$(dppe), Zn, DMF, THF, KI, 67°	(63), (86)^a	192, 193
CO$_2$Et, I	Ni(cod)$_2$, hv, toluene, N-methylimidazole	(90)	132
CO$_2$Et, I	CuI-P(Et)$_3$, lithium naphthalide, DME or THF, heat	(52)^b	85
i-PrO, Br	Pd/C, NaOH, H$_2$O, formic hydrazide, 95°	(78)^b	55

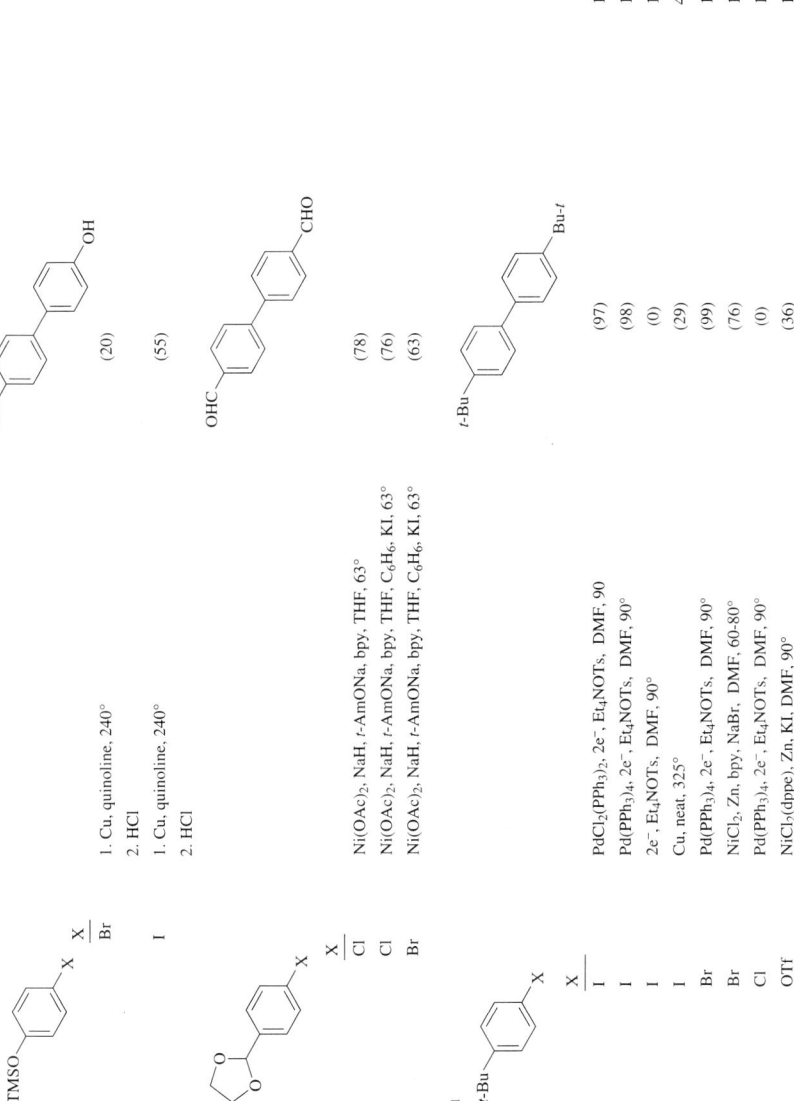

TABLE 4. 4,4'-DISUBSTITUTED BIARYLS (Continued)

Substrate	Conditions	Product(s) and Yield(s) (%)	Refs.
X–C₆H₄–[dioxolane-CH₃] X Br Cl	1. Ni(OAc)₂, NaH, t-AmONa, bpy, THF, C₆H₆, KI, 63° 2. acid 1. Ni(OAc)₂, NaH, t-AmONa, bpy, THF, C₆H₆, KI, 63° 2. acid	4,4'-diacetylbiphenyl (64) (72)	54 54
C₁₁ 4-iodo-(tert-pentyl)benzene	Cu, neat, 300°	4,4'-di(tert-pentyl)biphenyl (40)	470
C₁₂ (Me₃Si)₂N–C₆H₄–I	Cu, quinoline, 240°	4,4'-diaminobiphenyl (60)	433
4-X-biphenyl X Cl Br I I I	NiCl₂, bpy, LiBF₄, DMAc, 2e⁻ Pd/C, HCO₂Na, NaOH, H₂O, CTAB, 95° Cu, neat, 250–270° Cu, neat, 290° Cu, neat, 300°	p-quaterphenyl (22) (48) (66) (42) (97)	471 179 472 473 437

Substrate	Conditions	Product (Yield %)	Ref.
I	Cu, neat, 250-270°	(82)	2
OMs	NiCl₂(PPh₃)₂, Zn, Et₄NI, THF, 67°	(99)[b]	133
OFs	NiCl₂(PPh₃)₂, Zn, Et₄NI, THF, 67°	(60)	133

4'-iodo-2-nitrobiphenyl → 2,2''-dinitro-p-terphenyl

Substrate	Conditions	Product (Yield %)	Ref.
	Cu, neat, 225-235°	(42)	438
	Cu, neat, 200-220°	(47)	238

4'-iodo-3-nitrobiphenyl → 3,3''-dinitro-p-terphenyl

Substrate	Conditions	Yield	Ref.
	Cu, neat, 200-220°	(50)	238

4-R-4'-X-biphenyl → p-quaterphenyl derivative

R	X	Conditions	Yield	Ref.
NO₂	I	Cu, neat, heat	(14)	474
OH	Br	Pd/C, Na₂CO₃, H₂O, MeOH, formic hydrazide, 70°	(75)	55

359

TABLE 4. 4,4'-DISUBSTITUTED BIARYLS (Continued)

Substrate	Conditions	Product(s) and Yield(s) (%)	Refs.
4-iodophenyl phenyl ether	Cu, DMF, reflux	bis(4-phenoxyphenyl) (45)	410
4-(phenylsulfonyl)phenyl tosylate	NiCl$_2$(PPh$_3$)$_2$, PPh$_3$, NaBr, Zn, DMF, 100°	4,4'-bis(phenylsulfonyl)biphenyl (78)	137
4-R-4'-X-biphenyl (R, X table below)		4,4'''-di-R-p-quaterphenyl	
R = F, X = Cl	NiBr$_2$, Zn, DMAc, PPh$_3$, 50-90°	(82)	475
R = F, X = Br	Cu, neat, 200°	(30)	55
R = Cl, X = Br	Cu, neat, 200°	(0)	405
C$_{13}$ 4'-iodo-2-methoxybiphenyl	Cu, neat, 280°	2,2'''-dimethoxy-p-quaterphenyl (70)	474

Substrate	Conditions	Product	Yield (%)	Ref.
4-iodo-4'-methoxybiphenyl	Cu, neat, 280°	MeO-C6H4-C6H4-C6H4-C6H4-OMe (45)	—	474
4-X-benzophenone (X = Br, I, OTs)	Br: Pd/C, Zn, air, acetone, H2O, 25° I: Cu, neat, 250° OTs: NiCl2(PEt3)2, PPh3, NaBr, Zn, DMF, 100°	4,4'-dibenzoylbiphenyl	(0) (55) (—)	185 2 137
n-C6H13O2C-C6H4-Br	Pd(OAc)2, n-Bu3N, DMF, AsPh3, 140°	n-C6H13O2C-C6H4-C6H4-CO2C6H13-n (38)		172
PhCH2C(O)-C6H4-Br (C14)	Ni(cod)2, DMF, 40°	PhCH2C(O)-C6H4-C6H4-C(O)CH2Ph (37)		475

TABLE 4. 4,4'-DISUBSTITUTED BIARYLS (Continued)

Substrate	Conditions	Product(s) and Yield(s) (%)	Refs.
HO₂C–C₆H₃(CO₂H)–O–C₆H₄–Br	Pd/C, NaOH, H₂O, formic hydrazide, 80°	biaryl diether bis(dicarboxylic acid) (92)^c	55
(n-BuO)₂P(=O)–O–C₆H₄–Br	Ni(OAc)₂, bpy, Et₄NBr, 2e⁻, CH₃CN	bis[4-(di-n-butylphosphate)phenyl]-biphenyl (74)	405
4-acetyl-4'-iodobiphenyl	Ni(cod)₂, DMF, 42°	4,4'''-diacetyl-p-quaterphenyl (45)	475
C₁₅ 2-phenyl-5-(4-iodophenyl)oxazole	Cu, neat, 120-250°	bis[2-phenyloxazol-5-yl]biphenyl (25) lower yield with DMF as solvent	476

C₁₈	![structure] PhCH=CH-C(O)-C₆H₄-Br	Ni(cod)₂, DMF, 50-65°	(62) 49
	Ar–X, Ar = biphenyl-4-yl, X = I / OMs	Cu, neat, 280° / NiCl₂(PPh₃)₂, Zn, Et₄NI, THF, 67°	(40) 62 / (60) 133
C₁₉₋₂₅	X–C₆H₄–C₆H₄–I, X = (CH₂)₅COR; R = Me, Et, n-Pr, n-Heptyl	Cu, biphenyl, 255-260°	(11), (13), (12), (7) 477
C₂₀	X–C₆H₄–C₆H₄–I, X = PhC(O)C(O)–	PdCl₂, Bu₃N, 220°	(70) 478

363

TABLE 4. 4,4'-DISUBSTITUTED BIARYLS (Continued)

Substrate	Conditions	Product(s) and Yield(s) (%)	Refs.
Ar—⟨⟩—⟨⟩—Br, Ar = biphenyl-4-yl	NiCl$_2$, bpy, LiBF$_4$, DMAc, 2e$^-$	Ar—⟨⟩—⟨⟩—⟨⟩—⟨⟩—Ar (35)	471
(biphenyl-2-yl iodide structure), Ar = biphenyl-2-yl	Cu, neat, 225-235°	quaterphenyl with Ar substituents (61)	438
C$_{29}$ R—⟨⟩—⟨⟩—I with R = oxazole group	Cu, neat, 270-280°	R—⟨⟩—⟨⟩—⟨⟩—⟨⟩—R (27)	479

[a] The yield was determined by NMR spectroscopy.
[b] The yield was determined by gas chromatography.
[c] The yield was determined by HPLC.

TABLE 5. 2,2',3,3'-TETRASUBSTITUTED BIARYLS

Substrate			Conditions	Product(s) and Yield(s) (%)	Refs.
R^2	R^3	X		(biaryl with R^2, R^3)	
NO_2	NO_2	I	Cu, p-nitrotoluene, heat	(53)	58
Cl	Cl	I	Cu	(63)	480
OMe	F	I	Cu, heat	(67)	481
Me	Me	I	Cu, neat, 150°	(45)	482
Me	Me	I	$PdCl_2$, $HgCl_2$, $N_2H_4 \cdot H_2O$, MeOH, reflux	(64)	47
Me	OMe	I	Cu, neat, 270°	(92)	483
OMe	OMe	I	Cu, neat, 250°	(70-83)	484
$(CH_2)_2CO_2Et$	Cl	I	Cu, neat, 240°	(20) yield includes iodide preparation and subsequent ester hydrolysis	485
Ph	Ph	I	Cu, neat, 260-270°	(64)	486

C_{6-18}

TABLE 6. 2,2',4,4'-TETRASUBSTITUTED BIARYLS

Substrate	Conditions	Product(s) and Yield(s) (%)	Refs.
C_6 with R^4, R^2, X (R² / R⁴ / X: Cl / NO₂ / I)	Cu, neat, 210°	biaryl with R^2, R^4 (44)	487
Br / NO₂ / I	Cu, neat, 210°	(23)	487
Cl / Cl / I	Cu, neat, 220-270°	(53)	2
F / F / Br	Pd/C, NaOH, diethylene glycol dimethyl ether, polyethylene glycol dimethyl ether, glycerol, 100°	(87)[a]	462
O_2N–phenyl–NO_2, X = I	Cu, xylene, 110-140°	2,2',4-trinitrobiphenyl (94)	238
X = Br	Cu, nitrobenzene, reflux	(65)	1
X = Cl	Cu, nitrobenzene, reflux	(60)	1
X = Cl	Cu, methyl benzoate, 190°	(18)	46
X = Cl	Cu, 1,2,4-trichlorobenzene, 190°	(16)	46
X = Cl	Cu, neat, 190°	(0)	32
R^4-phenyl-NO_2, X (R⁴ / X: F / Br)	Cu, neat, 240-250°	biaryl with R^4, NO_2 (46)	488
Br / Br	Cu, neat, 190-250°	(65)	1
Br / Br	Cu, DMF, reflux	(76)	489
Br / Br	Cu, DMF, 120°	(70)	490

| | | Cl | Cl | Cu, neat, 240° | (42) | 1 |
| | | Cl | Cl | Cu, DMF, reflux | (75) | 57 |

C7

R^2	R^4	X			
NO_2	Me	I	Cu, sand, 180°	(46)	491
NO_2	CF_3	Cl	Cu, neat, reflux	(30)	492
NO_2	CF_3	Cl	Cu, neat, reflux	(24)	493
NO_2	OMe	I	Cu, neat, 130-170°	(82)	494
Me	NO_2	I	Cu, sand, 205-210°	(10)	495
Me	NO_2	I	Cu, neat, 280°	(25)	491
Me	F	Br	Pd/C, NaOH, diethylene glycol dimethyl ether, polyethylene glycol dimethyl ether, glycerol, 100°	$(73)^a$	462
CF_3	NO_2	I	Cu, neat, 250-300°	(17)	496
CF_3	F	I	Cu, heat	(63)	497
CF_3	Cl	I	Cu, heat	(56)	497
OMe	F	I	Cu, sand, 200°	(43)	498
OMe	NO_2	I	Cu, neat, 230°	(50)	499
OMe	NO_2	I	Cu, heat	(<10)	238
OMe	NO_2	I	Cu, biphenyl, 200°	(54)	238
OMe	NO_2	I	Cu, xylene, 200°	(91)	238
OMe	NO_2	I	Cu, nitrobenzene, 210-220°	(83)	500

TABLE 6. 2,2',4,4'-TETRASUBSTITUTED BIARYLS (Continued)

Substrate			Conditions	Product(s) and Yield(s) (%)	Refs.
R^2	R^4	X			
C₈					
NO₂	CO₂Me	Br	Cu, neat, 185°	(81)	1
NO₂	CO₂Me	I	Cu, neat, 170-190°	(78)	501
NO₂	OAc	I	Cu, neat, 140-180°	(77)	502
Me	Me	I	Cu, neat, sealed tube, 230-260°	(86)	2
Me	Me	I	Ni powder, DMF, 140°	(54)	120
Me	Me	I	NiCl₂, NaBr, Zn, bpy, DMF, 60-80°	(79)	146
Me	OMe	I	Cu, neat, 260-290°	(62)	503
CF₃	CF₃	I	Cu, heat	(32)	504
CO₂Me	NO₂	Br	Cu, neat, 200-205°	(51)	505
CO₂Me	Cl	Br	Cu, neat, 200°	(61)	505
CO₂Me	Br	I	Cu, neat, 180-230°	(75)	506
OMe	OMe	I	Cu, neat, 260°	(71)	507
OMe	OMe	I	Cu, neat, 190°	(65)	32
OMe	OMe	I	Cu₂O, heat	(47)	65
OMe	OMe	I	Cu, heat	(72-73)	65
C₉₋₁₂					
NO₂	CO₂Et	I	Cu, nitrobenzene, reflux	(69)	508
CO₂Me	Me	I	NiCl₂, KI, Zn, HMPA, 20°	(98)	129
CO₂Me	OMe	Br	Cu, neat, 220-250°	(100)	509

OMe	CO₂Me	I	Cu, neat, 210-220°	(91)	510
OMe	CO₂Me	Br	Cu, neat, 210-220°	(<30)	510
OMe	CO₂Me	Br	Cu, neat, 255-260°	(34)	511
NHCOMe	NHCOMe	Br	Cu, DMF, 100-120°	(95)	34
SO₃Ph	NO₂	I	Cu, neat, 205°	(68)	512

C₁₂

[structure: 4-iodo-3-nitrobiphenyl] + Cu, xylene, 110-140° / Cu, neat, 190-205° → [2,2'-dinitro-4,4'-diphenylbiphenyl structure] (92-98) / (83)   238 / 513

[structure: 4'-bromo-3',4-dinitro-2-nitrobiphenyl] + Cu, biphenyl, 180° → [quaterphenyl tetranitro structure] (48)   238

[structure: 2-iodo-5-nitrobiphenyl] + Cu, neat, 215-225° → [terphenyl dinitro structure] (70)   514

TABLE 6. 2,2',4,4'-TETRASUBSTITUTED BIARYLS (Continued)

Substrate	Conditions	Product(s) and Yield(s) (%)	Refs.
C₁₃			
R² = Me	Cu, 200-270°	(45)	515
R² = OMe	Cu, neat, 190-210°	(54)	513
C₁₄ (PhO₂S-aryl-I with Me)	Cu, neat, 180-210°	(92)	421
C₁₄ (benzoyl-aryl-I with Me)	Cu, neat, 230°	(77)	516
C₁₅	NiCl₂(PPh₃)₂, PPh₃, Zn, Et₄NI, THF, 67°	R⁴' = H (65)	134
	NiCl₂(PPh₃)₂, PPh₃, Zn, Et₄NI, THF, 67°	R⁴' = F (62)	134

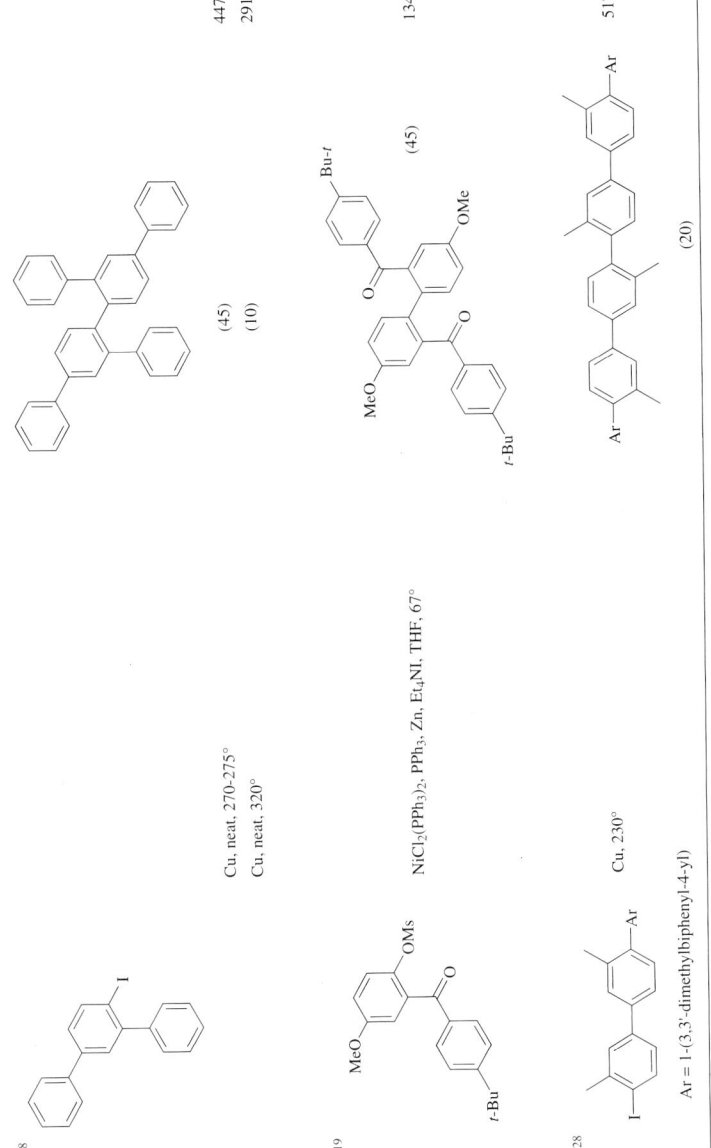

a The yield was determined by gas chromatography.

TABLE 7. 2,2′,5,5′-TETRASUBSTITUTED BIARYLS

Substrate			Conditions	Product(s) and Yield(s) (%)	Refs.
R^2	R^5	X			
NO_2	Me	Br	Cu, neat, 215-235°	(73)	518
NO_2	OMe	I	Cu, neat, 140-170°	(80)	502
NHCHO	Cl	Br	Cu, DMSO, 70°	(97)	34
NHCHO	NO_2	Br	Cu, DMF, 100°	(44)	34
Me	NO_2	I	Cu, neat, 220-270°	(22)	519
Me	SO_2F	I	Cu, neat, 220°	(41)	452
Me	Cl	I	Cu, neat, 280-290°	(50)	520
CHO	NO_2	I	Cu, DMF, reflux	(65)	428
CHO	Cl	I	Cu, DMF, reflux	(42)	521
SMe	Cl	I	Cu, neat, 200-280°	(25-30)	522
NO_2	CO_2Me	Br	Cu, nitrobenzene, reflux	(81)	518
NHCOMe	NO_2	Br	Cu, DMF, 110°	(94)	34
NHCOMe	Cl	Br	Cu, DMF, 100°	(76)	34
NHCOMe	Cl	Br	Cu, DMSO, 50°	(88)	34
NHCOMe	Br	Br	Cu, DMSO, 60°	(84)	34
Me	Me	I	Cu, neat, 265-300°	(46)	2
Me	Me	I	Ni powder, DMF, 140°	(42)	120
Me	Me	I	Pd(OAc)$_2$, Et$_3$N, 100°	(3)	169
Me	Me	I	Pd(OAc)$_2$, n-Bu$_3$N, 100°	(trace)	169
Me	Me	I	PdCl$_2$, HgCl$_2$, N$_2$H$_4$•H$_2$O, MeOH, reflux	(74)	47
Me	Me	I	PdCl$_2$, N$_2$H$_4$•H$_2$O, MeOH, reflux	(19)	47
Me	Me	Br	NiCl$_2$, bpy, NaBr, Zn, DMF, 60-80°	(76)	146

C$_9$

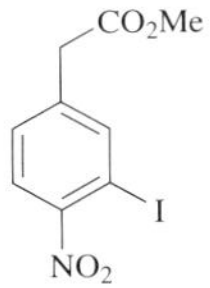

[structure: CO$_2$Me-CH$_2$-C$_6$H$_3$(I)(NO$_2$)]

Cu, DMF, reflux

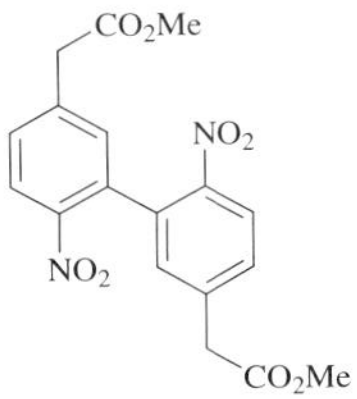

[structure (45): dimethyl 2,2'-dinitro-[1,1'-biphenyl]-4,4'-diacetate]

523

C$_{8,9}$

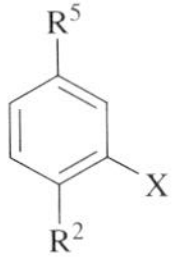

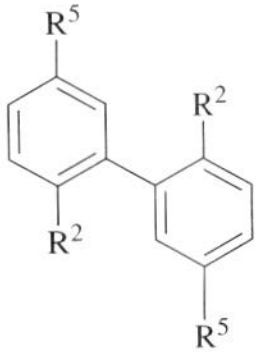

[structure: 2,5-disubstituted biphenyl with R^2, R^5, X]

[structure: biphenyl product with R^2, R^5]

R^2	R^5	X		
Me	Me	Br	NiCl$_2$, bpy, NaBr, Zn, DMF, 60-80°	(76)	146
Me	Me	I	PdCl$_2$, HgCl$_2$, N$_2$H$_4$•H$_2$O, MeOH, reflux	(74)	47
Me	Me	I	PdCl$_2$, N$_2$H$_4$•H$_2$O, MeOH, reflux	(19)	47
CF$_3$	CF$_3$	I	Cu, sand, reflux	(42)	524
CO$_2$Me	NO$_2$	I	Cu, neat, 200°	(75)	525
OMe	Me	Br	PdCl$_2$(PPh$_3$)$_2$, 2e$^-$, Et$_4$NOTs, DMF	(0)	190
OMe	OMe	I	Cu, neat, 215-260°	(93)	2
OMe	OMe	I	Cu, neat, heat	(80-90)	526
OMe	OMe	Br	NiBr$_2$, bpy, NaBr, 2e$^-$, EtOH, MeOH	(29)	53
OMe	OMe	I	NiCl$_2$(PPh$_3$)$_2$, PPh$_3$, Zn, DMF, 54°	(16)	457
OMe	CHO	I	Cu, neat, sealed tube, 240-250°	(7)	527
OMe	CO$_2$Me	Br	Cu, neat, 175°	(38)	37
NO$_2$	CH$_2$CO$_2$Me	Br			
Me	CO$_2$Me	I	Cu, neat, 260-310°	(65) crude	528

TABLE 7. 2,2',5,5'-TETRASUBSTITUTED BIARYLS (Continued)

Substrate	Conditions	Product(s) and Yield(s) (%)	Refs.

C_{10}

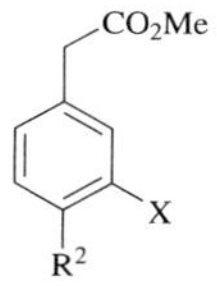

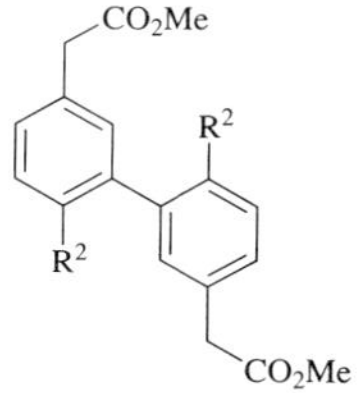

R^2	X	
OMe	I	Cu, neat, 175°
OMe	I	Cu, neat, 230°
OSO$_2$Me	Br	Cu, neat, 230°

(17) 37
(26) 523
(48) 37

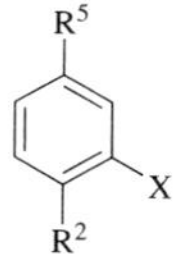 R^2	R^5	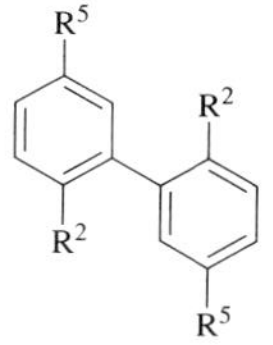 X	
Me	CO$_2$Et	I	Cu, neat, 260-310°
CO$_2$Me	CO$_2$Me	I	Cu, neat, 260-310°

(55) 528
(70) 528

C_{11}

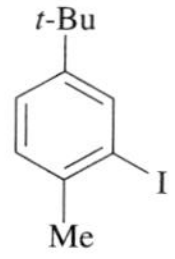

Cu, neat, 300°

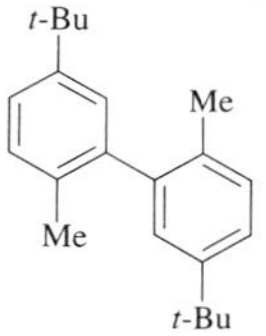

(68) 529

Starting material	Conditions	Product	Yield (%)	Ref.
3-iodo-4-methoxy aryl with CH2CH2CO2Me	Cu, neat, 225-285°	biaryl dimer with OMe, CH2CH2CO2Me groups	(58)	530
methyl 3-iodo-4-isopropylbenzoate	Cu, heat, 270-275°	biaryl with i-Pr, CO2Me	(81)	431
methyl (E)-3-(3-iodo-4-methoxyphenyl)acrylate	Cu, heat	biaryl divinyl diester with OMe	(60-70)	531
C12; methyl 3-iodo-4-tert-butylbenzoate	Cu, neat, 235°	biaryl with t-Bu, CO2Me	(62)	532

TABLE 7. 2,2',5,5'-TETRASUBSTITUTED BIARYLS (Continued)

Substrate	Conditions	Product(s) and Yield(s) (%)	Refs.
C₁₃ (2-chloro-4-fluorophenyl phenyl ketone)	NiBr₂, PPh₃, Zn, DMAc, 80-90°	(88)	440
(3-iodo-4-methoxybiphenyl)	Cu, neat, 260°	(59)	533
C₁₄ (5-formyl-2-benzyloxy-iodobenzene)	NiCl₂(PPh₃)₂, PPh₃, Zn, DMF, 54°	(16)	457

C_{16} (structure: 3-iodo-4-methoxybenzyl CO₂Bn)	Cu, neat, 200–220°	(structure: biaryl with OMe, OMe, CH₂CO₂Bn groups)	(75)	37
C_{18} (structure: 2-iodo-1,4-diphenylbenzene)	Cu, neat, 270–275°	(structure: tetraphenyl biaryl)	(64)	447
C_{20} (structure: 2-iodo-4,4′-di-t-Bu biphenyl)	Cu, neat, 300°	(structure: tetra-t-Bu tetraaryl)	(60)	529

TABLE 8. 2,2',6,6'-TETRASUBSTITUTED BIARYLS

Substrate			Conditions	Product(s) and Yield(s) (%)	Refs.

C_{6-7}

R²	R⁶	X	Conditions	R⁶—R² / R²—R⁶ biaryl yield	Refs.
NO₂	NO₂	Cl	Cu, DMF, 145°	(51)	534
NO₂	NO₂	Cl	Cu, nitrobenzene, heat	(54)	58
NO₂	NO₂	Cl	Cu, DMF, reflux	(0)	61
NO₂	Br	Br	Cu, neat, 140-160°	(72)	535
NO₂	I	I	Cu, neat, 150-160°	(75)	536
Cl	Cl	I	Cu, sealed tube, 230°	(59)	537
Cl	Cl	I	Cu, sealed tube, 200°	(quantitative) crude	538
NO₂	Me	I	Cu, neat, 180-235°	(81)	539
NO₂	Me	I	Cu, neat, 200°	(79)	491
NO₂	Me	I	Cu, neat, 200-235°	(80)	540
NO₂	Me	I	Cu, neat, 240-280°	(68)	541
NO₂	Me	I	Cu, neat, 180-230°	(67), yield includes preparation of precursor	542
NO₂	Me	I	Cu, DMF, reflux	(>66)	543
NO₂	OMe	I	Cu, DMF, reflux	(>66)	543
NO₂	OMe	Cl	Cu, nitrobenzene, 180-210°	(70)	544
NO₂	OMe	Cl	Cu, DMF, reflux	(84)	57
F	OMe	I	Cu, neat, 180-200°	(65)	545

C₈

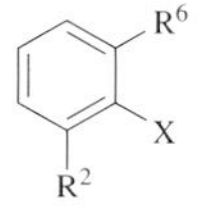

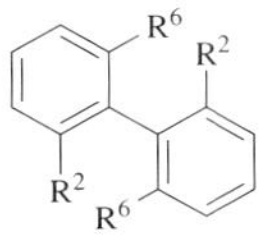

R²	R⁶	X	Conditions	Yield	Ref
NO_2	CO_2Me	I	Cu, neat, 165°	(79)	546
NO_2	CO_2Me	I	Cu, neat, 165-175°	(83)	547
NO_2	CO_2Me	I	Cu, neat, 100-180°	(91)	548
NO_2	CO_2Me	Br	Cu, heat	(50)	549
NO_2	CO_2Me	Cl	Cu, neat, 225-235°	(60)	550
I	CO_2Me	I	Cu, neat, 110-115°	(61) and 31% recovered starting material	33
Br	CO_2Me	I	Cu, neat, 160-170°	(81)	542
F	CO_2Me	I	Cu, neat, 180-240°	(57)	551
Me	CO_2Me	I	Cu, neat, 240-265°	(21)	552
Me	Me	Br	Ni(cod)₂, DMF, 54°	(0)	49, 50
Me	Me	Br	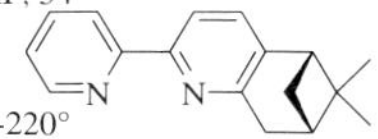 NiCl₂, CrCl₂, Mn, THF, rt	(0)	406
Me	CHO	I	Cu, neat, 190-220°	(50)	428
Me	OMe	I	Cu, neat, heat	(70)	553
OMe	OMe	I	Cu, neat, 170-210°	(85-90)	554
OMe	OMe	Br	Cu, neat, 240-260°	(48)	555
CF_3	CF_3	I	Cu, neat, 270°	(33)	556

C₁₀

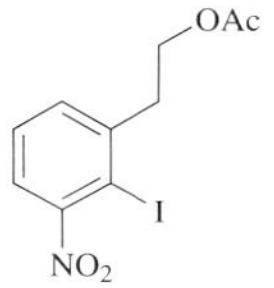

Cu, neat, 100-160° 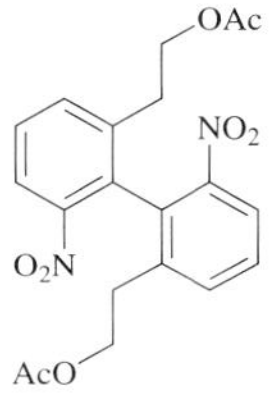   (88)   548

TABLE 8. 2,2′,6,6′-TETRASUBSTITUTED BIARYLS (*Continued*)

Substrate	Conditions	Product(s) and Yield(s) (%)	Refs.

C$_{9-10}$

R^2	R^6			
NO$_2$	CO$_2$Et	Cu, neat, 155-235°	(80)	557
Me	CO$_2$Me	Cu, neat, 260-270°	(74)	558
Me	CO$_2$Me	Cu, DMF, heat	(80)	559
Me	CO$_2$Me	Cu, neat, 270°	(90) after saponification to diacid	560
OMe	CO$_2$Me	Cu, neat, 205-215°	(81)	551
Et	CO$_2$Me	Cu, DMF, reflux	(41)	33

C$_{11}$ Cu, neat, 160°   (51)   561

C$_{12}$ Cu, neat, 210-215°   (95)   562

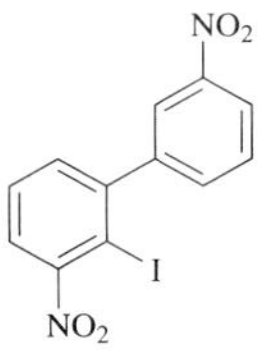

Cu, neat, 190-195° 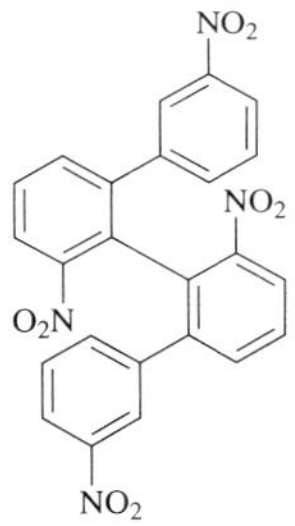 (91) 563

C₁₈₋₁₉

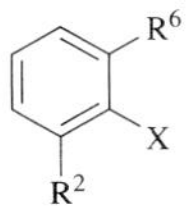

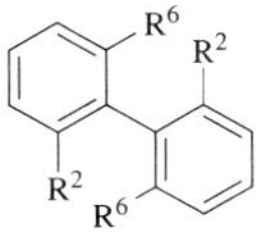

R²	R⁶	X		
F	POPh₂	Cl	Cu, DMF, 140°	(71) 564
Ph	Ph	I	Cu, neat, 260-270°	(3) 486
OMe	POPh₂	Br	Cu, DMF, 140°	(91) 565
OMe	POPh₂	I	Cu, DMF, 140°	(91) 565

381

TABLE 9. 3,3',4,4'-TETRASUBSTITUTED BIARYLS

Substrate	Conditions	Product(s) and Yield(s) (%)	Refs.
C_{6-7}			
R³–C₆H₃(R⁴)–X R³ / R⁴ / X		3,3',4,4'-biaryl with R³, R⁴	
F / OH / Br	NiBr₂, EtOH, MeOH, 2e⁻, bpy, NaBr	(75)	53
NO₂ / NO₂ / I	Cu, neat, 250°	(47)	1
Cl / Cl / SO₂Cl	PdCl₂(PhCN)₂, Ti(OPr-i)₄, m-xylene, 140°	(70)	168
NO₂ / Me / I	Cu, nitrobenzene, reflux	(39)	566
Cl / Me / I	Cu, neat, 220°	(90)	567
Br / OMe / Br	Pd(OAc)₂, Bu₃N, DMF, AsPh₃, 140°	(46)	172
Me / Cl / SO₂Cl	PdCl₂(PhCN)₂, Ti(i-PrO)₄, m-xylene, 140°	(75)	168
CF₃ / NO₂ / I	Cu, neat, 265–300°	(35)	496
C₇			
methylenedioxyphenyl–X, X = I	Cu, neat, 200°	bis(methylenedioxyphenyl) (21)	568
X = Br	NiBr₂, EtOH, DMF, 2e⁻, bpy, Bu₄NBF₄	(48)	53
X = Br	NiBr₂, EtOH, MeOH, 2e⁻, NaBr, bpy	(36)	53
C₈			
3,4-dimethylphenyl–X, X = I	Cu, neat, 270°	3,3',4,4'-tetramethylbiphenyl (23)	569
X = I	Cu, neat, heat	(60)	570
X = I	Ni powder, DMF, 140°	(70)	120
X = I	PdCl₂, N₂H₄·H₂O, MeOH, reflux	(56)	47

R³	R⁴	X	Conditions	(Yield)	Ref
CN	CN	I	PdCl$_2$, HgCl$_2$, N$_2$H$_4$•H$_2$O, MeOH, reflux	(78)	47
Me	OMe	Br	Pd(PPh$_3$)$_4$, 2e⁻, Et$_4$NOTs, DMF	(78)	190
CO$_2$H	CO$_2$H	Cl	NiCl$_2$, bpy, NaBr, DMF, 60-80°	(75)	146
CO$_2$H	CO$_2$H	Br	PdCl$_2$, HgCl$_2$, 3,4-xylylhydrazine hydrochloride, NaOH, MeOH, reflux 60-80°	(88)	165
		Cl	Pd/C, H$_2$, PEG-400, H$_2$O, NaOH, 110°	(39)[a]	186
		Cl	Pd/C, HCO$_2$Na, CTAB, H$_2$O, NaOH, 110°	(20), (25)[a]	181
			Ni I$_2$, lithium naphthalide, DME, 30-35°	(78)	571
			NiBr$_2$(PPh$_3$)$_2$, Zn, Et$_4$NI, THF, 50°	(57)	139
			Pd/C, NaOH, H$_2$O, formic hydrazide, 80-85°	(52)[a]	55
			Pd/C, NaOH, H$_2$O, formic hydrazide, 85°	(86-93)[b]	55

X	Conditions	(Yield)	Ref
I	Cu, neat, heat	(51)	572
I	Cu, neat, 260-270°	(86)	507
I	Cu, neat, 260-280°	(87)	490
I	Cu, neat, 190°	(55)	32
I	Cu, neat, CO$_2$, 235°	(77)	573
I	PdCl$_2$, HgCl$_2$, N$_2$H$_4$•H$_2$O, MeOH, reflux	(51)	47
Br	NiBr$_2$(PPh$_3$)$_2$, Zn, Et$_4$NI, THF, 50°	(70)	574
Br	Ni(OAc)$_2$, t-BuOLi, LiH, bpy, THF, 63°	(75), (91)[a]	110

TABLE 9. 3,3',4,4'-TETRASUBSTITUTED BIARYLS (*Continued*)

	Substrate	Conditions	Product(s) and Yield(s) (%)	Refs.
C_{10}	EtO–C$_6$H$_3$(OEt)–Br	Cu, neat, 250°	EtO–C$_6$H$_3$(OEt)–C$_6$H$_3$(OEt)–OEt (3)	575
	MeO$_2$C–C$_6$H$_3$(CO$_2$Me)–X		MeO$_2$C–C$_6$H$_3$(CO$_2$Me)–C$_6$H$_3$(CO$_2$Me)–CO$_2$Me	
	X			
	Cl	NiCl$_2$(PPh$_3$)$_2$, Zn, NaBr, DMAc, 80°	(>97)	576
	Cl	NiCl$_2$(PPh$_3$)$_2$, Zn, KI, NMP, 80°	(67)	576
	Cl	NiCl$_2$(PEt$_3$)$_2$, Zn, KI, NMP, 80°	(>64)	576
	Cl	NiCl$_2$(PBu$_3$)$_2$, Zn, NaBr, DMAc, 70°	(71)[a]	576
	Cl	NiCl$_2$, PBu$_3$, Zn, NaBr, DMAc, 70°	(>21)	576
	Cl	NiCl$_2$(PEt$_3$)$_2$, Zn, NaCl, DMF, 80°	(87)[a]	576
	Cl	NiCl$_2$, PEt$_3$, Zn, NaCl, DMF, 77-80°	(13)[a]	576
	Br	NiCl$_2$(PPh$_3$)$_2$, Zn, NaBr, DMAc, 80°	(>84)	576
C_{11}	t-Bu–C$_6$H$_3$(OMe)–Br	Ni(OAc)$_2$, NaH, t-AmOH, bpy, THF, reflux	t-Bu–C$_6$H$_3$(OMe)–C$_6$H$_3$(OMe)–Bu-t (50)	577
C_{13}	R^4–C$_6$H$_3$(R^3)–X		R^4–C$_6$H$_3$(R^3)–C$_6$H$_3$(R^3)–R^4	
	R^3 R^4 X			
	OMe OC$_6$H$_{13}$ I	Cu, neat, 270°	(66)	578
	OC$_6$H$_{13}$ OMe I	Cu, neat, 270°	(48)	578

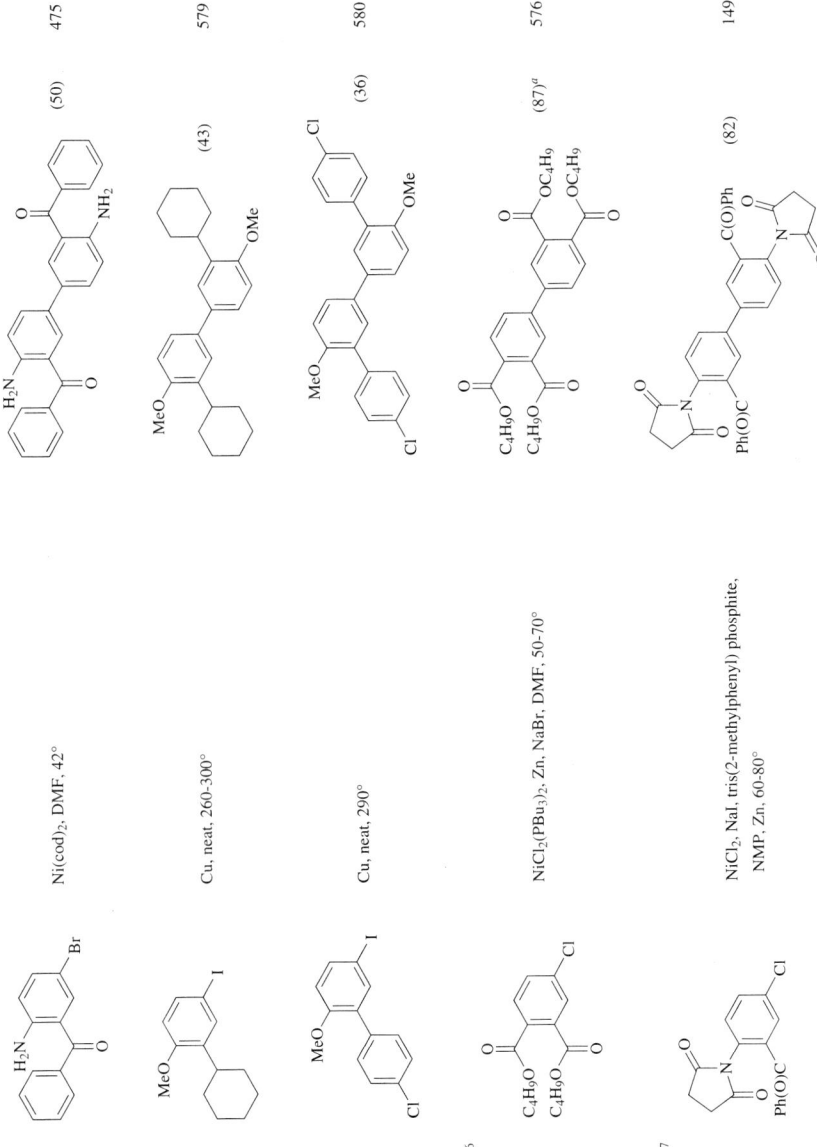

TABLE 9. 3,3',4,4'-TETRASUBSTITUTED BIARYLS (Continued)

Substrate	Conditions	Product(s) and Yield(s) (%)	Refs.
C_{18}			
(4-iodo-2-phenylbiphenyl)	Cu, neat, 260°	(tetraphenyl-substituted p-terphenyl) (67)	447
4-iodo-1,2-bis(hexyloxy)benzene	Cu, neat, 270-310°	3,3',4,4'-tetrakis(hexyloxy)biphenyl (50)	581

[a] The yield was determined by gas chromatography.
[b] The yield was determined by HPLC.

TABLE 10. 3,3',5,5'-TETRASUBSTITUTED BIARYLS

Substrate	Conditions	Product(s) and Yield(s) (%)	Refs.
C$_{6-8}$ 3,5-disubstituted aryl halide (R^3, R^5, X)		3,3',5,5'-tetrasubstituted biaryl (R^3, R^5)	
R^3=F, R^5=F, X=Br	NiBr$_2$, EtOH, DMF, 2e$^-$, Bu$_4$NBF$_4$, bpy	(54)	53
R^3=NO$_2$, R^5=NO$_2$, X=I	Cu, neat, 270°	(15)	582
R^3=Me, R^5=NO$_2$, X=I	Cu, DMF, reflux	(71)	583
R^3=CF$_3$, R^5=CF$_3$, X=I	Cu, sand, reflux	(30)	524
R^3=CO$_2$Me, R^5=Cl, X=I	Cu, neat, 265-270°	(25)	584
R^3=OMe, R^5=OMe, X=I	Cu, neat, 200-275°	(62-70)	585
R^3=OMe, R^5=OMe, X=I	Cu, neat, 230°	(37)	586
C$_{16}$ 3-bromo-5-(pyridin-2-yl) with 2-pyridyl substituent	NiCl$_2$(PPh$_3$)$_2$, Zn, DMF, 50°	tetra(2-pyridyl)biphenyl (70)	584, 587
C$_{30}$ 3,5-di(biphenyl-2-yl)iodobenzene	Cu, neat, heat	3,3',5,5'-tetra(Ar)biphenyl (35)	588

Ar = biphenyl-2-yl

TABLE 11. 2,2',3,3',4,4'-HEXASUBSTITUTED BIARYLS

Substrate	Conditions	Product(s) and Yield(s) (%)	Refs.
C9	Cu, neat, heat	(59)	589
C10	Cu, 2,4-dimethylsulfolane, 250°	(85)	590
C11-12	Cu, neat, 135-190°	(20)	591
R² = CO₂Me, R³ = CO₂Me, R⁴ = OMe	Cu, neat, 220-250°	(56)	592
R² = OMe, R³ = CO₂Me, R⁴ = CO₂Me	Cu, neat, 220-250°	(92)	592
R² = OMe, R³ = Et, R⁴ = OEt	Cu, neat, 260°	(67)	593
R² = OEt, R³ = Et, R⁴ = OMe	Cu, neat, 260°	(49)	593
R² = OEt, R³ = Et, R⁴ = OEt	Cu, neat, 200°	(62)	593
C13	Cu, neat, 150-200°	(73)	594
	CuOTf, DMSO, acetone, CH₃CN, aq. NH₃	(62)	420

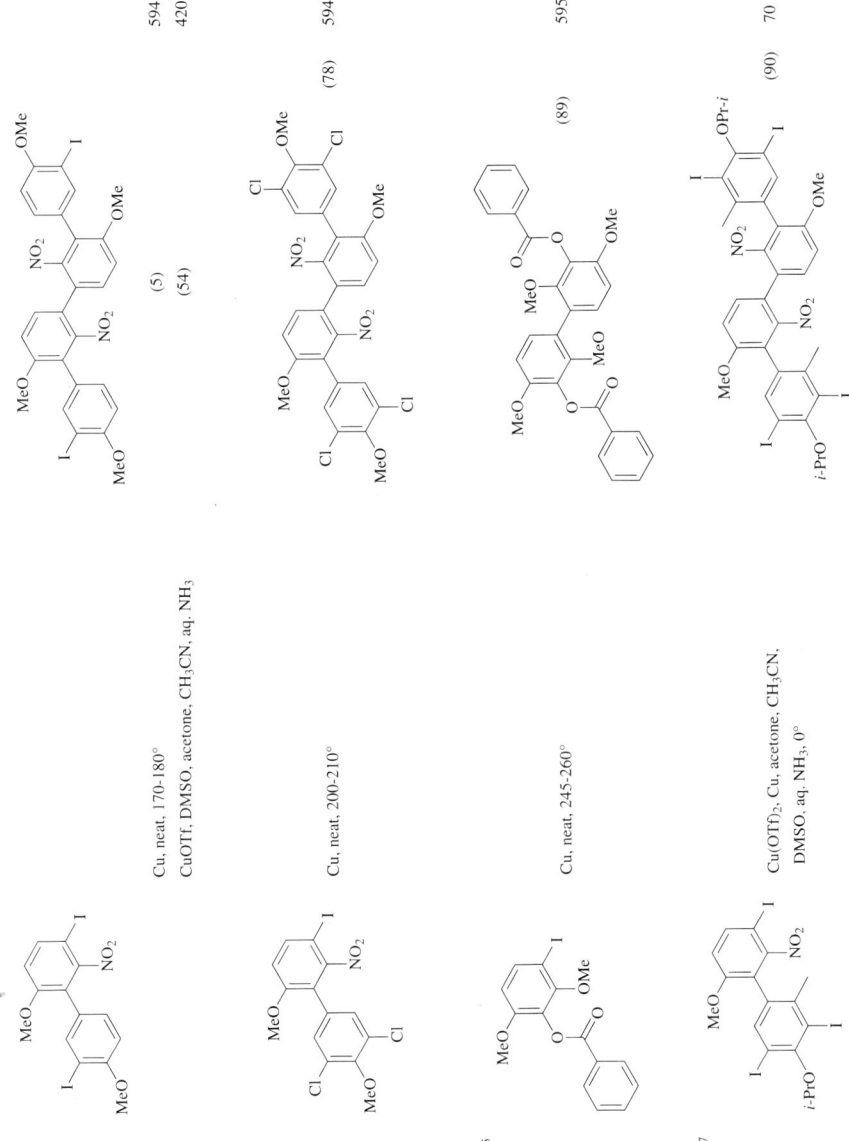

C$_{14}$	Cu, neat, 170-180°	(5) 594
	CuOTf, DMSO, acetone, CH$_3$CN, aq. NH$_3$	(54) 420
	Cu, neat, 200-210°	(78) 594
C$_{15}$	Cu, neat, 245-260°	(89) 595
C$_{17}$	Cu(OTf)$_2$, Cu, acetone, CH$_3$CN, DMSO, aq. NH$_3$, 0°	(90) 70

TABLE 11. 2,2',3,3',4,4'-HEXASUBSTITUTED BIARYLS (Continued)

Substrate	Conditions	Product(s) and Yield(s) (%)	Refs.
C$_{18}$	Cu(OTf)$_2$, Cu, acetone, CH$_3$CN, DMSO, aq. NH$_3$, 0°	(84)	70
C$_{22}$	Cu, neat, 180–200°	(87)	594

TABLE 12. 2,2',3,3',5,5'-HEXASUBSTITUTED BIARYLS

Substrate	Conditions	Product(s) and Yield(s) (%)	Refs.
C₇ (2,5-dibromobenzaldehyde, Br/CHO/Br)	Cu, DMF, reflux	biaryl with Br, CHO, Br, Br, CHO, Br (42)	596
C₇₋₉ (arene with R³, R⁵, X, OMe)			

R³	R⁵	X			
NO₂	NO₂	I	Cu, nitrobenzene, reflux	(13)	597, 598
OMe	Me	Br	Cu, neat, 230–290°	(19)	599
OMe	Me	Br	Cu, neat, 300°	(19-20)	484
OMe	CHO	I	Cu, neat, 250°	(65)	600
OMe	CHO	I	NiCl₂(PPh₃)₂, PPh₃, Zn, DMF, 54°	(8)	457
OMe	CHO	Br	Cu, nitrobenzene, reflux	(0)	601
OMe	CHO	Br	Cu, methyl benzoate, reflux	(0)	601
OMe	CHO	Br	Cu, DMF, reflux	(0)	601
OMe	CHO	Br	Cu, neat, heat	(0)	601

| C₁₂ (t-Bu, Br, Me, Me arene) | Cu, neat, 300° | tetramethyl-di-t-Bu biaryl (56) | 529 |

TABLE 12. 2,2',3,3',5,5'-HEXASUBSTITUTED BIARYLS (Continued)

Substrate	Conditions	Product(s) and Yield(s) (%)	Refs.
C₁₈ (substrate structure: mesityl-Br with ArC(O)- group; Ar = mesityl)	Cu, neat, 200°	biaryl product (63)	602

TABLE 13. 2,2',3,3',6,6'-HEXASUBSTITUTED BIARYLS

Substrate	Conditions	Product(s) and Yield(s) (%)	Refs.
C₈	Cu, DMF, reflux	(81)	603
C₉	Cu, DMF, heat	(>64 but <73)	604
C₉,₁₀			
R⁶ = CHO, X = Br	Cu, neat, 220–240°	(60)	605
R⁶ = CHO, X = I	Cu, DMF, reflux	(63)	601
R⁶ = CHO, X = I	Cu, neat, 250°	(45)	600
R⁶ = OMe, X = I	Cu, neat, 180–215°	(76)	606
R⁶ = CO₂Me, X = Br	Cu, neat, heat	(70)	605
C₁₀	Cu, neat, 200–235°	(47)	607
C₁₁	Cu, neat, sealed tube, 205–210°	(13)	608

393

TABLE 13. 2,2',3,3',6,6'-HEXASUBSTITUTED BIARYLS (Continued)

Substrate	Conditions	Product(s) and Yield(s) (%)	Refs.
C₁₃	Cu, DMF, reflux	(66)	56
C₁₅₋₁₆			
	Cu, neat, 220-240°	R⁶ = CHO (56)	609
	Cu, neat, 220-240°	R⁶ = CO₂Me (65)	609
C₂₀	Cu, DMF, reflux	(85)	610
	Cu, DMF, 140°	(82)	565

TABLE 14. 2,2',4,4',5,5'-HEXASUBSTITUTED BIARYLS

Substrate	Conditions	Product(s) and Yield(s) (%)	Refs.
C₆	Cu, nitrobenzene, reflux	(41)	598
C₇	Cu, DMF, 70°	(77)	611
	Cu, nitrobenzene, 160° (X = I)	(57)	568
	Cu, nitrobenzene, 160° (X = Br)	(79)	568
C₈	Cu, nitrobenzene, 220-230° (R² = Br)	(30)	612
	Cu neat, 230-250° (R² = Cl)	(31-36)	484
	Cu neat, 210-240° (R² = NO₂)	(70)	613

TABLE 14. 2,2',4,4',5,5'-HEXASUBSTITUTED BIARYLS (*Continued*)

Substrate	Conditions	Product(s) and Yield(s) (%)	Refs.
C₈			
(substrate 1)	Cu, neat, 150-225°	(40)	611
(substrate 2)	Cu, DMSO, 90°	(79)	34
(substrate 3)	Cu, DMF, reflux	(60)	583
(substrate 4)	Ni(cod)₂, 34°	(0)	49, 50

Substrate	R groups	Conditions	Product	Yield (%)	Refs.
C₉ (2-acetamido-bromoarene with R⁴, R⁵)	R⁴=Me, R⁵=Cl; R⁴=OMe, R⁵=Br	Cu, DMSO, 50°; Cu, DMF, 80°	bis-acetamido biaryl	(61); (95)	34; 34
iodoxylene with R²	R²=Me; R²=OMe	Cu, neat, 230-250°; Cu, neat, 200-260°	tetramethylbiaryl	(50); (69)	2; 589
dimethoxy haloarene (X=I, Br)	R²=Me, X=I; R²=CHO, X=Br	Cu, neat, 200-210°; Cu, neat, sealed tube, 200-220°	tetramethoxybiaryl	(55); (68)	614; 615
methylenedioxy haloarene	R₂=CH₂OMe, X=Br; R₂=CO₂Me, X=I/Br (not defined)	Ni(cod)₂, 40-60°; Cu, DMF, reflux	bis(methylenedioxy)biaryl	(67); (>64 but <73)	49, 50; 604

TABLE 14. 2,2',4,4',5,5'-HEXASUBSTITUTED BIARYLS (Continued)

Substrate	Conditions	Product(s) and Yield(s) (%)	Refs.
C₁₀ (tetrahydronaphthalene with I and NO₂)	Cu, neat, 110–140°	(bis-tetrahydronaphthyl with NO₂ groups) (75)	591
	Cu, neat, <140°	(73)	616
	Cu, DMF, reflux	(23)	61
(aryl with CO₂Me, I, OMe, OMe, OMe)	Cu, neat, 210–220°	(biaryl with CO₂Me, OMe groups) (75)	607
C₁₁ (aryl with OMe, Br, NHAc, NHAc, OMe)	Cu, DMSO, 90°	(biaryl with NHAc, OMe groups) (86)	34
(aryl with OMe, I, CH₂CO₂Me, OMe)	Cu, neat, 200–220°	(biaryl with OMe, CH₂CO₂Me groups) (38)	614

398

C_{12}	Cu, neat, 240-250°	(78)	617
C_{15}	Cu, neat, 230°	(38)	618
C_{16}	NiCl$_2$(PPh$_3$)$_2$, Zn, PPh$_3$, DMF, 50°	(68)	122
C_{17}	Cu, DMF, reflux	(29)	575

TABLE 15. 2,2',4,4',6,6'-HEXASUBSTITUTED BIARYLS

Substrate	Conditions	Product(s) and Yield(s) (%)	Refs.
C₆ (2,6-dinitro-4-nitro-X-benzene) X = Cl	Cu, nitrobenzene, heat	2,2',6,6'-tetranitro-4,4'-dinitrobiphenyl (55)	1
Cl	Cu, ethylene dichloride, 84°	(71)	619
Cl	Cu, nitrobenzene, heat	(58)	620
Cl	Ti, neat, 140-150°	(11)	73
Br	Cu, nitrobenzene, ultrasound	(77)[a]	30
(2,4-dichloro-6-chloro-iodobenzene)	Cu, neat, 220-230°	2,2',4,4',6,6'-hexachlorobiphenyl (53)	2
(2,6-dibromo-4-bromo-iodobenzene)	Cu, neat, sealed tube, 200°	2,2',4,4',6,6'-hexabromobiphenyl (75) crude	538
(4-nitro-2-chloro-6-R²-iodobenzene) R² = Cl	Cu, neat, 240-260°	(37)	621
R² = NO₂	Cu, neat, 240°	(46)	598

Substrate	Conditions	Product (Yield %)	Ref.
3-bromo-4-iodo-5-nitro substrate (O₂N, Br, I, Br)	Cu, sand, 180-200°	2,2',6,6'-tetrabromo-4,4'-dinitrobiphenyl (30)	460
3,5-difluoro-2-iodonitrobenzene (F, I, NO₂, F)	Cu, neat, 115-260°	2,2'-dinitro-4,4',6,6'-tetrafluorobiphenyl (75)	622
C₇ 2-X-3-R²-5-methyl-R⁶-benzene; R²=NO₂, R⁶=Br, X=I; R²=NO₂, R⁶=NO₂, X=Cl	Cu, DMF, reflux; Cu, nitrobenzene, heat	biphenyl products (58), (40)	623, 624
2-X-3-R²-5-MeO-R⁶-benzene; R²=NO₂, R⁶=Cl, X=Cl; R²=Cl, R⁶=Cl, X=I	Cu, nitrobenzene, heat; Cu, neat, heat	biphenyl products (90), (79)	624, 625
2-chloro-3-nitro-6-methyl substrate	Cu, neat, 205-230°	dimethyl dinitro biphenyl (52)	626

401

TABLE 15. 2,2',4,4',6,6'-HEXASUBSTITUTED BIARYLS (Continued)

Substrate	Conditions	Product(s) and Yield(s) (%)	Refs.
C₈ (2,6-dimethyl-1-iodo-3-nitrobenzene type)	Cu, neat, 190–230°	2,2'-dinitro-4,4',6,6'-tetramethylbiphenyl (50)	542
MeO-C₆H₂(R²)(R⁶)-X; R²=NO₂, R⁶=Me, X=Br	Cu, nitrobenzene, 190°	biaryl product (59)	627
R²=NO₂, R⁶=Me, X=Br	Cu, neat, 200–230°	(65)	627
R²=OMe, R⁶=Cl, X=I	Cu, neat, 190°	(85–89)	625
MeO₂C-C₆H₂(R²)(R⁶)-X; R²=NO₂, R⁶=NO₂, X=Br	Cu, p-xylene, heat	biaryl product (90) crude	624
R²=NO₂, R⁶=NO₂, X=Cl	Cu, nitrobenzene, heat	(54)	1
R²=Cl, R⁶=Cl, X=I	Cu, neat, 280°	(40)	628
2-Cl-3-NO₂-5-O₂N-C₆H₂-CO₂Me	Cu, neat, 160°	biaryl (62)	629

R^2	R^6	X			
Me	Me	I	Cu, neat, 260-270°	(13)	2
Me	Me	I	PdCl$_2$, N$_2$H$_4$·H$_2$O, MeOH, reflux	(0)	47
Me	Me	I	PdCl$_2$, HgCl$_2$, N$_2$H$_4$·H$_2$O, MeOH, reflux	(0)	47
Me	Me	I	Pd(OAc)$_2$, Et$_3$N, 100°	(0)	169
Me	Me	I	Pd(OAc)$_2$, n-Bu$_3$N, 100°	(0)	169
Me	Me	OTf	NiCl$_2$, Zn, PPh$_3$, NaI, DMF, sonication, 60°	(0)[b]	135
OMe	OMe	I	Cu, neat, 170-210°	(74)	630
OMe	OMe	Br	Cu, neat, 290°	(88)	631

R^2	R^6	X			
NO$_2$	CO$_2$Me	I	Cu, DMF, reflux	(72)	632
Me	Me	I	Cu, heat	(18)	633
Me	OMe	I	Cu, neat, 200°	(90)	634
OMe	OMe	I	Cu, neat, 230-270°	(64)	635
OMe	OMe	Br	Cu, neat, 270-305°	(24)	635
CHO	OMe	I	Cu, neat, 270°	(64)	636

Cu, neat, 190° (60-63) 625

TABLE 15. 2,2',4,4',6,6'-HEXASUBSTITUTED BIARYLS (Continued)

Substrate	Conditions	Product(s) and Yield(s) (%)	Refs.
C₈ (MeO₂C–Ar(Me)(NO₂)–Br)	Cu, neat, 170-220°	biaryl (70)	637
C₉ (O₂N–Ar(Me)(CO₂Me)–I)	Cu, DMF, reflux	biaryl (65)	638
C₁₀ (OEt, Br, OMe substituted)	Cu, neat, 290°	biaryl (63)	631
C₁₀ (MeO, CO₂Me, X, OMe substituted) X = Br	Cu, neat, 250-260°	biaryl (64), (80)[b]	639
X = Br	Cu, neat, 230-240°	(84)	640
X = I	Cu, neat, 220-225°	(89)	640

404

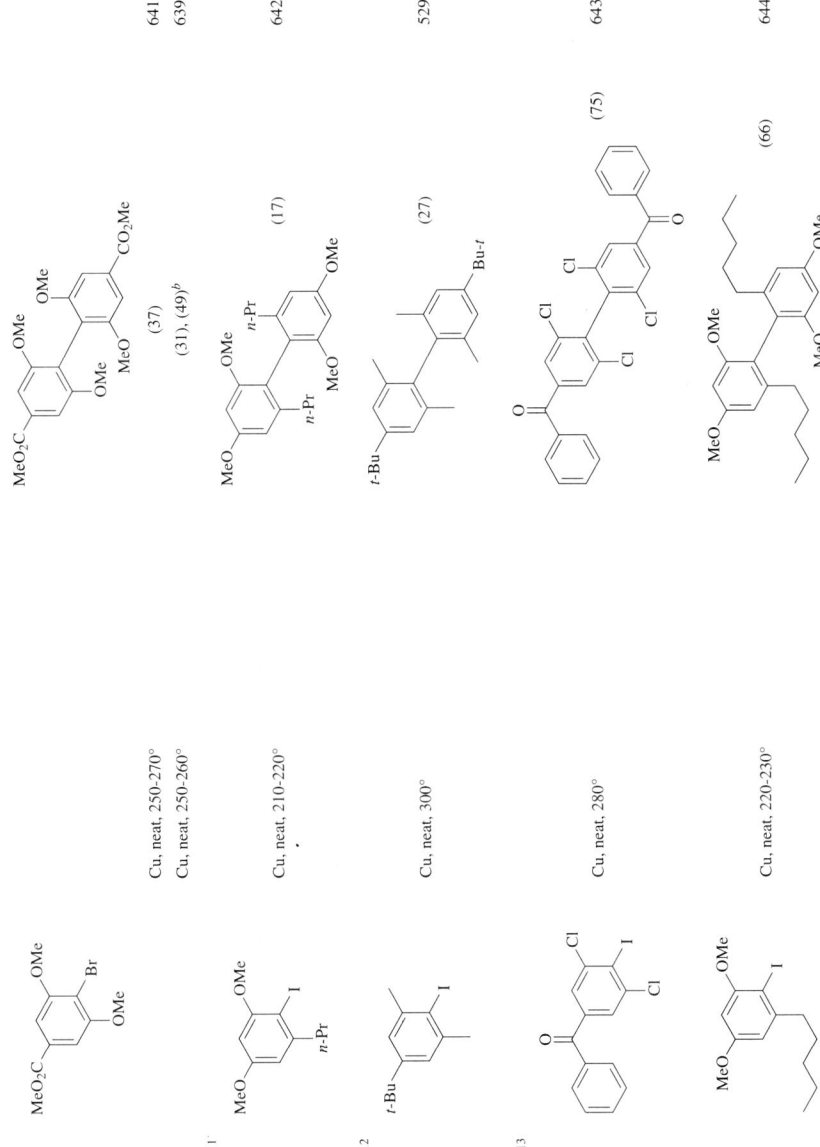

TABLE 15. 2,2',4,4',6,6'-HEXASUBSTITUTED BIARYLS *(Continued)*

Substrate	Conditions	Product(s) and Yield(s) (%)	Refs.
C$_{13-14}$ (R-C$_6$H$_4$-CO-aryl with Cl, NO$_2$, O$_2$N substituents)			
R = H	Cu, neat, AcOH, 175-190°	(65)	59
R = H	Cu, neat, nitrobenzene, heat	(21)	59
R = Me	Cu, neat, AcOH, 130-150°	(63)	59
R = NO$_2$	Cu, neat, AcOH, 150-170°	(54)	59
R = OMe	Cu, neat, AcOH, 155°	(29)	59
C$_{14}$ (mesityl-I-SO$_3$Ph)	Cu, neat, 170-180°	(80)	645
(dimethyl-iodo-phenyl)	Cu, neat, heat	(trace)	515

C₁₆ (aryl bromide with oxazoline, isopropyl)	Cu, DMF, reflux	biaryl product, dr 95 : 5, (>60)ᶜ prior to purification 230
(aryl bromide with oxazoline, isopropyl, methyl)	Cu, DMF, reflux	(43) RSS + (31) SSS 225, 226
aryl iodide/bromide with dimethyl malonate side chain	Cu, neat, 240° (X=I); Cu, neat, 240° (X=Br)	coupled biaryl bis-malonate (66) ; (0) 646; 646
C₁₇ (aryl bromide with phenyl-oxazoline)	Cu, DMF, reflux	biaryl product, dr 91.5 : 8.5, (>51)ᶜ prior to purification 230

TABLE 15. 2,2',4,4',6,6'-HEXASUBSTITUTED BIARYLS (Continued)

Substrate	Conditions	Product(s) and Yield(s) (%)	Refs.
C₂₀ ![MeO, P(O)Ph₂, I, OMe substrate]	Cu, DMF, 140°	[biaryl product] (66)	565
C₂₄ ![Ph, I, Ph substrate]	Cu, neat, 270–280°	[biaryl product] (50)	647

[a] The yield was determined by NMR spectroscopy.
[b] The yield was determined by gas chromatography.
[c] The yield includes oxazoline opening and acetylization.

TABLE 16. 3,3',4,4',5,5'-HEXASUBSTITUTED BIARYLS

Substrate	Conditions	Product(s) and Yield(s) (%)	Refs.
$C_{8,9}$			
(3-bromo-4-R⁴-dimethylbenzene)		(R⁴-tetramethylbiphenyl)	
R⁴ = OH	Pd/C, formic hydrazide, NaOH, H₂O, MeOH, 70°	(86)	55, 648
R⁴ = OMe	NiCl₂, Zn, PPh₃, DMF, 50-60°	(58)	490

TABLE 17. 2,2',3,3',4,4',5,5'-OCTASUBSTITUTED BIARYLS

Substrate	Conditions	Product(s) and Yield(s) (%)	Refs.
C₆			
(perfluorobromobenzene structure)	Cu, DMF, reflux	(perfluorobiphenyl) (73)	649
C₁₀			
(iodo-methyl-methoxy arene with R⁴, R⁵)		(biaryl product with R⁴, R⁵, OMe, Me)	
R⁴ R⁵			
Me OMe	Cu, neat, >220°	(76)	650
Me OMe	Cu, neat, 235°	(36)	651
OMe Me	Cu, neat, 220°	(86)	652
(iodo-dimethoxy arene with R², R³)		(biaryl with R², R³, OMe)	
R² R³			
CHO OMe	Cu, DMF, heat	(77)	653
OMe OMe	Cu, neat, 220°	(69)	654

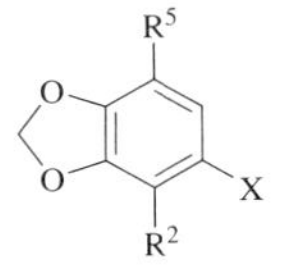

C₁₁₋₁₄	X	R²	R⁵	
	I	CO₂Me	OMe	Cu, DMF, reflux
	Br	OMe	CO₂Me	Cu, DMF, reflux

(44) 655
(10) 655

R	X	
Me	I	Cu, neat, 210°
Me	I	Cu, neat, 240-250°
Me	Br	Cu, neat, 240-250°
n-Bu	I	Cu, neat, 250°

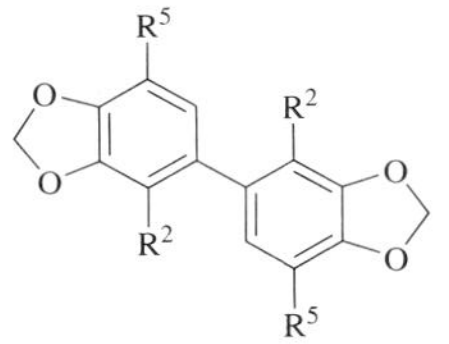

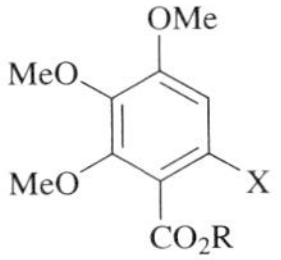

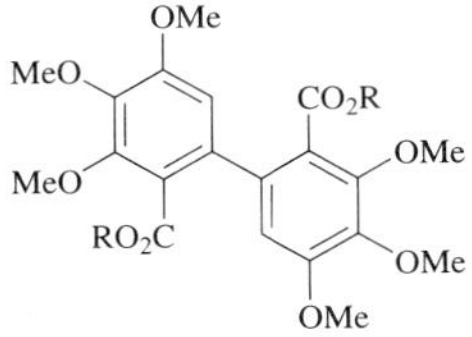

(52) 458
(45) 617
(63) 617
(74) 617

C₁₉

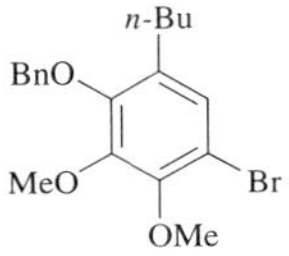

NiCl₂(PPh₃)₂, PPh₃, Zn, DMF, 54°   (5) 457

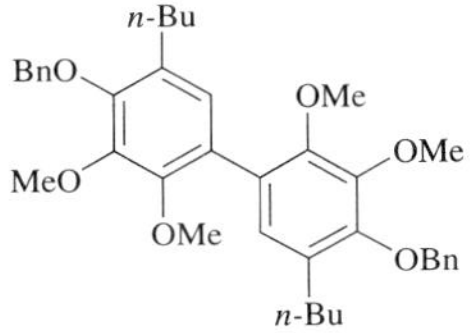

TABLE 18. 2,2',3,3',4,4',6,6'-OCTASUBSTITUTED BIARYLS

	Substrate	Conditions	Product(s) and Yield(s) (%)	Refs.
C_7				
	R³ | X | Me | Br	Cu, nitrobenzene, 160-183°	(70)	656
	Me | Br	Cu, nitrobenzene, heat	(80)	657
	OMe | Br	Cu, 1,2-dichloroethane, 84°	(77) as derivative	619
	OMe | Br	Cu, toluene, reflux	(83)	658
	OMe | Br	Cu, toluene, reflux	(83)	659
	OMe | Cl	Cu, xylene, reflux	(71) as derivative	658
C_{10}		Cu, neat, 200°	(75)	660
		Cu, neat, 250°	(81)	661
		Cu, DMF, heat	(92)	653
$C_{9,10}$	R³ | R⁴			
	OMe | OMe	Cu, neat, 220°	(63)	652
	NO₂ | Me	Cu, neat, 200-300°	(0)	662

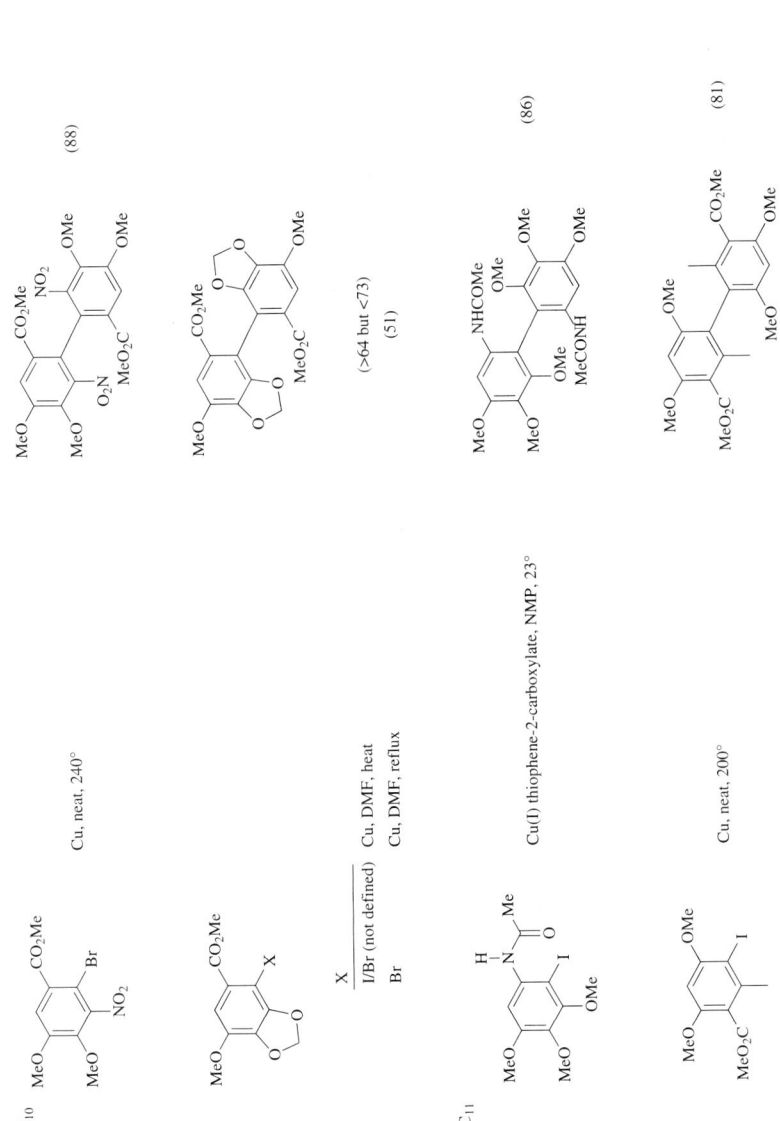

TABLE 18. 2,2',3,3',4,4',6,6'-OCTASUBSTITUTED BIARYLS (*Continued*)

Substrate	Conditions	Product(s) and Yield(s) (%)	Refs.
(structure with CO$_2$Me, X, OMe, MeO, MeO) X = I/Br (not defined)	Cu, DMF, heat	(structure, >64 but <73)	604
X = Br	Cu, neat, 240°	(36)	458
X = Br	Cu, DMF, reflux	(40)	666
X = Br	Cu, neat, 210–220°	(60)	667
X = Br	Cu, heat	(47)	640
C$_{12}$ (structure with OAc, I, OAc, MeO, MeO)	Cu, neat, heat	(structure, 41)	630
(structure with OMe, I, EtO$_2$C, MeO, O$_2$N)	Cu, 210–220°	(structure, 72)	668
(structure with O$_2$N, NO$_2$, X, NO$_2$) X = Cl	Cu, nitrobenzene, 200°	(84)	659
X = Br	Cu, nitrobenzene, 200°	(67)^a	659

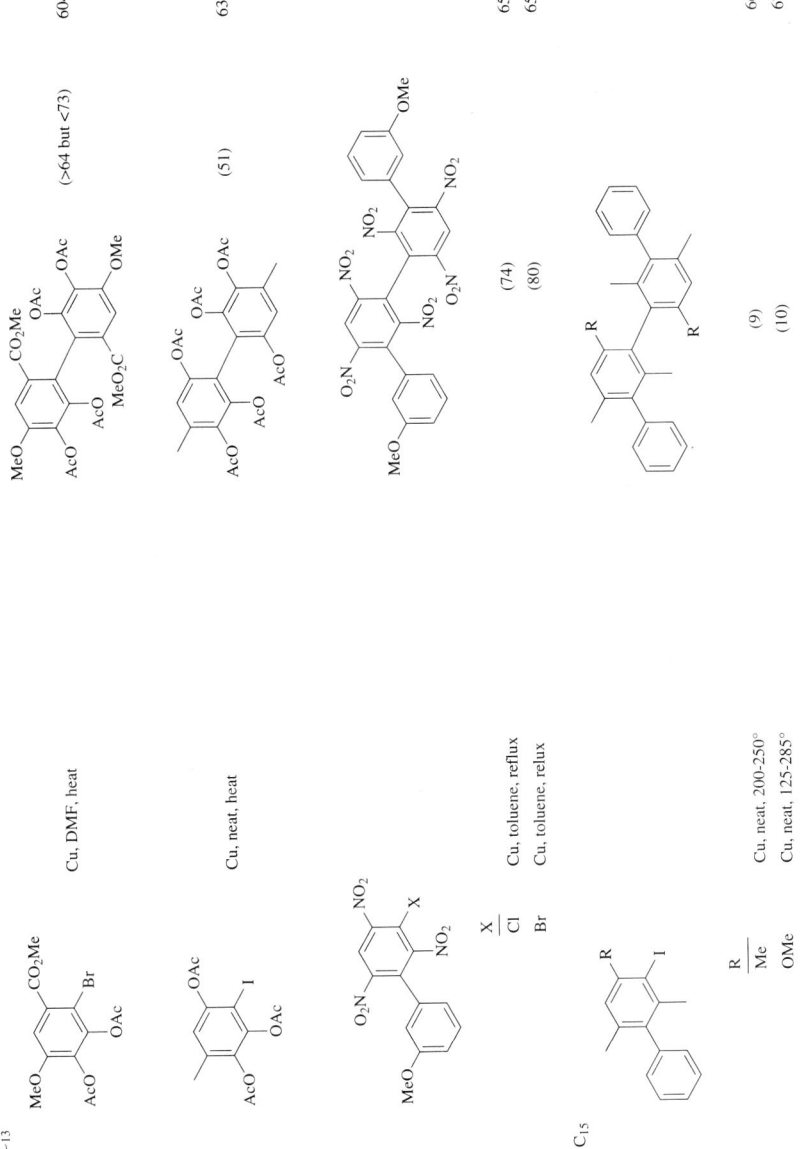

TABLE 18. 2,2',3,3',4,4',6,6'-OCTASUBSTITUTED BIARYLS (Continued)

Substrate	Conditions	Product(s) and Yield(s) (%)	Refs.
C₁₆ (bromo/X aryl with oxazoline, OMe groups)	Cu, DMF, reflux	(58) dr 93:7, prior to purification	214, 215, 218
C₁₆ (BnO, I, OMe, Me aryl)	Cu, neat, 180°	(56)	671
(OMe, I, OBn, Me aryl)	Cu, neat, 180°	(42)	671
C₁₈ (mesityl iodide)	Cu, neat, 250–270°	(48)	669
C₂₀₋₂₃ (oxazoline/Br aryl with neopentyl group) R = Me	Cu, DMF, reflux	(66) of S atropisomer dr 7.2:1 (S/R biaryl axis)	224
R = t-Bu	Cu, DMF, reflux	(64) of S atropisomer dr 3:1 (S/R biaryl axis)	224

C$_{21}$ — Cu, DMF, 140° — (79-90) — 565

C$_{22}$ — Cu, DMF, reflux — dr 1 : 6.7 (S/R biaryl axis) (74) of R atropisomer — 224

---

[a] The yield was determined by NMR spectroscopy.

TABLE 19. 2,2',3,3',5,5',6,6'-OCTASUBSTITUTED BIARYLS

Substrate	Conditions	Product(s) and Yield(s) (%)	Refs.
C₆ (tetrachloroiodobenzene)	Cu, neat, sealed tube, 190°	(decachlorobiphenyl) (65)	672
C₇ (chloro-difluoro-methyl-iodobenzene)	Cu, neat, 200°	(34)	16
C₈ (dichloro-dimethoxy-iodobenzene)	Cu, heat	(30)	673
C₈ (difluoro-dimethyl-iodobenzene)	Cu, neat, 200°	(60)	16
C₈ (fluoro-dimethyl-nitro-iodobenzene)	Cu, nitrobenzene, 200°	(75)	674

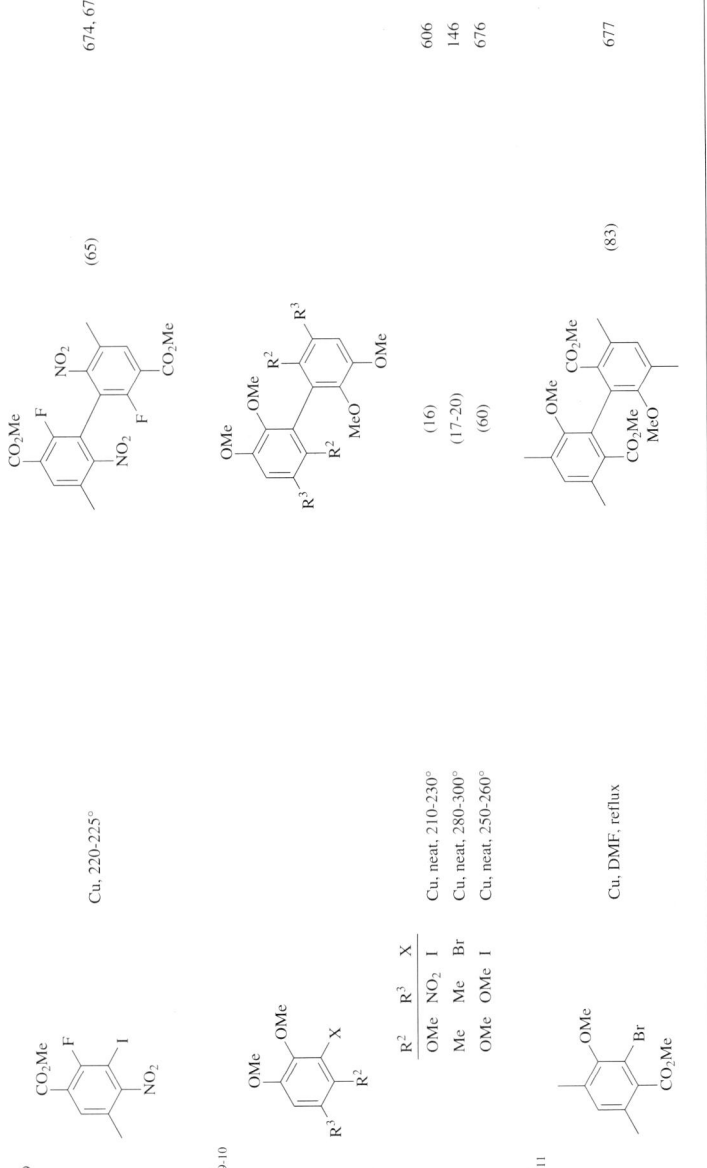

TABLE 20. 2,2',3,3',4,4',5,5',6,6'-DECASUBSTITUTED BIARYLS

Substrate			Conditions	Product(s) and Yield(s) (%)	Refs.
C₆₋₇					

R²	X	Conditions		Refs.
NO₂	Br	Cu, DMF, reflux	(57)	649
F	I	Cu, neat, sealed tube, 300°	(72)	678
F	I	CuI, potassium naphthalide, DME, 85°	(83)	24, 25
F	I	CuCl, lithium naphthalide, dioxane, 101°	(91)[a]	26
F	I	NiI₂, lithium naphthalide, DME, 20°	(100)[a]	112
F	I	Cu, neat, sealed tube, 180-290°	(87)	679
F	I	NiI₂, lithium naphthalide, DME, 80°	(100)[a]	111
F	I	CuI-PEt₃, lithium naphthalide, THF, mesitylene, 151°	(91)	25
F	I	CuCl, lithium naphthalide, DME, 85°	(92)[a]	25
F	Br	Cu, neat, sealed tube, 200°	(91)	680
F	Br	Cu, neat, sealed tube, 180-240°	(87)	681
F	Br	Cu, DMF, reflux	(96)	682
F	Br	NiI₂, lithium naphthalide, DME, 80°	(49)[a]	111
F	Br	NiBr₂, lithium naphthalide, DME, 80°	(37)[a]	111
F	Cl	Cu, neat, 320-360°	(73)	683
F	Cl	Cu, neat, sealed tube, 230°	(68)	684
F	OMs	NiCl₂(PPh₃)₂, Zn, Et₄NI, THF, 67°	(0)[a]	133
Me	Br	Cu, DMF, reflux	(61)	649
CN	Br	Cu, DMF, reflux	(77)	649

420

C$_7$	Cu, neat, 270-290°	(73)	685
C$_8$	Cu, nitrobenzene, 190°	(69)	674
C$_{11}$	NiCl$_2$, bpy, NaBr, Zn, DMF, 60-80°	(65)	146
	Cu, neat, 250°	(19)	676
C$_{12}$	Cu, neat, 250-260°	(29)	677

TABLE 20. 2,2',3,3',4,4',5,5',6,6'-DECASUBSTITUTED BIARYLS (*Continued*)

Substrate	Conditions	Product(s) and Yield(s) (%)	Refs.

C$_{22}$

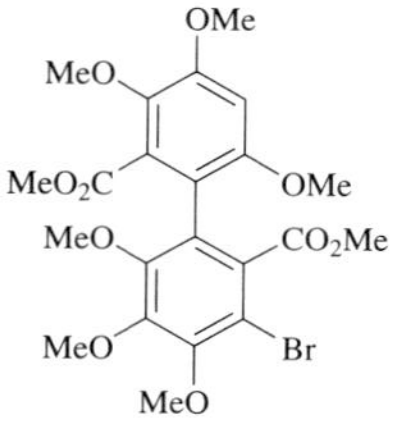

[Substrate: biaryl with OMe, OMe, MeO, MeO₂C, CO₂Me, MeO, MeO, Br, MeO substituents]

1. Cu, neat, 220°
2. HI, Ac$_2$O

[Product: bis-lactone polyphenol structure] (88)

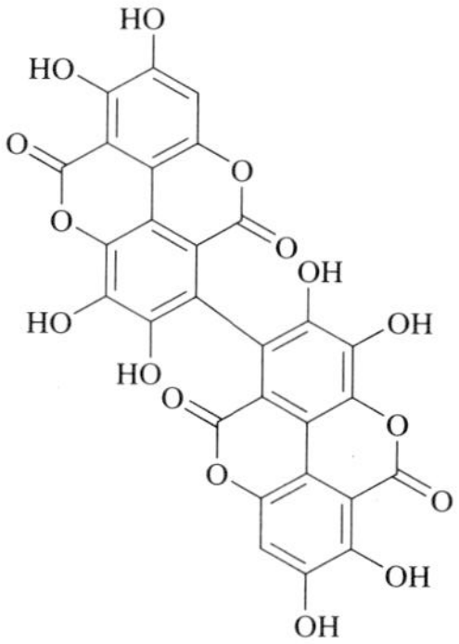

667

[a] The yield was determined by gas chromatography.

TABLE 21. 1,1'-BINAPHTHYLS

Substrate	Conditions	Product(s) and Yield(s) (%)	Refs.
C$_{10}$			
(perfluoro-1-bromonaphthalene)	Cu, 200°	(perfluoro-1,1'-binaphthyl) (80)	686
(1-bromo-4-iodo-2-nitronaphthalene)	Cu, DMF, 150-160°	(4,4'-dibromo-2,2'-dinitro-1,1'-binaphthyl) (8)	61
(1,4-diiodo-2-nitronaphthalene)	Cu, PhNO$_2$, reflux	(4,4'-diiodo-2,2'-dinitro-1,1'-binaphthyl) (11)	687

TABLE 21. 1,1'-BINAPHTHYLS (Continued)

Substrate 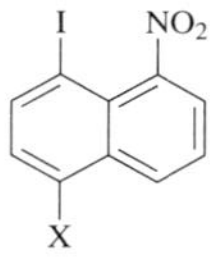	Conditions	Product(s) and Yield(s) (%) 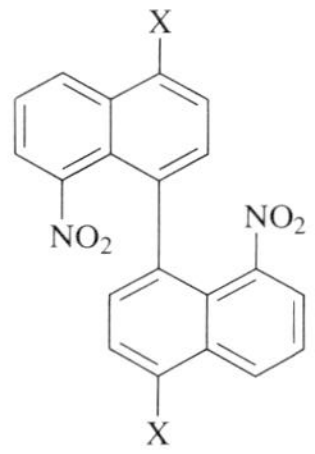	Refs.

X		
NO₂	Cu, nitrobenzene, reflux	(36)	688
Br	Cu, nitrobenzene, reflux	(29)	688

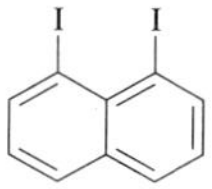

Pd(OAc)₂, K₂CO₃, Bu₄NBr, DMF, 100° — (39) + (42)   173

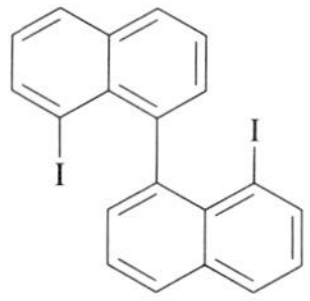

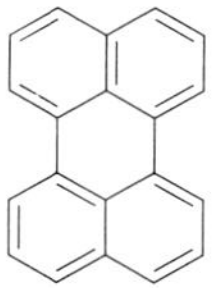

C₁₀₋₁₁

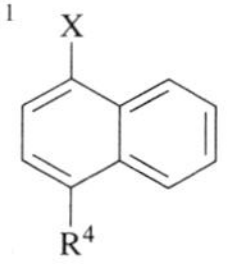

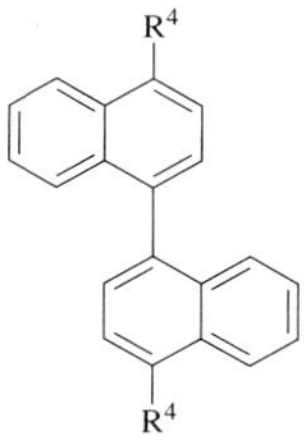

R⁴	X		
NO₂	I	Cu, naphthalene, 220-230°	(52)	689
Me	Br	Cu, I₂, 1-methylnaphthalene, 230-270°	(3)	690
OMe	I	Cu, 220-230°	(15-20)	691

424

Pd(OAc)$_2$, K$_2$CO$_3$, Bu$_4$NBr, DMF, 100°	(39)	173
Cu, nitrobenzene, reflux	(44)	688
Cu, H$_2$O, 100°	(8)	411

R^3	X			
NO$_2$	I	Cu, 215-280°	(7-20)	591, 692
CO$_2$Me	Br	Ni(cod)$_2$, DMF, 60°	(50)	693

X			
Cl	Ni(OAc)$_2$, NaH, t-AmOH, bpy, THF, 63°	(88-90)	131
Cl	Ni(OAc)$_2$, LiH, t-BuOH, bpy, THF, 63°	(87)	110
Br	Cu, I$_2$, 180-285°	(50)	689
Br	Ni(bpy)Br$_2$, bpy, NMP, −1.3 V	(75)	128
Br	Ni(OAc)$_2$, LiH, t-BuOH, bpy, THF, 65°	(88-94)	110, 400
Br	Ni(OAc)$_2$, NaH, t-AmOH, bpy, THF, 63°	(70-74)	131
Br	Ni(OAc)$_2$, NaH, t-AmOH, PPh$_3$, THF, 63°	(85)	131

TABLE 21. 1,1'-BINAPHTHYLS (Continued)

Substrate	Conditions	Product(s) and Yield(s) (%)	Refs.
Br	NiCl$_2$, CrCl$_2$, Mn, bpy-like ligand, THF, 25°	(98)	406
Br	Pd(OAc)$_2$, Et$_3$N, DMF, 115°	(40)	175
Br	Pd(OAc)$_2$, As(o-tolyl)$_3$, hydroquinone, DMA, 100°	(80)	178
Br	PdCl$_2$, HgCl$_2$, PhNHNH$_2$, NaOH, MeOH, reflux	(71)	165
Br	[PdCl(π-C$_3$H$_5$)]$_2$, TBAF, DMSO, 120°	(48)	198
Br	Pd/CaCO$_3$, N$_2$H$_4$, KOH, MeOH, heat	(14)	48
I	Cu, 260-285°	(74-92)	1, 694
I	Cu, DMF, reflux	(76)	57
I	NiCl$_2$, Ph$_3$P, Zn, NaI, DMF, 60°, ultrasound	(80)	135
I	Ni(OAc)$_2$, NaH, t-AmOH, bpy, THF, 63°	(62-66)	131
I	Pd/C, Zn, acetone, H$_2$O, rt	(70)	185
I	Pd/C, Zn, H$_2$O, 18-crown-6, rt	(40)	695
I	In, DMF, reflux	(80)	74
SO$_2$Cl	PdCl$_2$(PhCN)$_2$, Ti(OPr-i)$_4$, m-xylene, 140°	(26)	168
OTf	NiCl$_2$, Ph$_3$P, Zn, NaI, DMF, 60°, ultrasound	(80)	135
OTf	NiCl$_2$(dppe), Zn, KI, DMF, THF, 90°	(3-85)	192
OTf	NiCl$_2$(PPh$_3$)$_2$, Zn, KI, DMF, 90°	(92)	192
OTf	NiCl$_2$(dppf), Zn, KI, DMF, THF, 90°	(93)	192
OTf	PdCl$_2$(PPh$_3$)$_2$, DMF, 20°, 2e$^-$	(20)	191
OTf	PdCl$_2$(PPh$_3$)$_2$, DMF, 60°, 2e$^-$	(40)	191
OTf	PdCl$_2$(PPh$_3$)$_2$, DMF, 90°, 2e$^-$	(50)	191
OTf	PdCl$_2$(PPh$_3$)$_2$, Bu$_4$NBF$_4$, DMF, 2e$^-$	(7-50)	191
OTf	PdCl$_2$(MePPh$_2$)$_2$, Bu$_4$NBF$_4$, DMF, 2e$^-$	(30)	192
OTf	PdCl$_2$(dppm), Bu$_4$NBF$_4$, DMF, 2e$^-$	(35)	192
OTf	PdCl$_2$(dppe), Bu$_4$NBF$_4$, DMF, 2e$^-$	(33)	192
OTf	PdCl$_2$(dppp), Bu$_4$NBF$_4$, DMF, 2e$^-$	(31)	192
OTf	PdCl$_2$(dppb), Bu$_4$NBF$_4$, DMF, 2e$^-$	(40)	192
OTf	NiCl$_2$(PPh$_3$)$_2$, Zn, KI, THF, 67°	(92)	193
OTf	NiCl$_2$(dppe), Zn, KI, THF, DMF, 67°	(82)	193
OTf	PdCl$_2$(PPh$_3$)$_2$, Zn, DMF, 90°	(59)	192, 193

C₁₁₋₁₇

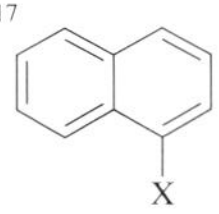

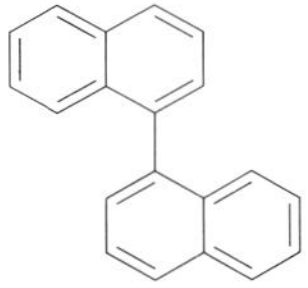

X			
OTf	Pd(OAc)₂, Zn, BINAP, DMF, 90°	(61-98)	192
OTf	Pd(OAc)₂, Zn, PPh₃, DMF, 90°	(51-67)	192
OTf	Pd(OAc)₂, Zn, dppe, DMF, 90°	(58)	192
OTf	Pd(OAc)₂, Zn, dppf, DMF, 90°	(64)	192
OTf	Pd(OAc)₂, Zn, DIOP, DMF, 90°	(13)	192
OTs	NiCl₂(dppe), Zn, KI, DMF, 140°	(88)	192, 193
OTs	NiCl₂(PPh₃)₂, Zn, KI, DMF, 140°	(70)	192, 193
OTs	NiCl₂(PPh₃)₂, Zn, PPh₃, NaBr, DMF, 100°	(79)	137

C₁₀₋₁₆

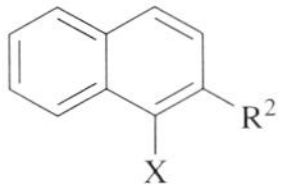

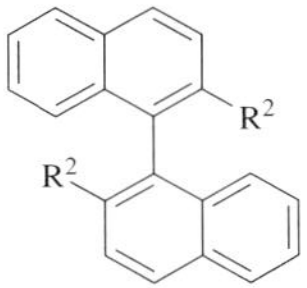

R^2	X			
NO₂	I	Cu, 120-130°	(17-70)	591, 692
NO₂	I	Cu, DMF, reflux	(77)	61
OH	Br	PdCl₂(PPh₃)₂, Et₄NOTs, DMF, 2e⁻	(16)	190
CN	Br	Cu, DMF, reflux	(63)	397
OMe	Br	PdCl₂(PPh₃)₂, Et₄NOTs, DMF, 2e⁻	(53)	190
OMe	Br	Pd(OAc)₂, K₂CO₃, i-PrOH, DMF, 115°	(3)	74
OMe	I	Pd(OAc)₂, As(o-tolyl)₃, hydroquinone, DMA, 125°	(91)	178
Me	OTf	NiCl₂(dppf), Zn, KI, DMF, 100°	(66)	192
Me	OTf	Pd(OAc)₂, Zn, BINAP, DMF, 100°	(16)	192
CO₂Me	Cl	Cu, I₂, 250-300°	(25-48)	696
CO₂Me	Br	Cu, 190°	(61-87)	697-700
CO₂Me	Br	Cu, 270-290°	(46-78)	701-703
CO₂Me	Br	Cu, DMF, reflux	(75-85)	704-706
SO₃Ph	I	Cu, <300°	(83)	707

TABLE 21. 1,1'-BINAPHTHYLS (Continued)

Substrate	Conditions	Product(s) and Yield(s) (%)	Refs.
C$_{12}$ (naphthyl with CO$_2$Me and I)	Cu, CO$_2$, 220-240° Cu, CO$_2$, heat	binaphthyl-(CO$_2$Me)$_2$ (75) (>50)	708 709
(dibromo-CO$_2$Me naphthyl)	Cu, 210-220°	tetrasubstituted binaphthyl (Br, CO$_2$Me)$_2$ (55)	36
(AcNH, Br naphthyl)	Cu, DMSO, 65°	bis(AcNH, Br) binaphthyl (96)	34

Substrate	Conditions	Product (Yield %)	Ref.
C₁₃ naphthalene with CO₂Me, Br, Br substituents	Cu, 160–170°	binaphthyl with CO₂Me, Br (41)	36
C₁₃ naphthalene with AcNH, Me, Br	Cu, tetralin, reflux	binaphthyl with AcNH, Me (39)	710
naphthalene with oxazoline, I	Cu, pyridine, 120°	bi-oxazoline binaphthyl (69)	711
C₁₄₋₁₅ naphthalene with OMe, OMe, OMe, X; R/X: OMe/Br, CO₂Me/I	Cu, 260–280°; Cu, 120°	hexamethoxy binaphthyl (14), (80)	712, 713

TABLE 21. 1,1'-BINAPHTHYLS (Continued)

Substrate	Conditions	Product(s) and Yield(s) (%)	Refs.
C₁₅ [1-chloro-2-(2-pyridyl)naphthalene]	Ni(PPh₃)₂Cl₂, Zn, Et₄NI, THF, rt	[2,2'-bis(2-pyridyl)-1,1'-binaphthyl] (69)	714
C₁₆₋₁₉ [1-bromo-2-Ox*-naphthalene]; Ox* = oxazoline with R = *i*-Pr, *t*-Bu, Ph	Cu, pyridine, reflux	[2,2'-bis(Ox*)-1,1'-binaphthyl] (60), (77–79), (75)	216; 212, 216, 220; 216
C₁₇ [8-bromo-1-(4-*t*-Bu-oxazolinyl)naphthalene]	Cu, DMF, reflux	[8,8'-bis(4-*t*-Bu-oxazolinyl)-1,1'-binaphthyl] (91)	220

C$_{19-38}$ structure with CO$_2$R and I on naphthalene	Cu, DMF, reflux	

R		
(+)-1-phenethyl	(80-95)	213
(−)-1-phenethyl	(80-95)	213
(−)-2-octyl	(80-95)	213
(−)-menthyl	(89)	213
cholesteryl	(80-95)	213

C$_{21}$ — Cu, 220–230° — (45-50) 715

C$_{32}$ — Cu, biphenyl, 255° — (41) 716

TABLE 21. 1,1'-BINAPHTHYLS (Continued)

Substrate	Conditions	Product(s) and Yield(s) (%)	Refs.
C_{42}			
R	Cu, DMF, reflux	(66) R,R,R	717

TABLE 22. 2,2'-BINAPHTHYLS

Substrate	Conditions	Product(s) and Yield(s) (%)	Refs.
C$_{10}$ perfluoro-2-bromonaphthalene	Cu, 250-280°	perfluoro-2,2'-binaphthyl (65)	686
2-X-naphthalene, X = Br	Pd/CaCO$_3$, N$_2$H$_4$, KOH, MeOH, heat	2,2'-binaphthyl (50)	48
X = Br	PdCl$_2$, HgCl$_2$, PhNHNH$_2$, NaOH, MeOH, reflux	(50)	165
X = I	Cu, 225-260°	(68)	2
X = I	[PdCl(π-C$_3$H$_5$)]$_2$, TBAF, DMSO, 120°	(58)	19
1-NO$_2$-3-I-naphthalene	Cu, nitrobenzene, reflux	1,1'-dinitro-3,3'-binaphthyl (32)	687
C$_{10\text{-}12}$ 3-R^3-2-X-naphthalene; R^3=NO$_2$, X=I	Cu, 135-190°	(31)	718
R^3=CO$_2$Me, X=Br	Cu, 190°	(90)	698

TABLE 22. 2,2'-BINAPHTHYLS (Continued)

Substrate		Conditions	Product(s) and Yield(s) (%)	Refs.
C_{10-13} R^6-naphthyl-X			R^6-binaphthyl-R^6	
R^6	X			
OH	Br	Pd, SrCO$_3$, N$_2$H$_4$, EtOCH$_2$CH$_2$OH, 135-140°	(10)	719
OH	Br	Pd/CaCO$_3$, NaOH, H$_2$O, HCONHNH$_2$, 80°	(63)	55
OMe	Br	Pd, SrCO$_3$, N$_2$H$_4$, EtOCH$_2$CH$_2$OH, 135-140°	(10)	719
CO$_2$Me	OMs	NiCl$_2$(PPh$_3$)$_2$, Zn, Et$_4$NI, THF, 67°	(91)	133
C_{10-16} 1-iodo-naphthyl-R^1			binaphthyl-R^1,R^1	
R^1				
NO$_2$		Cu, 135-190°	(52)	591
NO$_2$		Cu, nitrobenzene, reflux	(35-84)	692, 720
NO$_2$		Cu, DMF, reflux	(78)	61
OMe		Cu, I$_2$, 200-210°	(36)	721
SO$_3$Ph		Cu, 170-175°	(95)	722
C_{12} 4-Br, 2-Br, 1-NHAc naphthalene		Cu, DMSO, 60°	(96)	34
C_{13} 1,4-diOMe, 2-Me, 3-I naphthalene		Cu, 210°	(46)	723

C$_{23}$	Cu, pyridine, reflux	(81)	227, 229
C$_{25}$	Cu, DMF, reflux	(80)	227, 228

TABLE 23. BISTETRAHYDRONAPHTHYLS

Substrate	Conditions	Product(s) and Yield(s) (%)	Refs.
C_{10}			
(tetrahydronaphthyl with I and NO_2)	Cu, 135-190°	(bis-tetrahydronaphthyl with NO_2, NO_2) (20)	591
(tetrahydronaphthyl with NO_2 and I)	Cu, 110-140° Cu, 140° Cu, DMF, reflux	(bis-tetrahydronaphthyl with NO_2, O_2N) (75) (73) (23)	591 616 61

TABLE 24. MISCELLANEOUS BIARYLS

Substrate	Conditions	Product(s) and Yield(s) (%)	Refs.
C₁₀	Cu, 200°	(83)	724
	Cu, 180°	(54)	724
C₁₁	Ni(PPh₃)₂Br₂, Zn, Et₄NI, 50°	(43) + (45)	404, 725
C₁₂	Cu, 220°	(40)	724
	Cu, (Ph₃P)₄Pd, DMF, 130°	(>43)	726

TABLE 24. MISCELLANEOUS BIARYLS (Continued)

Substrate	Conditions	Product(s) and Yield(s) (%)	Refs.
C₁₃			
2-iodo-9-fluorenone	Cu, 225-230°	bis-fluorenone biaryl (31)	727
3-bromo-2-nitro-9-fluorenone	Cu, xylene, reflux	dinitro bis-fluorenone biaryl (6)	728
2-iodofluorene	Cu, biphenyl, 190-220°	2,2'-bifluorene (44)	729
1-carbethoxy-2-iodoazulene	Cu, 220°	bis(1-carbethoxyazulen-2-yl) (82)	724
C₁₄			
9-bromo-10-nitroanthracene	Cu, 220-235°	10,10'-dinitro-9,9'-bianthryl (41)	730

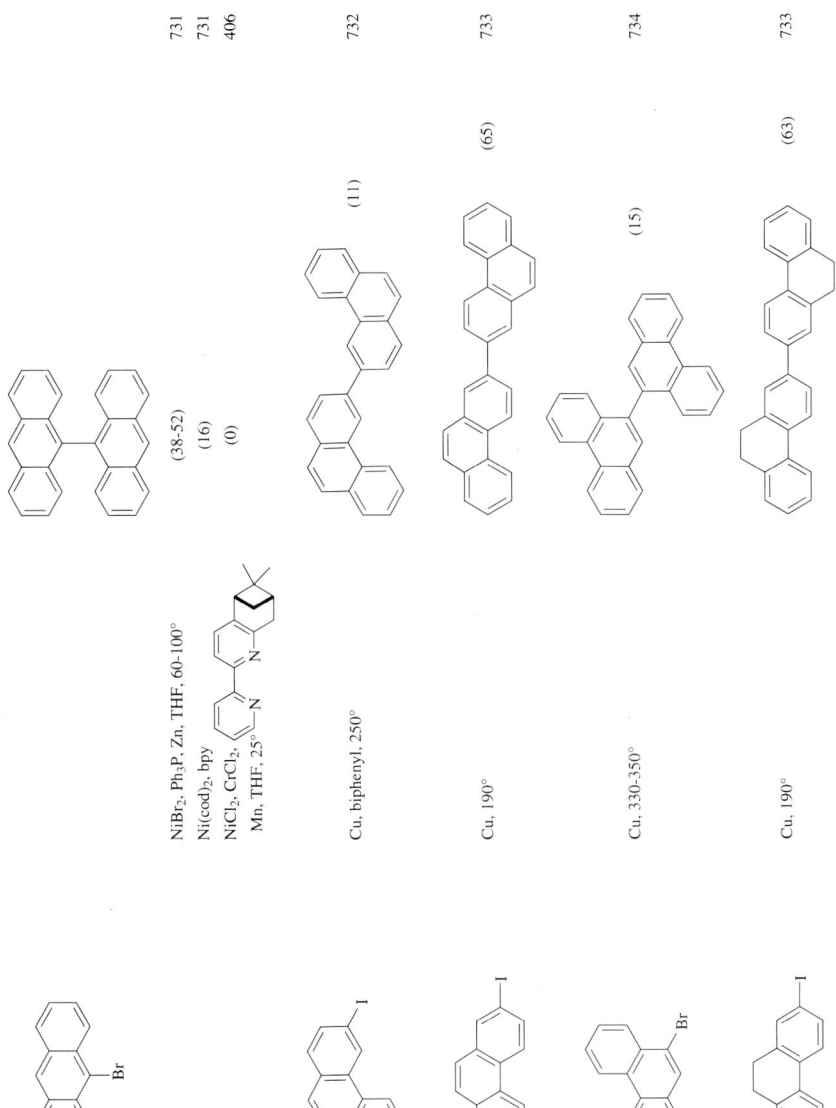

TABLE 24. MISCELLANEOUS BIARYLS (Continued)

Substrate	Conditions	Product(s) and Yield(s) (%)	Refs.
C_{14-16}, phenanthrene-Br with R: NO_2; NO_2; CO_2Me; CO_2Me	Cu, 200°; Cu, DMF, reflux; Cu, 265°; Cu, DMF, reflux	biphenanthryl with R (5); (0); (80); (70)	61; 61; 735; 736
C_{15}, tetralone with OMe, OMe, I, CH_2CO_2Me	Cu, 280-300°	coupled bi-tetralone product (26)	737
C_{16}, 1-bromopyrene	Cu, 280-290°	1,1'-bipyrene (30)	738
C_{16-19}, azulene with CO_2Et, Cl, CO_2Et, R (R = H, Me, i-Pr)	Cu, 220°; Cu, 220°; Cu, 235°	biazulene product (83); (53); (11)	724; 724; 724

C_{17}

	Cu, naphthalene, heat	(95)	739
	Cu, naphthalene, heat	(8)	739
X: Cl, Br	Cu, 280–300°	(5), (13)	740, 740

TABLE 24. MISCELLANEOUS BIARYLS (Continued)

Substrate	Conditions	Product(s) and Yield(s) (%)	Refs.
C₁₈	Ni(PPh₃)₂Cl₂, Ph₃P, Zn, DMF, 50°	(41)	741
C₂₀	Cu, 250°	(4)	742
C₃₂	NiCl₂, Ph₃P, Zn, NaBr, DMAc, 80°	(40)	743
	Ni(cod)₂, cod, bpy, toluene, DMF, 75°	(95 - 99)	247, 744
C₃₄	Ni(cod)₂, DMF 60°	(89)	745
	Ni(cod)₂, DMF 60°	(83)	745

R = n-C₁₂H₂₅, i-Pr (2-methylphenyl), i-Pr

C_{62-102}

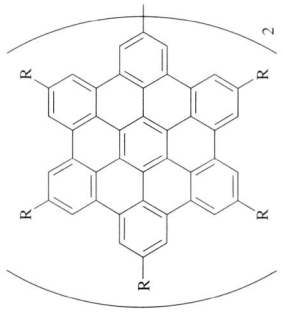

R		
t-C$_4$H$_9$	Ni(cod)$_2$, cod, bpy, toluene, 60°	(75) 746
n-C$_{12}$H$_{25}$	Ni(cod)$_2$, cod, bpy, toluene, 60°	(85) 746

TABLE 25. 1,1'-BIANTHRAQUINONES

Substrate				Conditions	Product(s) and Yield(s) (%)	Refs.
R^2	R^3	R^4	X			
Br	H	H	Br	Cu, PhNO$_2$, 200-210°	(68)	747
H	Br	H	Br	Cu, PhNO$_2$, reflux	(65)	748
H	Cl	H	Cl	Cu, PhNO$_2$, reflux	(86)	749
H	H	H	Cl	Cu	(77)	750
H	H	H	Cl	Cu, PhNO$_2$, reflux	(70-80)	751
H	H	H	I	Cu, CO$_2$, 210-275°	(20)	752
H	SO$_3$H	NH$_2$	Br	Cu, H$_2$O	(78)	753
Me	H	H	I	Cu, PhNO$_2$, reflux	(50)	754
Me	H	H	I	Cu, 210-290°	(50)	755
Me	H	H	I	Cu	(30-45)	706, 756
H	Me	H	Br	Cu, PhNO$_2$, reflux	(50-70)	757
H	H	Me	Cl	Cu, 290-300°	(77)	758
H	H	Me	Cl	Cu, PhNO$_2$, reflux	(70)	758
OMe	H	H	Cl	Cu, PhNO$_2$, reflux	(0)	759
OMe	H	H	I	Cu, 360°	(20)	760
OMe	H	H	I	Cu	(80)	761
Et	H	H	I	Cu, 240°	(50)	762
CO$_2$Me	H	H	Br	Cu, DMF, reflux	(75)	706
H	H	CO$_2$Me	Cl	Cu, 265-300°	(>62)	763
OMe	H	OMe	Br	Cu, naphthalene, 225°	(59)	764
Me	H	OMe	Cl	Cu, CO$_2$, 300°	(10-40)	765

C_{14-22}

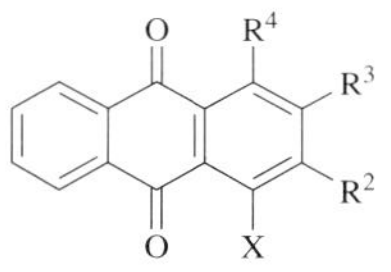

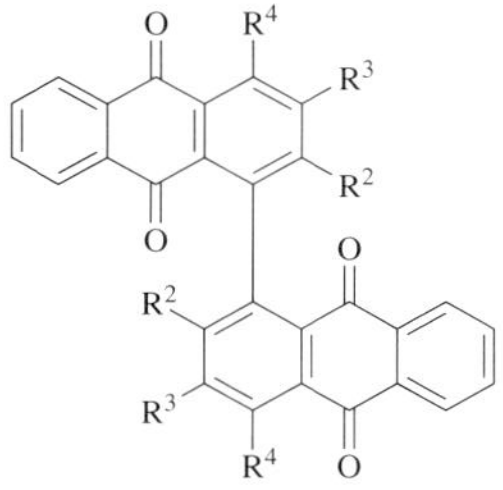

R^2	R^3	R^4	X			
H	OMe	OMe	I	Cu, 310°	(53)	766
Me	H	Me	I	Cu, 210-250°	(35)	767
i-Pr	H	H	I	Cu, 200°	(75)	762
CO_2Et	H	H	Cl	Cu, PhNO$_2$, reflux	(75)	768
N=CHPh	Br	H	Br	Cu, naphthalene, 220-260°	(66)	769
H	OMe	OCOPh	Br	Cu, PhNO$_2$, reflux	(69)	770
H	OMe	OCOPh	Br	Cu, naphthalene, reflux	(83)	770
H	H	NHCOPh	Cl	Cu, DMF, reflux	(90)	771

C_{16}

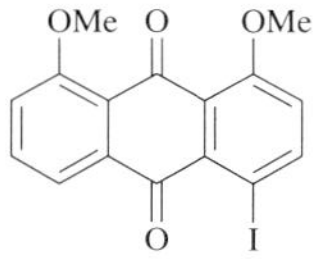

Cu, CO$_2$, 310° 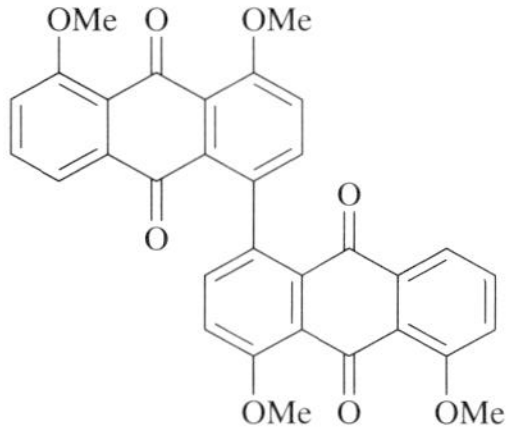   (87)   772

TABLE 25. 1,1'-BIANTHRAQUINONES (Continued)

Substrate	Conditions	Product(s) and Yield(s) (%)	Refs.
C_{18} (structure)	Cu, naphthalene, 230–240°	(structure) (40–90)	773–775
C_{21} (structure)	Cu, naphthalene, 240°	(structure) (83)	776
C_{16} (structure)	Cu, PhNO$_2$, reflux	(structure) (37)	777

TABLE 26. 2,2'-BIANTHRAQUINONES

Substrate	Conditions	Product(s) and Yield(s) (%)	Refs.
C₁₄ (2-iodoanthraquinone)	Cu, 230-330°	2,2'-bianthraquinone (49)	778
C₁₅ (2-bromo-1-cyanoanthraquinone)	Cu, PhNO₂, reflux	1,1'-dicyano-2,2'-bianthraquinone (58)	779
C₁₅ (3-iodo-2-methoxyanthraquinone)	Cu, 360°	3,3'-dimethoxy-2,2'-bianthraquinone (2)	760

TABLE 27. BICOUMARINS, BICHROMONES, AND BIFLAVONES

Substrate	Conditions	Product(s) and Yield(s) (%)	Refs.

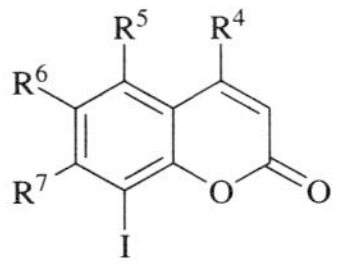

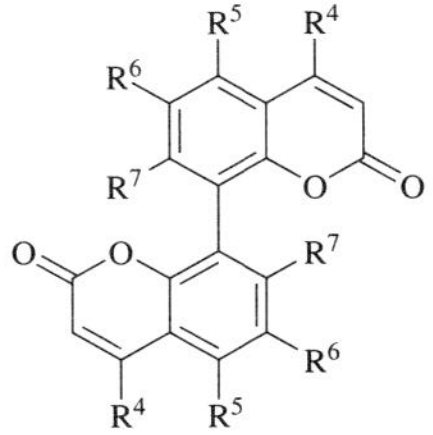

C_{10-13}

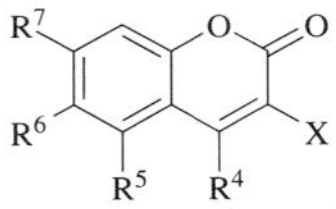

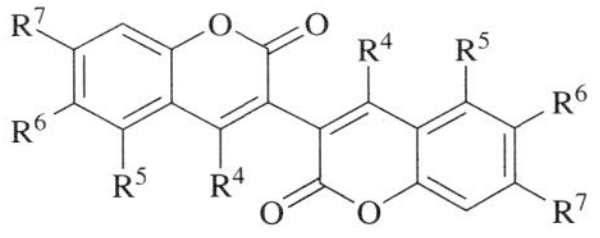

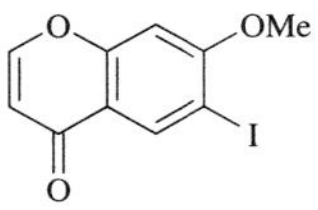

Substrate (coumarin with R^4, R^5, R^6, R^7, X):

R^4	R^5	R^6	R^7	X	Conditions	Yield	Refs.
OMe	H	H	H	Br	Cu, 210°	(35)	780
H	H	OMe	OMe	I	Cu, Ph_2O, reflux	(25)	781
Me	H	H	OMe	I	Cu, Ph_2O, reflux	(17)	781
H	H	CO_2Me	OMe	I	Cu, Ph_2O, reflux	(16)	781
Me	H	CO_2Me	OMe	I	Cu, Ph_2O, reflux	(20)	781

Product: bicoumarin linked at 3,3′ with substituents R^4, R^5, R^6, R^7.

Substrate (8-iodocoumarin with R^4, R^5, R^6, R^7):

R^4	R^5	R^6	R^7	Conditions	Yield	Refs.
H	H	H	OMe	Cu, Ph_2O, reflux	(48)	781
Me	OMe	H	H	Cu, Ph_2O, reflux	(8)	781
Me	H	H	OMe	Cu, Ph_2O, reflux	(68)	781
H	H	CO_2Me	OMe	Cu, Ph_2O, reflux	(63)	781
Me	OMe	CO_2Me	H	Cu, Ph_2O, reflux	(12)	781
Me	H	CO_2Me	OMe	Cu, Ph_2O, reflux	(64)	781

Product: 8,8′-bicoumarin with substituents.

C_{10}

Substrate: 7-OMe-6-iodochromone; Conditions: Cu; Product: 6,6′-bichromone (7,7′-diOMe) (<15); Refs. 782

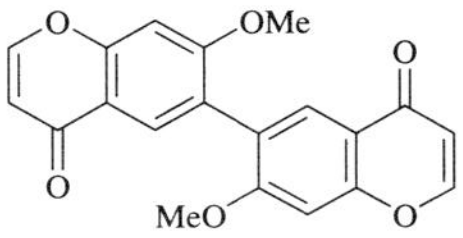

C	Cu	(<15)	782
C₁₂	Cu, DMF, 155-160°	(14)	783
C₁₆	Cu	(<15)	782
C₁₈	Cu, DMF, reflux	(40)	784

TABLE 27. BICOUMARINS, BICHROMONES, AND BIFLAVONES (Continued)

Substrate	Conditions	Product(s) and Yield(s) (%)	Refs.
C_{23} [substrate structure: 7-OMe, 6-I chromone with 2-Ph and 3-benzoyl]	Cu, heat	[biflavone dimer structure] (<15)	782

TABLE 28. BIPHENYLENES VIA INTERMOLECULAR COUPLING

Substrate	Conditions	Product(s) and Yield(s) (%)	Refs.
C$_6$ 2,3-dihalonitrobenzene X/Cl, I	Cu, DMF, reflux; Cu, DMF, 152°	1,8-dinitrobiphenylene (1), (22-43)	785, 536
1,2-dichlorobenzene	(i-Pr$_2$P)$_2$NiCl$_2$, DMSO, 2 e$^-$, 65°	biphenylene (0)	154
C$_7$ 2,3-diiodoanisole	Cu, DMF, reflux	1,8-dimethoxybiphenylene (9) + 1,2-dimethoxybiphenyl (2)	603
1-bromo-2-halo-3-nitro-5-methylbenzene X/Cl, Br, I	Cu, DMF, reflux; Cu, DMF, reflux; Cu, DMF, reflux	2,7-dimethyl-1,8-dinitrobiphenylene (7), (48), (3)	623, 623, 623

TABLE 28. BIPHENYLENES VIA INTERMOLECULAR COUPLING (Continued)

Substrate	Conditions	Product(s) and Yield(s) (%)	Refs.
C₈ (2,3-diiodo-methylbenzoate)	Cu, DMF, 55°	dimethyl biphenyl-dicarboxylate (75)	33
C₁₀ (1-nitro-2,3-dibromonaphthalene)	Cu, DMF, reflux	dinitro-dinaphthylene (2) + tetranaphthylene	786
(2,3-dihalonaphthalene) X = Br; X = I	Cu, DMF, reflux; Cu, DMF, reflux	(1); (1)	787; 786
1,8-diiodonaphthalene	Pd(OAc)₂, K₂CO₃, Bu₄NBr, DMF, 100°	1,1'-binaphthyl-8,8'-diiodide (39) + perylene (42)	173

452

TABLE 29. TRIARYLENES

Substrate		Conditions	Product(s) and Yield(s) (%)	Refs.
C₆		Cu₂O, DMF, reflux	(2)	785
	X			
	Cl	Ni(cod)₂, bpy, DMF, 70°	(9)	731
	Cl	Ni(cod)₂, bpy, THF, reflux	(11)	731
	Cl	[Ni complex with i-Pr phosphine], DMSO, −2.2 V, 65°	(0)	154
	Br	Ni(0), DMF, 140°	(25)	120
	Br	Ni(cod)₂, bpy, THF, reflux	(37)	731
	Br	Ni(cod)₂, PPh₃, DMF, 70°	(60)	731
C₁₄		Cu, biphenyl, 250°	(11)	732, 788
		Cu, DMF, reflux	(0)	788

TABLE 30. TETRAARYLENES

Substrate	Conditions	Product(s) and Yield(s) (%)	Refs.
C_4 3,4-dibromothiophene	Ni(cod)$_2$, PPh$_3$, DMF, 70° Ni(cod)$_2$, bpy, DMF, 70°	tetrathienocyclooctatetraene (64) (70)	731 731
C_6 1,2-dibromobenzene	Ni(0), DMF, 140°	tetraphenylene (25)	120
C_{12} 2,2'-diiodo-6,6'-dinitrobiphenyl	Cu$_2$O, DMF, 152°	tetranitrotetraphenylene (4-20)	536

TABLE 31. INTRAMOLECULAR COUPLINGS FORMING SYMMETRIC BIPHENYLS AND BINAPHTHYLS

Substrate	Conditions	Product(s) and Yield(s) (%)	Refs.
C$_{12}$ (perfluoro bis(bromophenyl) sulfide)	Cu, 200°	(perfluorodibenzothiophene) (100)	789
	Cu, DMF, 150°	(0)	789
bis(2-iodophenyl) disulfide	Cu, 180°	dibenzo[c,e][1,2]dithiine (60)	422
C$_{13}$ bis(2-halo-4-nitrophenyl)methane, X = Br, I	Cu, sand, PhNO$_2$, DMF, heat	2,7-dinitrofluorene (0)	790
bis(2-iodo-4-nitrophenyl) ketone	Cu, DMF, reflux	2,7-dinitrofluorenone (0)	790

TABLE 31. INTRAMOLECULAR COUPLINGS FORMING SYMMETRIC BIPHENYLS AND BINAPHTHYLS (*Continued*)

Substrate	Conditions	Product(s) and Yield(s) (%)	Refs.
C$_{14}$			
(2,4-dichlorobenzoic anhydride)	Cu, pyridine, 60-70°	(89)	791
	Cu, DMF, 60-70°	(88)	791
(2-bromobenzoic anhydride)	Cu, complexing agent, benzene, reflux		
	complexing agent		
	DMF, heat	(85)	791
	HMPA	(87)	791
	TMED	(28)	791
	4,5-phenanthroline	(95)	791
	bpy	(98)	791
(2-chlorobenzoic anhydride)	Cu, TMED, 60-70°	(36)	791
	Cu, DMF, 60-70°	(34)	791

X		
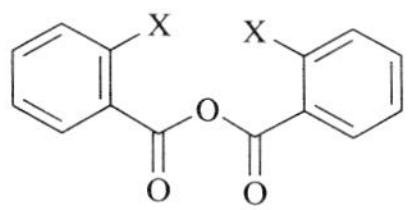 Br	Cu, solvent, 60-70°	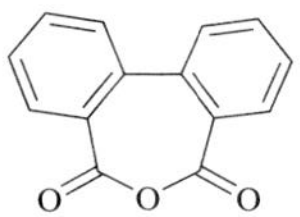 (structure: dibenzo dilactone)

X	solvent		
Br	TMED	(22)	791
Br	NMP	(65)	791
Br	HMPA	(61)	791
Br	DMF	(90)	791
Br	pyridine	(98)	791
I	TMED	(18)	791
I	NMP	(65)	791
I	HMPA	(58)	791
I	DMF	(56)	791
I	pyridine	(98)	791

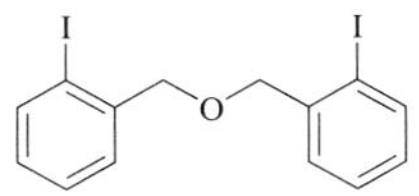

Pd(OAc)$_2$, As(o-tolyl)$_3$, hydroquinone, Cs$_2$CO$_3$, DMA, 100°   (61)   178

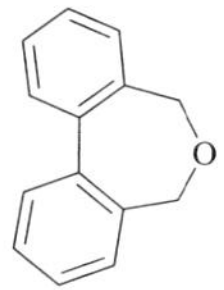

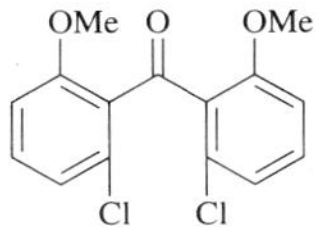

Cu, heat   (0)   577

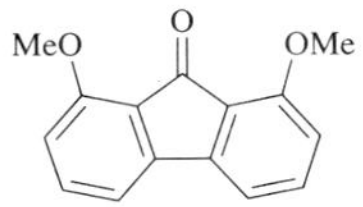

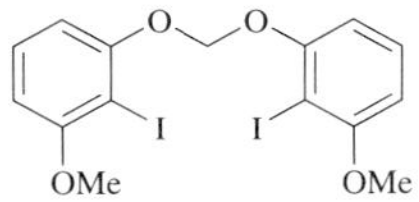

Cu, 200°   (15)   792

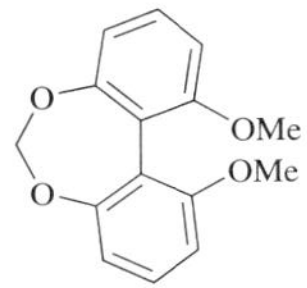

TABLE 31. INTRAMOLECULAR COUPLINGS FORMING SYMMETRIC BIPHENYLS AND BINAPHTHYLS (*Continued*)

Substrate	Conditions	Product(s) and Yield(s) (%)	Refs.
C_{16}	OCu, NMP, rt	(88)	71
C_{16}	Cu, 210°	(8)	432
$C_{16\text{-}20}$			
	n		
	2 Ni(PPh$_3$)$_2$Cl$_2$, Ph$_3$P, Zn, DMF, 50°	(98)	793
	2 (Ph$_3$P)$_4$Ni, DMF, 55°	(81)	49, 138
	3 Cu, 240°	(74)	49
	3 (Ph$_3$P)$_4$Ni, DMF, 55°	(83)	49, 138
	4 (Ph$_3$P)$_4$Ni, DMF, 55°	(76)	49, 138
	5 (Ph$_3$P)$_4$Ni, DMF, 55°	(85)	49, 138
	6 (Ph$_3$P)$_4$Ni, DMF, 55°	(38)	49, 138

C_{16-20}	Cu, DMF, reflux	(n=2) (21) (n=3) (86) (n=4) (41) (n=5) (26) (n=6) (19)	794 794 794 794 794
C_{17}	Ni(OAc)$_2$, NaH, t-AmOH, bpy, THF, reflux	(52), including accompanying rearrangements	577
	Cu	(0)	577
	(Ph$_3$P)$_4$Ni, DMF, 55°	(80-85)	49, 138
C_{18}	Cu, DMF, reflux	(90)	33

TABLE 31. INTRAMOLECULAR COUPLINGS FORMING SYMMETRIC BIPHENYLS AND BINAPHTHYLS (*Continued*)

Substrate	Conditions	Product(s) and Yield(s) (%)	Refs.
C_{19}			
(substrate with MeO, OMe, NMe linker, two aryl iodides)	$(Ph_3P)_4Ni$, DMF, 55°	(biphenyl product with OMe, OMe, Me–N bridge) (60)	795
(substrate with two OMe, OMe aryl iodides, propyl linker)	Cu, Hg, 250°	(dibenzo product with OMe, OMe, MeO, MeO) (23)	796
$C_{19\text{-}24}$			
(substrate with OMe, MeO, two aryl iodides, $(CH_2)_n$ linker)	$Ni(PPh_3)_2Cl_2$, Ph_3P, Zn, DMF, 50°	cyclophane product $(CH_2)_n$, MeO, MeO	

n	yield
5	(0)	793
6	(7)	793
7	(10)	793
8	(7)	793
9	(11)	793
10	(9)	793

C20	Pd(OAc)2, K2CO3, Bu4NBr, DMF, 100°	(85)	173
C21	Pd(OAc)2, As(o-tolyl)3, hydroquinone, Cs2CO3, DMA, 100°	(82)	178
	Pd(OAc)2, As(o-tolyl)3, hydroquinone, Cs2CO3, DMA, 100°	(65)	178
	Cu, 230°	(95)	797
	Ni(PPh3)2Cl2, Ph3P, Zn, DMF, 70-75°	(85)	798
	NiCl2•6H2O, PPh3, Zn, DMF, 70-75°	(76)	798

TABLE 31. INTRAMOLECULAR COUPLINGS FORMING SYMMETRIC BIPHENYLS AND BINAPHTHYLS (*Continued*)

Substrate	Conditions	Product(s) and Yield(s) (%)	Refs.
C_{22}	Ni(PPh$_3$)$_2$Cl$_2$, Ph$_3$P, Zn, NaBr, DMF, 80°	(62)	577
C_{23}	NiCl$_2$, Ph$_3$P, Zn, NaI, DMF, 70-80°	(33)	209
C_{24}	(Ph$_3$P)$_4$Ni, DMF, 45°	(58)	49, 138

C27			
X			(9) (67)
Br	Cu, DMF, reflux		799
I	Cu, DMF, reflux		799

C34			
X			(42) (70)
Cl	Cu, DMF, reflux		206
I	Cu, DMF, reflux		206

463

TABLE 31. INTRAMOLECULAR COUPLINGS FORMING SYMMETRIC BIPHENYLS AND BINAPHTHYLS (*Continued*)

Substrate	Conditions	Product(s) and Yield(s) (%)	Refs.
C₃₈	Ni(PPh₃)₂Cl₂, Ph₃P, Zn, DMF, 50°	(55)	800
C₃₈	Ni(PPh₃)₂Cl₂, Ph₃P, Zn, DMF, 54°	(6)	457
C₄₂	Cu, DMF, reflux	(36-98)	206, 801, 802

Cu, DMF, reflux

(11)

TABLE 32. INTRAMOLECULAR COUPLINGS FORMING UNSYMMETRIC BIPHENYLS AND BINAPHTHYLS

Substrate	Conditions	Product(s) and Yield(s) (%)	Refs.
C14	Cu, DMF, reflux	(56)	207
	Cu, DMF, reflux	(51)	207
C18	Cu, DMF, reflux	(61)	207
	Cu, DMF, reflux	(38)	207

C19	Cu, DMF, reflux	(43)	207
	Cu, DMF, reflux	(28)	207
C20	Cu, DMF, reflux	(31)	207
	Cu, DMF, reflux	(84)	604

TABLE 32. INTRAMOLECULAR COUPLINGS FORMING UNSYMMETRIC BIPHENYLS AND BINAPHTHYLS (Continued)

Substrate	Conditions	Product(s) and Yield(s) (%)	Refs.

C_{21-26}

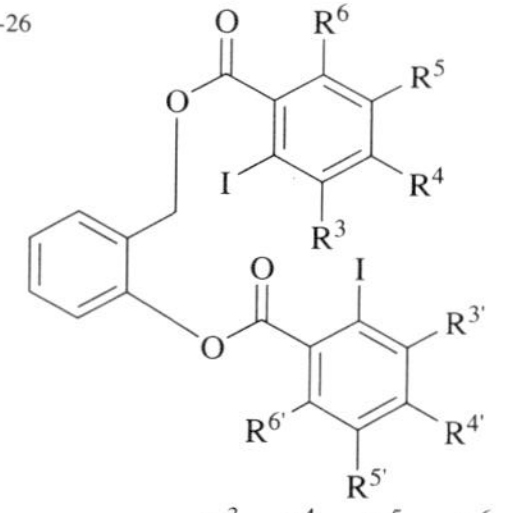

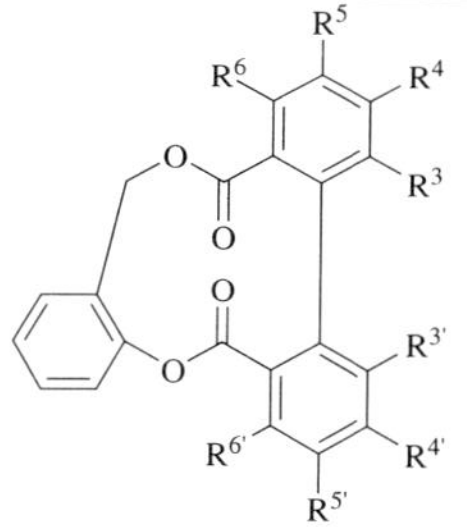

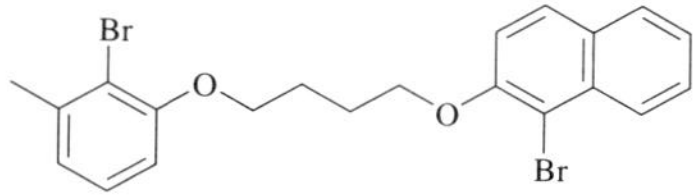

R^3	R^4	R^5	R^6	$R^{3'}$	$R^{4'}$	$R^{5'}$	$R^{6'}$		
H	H	NO$_2$	H	Cl	H	H	Cl	Cu, DMF, reflux	(84) 804
H	OMe	OMe	H	H	H	H	NO$_2$	Cu, DMF, reflux	(90) 804
H	OMe	OMe	H	H	H	H	H	Cu, DMF, reflux	(71) 805
-OCH$_2$O-		H	H	H	-OCH$_2$O-		H	Cu, DMF, reflux	(75) 805
H	OMe	OMe	H	H	H	H	NO$_2$	Cu, DMF, reflux	(87) 804
H	Cl	H	H	Cl	H	OMe	OMe	Cu, DMF, reflux	(89) 804
-OCH$_2$O-		H	H	H	OMe	OMe	H	Cu, DMF, reflux	(89) 805
H	-OCH$_2$O-		H	OMe	OMe	H	H	Cu, DMF, reflux	(78) 805
H	OMe	OMe	H	OMe	OMe	H	H	Cu, DMF, reflux	(83) 805

C_{21}

	Cu, DMF, reflux	(49)	207
	Ni(PPh$_3$)$_2$Cl$_2$, Ph$_3$P, Zn, DMF	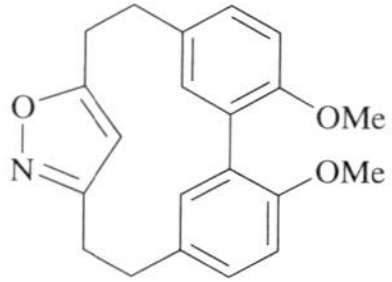 (31)	766

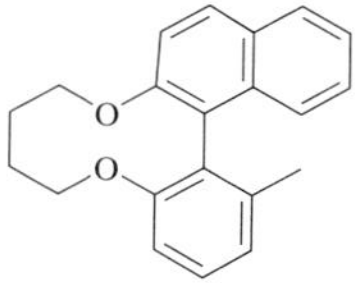

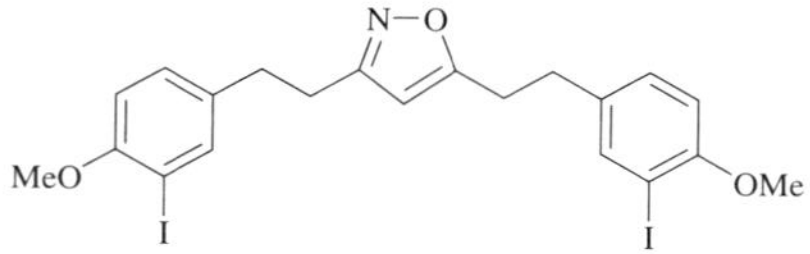

Substrate	Conditions	Product	(Yield %)	Refs.
	(PPh$_3$)$_4$Ni, DMF, 50°		(52)	49, 138
C$_{23}$	Ni(PPh$_3$)$_2$Cl$_2$, Ph$_3$P, Zn, DMF, 55°		(57)	807
	(PPh$_3$)$_4$Ni, DMF, 50°		(46)	49
	Ni(PPh$_3$)$_2$Cl$_2$, Ph$_3$P, Zn, DMF, 50°		(49)	806
	Ni(PPh$_3$)$_2$Cl$_2$, Ph$_3$P, Zn, DMF, 54°		(8)	457

TABLE 32. INTRAMOLECULAR COUPLINGS FORMING UNSYMMETRIC BIPHENYLS AND BINAPHTHYLS (*Continued*)

Substrate	Conditions	Product(s) and Yield(s) (%)	Refs.
C$_{25}$ (substrate structure)	Ni(PPh$_3$)$_2$Cl$_2$, Ph$_3$P, Zn, DMF, 54°	(9)	457
C$_{32}$ (substrate structure)	Cu, DMF, reflux	(80)	207
C$_{33}$ (substrate structure) X = Br; X = I	Ni(cod)$_2$; Ni(cod)$_2$	(0); (0)	808; 808

C_{35}	Ni(PPh$_3$)$_2$Cl$_2$, Ph$_3$P, Zn, DMF, 54°	(10)	457, 809
C_{36}	Cu, DMF, reflux	(44)	207
	Cu, DMF, reflux	(33)	207
	(Ph$_3$P)$_4$Ni, DMF, 55°	(20)	810

TABLE 32. INTRAMOLECULAR COUPLINGS FORMING UNSYMMETRIC BIPHENYLS AND BINAPHTHYLS (Continued)

Substrate	Conditions	Product(s) and Yield(s) (%)	Refs.
C_{37}	Ni(PPh$_3$)$_2$Cl$_2$, Ph$_3$P, Zn, DMF, 54°	(7)	457, 809
C_{41}	Cu, DMF, reflux	(77)	208
C_{41-47}	Cu, DMF, reflux	(50)	208
	Cu, DMF, reflux	(50)	208

C_43   Cu, DMF, reflux   208   (69)

TABLE 33. BIPHENYLENES VIA INTRAMOLECULAR COUPLING

Substrate	Conditions	Product(s) and Yield(s) (%)	Refs.
C_{12} (iodonium salt with 2,2'-NO$_2$ groups)	Cu$_2$O, 350°	2,2'-dinitrobiphenylene (<1)	456
C_{12-13} (2,2'-diiodobiphenyl with R)		R-biphenylene	
R = NO$_2$	Cu$_2$O, 350°	(17)	456
R = OMe	Cu$_2$O, 350°	(23)	456
C_{12-14} (iodonium salt, 2,2'-R)		2,7-R$_2$-biphenylene	
R = H	Cu$_2$O, heat	(15)	811
R = H	Cu$_2$O, 350-360°	(21-28)	415, 812
R = Me	Cu$_2$O, heat	(5)	811
R = Me	Cu$_2$O, CuC$_2$O$_4$, heat	(35)	813
C_{12-18} (iodonium salt with R)		R-biphenylene	
R = NO$_2$	Cu$_2$O, heat	(16-17)	456
R = Ph	Cu$_2$O, 450°	(17)	813

C_{12-20}

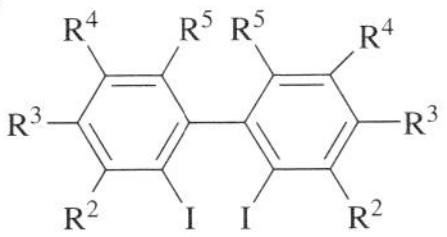

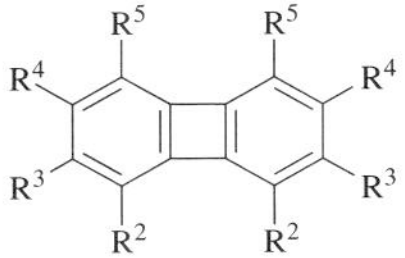

R^2	R^3	R^4	R^5			
F	F	F	F	$Cu, 200°$	(5)	814
H	H	H	H	Cu, heat	(38-44)	65
H	H	H	H	Cu_2O, heat	(31)	65
H	H	OMe	H	Cu_2O, 330-360°	(42)	612
H	H	H	CO_2Me	Cu, pyridine, reflux	(11)	33
H	Me	Me	H	Cu_2O, 380-400°	(5)	813
H	H	Me	H	Cu_2O, 350°	(31)	603
H	OMe	OMe	OMe	Cu_2O, 330-360°	(31)	612
H	H	OMe	H	Cu_2O, 340°	(27)	654
Me	OMe	Me	H	Cu_2O, 330-360°	(28)	612
Me	OMe	OMe	H	Cu_2O, CuC_2O_4, 360-400°	(18)	651
Me	OMe	OMe	OMe	Cu_2O, 350°	(18)	654
H	OMe	OMe	OMe	Cu, heat	(46)	65, 654
H	OMe	OMe	H	Cu_2O, 340°	(22)	654
OMe	OMe	OMe	H	Cu_2O, 330-350°	(30)	654
OMe	H	OMe	OMe			
Me	OMe	Me	Me	Cu_2O, CuC_2O_4, 360-400°	(7)	651

C_{13}

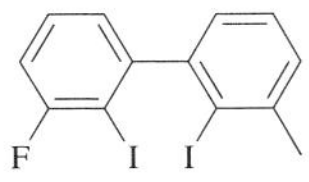

Cu_2O, heat (27) 815

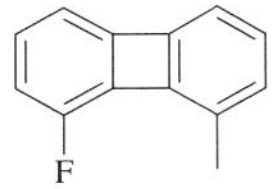

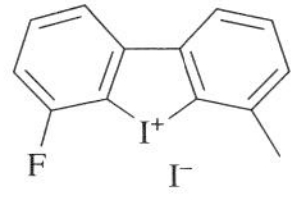

Cu_2O, heat (74) 815

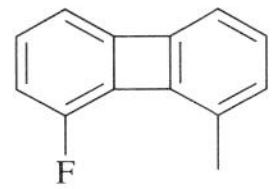

TABLE 33. BIPHENYLENES VIA INTRAMOLECULAR COUPLING (Continued)

Substrate	Conditions	Product(s) and Yield(s) (%)	Refs.
C_{13}	Cu_2O, 350°	(41)	456
C_{14-26}			
R = Me	Cu, DMF, reflux	(79)	816
R = Me	Cu_2O, 270°	(28)	816
R = PhCH$_2$	Cu, DMF, reflux	(42)	816
C_{20}	Cu, 250-350°	(9)	817
	Cu_2O, 230-330°	(6)	818
	Cu_2O, 340°	(3-5)	616

| C$_{32}$ starting material | Cu, DMF, reflux | product (76) | 819 |

TABLE 34. BIPYRROLES

Substrate	Conditions	Product(s) and Yield(s) (%)	Refs.
C₉ 2,4-dimethyl-3-iodo-5-(ethoxycarbonyl)pyrrole	Cu, DMF, reflux	bipyrrole diester (25)	820
1-methyl-2-iodoindole	Cu, 205°	2,2'-bi(1-methylindole) (66)	13

C₉₋₂₁ Substrate: 2-iodopyrrole with R³, R⁴, R⁵ substituents; Product: 2,2'-bipyrrole with R³, R⁴, R⁵ substituents

R³	R⁴	R⁵	Conditions	Yield (%)	Refs.
CO₂Me	Me	CO₂Me	Cu, DMF, rt	(60)	821
Me	Me	CO₂Et	Cu, 240°	(15)	822
Me	Me	CO₂Et	Cu, DMF, 100°	(35)	823
Me	CO₂Et	Me	Cu, DMF, rt	(32)	823
Me	CO₂Et	CO₂Me	Cu, DMF, rt	(21)	821
Et	Me	CO₂Et	Cu, DMF, 100-140°	(29-50)	823-825
Me	Et	CO₂Et	Cu, DMF, 110°	(45-50)	824, 826, 827
Et	Et	CO₂Et	Cu, DMF, 110°	(35-40)	827
CO₂Et	Me	CO₂Et	Cu, DMF, rt	(61-63)	821, 823
CO₂Et	Me	CO₂Et	Cu, benzene, 80°	(77)	821
CO₂Et	Me	CO₂Et	Cu, PhNO₂, 100°	(38)	821
CO₂Et	Me	CO₂Et	Cu, 210°	(54)	828

CO₂Et	Et	CO₂Et		Cu, DMF, reflux	(65)	823
CO₂Et	n-C₃H₇	CO₂Et		Cu, DMF, 20°	(62–71)	829
CH₂CH₂CO₂Et	Me	CO₂Et		Cu, DMF, rt	(30)	823
CO₂Et	CH₂CH₂OMe	CO₂Et		Cu, 1,3-dimethyl-2-imidazolidinone, rt	(67)	830
Me	Me	CO₂CH₂Ph		Cu, DMF, rt	(57)	821
Me	CO₂CH₂Ph	Me		Cu, DMF, rt	(18)	821
CO₂Et	Me	CO₂CH₂Ph		Cu, DMF, 100°	(77)	821
CO₂Et	Ph	CO₂Et		Cu, DMF, reflux	(44)	831
CO₂CH₂Ph	Me	CO₂Et		Cu, DMF, rt	(27)	821
CO₂CH₂Ph	Me	CO₂CH₂Ph		Cu, DMF, rt	(56)	821

C₁₀ [structure] Cu, 220° (59) 820

C₁₁ [structure] [reagent structure], NMP, rt (69) 71

C₁₂ [structure] Cu, I₂, DMF, reflux (35) 832

C₁₉ [structure] Cu, DMF, 140° (69) 833

TABLE 35. UNSUBSTITUTED 2,2'-BIPYRIDYL

Substrate	Conditions	Product(s) and Yield(s) (%)	Refs.
C₅ 2-fluoropyridine	NaH, Ni(OAc)₂, t-BuOH, Ph₃P, DME, 65°	bipyridyl (51)	398
2-chloropyridine	NaH, Ni(OAc)₂, t-BuOH, Ph₃P, DME, 45°	I (66)	398, 834
	NiCl₂, Ph₃P, Zn, DMAc, 50-80°	(70)	51
	NiBr₂(PPh₃)₂, Zn, Et₄NI, THF, 50°	(60)	139
	HCO₂Na, Pd/C, TEBAC, NaOH(aq)	(52)	179
2-bromopyridine	Cu, cymene, reflux	I (60-63)	835-837
	Cu, biphenyl, 230°	(72)	835, 838
	Cu, DMF, reflux	(87)	6, 835
	Cu, tetralin, reflux	(26)	6, 835
	Cu, decane, reflux	(51)	6, 835
	Cu, decalin, reflux	(61)	6, 835
	Cu, pseudocumene, reflux	(63)	6, 835
	NiCl₂, Ph₃P, Zn, DMF, 50°	(68)	142, 839
	Ni(bpy)₂(OAc)₂, Et₄NBr, CH₃CN, 2 e⁻	(88)	405
	NiBr₂(PPh₃)₂, Zn, Et₄NI, THF, 50°	(72)	139
	NaH, Ni(OAc)₂, t-BuOH, Ph₃P, DME, 30-45°	(65-70)	398, 834

Ni(CO)₂(PPh₃)₂, toluene, DMF, 70°	(75)		145
NiCl₂, CrCl₂, Mn, THF, 25°	(0)		406
Pd(OAc)₂, P(o-tol)₃, KF, DMF, 120°	(63)		196
Pd(OAc)₂, As(o-tolyl)₃, hydroquinone, Cs₂CO₃, DMAc, 100°	(72)		178
PdCl₂(PPh₃)₂, Et₄NOTs, DMF, 2e⁻	(91)		190
Pd(OAc)₂, Bu₄NBr, DMF, H₂O, i-PrOH, K₂CO₃, 50°	(92)		174, 175
PdCl₂, HgCl₂, PhNHNH₂, NaOH, MeOH, reflux	(42)		165
Cu, cymene, 220°	(20)		836
NiCl₂(PEt₃)₂, Zn, KI, HMPA, 40°	(0)		401
Pd(OAc)₂, Ph₃P, styrene, Et₃N, 100°	(66-71)		840
Pd(OAc)₂, styrene, Et₃N, 100°	(68)		840
PdCl₂(PPh₃)₂, Bu₄NBF₄, DMF, 2e⁻	(78)		192

TABLE 36. 3,3'-DISUBSTITUTED-2,2'-BIPYRIDYLS

Substrate		Conditions	Product(s) and Yield(s) (%)	Refs.
R	X			
OH	I	Cu, DMF, reflux	(25)	841
OH	Br	NiCl$_2$•6 H$_2$O, Zn, PPh$_3$, DMF, 50°	(55)	140
NO$_2$	Cl	Cu, DMF, 100-150°	(51-85)	842-845
OMe	I	Cu, DMF, reflux	(39)	841
OMe	I	NiCl$_2$, Zn, DMF, 50°	(60)	841
OMe	I	NiCl$_2$, Zn, PPh$_3$, DMF, 50°	(65)	841
OMe	Br	NiCl$_2$•6 H$_2$O, PPh$_3$, Zn, DMF, 50°	(75)	142, 839
Me	Br	Cu, neat, 240°	(40)	846
Me	Br	Cu, NaCl, p-cymene, reflux	(19)	847
CF$_3$	Cl	Ni(PPh$_3$)$_2$Cl$_2$, Zn, Et$_4$NI, THF, 60°	(0)	848
CF$_3$	Cl	HCO$_2$Na, PhCH$_2$NEt$_3$Cl, Pd/C, NaOH, ethylene glycol, H$_2$O, toluene, reflux	(trace)	849
CF$_3$	Cl	Cu, DMF, 150°	(39)	850
CN	Br	NiBr$_2$(PPh$_3$)$_2$, Zn, Et$_4$NI, THF, 50°	(53)	139
CO$_2$Me	Cl	Cu, DMF, 100°	(34)	842
NHAc	Br			

C$_{10}$

NiCl$_2$•6 H$_2$O, Zn, PPh$_3$, DMF, 50-60°   (79)   851

TABLE 37. 4,4'-DISUBSTITUTED-2,2'-BIPYRIDYLS

Substrate	Conditions	Product(s) and Yield(s) (%)	Refs.

C$_{6-12}$

R	X		
CF$_3$	Cl	NiCl$_2$(PPh$_3$)$_2$, Zn, Et$_4$NI, THF, 60°	(34)	848
Me	Br	Cu, neat, 240°	(33)	846
OMe	Cl	NiCl$_2$•6H$_2$O, PPh$_3$, Zn, DMF, 50°	(88)	852
Et	Br	Cu, neat, 220°	(25)	853
Ph	Br	Cu, neat, 250°	(18)	853
p-Tol	Br	Ni(PPh$_3$)$_2$Cl$_2$, PPh$_3$, Zn, DMF, 50°	(27)	854

TABLE 38. 5,5'-DISUBSTITUTED-2,2'-BIPYRIDYLS

Substrate: R-pyridyl-X (C$_{5-12}$)

Product: 5,5'-R-substituted-2,2'-bipyridyl

R	X	Conditions	Product(s) and Yield(s) (%)	Refs.
NO_2	Br	Pd(OAc)$_2$, K$_2$CO$_3$, toluene, DMF, i-PrOH, 105°	(8-19)	175
NO_2	Br	Pd(OAc)$_2$, K$_2$CO$_3$, toluene, DMF, 115°	(8)	175
NO_2	I	Cu, neat, 180°	(2)	846
NH_2	Cl	NiCl$_2$·6 H$_2$O, Zn, Ph$_3$P, DMF, 50°	(60)	855-856
Cl	Br	Cu, neat, 225°	(8)	846
Br	Br	Cu, neat, 225°	(1)	846
CF_3	Cl	NaH, Ni(OAc)$_2$, t-AmOH, bpy, THF, reflux	(57)	398
CF_3	Cl	Ni(PPh$_3$)$_2$Cl$_2$, Zn, Et$_4$NI, THF, 60°	(32)	848
CF_3	Cl	HCO$_2$Na, Pd/C, Et$_3$NCH$_2$PhCl, H$_2$O, reflux	(32)	848
Me	Br	Pd(OAc)$_2$, Bu$_4$NBr, DMF, H$_2$O, i-PrOH, K$_2$CO$_3$, 115°	(95)	174, 175
Me$_2$N	Br	NiCl$_2$·6 H$_2$O, Zn, Ph$_3$P, DMF, 50°	(64)	855
(N-Me pyrrolidinyl)	Cl	Pd(PPh$_3$)$_4$, (n-Bu$_3$Sn)$_2$, Et$_3$N, DMF, 100°	(58)	857
PhN=CH	Cl	Ni(PPh$_3$)$_2$Br$_2$, Zn, Et$_4$NI, THF, 50-80°	(83)	858

TABLE 39. 6,6'-DISUBSTITUTED-2,2'-BIPYRIDYLS

Substrate		Conditions	Product(s) and Yield(s) (%)	Refs.
C$_{5-7}$				
R—[pyridine]—X with R at 6-position			6,6'-R$_2$-2,2'-bipyridyl	
R	X			
NO$_2$	Cl	Cu, DMF, 100°	(51)	844
CF$_3$	Cl	Ni(PPh$_3$)$_2$Cl$_2$, Zn, Et$_4$NI, THF, 60°	(29)	848
CF$_3$	Cl	Pd/C, HCO$_2$Na, TEBAC, NaOH, H$_2$O, reflux	(9)	848
Me	Br	NaH, Ni(OAc)$_2$, t-BuOH, PPh$_3$, DME, 45°	(73)	398
Me	Br	Raney Ni, toluene, reflux	(64-68)	859, 860
Me	Br	Ni(cod)$_2$, DMF	(95)	861
Me	Br	Ni(dmpb)Cl$_2$, Zn, Bu$_4$NBF$_4$, DMF, 2e$^-$	(67)	862
Me	Br	Pd/C, HCO$_2$Na, TEBAC, NaOH, H$_2$O, reflux	(50-67)	183, 863, 864
Me	Br	Pd(OAc)$_2$, K$_2$CO$_3$, Bu$_4$NBr, i-PrOH, H$_2$O, 110°	(74)	862
MeO	Br	NiCl$_2$, PPh$_3$, Zn, DMF, 50°	(70-87)	142, 839, 852
MeO	Cl	NaH, Ni(OAc)$_2$, t-BuOH, PPh$_3$, DME, 45°	(79-80)	398, 834
CO$_2$Me	Cl	NiBr$_2$(PPh$_3$)$_2$, Zn, Et$_4$NI, THF, 50°	(90)	139
C$_8$				
2-Br-6-(i-PrS)-pyridine		NiCl$_2$·6H$_2$O, PPh$_3$, Zn, KI, DMF, 50°	6,6'-bis(i-PrS)-2,2'-bipyridyl (50)	865
		NiCl$_2$·6H$_2$O, PPh$_3$, Zn, DMF, 50°	(45)	865
2-Br-6-(1,3-dioxolan-2-yl)-pyridine		Ni(PPh$_3$)$_2$Cl$_2$, Zn, Et$_4$NI, THF, 50°	6,6'-bis(1,3-dioxolan-2-yl)-2,2'-bipyridyl (65)	866

TABLE 39. 6,6'-DISUBSTITUTED-2,2'-BIPYRIDYLS (Continued)

Substrate	Conditions	Product(s) and Yield(s) (%)	Refs.
C₉ (2-Br-6-(t-BuS)-pyridine)	NiCl₂·6 H₂O, PPh₃, Zn, KI, DMF, 50°	t-BuS-bipyridyl-SBu-t (60)	865
	NiCl₂·6 H₂O, PPh₃, Zn, DMF, 50°	(40)	865
C₁₀ (2-Br-6-(CH(OAc)Me)-pyridine)	NiCl₂, Zn, PPh₃, DMF, 72°	bipyridyl-(CH(OAc)Me)₂ (39)	867
(2-Br-6-(CH(OH)CMe₃)-pyridine)	NiCl₂, Zn, PPh₃, DMF, 72°	bipyridyl-(CH(OH)CMe₃)₂ (50-60)	867, 868
(2-X-6-(2-pyridyl)-pyridine)		terpyridyl	
X = Br	Cu, biphenyl, 230°	(30)	838
X = Br	NiCl₂·6 H₂O, PPh₃, Zn, DMF, 50°	(43)	866
X = Br	Ni(CO)₂(PPh₃)₂, toluene, DMF, 70°	(70)	145
X = Cl	Ni(PPh₃)₂Cl₂, PPh₃, Zn, DMF, 50°	(27-38)	866, 867
C₁₁ (2-Br-6-(CH(OMe)CMe₃)-pyridine)	NiCl₂, Zn, PPh₃, DMF, 72°	bipyridyl-(CH(OMe)CMe₃)₂ (65)	868, 869

Substrate	Conditions	Product (Yield %)	Refs.
C₁₄ (pyridine-Br with Ph)	NiCl₂·6H₂O, Zn, PPh₃, DMF, 50°	bipyridine-pyrazole-MOMO (40)	870, 871
	Ni(CO)₂(PPh₃)₂, toluene, DMF, 70°	Ph-bipyridine-Ph (70)	145
C₁₅ (chloro-pyridyl pinane)	NiCl₂·6H₂O, PPh₃, Zn, NaI, DMF, 70°	bipyridine-dipinanyl (60)	872
(bromo-bipyridine)	Cu, biphenyl, 230°	terpyridine (40)	870
	NiCl₂·6H₂O, PPh₃, Zn, DMF, reflux	(48)	873
C₁₇ (bromo-pyridyl sugar)	NiCl₂·6H₂O, PPh₃, Zn, DMF, 80°	bipyridine-bis(sugar) (75)	874

487

TABLE 39. 6,6'-DISUBSTITUTED-2,2'-BIPYRIDYLS (*Continued*)

Substrate	Conditions	Product(s) and Yield(s) (%)	Refs.
C$_{18}$	NiCl$_2$•6 H$_2$O, PPh$_3$, Zn, DMF, 60°	(51)	875
C$_{21}$	NiCl$_2$•6 H$_2$O, PPh$_3$, Zn, DMF, reflux	(23)	873

TABLE 40. POLYSUBSTITUTED-2,2'-BIPYRIDYLS

Substrate: 3,4,5,6-tetrasubstituted-2-halopyridine (R^3, R^4, R^5, R^6, X)
Product: 3,3',4,4',5,5',6,6'-octasubstituted-2,2'-bipyridyl

C$_{5-12}$

R^3	R^4	R^5	R^6	X	Conditions	Product(s) and Yield(s) (%)	Refs.
NO_2	H	H	Cl	Cl	Cu, DMF, 100°	(54)	842
NO_2	H	NO_2	H	Cl	Cu, DMF, 100°	(<10)	842
NO_2	H	Me	H	Cl	Cu, DMF, 100-105°	(63)	396
NO_2	H	OMe	Br	Br	Cu, DMF, 80°	(75)	842
NO_2	H	OMe	Br	Br	excess Cu, DMF, 100°	(0)	842
OMe	H	H	NO_2	I	Cu, DMF, 100°	(42)	841
H	H	OMe	Br	Br	Cu, DMF, 100°	(0)	842
F	OMe	H	H	Br	NiCl$_2$, Zn, PPh$_3$, DMF, 50°	(42)	876
Me	H	H	OH	Br	NiCl$_2$, Zn, PPh$_3$, DMF, 50°	(42)	140
Me	H	H	OH	I	Cu(I) thiophenecarboxylate, NMP, rt	(60)	71
CF$_3$	H	CF$_3$	H	Cl	Ni(PPh$_3$)$_2$Cl$_2$, Zn, Et$_4$NI, THF, 60°	(33)	848
CF$_3$	OMe	CF$_3$	H	Cl	Pd/C, PhCH$_2$NEt$_3$Cl, HCO$_2$Na, H$_2$O, reflux	(7)	848
OMe	OMe	H	H	Br	Cu, DMF, 150°	(<10)	144
OMe	OMe	H	H	Br	NiCl$_2$, Zn, PPh$_3$, DMF, 50°	(25-87)	144, 876-878
OMe	OMe	H	H	Br	Pd/C, HCO$_2$NH$_4$, CTAB, NaOH(aq), reflux	(15-20)	144
OMe	OMe	H	H	I	NiCl$_2$, Zn, PPh$_3$, DMF, 50°	(80)	879, 880
H	OMe	OMe	H	I	NiCl$_2$, Zn, PPh$_3$, DMF, 50°	(79)	881
Me	H	H	(2,6-lutidinyl)		Cu, 150°	(70)	882

C$_{10}$

Substrate (6-bromo-3-isopropyl-2,3-dihydrofuro[3,2-b]pyridine)	Conditions	Product and Yield	Refs.
	NiCl$_2$, PPh$_3$, Zn, DMF, 50°	bis(furopyridyl) dimer (63)	883

TABLE 40. POLYSUBSTITUTED-2,2'-BIPYRIDYLS (Continued)

Substrate	Conditions	Product(s) and Yield(s) (%)	Refs.
C₁₂	NiCl₂, PPh₃, Zn, DMF, 50°	(46)	884
	NiCl₂, PPh₃, Zn, DMF, 60°	(50)	885
	NiCl₂, PPh₃, Zn, DMF, 50°	(65)	884
C₁₃	NiCl₂, PPh₃, Zn, DMF, 50°	(66)	884
C₁₅	NiCl₂, PPh₃, Zn, DMF, 50°	(50)	884

| C17 | NiCl$_2$, Zn, PPh$_3$, DMF, 50° | (91) | 886, 887 |

| C20 | Ni(PPh$_3$)$_2$Cl$_2$, Zn, Et$_4$NI, THF, 60° | (89) | 85 |

TABLE 41. 3,3'-BIPYRIDYLS

C_{5-19}

Substrate							Conditions	Product(s) and Yield(s) (%)	Refs.
R^2	R^4	R^5	R^6	X					
F	F	Cl	F	Cl			Cu, DMF, reflux	(15)	888
Cl	Cl	Cl	Cl	I			Cu, DMF, reflux	(71)	888
H	H	H	H	Cl			Ni(OAc)$_2$, NaH, t-BuOH, Ph$_3$P, DME, 65°	(90)	398, 834
H	H	H	H	Br			Cu, biphenyl, reflux	(34)	889
H	H	H	H	Br			Ni(OAc)$_2$, NaH, t-BuOH, Ph$_3$P, DME, 45°	(78)	398, 834
H	H	H	H	Br			NiCl$_2$, Zn, PPh$_3$, DMF, 50°	(80)	142, 839
H	H	H	H	Br			NiBr$_2$, Zn, PPh$_3$, Et$_4$NI, DMF, 50°	(90)	890
H	H	H	H	I			Cu, biphenyl, reflux	(7-8)	889
H	H	H	H	I			Cu, n-C$_4$H$_9$Ph, reflux	(2)	891
H	H	H	H	I			NiCl$_2$, Zn, PPh$_3$, DMF, 50°	(39)	892
H	H	H	H	I			NiCl$_2$(PEt$_3$)$_2$, Zn, KI, HMPA, 40°	(0)	401
NH$_2$	H	H	H	Cl			Ni(OAc)$_2$, NaH, t-BuOH, Ph$_3$P, DME, 65°	(40)	398
H	H	H	NH$_2$	Cl			CuSO$_4$, Na$_2$SO$_3$, benzene, 300°	(2)	893
H	H	H	NH$_2$	Cl			CuSO$_4$, EtONa, 300°	(2)	893
OMe	H	H	NO$_2$	I			Cu, DMF, reflux	(42)	841
OMe	H	H	H	Cl			NiCl$_2$, Zn, PPh$_3$, DMF, 50°	(51)	142, 839
OMe	H	H	H	Br			NiCl$_2$·6 H$_2$O, Ph$_3$P, Zn, DMF, 50°	(16)	852
OMe	H	H	H	Br			NiCl$_2$·6 H$_2$O, Ph$_3$P, Zn, DMF, 50°	(12)	852
H	OMe	H	H	Br			NiCl$_2$·6 H$_2$O, Ph$_3$P, Zn, DMF, 50°	(42)	852
H	H	OMe	H	Cl			NiCl$_2$, Zn, PPh$_3$, DMF, 50°	(88-89)	142, 839
H	H	OMe	H	Cl			NiCl$_2$, Zn, PPh$_3$, DMF, 50°	(56)	142, 839
H	H	H	OMe	Br			NiCl$_2$, Zn, PPh$_3$, DMF, 50°	(56)	142, 839
H	H	CO$_2$Me	H	Br			NiBr$_2$(PPh$_3$)$_2$, Zn, Et$_4$NI, THF, 50°	(69)	139

H	H	t-BuS	H	Br	NiCl$_2$•6 H$_2$O, Zn, PPh$_3$, DMF, 50°	(40)	865
H	H	t-BuS	H	Br	NiCl$_2$•6 H$_2$O, Zn, PPh$_3$, KI, DMF, 50°	(74)	865
H	H	H	t-BuS	Br	NiCl$_2$•6 H$_2$O, Zn, PPh$_3$, DMF, 50°	(30)	865
H	H	H	t-BuS	Br	NiCl$_2$•6 H$_2$O, Zn, PPh$_3$, KI, DMF, 50°	(50)	865
OMe	POPh$_2$	H	OMe	Br	Cu, DMF, reflux	(80)	894

TABLE 42. 4,4'-BIPYRIDYLS

C$_{5-15}$

Substrate					Conditions	Product(s) and Yield(s) (%)	Refs.
R^2	R^3	R^5	R^6	X			
F	F	F	F	Br	Cu, 230°	(50)	895
F	F	F	F	Br	Cu, DMF, reflux	(40)	895
F	F	F	F	I	Cu, 200°	(59)	896
F	Cl	F	F	I	Cu, 200°	(59)	897
F	Cl	Cl	F	I	Cu, 190°	(78)	897
Cl	Cl	Cl	Cl	Br	Cu, DMF, reflux	(8-18)	888
Cl	Cl	Cl	Cl	I	Cu, DMF, reflux	(52)	888
Cl	Cl	Cl	Cl	I	Cu, 200-210°	(40)	898
H	H	H	H	Cl	Ni(OAc)$_2$, NaH, t-BuOH, Ph$_3$P, DME, 45°	(86)	398, 834
H	H	H	H	Cl	NiCl$_2$, Zn, PPh$_3$, DMF, 50°	(82)	142, 839
H	H	H	H	Cl	HCO$_2$Na, Pd/C, TEBAC, NaOH aq.	(46)	179
H	H	H	H	Br	Ni(OAc)$_2$, NaH, t-BuOH, Ph$_3$P, DME, 45°	(78)	398, 834
H	H	H	H	Br	PdCl$_2$(PPh$_3$)$_2$, Et$_4$NOTs, DMF, 2e$^-$	(91)	190
Me	H	H	H	I	Cu, 210°	(52)	899
OMe	H	H	H	I	NiCl$_2$•6 H$_2$O, Zn, PPh$_3$, DMF, 50°	(23)	852
H	OMe	H	H	Br	NiCl$_2$•6 H$_2$O, Zn, PPh$_3$, DMF, 50°	(32)	852
Ph	H	H	H	Cl	Ni(PPh$_3$)$_2$Br$_2$, Zn, Et$_4$NI, THF, heat	(47)	900
2-pyridyl	H	H	H	Cl	Ni(PPh$_3$)$_2$Cl$_2$, PPh$_3$, Zn, DMF, rt	(20-70)	901, 902

TABLE 43. 2,2'-BIPYRIMIDYLS

Substrate				Conditions	Product(s) and Yield(s) (%)	Refs.

C$_{4-16}$

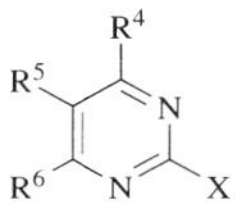

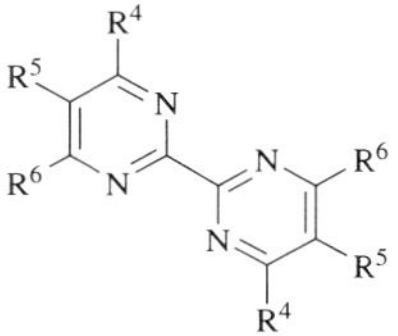

R^4	R^5	R^6	X	Conditions	Yield (%)	Refs.
H	H	H	Cl	NiCl$_2$, Zn, PPh$_3$, DMF, 50°	(60)	143
H	H	H	Cl	Ni(OAc)$_2$, NaH, t-BuOH, PPh$_3$, DME, 25°	(40)	398
H	H	H	Br	Cu, DMF, reflux	(10-50)	903, 904
H	H	H	Br	Cu, 110°	(10-20)	905
H	H	H	Br	Ni(OAc)$_2$, NaH, t-BuOH, PPh$_3$, DME, 25°	(30)	398
Me	H	H	Br	NiCl$_2$·6 H$_2$O, Zn, PPh$_3$, DMF, 50°	(12)	906
Me	H	Me	Cl	NiCl$_2$, Zn, PPh$_3$, DMF, 50°	(61)	143
Me	H	Me	Br	NiCl$_2$, Zn, PPh$_3$, DMF, 50°	(60)	143
Me	H	Me	Br	Cu, DMF, reflux	(26)	907
H	t-Bu	H	Cl	Cu, 190°	(25)	908
H	t-Bu	H	Cl	NiCl$_2$·6 H$_2$O, Zn, PPh$_3$, DMF, 50°	(43)	908
Ph	H	H	Br	Cu, DMF, reflux	(20)	909
H	n-C$_6$H$_{13}$	H	Cl	NiCl$_2$·6 H$_2$O, Zn, PPh$_3$, DMF, 50°	(10-34)	908
Me	H	PhCH$_2$	Cl	Cu, cumene, heat	(0)	910
Ph	H	Ph	Cl	NiCl$_2$, Zn, PPh$_3$, DMF, 50°	(85)	143
Ph	H	Ph	Br	NiCl$_2$, Zn, PPh$_3$, DMF, 50°	(80)	143

TABLE 44. 4,4'-BIPYRIMIDYLS

C$_{4-13}$

Substrate					Conditions	Product(s) and Yield(s) (%)	Refs.
R^2	R^5	R^6	X				
F	F	F	I		Cu, 180°	(58)	911
Me	H	Me	I		Pd/CaCO$_3$, N$_2$H$_4$, KOH, MeOH, reflux	(22)	912
Me	H	Me	I		Pd(OAc)$_2$, PPh$_3$, Et$_3$N, 160°	(50)	170
i-Pr	H	Me	I		Pd(OAc)$_2$, PPh$_3$, Et$_3$N, 160°	(97)	170
i-Pr	H	Me	I		Pd/C, acrylonitrile, Et$_3$N, 80°	(42)	840
n-Bu	H	Me	Cl		Cu, cumene, heat	(37)	910
Me	H	Ph	Cl		NiCl$_2$, Zn, PPh$_3$, DMF, 160°	(52)	913
Me	H	Ph	Cl		NiCl$_2$•6H$_2$O, Zn, PPh$_3$, DMF, 50°	(52)	914
Ph	H	Me	I		Cu, p-cymene, reflux	(28)	912
Et	H	Ph	Cl		NiCl$_2$, Zn, PPh$_3$, DMF, 160°	(30)	913
PhCH$_2$	H	Me	I		Cu, cumene, reflux	(63-64)	910, 915
i-Pr	H	Ph	Cl		NiCl$_2$, Zn, PPh$_3$, DMF, 160°	(67)	913

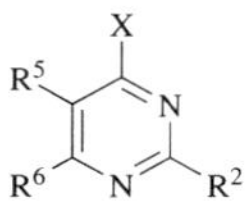

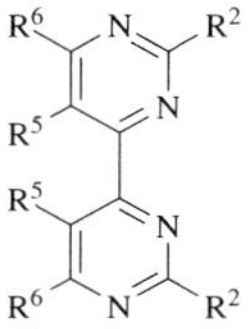

TABLE 45. 5,5'-BIPYRIMIDYLS

Substrate	Conditions	Product(s) and Yield(s) (%)	Refs.

C$_{4-10}$

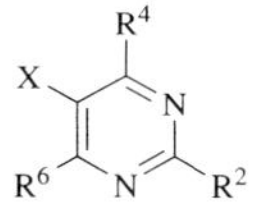

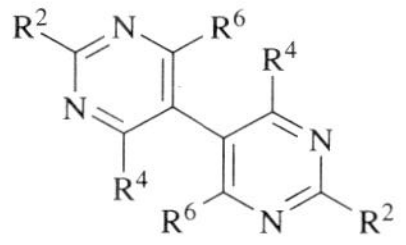

[substrate structure: pyrimidine with R^2, R^4, R^6, X]

R^2	R^4	R^6	X		
H	H	H	Br	Ni(OAc)$_2$, NaH, t-BuOH, PPh$_3$, DMF, 65°	(53) 398
Ph	H	H	Br	Pd/CaCO$_3$, N$_2$H$_4$, KOH, MeOH, reflux	(46) 912

C$_{6-11}$

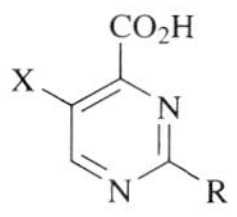

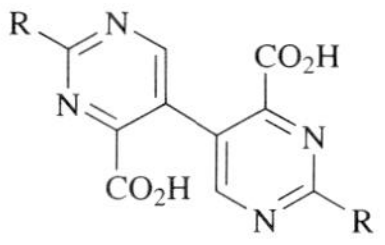

[substrate structure: pyrimidine with CO$_2$H, X, R]

R	X		
MeS	Br	Cu, DMF, reflux	(40), with accompanying decarboxylation 912
Ph	Br	Cu, DMF, reflux	(70), with accompanying decarboxylation 912

TABLE 46. 2,2'-BIQUINOLINES

C_{9-15}

Substrate: quinoline with R^3, R^4, and X (2-position)

R^3	R^4	X	Conditions	Product(s) and Yield(s) (%)	Refs.
H	H	Cl	Ni(OAc)_2, NaH, t-BuOH, PPh_3, DME, 45-60°	(68-70)	398, 834
H	H	Cl	NiBr_2(PPh_3)_2, Zn, Et_4NI, THF, 50°	(84)	139
H	H	Cl	NiCl_2, PPh_3, Zn, DMF, 50°	(83)	142, 839
H	H	Cl	Pd(OAc)_2, K_2CO_3, i-PrOH, DMF, 115°	(62)	175
H	H	Br	NiBr_2(PPh_3)_2, Zn, Et_4NI, THF, 50°	(61)	139
OH	H	Br	NiCl_2·6 H_2O, PPh_3, Zn, DMF, 50°	(2)	140
OH	H	Cl	Cu, 218°	(<1)	916
OMe	Me	Br	NiCl_2, PPh_3, Zn, DMF, 50°	(30)	140
H	Ph	Br	Cu, 260-280°	(3)	917
H	Ph	Br	Pd/CaCO_3, 5% KOH, MeOH, N_2H_4, heat	(11)	918

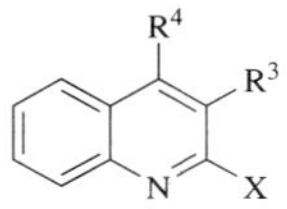

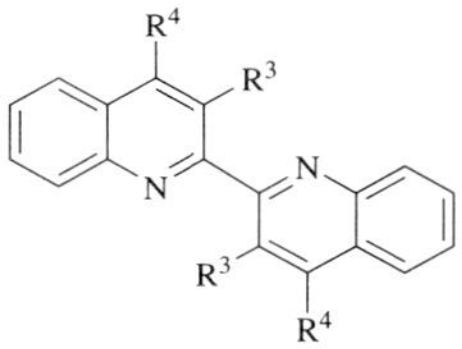

Product is 2,2'-biquinoline with R^3, R^4 substituents.

C_{11}

Substrate: 2-bromo-8-ethylquinoline

Conditions: Cu, 210-220°

Product: 8,8'-diethyl-2,2'-biquinoline (3)

Ref. 919

Substrate: 2-bromo-6-methoxy-4-methylquinoline

Conditions: Pd/CaCO_3, 5% KOH, MeOH, N_2H_4, heat

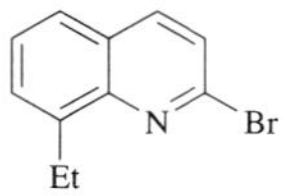

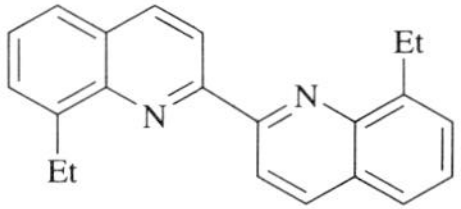

Product: 6,6'-dimethoxy-4,4'-dimethyl-2,2'-biquinoline (42)

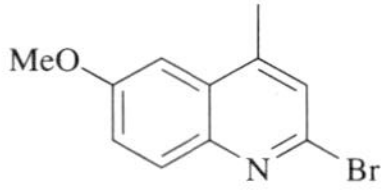

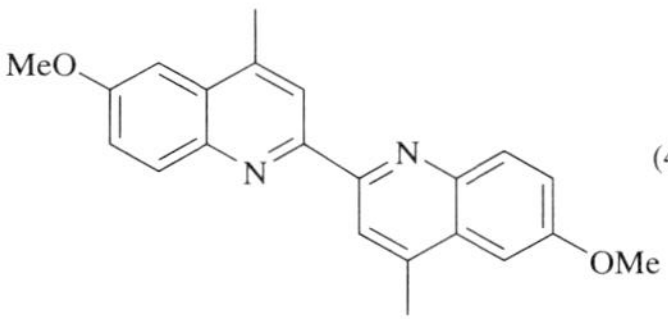

Ref. 918

C₁₆		Pd/CaCO₃, 5% KOH, MeOH, N₂H₄, heat	(20) 918
		Pd/CaCO₃, 5% KOH, MeOH, N₂H₄, heat	(20) 918
		Pd/CaCO₃, 5% KOH, MeOH, N₂H₄, heat	(39) 918

TABLE 47. OTHER BIQUINOLINES

Substrate			Conditions	Product(s) and Yield(s) (%)	Refs.
C_9					
R^4	X				
OH	Br		$NiCl_2 \cdot 6 H_2O$, PPh_3, Zn, DMF, 50°	(27)	140
H	Br		$NiBr_2(PPh_3)_2$, Zn, Et_4NI, THF, 50°	(61)	139
H	Br		$Ni(OAc)_2$, NaH, t-BuOH, PPh_3, DME, 65°	(72)	398, 834
H	Br		$Pd(OAc)_2$, K_2CO_3, i-PrOH, DMF, 115°	(79)	175
H	Br		$PdCO_3$, N_2H_4, KOH, MeOH, heat	(53)	163
H	I		$Pd(PPh_3)_4$, $BrZnCH_2CO_2Et$, DME, THF, reflux	(47)	920
C_{9-12}					
R^2	R^3	X			
H	H	Cl	$NiBr_2(PPh_3)_2$, Zn, Et_4NI, THF, 50°	(77)	139
H	H	Cl	$NiCl_2$, PPh_3, Zn, DMF, 50°	(70)	142, 839
H	H	Br	$Ni(OAc)_2$, NaH, t-BuOH, PPh_3, DME, 45°	(53)	398
H	H	Br	$NiBr_2(PPh_3)_2$, Zn, Et_4NI, THF, 50°	(60)	139
H	H	I	$Pd(PPh_3)_4$, $BrZnCH_2CO_2Et$, DME, THF, reflux	(10)	920
Me	H	Cl	$Ni(OAc)_2$, NaH, t-BuOH, PPh_3, DME, 45°	(71)	398, 834
Me	H	I	$Pd(PPh_3)_4$, $BrZnCH_2CO_2Et$, DME, THF, reflux	(32)	920
H	CO_2Et	Cl	$NiBr_2(PPh_3)_2$, Zn, Et_4NI, THF, reflux	(73)	921
C_{21}					
			$NiCl_2$, (o-tolyloxy)$_3$P, NaI, NMP, 60-80°	(73-75)	149, 922
			$NiCl_2(PPh_3)_2$, NaI, PPh_3, Zn, NMP, 70°	(81)	922

TABLE 48. BI-ISOQUINOLINES

Substrate	Conditions	Product(s) and Yield(s) (%)	Refs.
C₉			
(4-bromoisoquinoline)	Ni(PPh₃)₂Br₂, Zn, Et₄NI, THF, 50°	(60)	139
	Pd/CaCO₃, KOH, EtOH, N₂H₄·HCl	(33)	923
C₉₋₁₀			
R⁸ = H, X = Cl	Ni(PPh₃)₂Br₂, PPh₃, Zn, Et₄NI, THF, 50°	(20)	924
R⁸ = H, X = Cl	Ni(PPh₃)₂Br₂, Zn, Et₄NI, THF, 50°	(37)	139
R⁸ = H, X = Br	Cu, 200°	(18)	925
R⁸ = H, X = Br	NiCl₂, PPh₃, Zn, DMF, 50°	(85)	926
R⁸ = Me, X = Cl	NiCl₂, PPh₃, Zn, DMF, 50°	(56-61)	927, 928
R⁸ = OMe, X = Cl	NiCl₂, PPh₃, Zn, DMF, 50°	(70)	927

TABLE 48. BI-ISOQUINOLINES *(Continued)*

Substrate		Conditions	Product(s) and Yield(s) (%)	Refs.
C₉₋₂₀				
R¹	X			
H	Br	Cu, 200°	(14)	925
Ph	Cl	NiCl₂, PPh₃, Zn, NaI, THF, reflux	(93)	929
o-Tolyl	Cl	NiCl₂, PPh₃, Zn, NaI, THF, reflux	(75)	929
p-Tolyl	Cl	NiCl₂, PPh₃, Zn, NaI, THF, reflux	(73)	929
2-OMe-phenyl	Cl	NiCl₂, PPh₃, Zn, NaI, THF, reflux	(81)	929
8-Me-naphthyl	Cl	NiCl₂, PPh₃, Zn, NaI, THF, reflux	(82)	929
8-OMe-naphthyl	Cl	NiCl₂, PPh₃, Zn, NaI, THF, reflux	(64)	929
2-OMe-naphthyl	Cl	NiCl₂, PPh₃, Zn, NaI, THF, reflux	(73)	929

TABLE 49. 8,8'-BIQUINOLYL

Substrate	Conditions	Product(s) and Yield(s) (%)	Refs.
C₁₀	NiCl₂, PPh₃, Zn, NaI, DMF, reflux	(65)	930

TABLE 50. BIFURANS

Substrate	Conditions	Product(s) and Yield(s) (%)	Refs.
C₄ 3-bromofuran	Ni(OAc)₂, NaH, t-BuOH, bpy, THF, 25°	3,3'-bifuran (42)	398
	NiCl₂(PEt₃)₂, Zn, KI, HMPA, 40°	(80)	401
C₄₋₁₁ 2-X-furan with R³, R⁴, R⁵ substituents		2,2'-bifuran with R³, R⁴, R⁵	
R³ R⁴ R⁵ X			
H H NO₂ Br	Cu, 190°	(9)	931
H H NO₂ Br	Cu, DMF, reflux	(76)	822
H H H Br	Ni(OAc)₂, NaH, t-BuOH, bpy, THF, 25°	(10)	398
H H CHO I	Cu, DMF, reflux	(44–50)	822, 932
H H CO₂Me Br	Cu, DMF, reflux	(68)	822
H H CO₂Me Br	NiCl₂(PEt₃)₂, Zn, KI, HMPA, 40°	(90)	401
H Me CO₂Me Br	Cu, DMF, reflux	(62)	822
H H COPh I	Cu, DMF, reflux	(40)	822
C₈ 2-bromobenzofuran	Ni(OAc)₂, NaH, t-BuOH, bpy, THF, 25°	2,2'-bibenzofuran (70)	398

TABLE 51. BIS-DIBENZOFURANS AND BIS-DIBENZODIOXANES

Substrate	Conditions	Product(s) and Yield(s) (%)	Refs.
C$_{12}$			
(iodo-dibenzofuran)	Cu, 200-220°	(bis-dibenzofuran) (60)	933
(iodo-dibenzodioxin)	Cu, 250°	(bis-dibenzodioxin) (22)	934
(iodo-dibenzodioxin)	Cu, 240-250°	(bis-dibenzodioxin) (22)	934
C$_{13}$			
(iodo-methoxy-dibenzofuran)	Cu, 180-200°	(MeO-bis-dibenzofuran) (65)	933

TABLE 52. UNSUBSTITUTED 2,2'-BITHIENYL

Substrate		Conditions	Product(s) and Yield(s) (%)	Refs.
C$_{4-6}$	X			
	Cl	Ni(OAc)$_2$, NaH, t-AmOH, bpy, THF, 25°	(60)	398
	Cl	NiCl$_2$, PPh$_3$, Zn, DMAc, 50-80°	(98)	51
	Br	Ni(OAc)$_2$, NaH, t-AmOH, bpy, THF, 25°	(70)	398
	Br	NiI$_2$, Li, naphthalene, glyme, 85°	(45)	112
	Br	Ni(cod)$_2$, DMF, 42°	(30)	49, 50
	Br	Ni(PPh$_3$)$_2$Cl$_2$, Ph$_3$P, Zn, DMF, 50°	(41)	123
	Br	NiCl$_2$, CrCl$_2$, Mn, THF, 25°	(22)	406
	Br	NiCl$_2$(PEt$_3$)$_2$, Zn, KI, HMPA, 40°	(7)	401
	Br	PdCl$_2$, HgCl$_2$, PhNHNH$_2$, NaOH, MeOH, reflux	(42)	165
	I	Cu, 200-210°	(22)	935
	I	Cu, DMF, reflux	(67)	936
	I	NiCl$_2$(PEt$_3$)$_2$, Zn, KI, HMPA, 40°	(87)	401
	I	NiCl$_2$, CrCl$_2$, Mn, THF, 25°	(87)	406
	I	Pd/C, Zn, H$_2$O, acetone, rt	(64)	185
	I	[thienyl-C(O)-OCu], NMP, rt	(77)	71
	I	[Pd(OAc)(o-tol)(P(tol-o))]$_2$, DMF, 150°	(87)	194
	SiMe$_2$Cl	CuI, TBAF, MeCN, rt	(71)	72
	SiMe$_2$Br	CuI, TBAF, MeCN, rt	(75)	72

TABLE 53. SUBSTITUTED 2,2'-BITHIENYLS

	Substrate				Conditions	Product(s) and Yield(s) (%)	Refs.
	R^3	R^4	R^5	X			
C_{4-11}	Cl	Cl	Cl	I	Cu, DMF, reflux	(75)	888
	NO_2	H	NO_2	Cl	Cu, 200-215°	(43)	937
	H	H	NO_2	Br	Cu, 220-225°	(86)	938
	H	H	NO_2	Br	Pd(OAc)$_2$, i-Pr$_2$NEt, toluene, 105°	(58)	939
	H	H	NO_2	Br	Pd(OAc)$_2$, i-Pr$_2$NEt, n-Bu$_4$NBr, toluene, 105°	(68)	939
	H	H	Cl	Br	Pd(OAc)$_2$, i-Pr$_2$NEt, n-Bu$_4$NBr, toluene, 105°	(79)	940
	CHO	H	H	I	Cu, DMF, 130°	(85)	941
	H	H	CHO	Br	NiCl$_2$, PPh$_3$, Zn, DMF, reflux	(71)	939
	H	H	CHO	Br	Pd(OAc)$_2$, i-Pr$_2$NEt, n-Bu$_4$NBr, toluene, 105°	(26)	942
	Me	H	H	I	Cu, 200-210°	(92)	939
	H	H	H	I	Pd(OAc)$_2$, i-Pr$_2$NEt, toluene, 105°	(73)	398
	H	H	Me	Br	Ni(OAc)$_2$, NaH, t-AmOH, bipy, THF, 25°	(75-81)	943
	H	H	Me	I	Cu, 170-220°	(39)	937
	NO_2	H	COMe	Cl	Cu, 200-215°	(53)	937
	NO_2	H	CO$_2$Me	Cl	Cu, 205-225°	(24)	944, 945
	CO$_2$Me	H	H	Br	Cu, 210-225°	(73)	939
	H	H	COMe	Cl	Pd(OAc)$_2$, i-Pr$_2$NEt, n-Bu$_4$NBr, toluene, 105°	(80)	939
	H	H	COMe	Br	Pd(OAc)$_2$, i-Pr$_2$NEt, n-Bu$_4$NBr, toluene, 105°	(23)	942
	Me	Me	H	I	Cu, 200-210°	(22)	946
	H	Et	Me	I	Cu, 210-245°	(80)	947
	H	H	t-Bu	I	Cu, 190-200°	(37)	391
	Et	Et	H	I	Cu, 200°	(46)	164
	H	H	4-pyridyl		Pd(Hg), N$_2$H$_4$·H$_2$O, NaOH, MeOH	(65)	164
	H	Ph	H	I	NiCl$_2$, PPh$_3$, Zn, DMF, 50°	(63)	399
	H	Ph	H	Br	NiCl$_2$, PPh$_3$, Zn, DMF, 50°	(38)	942
	H	H	Ph	Br	Cu, 200-210°	(14)	948
	H	H	PhCH$_2$	I	Cu, 185-210°		

TABLE 53. SUBSTITUTED 2,2'-BITHIENYLS (Continued)

Substrate	Conditions	Product(s) and Yield(s) (%)	Refs.
C₈ (2-bromo-bithiophene)	NiCl₂, PPh₃, Zn, DMF, 50° Ni(PPh₃)₂Cl₂, Zn, Bu₄NI, THF, 50°	(terthiophene-thiophene dimer) (66) (87)	949 950
C₇₋₈ (R–thiophene–X, R = dioxolane/methyl dioxolane, X = Br)	Ni(OAc)₂, NaH, t-AmOH, bpy, THF, rt Ni(OAc)₂, NaH, t-AmOH, bpy, THF, rt	OHC–bithiophene–CHO (53) Ac–bithiophene–Ac (67)	398 398
C₉ (5-bromo-5'-formyl-bithiophene)	NiCl₂, PPh₃, Zn, DMF, reflux	OHC–(bithiophene)₂–CHO (78)	941
(5-bromo-5'-methyl-bithiophene)	Cu, 200°	H₃C–(bithiophene)₂–CH₃ (9)	951

C_{11-12}			
	R¹ Me MeO Et Me	R² H H H Me	NiCl₂, PPh₃, Zn, DMF, 50° · (75) 399 NiCl₂, PPh₃, Zn, DMF, 50° · (65) 399 NiCl₂, PPh₃, Zn, DMF, 50° · (65) 399 NiCl₂, PPh₃, Zn, DMF, 50° · (59) 399
C_{12}			NiCl₂, PPh₃, Zn, DMF, 70° · (58) 949
C_{13}			NiCl₂, PPh₃, Zn, DMF, reflux · (70) 941
C_{14}			NiCl₂, PPh₃, Zn, DMF, 50° · (82) 399
			NiCl₂, PPh₃, Zn, DMF, 50° · (85) 399

TABLE 53. SUBSTITUTED 2,2'-BITHIENYLS (*Continued*)

Substrate	Conditions	Product(s) and Yield(s) (%)	Refs.
C₁₆	NiCl₂, PPh₃, Zn, DMF, 70°	(18)	949
C₁₈	Ni(PPh₃)₃, DMF	(53)	952
C₂₃	Cu, 250°	(90)	953
C₂₄	NiCl₂, Zn, PPh₃, DMF, 70°	(33)	954
C₂₆	Ni(PPh₃)₃, DMF	(29)	952

C_{29-30}

R		
n-C$_{17}$H$_{35}$	Ni(PPh$_3$)Cl$_2$, Zn, PPh$_3$, KI, DMF, 60°	(56) 955
n-C$_{16}$H$_{33}$OCH$_2$	Ni(PPh$_3$)Cl$_2$, Zn, PPh$_3$, KI, DMF, 60°	(81) 955
n-C$_{15}$H$_{31}$	Ni(PPh$_3$)Cl$_2$, Zn, PPh$_3$, KI, DMF, 60°	(82) 955

TABLE 54. 3,3'-BITHIENYLS

C$_{4-8}$

Substrate: thiophene with R^2 (pos 2), R^4 (pos 4), R^5 (pos 5), X (pos 3)

Product: 3,3'-bithienyl with R^2, R^4, R^5 substituents

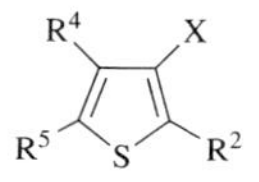

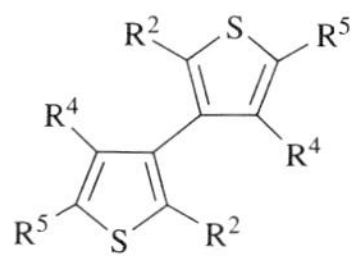

R^2	R^4	R^5	X	Conditions	Yield (%)	Refs.
H	NO$_2$	H	I	Cu, CuSO$_4$, NH$_3$, MeCN, acetone, 25°	(86)	956
H	H	H	Cl	Ni(OAc)$_2$, NaH, t-AmOH, bpy, THF, 25°	(52)	398
H	H	H	Br	Ni(OAc)$_2$, NaH, t-AmOH, bpy, THF, 25°	(60)	398
H	H	H	Br	Pd/C, NaOH, H$_2$O, HCONHNH$_2$, 95°	(67)	55
H	H	H	I	NiCl$_2$(PEt$_3$)$_2$, Zn, KI, HMPA	(83)	401
NO$_2$	NO$_2$	Me	I	Cu, 148-155°	(29)	937
CO$_2$Me	NO$_2$	H	Br	Cu, DMF, reflux	(73)	957
CO$_2$Me	Br	H	Br	Cu, DMF, reflux	(12)	958
Me	H	Me	I	Cu, 220-240°	(47)	959
Me	CO$_2$Me	Me	I	Cu, 230-235°	(8)	937

TABLE 55. BIBENZOTHIOPHENES

Substrate	Conditions	Product(s) and Yield(s) (%)	Refs.
C$_8$ (2-bromobenzothiophene)	Ni(OAc)$_2$, NaH, *t*-AmOH, bpy, THF, 25°	2,2'-bibenzothiophene (82)	398
3-X-benzothiophene		3,3'-bibenzothiophene	
X = Br	Ni(OAc)$_2$, NaH, *t*-AmOH, bpy, THF, 25°	(70)	398
X = I	Cu, 270–280°	(17)	960
C$_{12}$ (4-iododibenzothiophene)	Cu, 260°	4,4'-bidibenzothiophene (31)	961

TABLE 56. BISELENYLS

Substrate				Conditions	Product(s) and Yield(s) (%)	Refs.

C₄₋₆

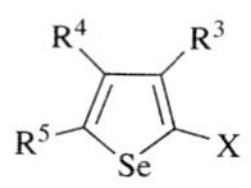

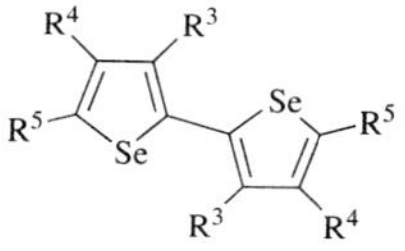

Substrate: thiophene-like selenophene with R³, R⁴, R⁵, X substituents (Se ring)

Product: biselenyl with R³, R⁴, R⁵ groups

R³	R⁴	R⁵	X	Conditions	Yield	Ref.
H	H	H	Br	Ni(PPh₃)₂Cl₂, Zn, Bu₄NI, THF, 50°	(19)	962
H	H	H	I	Cu, xylene, reflux	(76)	963
NO₂	H	H	Br	Cu, xylene, reflux	(70)	963
H	H	NO₂	I	Cu, xylene, reflux	(35)	963
NO₂	H	NO₂	I	Cu, xylene, reflux	(75)	963
H	H	OAc	I	Cu, xylene, reflux	(55)	963

C₆

Substrate: MeO₂C, Br, NO₂ substituted selenophene

Conditions	Product and Yield	Ref.
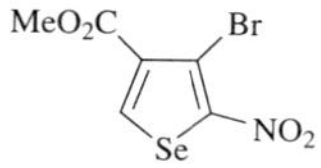 Cu, xylene, reflux	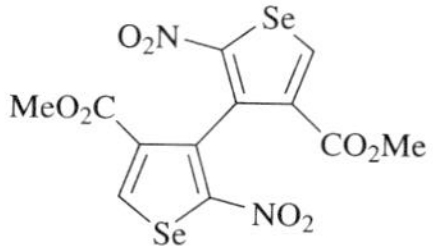 biselenyl product (40)	964

TABLE 57. BIMETALLOCENES AND POLYMETALLOCENYLENES

Substrate	Conditions	Product(s) and Yield(s) (%)	Refs.
C_{10}			
Fc–X (ferrocenyl-X)		biferrocene	
X = Cl	Cu, 140°	(65)	35
Br	Cu, 140°	(97)	35
I	Cu, 150°	(79)	965
I	Cu, 140-150°	(100)	35
I	Cu, biphenyl, 130-160°	(76)	35
I	Zn, 145-150°	(17)	35
X–M–X (1,1'-dihalometallocene)		[metallocenylene]$_n$	
M = Fe, X = Br	Ni(cod)$_2$, DMF	(20-28)	966
Fe, Br	Ni(PPh$_3$)$_4$, DMF	(50-60)	966
Fe, I	Ni(PPh$_3$)$_4$, DMF	(35-42)	966
Ru, I	Cu, 130-150°	(68)	966

TABLE 58. MISCELLANEOUS BI(HETEROCYCLE)

Substrate	Conditions	Product(s) and Yield(s) (%)	Refs.
C₃ (2-bromothiazole)	Ni(OAc)₂, NaH, t-BuOH, PPh₃, C₆H₆, 45° Pd(OAc)₂, i-Pr₂NEt, Bu₄NBr, toluene, 105°	2,2'-bithiazole (62) (86)	398
C₄ (2-chloropyrazine)	Ni(OAc)₂, NaH, t-BuOH, PPh₃, DME, 45°	2,2'-bipyrazine (43)	398
C₆ (2-bromo-4,5-dicyano-1-methylimidazole)	Cu, DMF, 85°	bi(4,5-dicyano-1-methylimidazol-2-yl) (48)	967
C₇ (2-bromo-4,5-dimethylphenylphosphine)	Ni(dppe)Cl₂, Zn, THF, 50°	Ni complex with biaryl diphosphine (30)	968
C₈ (5-bromoindole)	Ni(bpy)Br₂, NaBr, MeOH, EtOH, 2e⁻	5,5'-biindole (86)	53
C₉ (2-iodo-1-methylindole)	Cu, 205°	2,2'-bi(1-methylindole) (66)	969

C_{11}	(structure: 2-chloro-4,5-di-t-butylimidazole coupling)	Cu, heat	(0) 908
		NiCl$_2$, Zn, PPh$_3$, DMF, heat	(0) 908
C_{12}	(aryl bromide phosphine M(CO)$_5$)		
			M
		Ni(dppe)Cl$_2$, Zn, THF, 50°	Cr (55) 968
		Ni(dppe)Cl$_2$, Zn, THF, 50°	Mo (55) 968
		Ni(dppe)Cl$_2$, Zn, THF, 50°	W (35) 968
C_{13}	(9-chloroacridine)	Cu, 140°	(72) 970
C_{14-15}	(1-chlorothioxanthone)		
			R^2 R^4
		Cu, 180–250°	H Me (44) 971
		Cu, naphthalene, reflux	H Me (46) 971
		Cu, naphthalene, reflux	Me OMe (57) 971

517

TABLE 59. INTRAMOLECULAR COUPLINGS FORMING SYMMETRIC HETEROCYCLIC BIARYLS

Substrate	Conditions	Product(s) and Yield(s) (%)	Refs.
C₉	X = Br: Cu, DMF, reflux X = I: Cu, DMF, reflux	(49) (100)	972 972
C₁₁	Cu, DMF, reflux	(37)	972
C₁₁	Cu, DMF, reflux	(99-100)	972, 973
C₁₃	Cu, DMF, reflux	(75)	974
C₁₃	Cu, DMF, reflux	(89)	974
C₂₀	Cu, DMF, reflux	(28)	975

Substrate	Product	Conditions	Yield (%)	Ref.
C29 (diiodo isoquinoline ether)	macrocycle (18)	Cu, DMF, reflux	(18)	975
C28 (dipyrrole, X=Br / X=I)	Cu-porphyrin-like macrocycle	Cu, DMF, reflux / Cu, DMF, reflux	(50) / (50)	976 / 976
C29 (dibromo bipyridine)	pyridine macrocycle	Cu, DMF, reflux	(0)	977
		Cu, biphenyl, reflux	(0)	977
		Ni(cod)$_2$, bpy, DMF, heat	(0)	977
		Ni(PPh$_3$)$_2$Br$_2$, Zn, Et$_4$NI, THF, heat	(0)	977

TABLE 59. INTRAMOLECULAR COUPLINGS FORMING SYMMETRIC HETEROCYCLIC BIARYLS (*Continued*)

Substrate	Conditions	Product(s) and Yield(s) (%)	Refs.
C₃₄ (structure shown)	X = Br; Cu, DMF, reflux X = I; Cu, DMF, reflux	(50) (50)	976 976

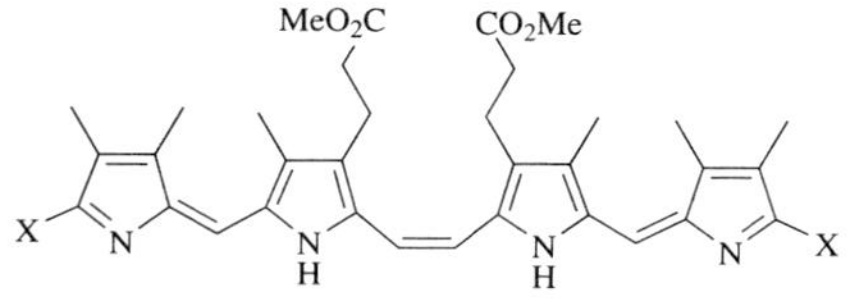

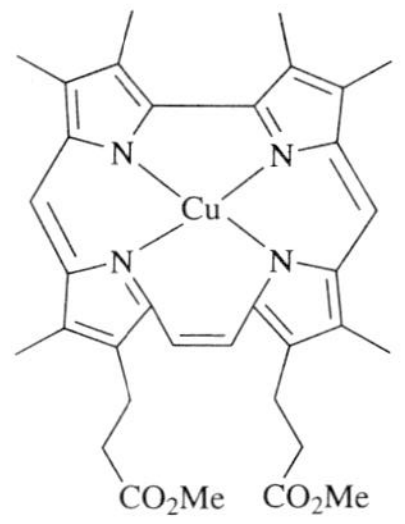

TABLE 60. INTRAMOLECULAR COUPLINGS FORMING UNSYMMETRIC HETEROCYCLIC BIARYLS

Substrate	Conditions	Product(s) and Yield(s) (%)	Refs.
C$_9$			
	Cu, DMF, reflux	(39)	972
	Cu, DMF, reflux	(57)	972
C$_{11}$			
		R / X: F/Cl (27); F/Br (1); Cl/Cl (4); Cl/Br (41)	
	Cu, DMF, reflux	(27)	888
	Cu, DMF, reflux	(1)	888
	Cu, DMF, reflux	(4)	888
	Cu, DMF, reflux	(41)	888
	Cu, DMF, reflux	(25)	888

TABLE 60. INTRAMOLECULAR COUPLINGS FORMING UNSYMMETRIC HETEROCYCLIC BIARYLS (*Continued*)

Substrate	Conditions	Product(s) and Yield(s) (%)	Refs.
C_{13}	Cu, DMF, reflux	(92)	974
C_{17}	Cu, DMF, reflux	(64)	978
	Cu, DMF, reflux	(81)	978, 979
C_{19}	Cu, DMF, reflux	(78)	978, 979

C$_{19}$	Cu, DMF, reflux	(83)	978, 979
	Cu, DMF, reflux	(87)	978
	Cu, DMF, reflux	(80)	978
C$_{20}$	Cu, DMF, reflux	(84)	978, 979

523

TABLE 60. INTRAMOLECULAR COUPLINGS FORMING UNSYMMETRIC HETEROCYCLIC BIARYLS (Continued)

Substrate	Conditions	Product(s) and Yield(s) (%)	Refs.
	Cu, DMF, reflux	(78)	978
$\dfrac{X}{Br}$ I	Cu, DMF, reflux Cu, DMF, reflux	(28) (67)	978 978, 979
C$_{21}$	Cu, DMF, reflux	(63–67)	978, 979

| | Cu, DMF, reflux | (87) | 978 |
| | Cu, DMF, reflux | (83) | 978 |

TABLE 61. POLY(PARAPHENYLENES)

Substrate		Conditions	Product(s) and Yield(s) (%)	Refs.
X—⟨C6H4⟩—X (C6-7)	X = Cl / Cl	[P(i-Pr)2(o-C6H4)P(i-Pr)2]NiCl2, DMSO, 65°, −2.2 V	—⟨C6H4⟩—n (quantitative)	154
	Br	Ni(cod)2, bpy, DMF, 60°	(92)	246
	Br	Ni(cod)2, Ph3P, DMF, 60°	(26)	246
	Br	Ni(cod)2, bpy, DMF, 30-60°	(95-99)	246
	Br	Ni(diphos)Cl2, LiClO4, HMPA, THF, −2.5 V	(75)	251
	I	Ni(cod)2, bpy, DMF, 60°	(91)	246
	I	[Pd(OAc)(P(o-tol)2(o-CH2C6H4))]2, i-Pr2NEt, DMF, 100°	(70)	194
	OMs	Ni(PPh3)2Cl2, Ph3P, Zn, Et4NI, THF, reflux	(49-50)	262, 263
R—⟨C6H3⟩—X, X (C8)	R / X		—⟨C6H3(R)—C6H3(R)⟩—n	
	CO2Me / Cl	Ni(PPh3)2Cl2, Zn, Et4NI, THF, 67°	(88)	261
	CO2Me / Cl	NiCl2, Ph3P, Zn, bpy, DMF, 90°	(72)	261
	CO2Me / Cl	NiCl2, Ph3P, Zn, DMF, 30-80°	(22-85)	233
	CO2Me / Br	Ni(PPh3)2Cl2, Zn, Et4NI, THF, 67°	(50)	261
	CO2Me / Br	NiCl2, Ph3P, Zn, bpy, DMF, 90°	(21)	261
	CN / OMs	Ni(PPh3)2Cl2, Ph3P, Zn, Et4NI, THF, reflux	(68)	263
	CO2Me / OTf	Ni(PPh3)2Cl2, Zn, Et4NI, THF, 67°	(16-65)	261
	CO2Me / OTf	NiCl2, Ph3P, Zn, bpy, DMF, 90°	(76)	261, 262
	CO2Me / OMs	Ni(PPh3)2Cl2, Ph3P, Zn, Et4NI, THF, reflux	(85)	263
	Me / OMs	Ni(PPh3)2Cl2, Ph3P, Zn, Et4NI, THF, reflux	(87)	263

$C_{8\text{-}18}$

R (structure)	R	X	Conditions	Yield	Ref.
1,4-C₆H₄ with R substituent	Ph	Cl	NiCl₂, Ph₃P, Zn, bpy, DMF, 90°	(57)	261
	Ph	Cl	Ni(PPh₃)₂Cl₂, Zn, Et₄NI, THF, 67°	(38-59)	261
	Ph	Br	NiCl₂, Ph₃P, Zn, bpy, DMF, 90°	(15)	261
	Ph	Br	Ni(PPh₃)₂Cl₂, Zn, Et₄NI, THF, 67°	(8-79)	261
	CO₂Pr-i	OMs	Ni(PPh₃)₂Cl₂, Ph₃P, Zn, Et₄NI, THF, reflux	(88)	263
	Ph	OTf	NiCl₂, Ph₃P, Zn, bpy, DMF, 90°	(8)	261
	Ph	OTf	Ni(PPh₃)₂Cl₂, Zn, Et₄NI, THF, 67°	(6-49)	261, 262
	Ph	OMs	Ni(PPh₃)₂Cl₂, Ph₃P, Zn, Et₄NI, THF, 67°	(82)	263
	t-Bu	OTf	Ni(PPh₃)₂Cl₂, Zn, Et₄NI, THF, 67°	(5-17)	262, 263
	t-Bu	OTf	NiCl₂, Ph₃P, Zn, bpy, DMF, 90°	(0)	261
	t-Bu	OMs	Ni(PPh₃)₂Cl₂, Ph₃P, Zn, Et₄NI, THF, 67°	(85)	263
morpholine-C(O)-		Cl	NiCl₂, Ph₃P, Zn, bpy, NaI, NMP, 50°	(100)	258
piperidine-C(O)-		Cl	NiCl₂, Ph₃P, Zn, NaI, NMP, 50°	(100)	258
4-F-C₆H₄-C(O)-		OMs	Ni(PPh₃)₂Cl₂, Ph₃P, Zn, Et₄NI, THF, reflux	(68)	263
4-Cl-C₆H₄-C(O)-		OMs	Ni(PPh₃)₂Cl₂, Ph₃P, Zn, Et₄NI, THF, reflux	(98)	263
Ph-C(O)-		Cl	NiCl₂, (o-tolyl-O)₃P, Zn, bpy, NaI, NMP, 90°	(92)	149
Ph-C(O)-		OMs	Ni(PPh₃)₂Cl₂, Ph₃P, Zn, Et₄NI, THF, reflux	(95)	263
4-t-Bu-C₆H₄-C(O)-		OMs	Ni(PPh₃)₂Cl₂, Ph₃P, Zn, Et₄NI, THF, reflux	(82)	263

TABLE 61. POLY(PARAPHENYLENES) (Continued)

Substrate	Conditions	Product(s) and Yield(s) (%)	Refs.
C_{12} Br–C6H4–C6H4–Br	Ni(cod)$_2$, bpy, DMF, 60°	[–C6H4–]$_n$ (95)	246
O_2N-substituted diiodobiphenyl	Cu, DMF, 140°	nitro-substituted poly(biphenylene) (95-97)	238
dimethyl diiodobiphenyl	Cu, 240°	dimethyl poly(biphenylene) (85)	237
MsO/EtO$_2$C/CO$_2$Et/OMs benzene	Ni(PPh$_3$)$_2$Cl$_2$, Ph$_3$P, Zn, Et$_4$NI, THF, 67°	CO$_2$Et/EtO$_2$C poly(phenylene) (81)	263
C_{15} X–C6H3(CO$_2$R)–X, R = 2-ethylhexyl	Ni(PPh$_3$)$_2$Cl$_2$, Zn, Et$_4$NI, solvent, reflux	OR–C(=O)–[C6H3]$_n$	

X	solvent	Yield	Ref.
Cl	THF	(87)	263
Cl	dioxane	(82)	263
Br	THF	(81)	263
Br	dioxane	(84)	263

528

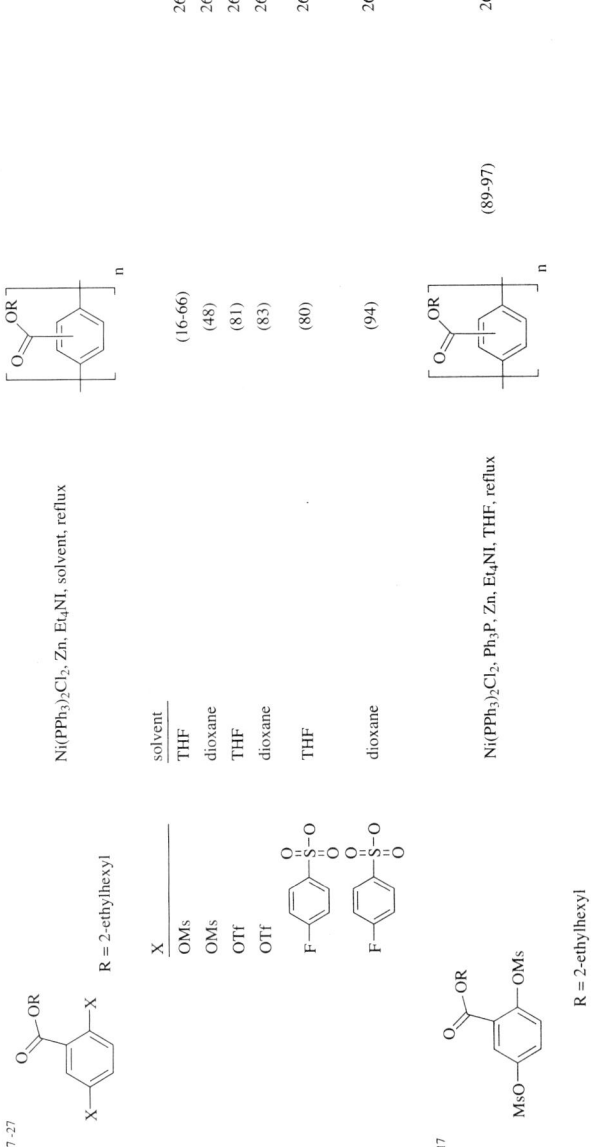

TABLE 62. POLY(METAPHENYLENES) AND POLY(BIPHENYLYLENES)

Substrate	Conditions	Product(s) and Yield(s) (%)	Refs.
C_{6-13}			
2,4-dichloro-C₆H₃R			
R = H	NiCl₂, Ph₃P, Zn, bpy, DMAc, 90°	poly(R-phenylene) (94)	257
CHO	NiCl₂, Ph₃P, Zn, bpy, DMAc, 90°	(77)	257
Me	NiCl₂, Ph₃P, Zn, bpy, DMAc, 90°	(91)	257
COMe	NiCl₂, Ph₃P, Zn, bpy, DMAc, 90°	(100)	257
COPh	NiCl₂, Ph₃P, Zn, bpy, DMAc, 90°	(100)	257
C_{19-25}			
3,3'-diiodo-Ar₂CHR	Cu, Hg, biphenyl, 225°	poly(biphenylylene-CHR)	240
R = Ph		(52)	
1-naphthyl		(65)	
p-tolyl, Ph		(70)	

TABLE 63. POLY(THIOPHENES)

Substrate		Conditions	Product(s) and Yield(s) (%)	Refs.
C_{4-14}				
R—[thiophene]—X			R—[thiophene]$_n$	
R	X			
H	Cl	Ni(cod)$_2$, Ph$_3$P, DMF, 60°	(55)	246
H	Br	Ni(cod)$_2$, Ph$_3$P, DMF, 25-100°	(74-100)	246
H	Br	Ni(cod)$_2$, bpy, DMF, 25-60°	(91-93)	246
H	Br	Ni(PPh$_3$)$_4$, DMF, 60°	(63)	246
CN	Br	Ni(cod)$_2$, Ph$_3$P, DMF, 60°	(91)	246
CN	Br	Ni(cod)$_2$, bpy, DMF, 25°	(80)	246
CN	Br	Ni(dppp)Cl$_2$, Zn, HMPT, 140°	(24)	250
CHO	Br	Ni(dppp)Cl$_2$, Zn, HMPT, 140°	(81)	250
CH=NOH	Br	Ni(dppp)Cl$_2$, Zn, HMPT, 140°	(18)	250
Me	Br	Ni(cod)$_2$, Ph$_3$P, DMF, 25-100°	(76)	246
n-C$_6$H$_{13}$	I	Ni(cod)$_2$, Ph$_3$P, DMF, 25-100°	(60)	246
n-C$_6$H$_{13}$	I	Ni(cod)$_2$, bpy, DMF, 60°	(70-94)	246
n-C$_6$H$_{13}$	I	Ni(cod)$_2$, bpy, toluene, 60°	(60)	246
CO$_2$C$_6$H$_{13}$-n	Br	Cu, DMF, 145-150°	(40)	242
n-C$_8$H$_{17}$	I	Ni(cod)$_2$, bpy, DMF, 30-60°	(80)	246
CO$_2$C$_8$H$_{17}$-n	Br	Cu, DMF, 145-150°	(56)	242
n-C$_{10}$H$_{21}$	I	Ni(cod)$_2$, bpy, DMF, 60°	(87)	246
C_{14}				
[crown-ether-dichlorothiophene]		Ni(cod)$_2$, bpy, DMF, 60°	[crown-ether-polythiophene]$_n$ (86)	249

TABLE 64. MISCELLANEOUS POLY(ARENES)

Substrate	Conditions	Product(s) and Yield(s) (%)	Refs.
C₄ 2,5-dibromoselenophene	Ni(cod)₂, bpy, cod, 60°	poly(selenophene) (63)	980
C₅ 2,5-dibromopyridine	Ni(cod)₂, PPh₃, bpy, DMF, rt	poly(pyridine) (64)	981
	Ni(cod)₂, PPh₃, bpy, DMF, 60°	(95)	981
	Ni(cod)₂, PPh₃, bpy, HMPA, 60°	(95)	981
	NiCl₂, Ph₃P, Zn, DMF, 60°	(59–66)	981
C₈₋₁₆ pyrrole-dione dimer, R = Bu, X = Br	Cu, DME, 200°	(56)	243, 982
R = Bu, X = I	Cu, DME, 200°	(67)	982
R = MeOCH₂CH₂OCH₂CH₂, X = Br	Cu, DME, 200°	(58)	982
R = n-C₁₂H₂₅, X = Br	Cu, DME, 200°	(62)	982
C₁₀ 5,5'-dibromo-2,2'-bipyridine	Ni(cod)₂, bpy, DMF, 70°	poly(bipyridine) (97)	983

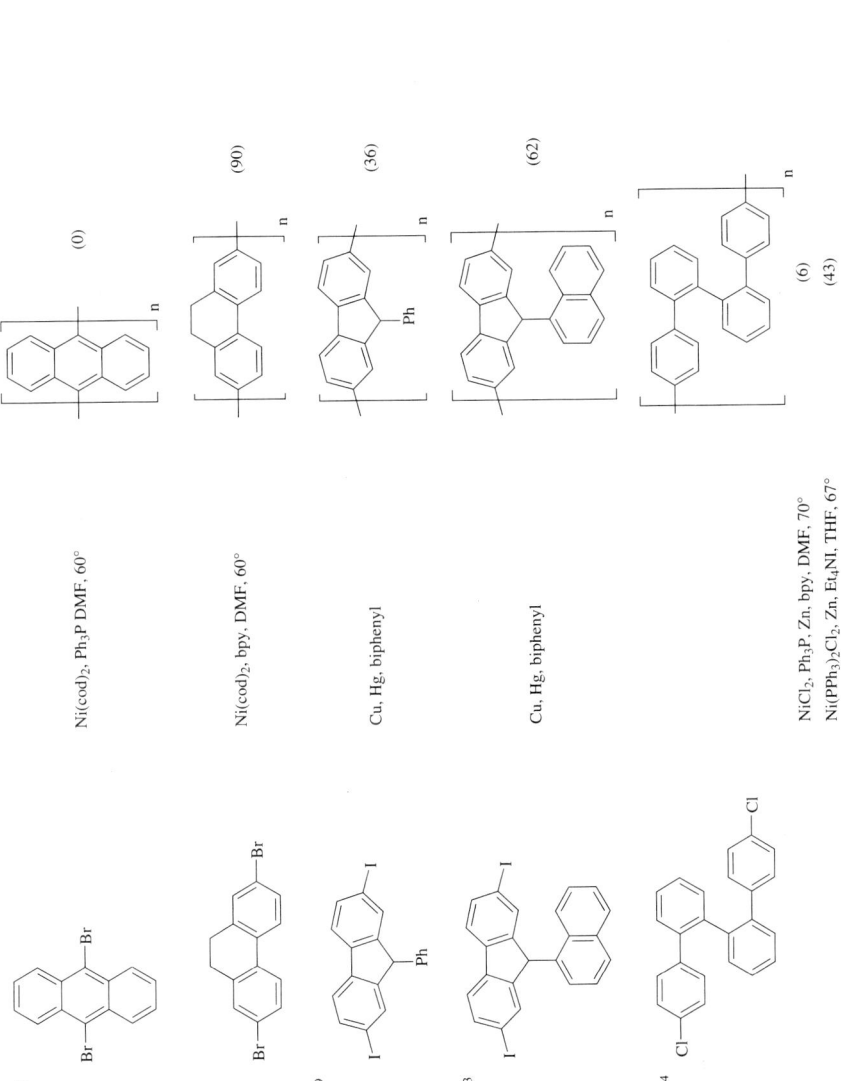

TABLE 64. MISCELLANEOUS POLY(ARENES) (Continued)

Substrate	Conditions	Product(s) and Yield(s) (%)	Refs.
(chloro-terphenyl substrate)	NiCl$_2$, Ph$_3$P, Zn, bpy, DMF, 70° Ni(PPh$_3$)$_2$Cl$_2$, Zn, Et$_4$NI, THF, 67°	poly(terphenylene), (45) (41)	260 260
(bis(4-chlorophenylsulfonyl) phenyl ether)	NiCl$_2$, Ph$_3$P, Zn, bpy, DMAc, 90-100°	poly(ether sulfone), (99-100)	465
C$_{32-34}$ (bis-naphthyl-phenyl substrate) X = Cl X = OMs	NiCl$_2$, Ph$_3$P, Zn, bpy, DMF, 70° Ni(PPh$_3$)$_2$Cl$_2$, Zn, Et$_4$NI, THF, 67°	(89-91) (2)	259 259

REFERENCES

[1] Ullmann, F.; Bielecki, J. *Chem. Ber.* **1901**, *34*, 2174.
[2] Ullmann, F.; Meyer, G. M.; Loewenthal, O.; Gilli, E. *Liebigs Ann. Chem.* **1904**, *332*, 38.
[3] Fanta, P. E. *Chem. Rev.* **1946**, *38*, 139.
[4] Fanta, P. E. *Chem. Rev.* **1964**, *64*, 613.
[5] Fanta, P. E. *Synthesis* **1974**, 9.
[6] Goshaev, M.; Otroshchenko, O. S.; Sadykov, A. S. *Russ. Chem. Rev.* **1972**, *41*, 1046.
[7] Naso, F.; Marchese, F. In *The Chemistry of Functional Groups, Supplement D;* Patai, S., Rappoport, Z., Eds.; Wiley: New York, 1983, p. 1353.
[8] Sainsbury, M. *Tetrahedron* **1980**, *36*, 3327.
[9] Knight, D. W. In *Comprehensive Organic Synthesis;* Trost, B. M., Fleming, I., and Pattenden, G., Eds.; Pergamon: Oxford, 1991; Vol. 3, p. 481.
[10] Jukes, A. E. *Adv. Organomet. Chem.* **1974**, *12*, 215.
[11] Bringmann, G.; Walter, R.; Weirich, R. *Angew. Chem., Int. Ed. Engl.* **1990**, *29*, 977.
[12] Bacon, R. G. R.; Hill, H. A. O. *Quart. Rev. (London)* **1965**, *19*, 95.
[13] Bergman, J.; Eklund, N. *Tetrahedron* **1980**, *36*, 1439.
[14] Paine, A. J. *J. Am. Chem. Soc.* **1987**, *109*, 1496.
[15] Fuson, R. C.; Cleveland, E. A. *Org. Synth. Collective Vol. 3* **1955**, 339.
[16] Kleiderer, E. C.; Adams, R. *J. Am. Chem. Soc.* **1933**, *55*, 4219.
[17] Lewin, A. H.; Zovko, M. J.; Rosewater, W. H.; Cohen, T. *J. Chem. Soc., Chem. Commun.* **1967**, 80.
[18] Lippert, T.; Wokaun, A.; Lenoir, D. *Environ. Sci. Technol.* **1991**, *25*, 1485.
[19] Gore, P. H.; Hughes, G. K. *J. Chem. Soc.* **1959**, 1615.
[20] Kulicki, A.; Karminski, W. *Zeszyty Nauk. Politech. Slask., Chem.* **1963**, *16*, 11; *Chem. Abstr.* **1965**, *62*, 4001c.
[21] Rieke, R. D.; Burns, T. P.; Wehmeyer, R. M.; Kahn, B. E. In *ACS Symposium Series;* Suslick, K. S., Ed.; 1987; Vol. 333, pp. 223–245.
[22] Rieke, R. D. *Science* **1989**, *246*, 1260.
[23] Fürstner, A. *Angew. Chem., Int. Ed. Engl.* **1993**, *32*, 164.
[24] Rieke, R. D.; Rhyne, L. D. *J. Org. Chem.* **1979**, *44*, 3445.
[25] Ebert, G. W.; Rieke, R. D. *J. Org. Chem.* **1988**, *53*, 4482.
[26] Ebert, G. W.; Rieke, R. D. *J. Org. Chem.* **1984**, *49*, 5282.
[27] Suslick, K. S.; Casadonte, D. J.; Doktycz, S. J. *Chem. Mater.* **1989**, *1*, 6.
[28] Lindley, J.; Lorimer, J. P.; Mason, T. J. *Ultrasonics* **1986**, *24*, 292.
[29] Lindley, J.; Mason, T. J.; Lorimer, J. P. *Ultrasonics* **1987**, *25*, 45.
[30] Nelson, K. A.; Adolph, H. G. *Synth. Commun.* **1991**, *21*, 293.
[31] Roberge, D. M.; Hölderich, W. F. *Applied Cat. A: Gen.* **2000**, *194–195*, 341; *Chem. Abstr.* **2000**, *132*, 293391.
[32] Forrest, J. *J. Chem. Soc.* **1960**, 592.
[33] Newman, M. S.; Logue, M. W. *J. Org. Chem.* **1971**, *36*, 1398.
[34] Kalk, W.; Bien, H.-S.; Schündehütte *Liebigs Ann. Chem.* **1977**, 329.
[35] Rausch, M. D. *J. Org. Chem.* **1961**, *26*, 1802.
[36] Rule, H. G.; Smith, F. R. *J. Chem. Soc.* **1937**, 1096.
[37] Radau, G.; Büllesbach, R.; Pachaly, P. *Tetrahedron* **1996**, *52*, 14735.
[38] Forrest, J. *J. Chem. Soc.* **1960**, 594.
[39] Sako, S.-i. *Bull. Chem. Soc. Jpn.* **1935**, *10*, 585.
[40] Nilsson, M. *Acta Chem. Scand.* **1966**, *20*, 423.
[41] Cohen, T.; Berninger, R. W.; Wood, J. T. *J. Org. Chem.* **1978**, *43*, 837.
[42] Moroz, A. A.; Shvartsberg, M. S. *Russ. Chem. Rev.* **1974**, *43*, 679; *Chem. Abstr.* **1975**, *82*, 15753.
[43] Lindley, J. *Tetrahedron* **1984**, *40*, 1433.
[44] Sawyer, J. S. *Tetrahedron* **2000**, *56*, 5045.
[45] Kalinin, A. V.; Bower, J. F.; Riebel, P.; Snieckus, V. *J. Org. Chem.* **1999**, *64*, 2986.
[46] Forrest, J. *J. Chem. Soc.* **1960**, 581.
[47] Nakajima, R.; Shintani, Y.; Hara, T. *Bull. Chem. Soc. Jpn.* **1980**, *53*, 1767.

[48] Busch, M.; Weber, W. *J. Prakt. Chem.* **1936**, *146*, 1.
[49] Semmelhack, M. F.; Helquist, P.; Jones, L. D.; Keller, L.; Mendelson, L.; Ryono, L. S.; Smith, J. G.; Stauffer, R. D. *J. Am. Chem. Soc.* **1981**, *103*, 6460.
[50] Semmelhack, M. F.; Helquist, P. M.; Jones, L. D. *J. Am. Chem. Soc.* **1971**, *93*, 5908.
[51] Colon, I.; Kelsey, D. R. *J. Org. Chem.* **1986**, *51*, 2627.
[52] Colon, I.; Maresca, L. M.; Kwiatkowski, G. T. U.S. Patent 4326989 (1982); *Chem. Abstr.* **1981**, *95*, 80437.
[53] Courtois, V.; Barhdadi, R.; Troupel, M.; Périchon, J. *Tetrahedron* **1997**, *53*, 11569.
[54] Lourak, M.; Vanderesse, R.; Fort, Y.; Caubère, P. *J. Org. Chem.* **1989**, *54*, 4840.
[55] Becker, A.; Ewenson, A. A.; Croitoru, B. U.S. Patent 5,177,258 (1993); *Chem. Abstr.* **1993**, *118*, 61912.
[56] Kelly, T. R.; Xie, R. L. *J. Org. Chem.* **1998**, *63*, 8045.
[57] Kornblum, N.; Kendall, D. L. *J. Am. Chem. Soc.* **1952**, *74*, 5782.
[58] Dieteren, H. M. L.; Koningsberger, C. *Recl. Trav. Chim. Pays-Bas* **1963**, *82*, 5.
[59] Fairfull, A. E. S.; Peak, D. A.; Short, W. F.; Watkins, T. I. *J. Chem. Soc.* **1952**, 4700.
[60] Nilsson, M. *Acta Chem. Scand.* **1958**, *12*, 537.
[61] Braithwaite, R. S. W.; Holt, P. F. *J. Chem. Soc.* **1959**, 3025.
[62] Nozaki, T.; Tamura, M.; Harada, Y.; Saito, K. *Bull. Chem. Soc. Jpn.* **1960**, *33*, 1329.
[63] Forrest, J. *J. Chem. Soc.* **1960**, 574.
[64] Cornforth, J.; Sierakowski, A. F.; Wallace, T. W. *J. Chem. Soc., Chem. Commun.* **1979**, 294.
[65] Salfeld, J. C.; Baume, E. *Tetrahedron Lett.* **1966**, 3365.
[66] Bacon, R. G. R.; Pande, S. G. *J. Chem. Soc. (C)* **1970**, 1967.
[67] Cohen, T.; Tirpak, J. G. *Tetrahedron Lett.* **1975**, 143.
[68] Cohen, T.; Cristea, I. *J. Am. Chem. Soc.* **1976**, *98*, 748.
[69] Cohen, T.; Tirpak, J. G. *Tetrahedron Lett.* **1975**, 143.
[70] Cornforth, J.; Kumar, A.; Stuart, A. S. *J. Chem. Soc., Perkin Trans. 1* **1987**, 859.
[71] Zhang, S.; Zhang, D.; Liebeskind, L. S. *J. Org. Chem.* **1997**, *62*, 2312.
[72] Kang, S.-K.; Kim, T.-H.; Pyun, S.-J. *J. Chem. Soc., Perkin Trans 1* **1997**, 797.
[73] Sharma, V. N.; Dutt, S. *J. Indian Chem. Soc.* **1935**, *12*, 774; *Chem. Abstr.* **1936**, *30*, 31624.
[74] Ranu, B. C.; Dutta, P.; Sarkar, A. *Tetrahedron Lett.* **1998**, *39*, 9557.
[75] Negrel, J. C.; Gony, M.; Chanon, M.; Lai, R. *Inorg. Chim. Acta* **1993**, *207*, 59.
[76] Rapson, W. S.; Shuttleworth, R. G. *Nature* **1941**, *147*, 675.
[77] Bell, F.; Morgan, W. H. D. *J. Chem. Soc.* **1954**, 1716.
[78] Nursten, H. E. *J. Chem. Soc.* **1955**, 3081.
[79] Xi, M.; Bent, B. E. *J. Am. Chem. Soc.* **1993**, *115*, 7426.
[80] Meyers, J. M.; Gellman, A. J. *Surf. Sci.* **1995**, *337*, 40.
[81] Douglass, S. E.; Massey, S. T.; Woolard, S. G.; Zoellner, R. W. *Transition Met. Chem.* **1990**, *15*, 317.
[82] Xi, M.; Bent, B. E. *Surf. Sci.* **1992**, *278*, 19.
[83] Cohen, T.; Cristea, I. *J. Org. Chem.* **1975**, *40*, 3649.
[84] Cohen, T.; Poeth, T. *J. Am. Chem. Soc.* **1972**, *94*, 4363.
[85] Ebert, G. W.; Cheasty, J. W.; Tehrani, S. S.; Aouad, E. *Organometallics* **1992**, *11*, 1560.
[86] Lewin, A. H.; Cohen, T. *Tetrahedron Lett.* **1965**, 4531.
[87] Nilsson, M.; Wennerström, O. *Tetrahedron Lett.* **1968**, 3307.
[88] Costa, G.; Camus, A.; Marsich, N.; Gatti, L. *J. Organomet. Chem.* **1967**, *8*, 339.
[89] Posner, G. H. *Org. React.* **1975**, *22*, 253.
[90] Lipshutz, B. H.; Sengupta, S. *Org. React.* **1992**, *41*, 135.
[91] Kauffmann, T. *Angew. Chem., Int. Ed. Engl.* **1974**, *13*, 291.
[92] Baxter, P.; Lehn, J.-M.; DeCian, A.; Fischer, J. *Angew. Chem., Int. Ed. Engl.* **1993**, *32*, 69.
[93] Garber, T.; Rillema, D. P. *Synth. Commun.* **1990**, *20*, 1233.
[94] Butler, I. R.; Soucy-Breau, C. *Can. J. Chem.* **1991**, *69*, 1117.
[95] Kishii, N.; Araki, K.; Shiraishi, S. *Bull. Chem. Soc. Jpn.* **1984**, *57*, 2121.
[96] Weber, E.; Josel, H.-P.; Puff, H.; Franken, S. *J. Org. Chem.* **1985**, *50*, 3125.
[97] Lipshutz, B. H.; Siegmann, K.; Garcia, E. *J. Am. Chem. Soc.* **1991**, *113*, 8161.
[98] Lipshutz, B. H.; Siegmann, K.; Garcia, E.; Kayser, F. *J. Am. Chem. Soc.* **1993**, *115*, 9276.

[99] Lipshutz, B. H.; Siegmann, K.; Garcia, E. *Tetrahedron* **1992**, *48*, 2579.
[100] Coleman, R. S.; Grant, E. B. *Tetrahedron Lett.* **1993**, *34*, 2225.
[101] Lipshutz, B. H.; Kayser, F.; Maullin, N. *Tetrahedron Lett.* **1994**, *35*, 815.
[102] Lipshutz, B. H.; Kayser, F.; Liu, Z.-P. *Angew. Chem., Int. Ed. Engl.* **1994**, *33*, 1842.
[103] Sugimura, T.; Yamada, H.; Inoue, S.; Tai, A. *Tetrahedron: Asymmetry* **1997**, *8*, 649.
[104] Lin, G.-Q.; Zhong, M. *Tetrahedron Lett.* **1997**, *38*, 1087.
[105] Lipshutz, B. H.; Liu, Z.-P.; Kayser, F. *Tetrahedron Lett.* **1994**, *35*, 5567.
[106] Lin, G.-Q.; Zhong, M. *Tetrahedron: Asymmetry* **1997**, *8*, 1369.
[107] Green, J. In *The Chemistry of Halides, Pseudo-halides, and Azides, Supplement D2*; Patai, S.; Rappoport, Z., Eds.; Wiley: New York, 1995; Part 2, Chapter 22, p. 1175.
[108] Semmelhack, M. F. *Org. React.* **1972**, *19*, 115.
[109] Bogdanovic, B.; Kroner, M.; Wilke, G. *Liebigs Ann. Chem.* **1966**, *699*, 1.
[110] Massicot, F.; Schneider, R.; Fort, Y. *J. Chem. Res. (S)* **1999**, 664.
[111] Matsumoto, H.; Inaba, S.-i.; Rieke, R. D. *J. Org. Chem.* **1983**, *48*, 840.
[112] Inaba, S.-i.; Matsumoto, H.; Rieke, R. D. *Tetrahedron Lett.* **1982**, *23*, 4215.
[113] Massicot, F.; Schneider, R.; Fort, Y.; Illy-Cherrey, S.; Tillement, O. *Tetrahedron* **2001**, *57*, 531.
[114] Marceau, P.; Beguin, F.; Guillaumet, G. *J. Organomet. Chem.* **1988**, *342*, 137.
[115] Lipshutz, B. H.; Tasler, S. *Adv. Synth. Catal.* **2001**, *343*, 327.
[116] Lipshutz, B. H. *Adv. Synth. Catal.* **2001**, *343*, 313.
[117] Kellogg, R. M. *Chemtracts* **2000**, *13*, 69.
[118] Lipshutz, B. H.; Blomgren, P. A. *J. Am. Chem. Soc.* **1999**, *121*, 5819.
[119] Lipshutz, B. H.; Scalafani, J. A.; Blomgren, P. A. *Tetrahedron* **2000**, *56*, 2139.
[120] Chao, C. S.; Cheng, C. H.; Chang, C. T. *J. Org. Chem.* **1983**, *48*, 4904.
[121] Yasuhara, A.; Kasano, A.; Sakamoto, T. *Organometallics* **1998**, *17*, 4754.
[122] Kende, A. S.; Liebeskind, L. S.; Braitsch, D. M. *Tetrahedron Lett.* **1975**, 3375.
[123] Zembayashi, M.; Tamao, K.; Yoshida, J.-i.; Kumada, M. *Tetrahedron Lett.* **1977**, 4089.
[124] Rollin, Y.; Troupel, M.; Perichon, J.; Fauvarque, J. F. *J. Chem. Res. (S)* **1981**, 322.
[125] Amatore, C.; Jutand, A. *Organometallics* **1988**, *7*, 2203.
[126] Bontempelli, G.; Fiorani, M. *Ann. Chim. (Rome)* **1985**, *75*, 303.
[127] Mori, M.; Hashimoto, Y.; Ban, Y. *Tetrahedron Lett.* **1980**, 631.
[128] Rollin, Y.; Troupel, M.; Tuck, D. G.; Perichon, J. *J. Organomet. Chem.* **1986**, *303*, 131.
[129] Takagi, K.; Hayama, N.; Inokawa, S. *Bull. Chem. Soc. Jpn.* **1980**, *53*, 3691.
[130] Puckette, T. A. U.S. Patent 4,939,309 (1990); *Chem. Abstr.* **1990**, *113*, 171646.
[131] Vanderesse, R.; Brunet, J.-J.; Caubère, P. *J. Organomet. Chem.* **1984**, *264*, 263.
[132] Knapp, S.; Albaneze, J.; Schugar, H. J. *J. Org. Chem.* **1993**, *58*, 997.
[133] Percec, V.; Bae, J.-Y.; Zhao, M.; Hill, D. H. *J. Org. Chem.* **1995**, *60*, 176.
[134] Percec, V.; Bae, J.-Y.; Zhao, M.; Hill, D. H. *J. Org. Chem.* **1995**, *60*, 1066.
[135] Yamashita, J.; Inoue, Y.; Kondo, T.; Hashimoto, H. *Chem. Lett.* **1986**, 407.
[136] Inoue, Y.; Yamashita, J.; Kondo, T.; Hashimoto, H. *Nippon Kagaku Kaishi* **1987**, 197; *Chem. Abstr.* **1987**, *107*, 197686k.
[137] Eilingsfeld, H.; Patsch, M.; Siegel, B. German Patent 3,941,494 A1 (1990); *Chem. Abstr.* **1991**, *114*, 23548.
[138] Semmelhack, M. F.; Ryono, L. S. *J. Am. Chem. Soc.* **1975**, *97*, 3875.
[139] Iyoda, M.; Otsuka, H.; Sato, K.; Nisato, N.; Oda, M. *Bull. Chem. Soc. Jpn.* **1990**, *63*, 80.
[140] Naumann, C.; Langhals, H. *Synthesis* **1990**, 279.
[141] Negishi, E.-I.; King, A. O.; Okukado, N. *J. Org. Chem.* **1977**, *42*, 1821.
[142] Tiecco, M.; Testaferri, L.; Tingoli, M.; Chianelli, D.; Montanucci, M. *Synthesis* **1984**, 736.
[143] Nasielski, J.; Standaert, A.; Nasielski-Hinkens, R. *Synth. Commun.* **1991**, *21*, 901.
[144] Tiecco, M.; Tingoli, M.; Testaferri, L.; Chianelli, D.; Wenkert, E. *Tetrahedron* **1986**, *42*, 1475.
[145] Leadbeater, N. E.; Resouly, S. M. *Tetrahedron Lett.* **1999**, *40*, 4243.
[146] Adonin, N. Y.; Ryabinin, V. A.; Starichenko, V. F. *Russ. J. Org. Chem.* **1998**, *34*, 286; *Chem. Abstr.* **1998**, *130*, 3636.
[147] Colon, I.; Maresca, L. M.; Kwiatkowski, G. T. U.S. Patent 4,326,989 (1981); *Chem. Abstr.* **1981**, *95*, 80437.

[148] Colon, I.; Maresca, L. M.; Kwiatkowski, G. T. U.S. Patent 4,263,466 (1981); *Chem. Abstr.* **1981**, *95*, 80437.
[149] Wang, Y.; Marrocco, M. L.; Trimmer, M. S. Intl. Patent WO 96/39455 (1996); *Chem. Abstr.* **1997**, *126*, 104556.
[150] Troupel, M.; Rollin, Y.; Sibille, S.; Fauvarque, J.-F.; Perichon, J. *J. Chem. Res. (S)* **1980**, 26.
[151] Troupel, M.; Rollin, Y.; Sibille, S.; Perichon, J.; Fauvarque, J.-F. *J. Organomet. Chem.* **1980**, *202*, 435.
[152] Troupel, M. *Ann. Chim. (Rome)* **1986**, *76*, 151.
[153] Courtois, V.; Barhdadi, R.; Condon, S.; Troupel, M. *Tetrahedron Lett.* **1999**, *40*, 5993.
[154] Fox, M. A.; Chandler, D. A.; Lee, C. *J. Org. Chem.* **1991**, *56*, 3246.
[155] Amatore, C.; Jutand, A. *Acta Chem. Scand.* **1990**, *44*, 755.
[156] Tsou, T. T.; Kochi, J. K. *J. Am. Chem. Soc.* **1979**, *101*, 7547.
[157] Yamamoto, T.; Wakabayashi, S.; Osakada, K. *J. Organomet. Chem.* **1992**, *428*, 223.
[158] Meyer, G.; Rollin, Y.; Perichon, J. *J. Organomet. Chem.* **1987**, *333*, 263.
[159] Schiavon, G.; Bontempelli, G.; Corain, B. *J. Chem. Soc., Dalton Trans.* **1981**, 1074.
[160] Tsou, T. T.; Kochi, J. K. *J. Am. Chem. Soc.* **1979**, *101*, 6319.
[161] Foà, M.; Cassar, L. *J. Chem. Soc., Dalton Trans.* **1975**, 2572.
[162] Durandetti, M.; Devaud, M.; Perichon, J. *New J. Chem.* **1996**, *20*, 659.
[163] Uyeda, K. *Yakugaku Zasshi* **1931**, *51*, 495; *Chem. Abstr.* **1931**, *2*, 5427.
[164] Nakajima, R.; Iida, H.; Hara, T. *Bull. Chem. Soc. Jpn.* **1990**, *63*, 636.
[165] Nakajima, R.; Kinosada, M.; Tamura, T.; Hara, T. *Bull. Chem. Soc. Jpn.* **1983**, *56*, 1113.
[166] Garves, K. *J. Org. Chem.* **1970**, *35*, 3273.
[167] Selke, R.; Thiele, W. *J. Prakt. Chem.* **1971**, *313*, 875.
[168] Miura, M.; Hashimoto, H.; Itoh, K.; Nomura, M. *Chem. Lett.* **1990**, 459.
[169] Clark, F. R. S.; Norman, R. O. C.; Thomas, C. B. *J. Chem. Soc., Perkin Trans. 1* **1975**, 121.
[170] Edo, K.; Sakamoto, T.; Yamanaka, H. *Chem. Pharm. Bull.* **1979**, *27*, 193.
[171] Heitz, W. German Patent 3,842,622 (1990); *Chem. Abstr.* **1990**, *113*, 77892.
[172] Brenda, M.; Knebelkamp, A.; Greiner, A.; Heitz, W. *Synlett* **1991**, 809.
[173] Dyker, G. *J. Org. Chem.* **1993**, *58*, 234.
[174] Penalva, V.; Hassan, J.; Lavenot, L.; Gozzi, C.; Lemaire, M. *Tetrahedron Lett.* **1998**, *39*, 2559.
[175] Hassan, J.; Penalva, V.; Lavenot, L.; Gozzi, C.; Lemaire, M. *Tetrahedron* **1998**, *54*, 13793.
[176] Penalva, V.; Lavenot, L.; Gozzi, C.; Lemaire, M. *Applied Cat. A: General* **1999**, *182*, 399; *Chem. Abstr.* **1999**, *131*, 73252.
[177] Yasui, S.; Nakamura, K.; Fujii, M.; Ohno, A. *J. Org. Chem.* **1985**, *50*, 3283.
[178] Hennings, D. D.; Iwama, T.; Rawal, V. H. *Org. Lett.* **1999**, *1*, 1205.
[179] Bamfield, P.; Quan, P. M. *Synthesis* **1978**, 537.
[180] Mukhopadhyay, S.; Rothenberg, G.; Gitis, D.; Sasson, Y. *Org. Lett.* **2000**, *2*, 211.
[181] Mukhopadhyay, S.; Rothenberg, G.; Gitis, D.; Wiener, H.; Sasson, Y. *J. Chem. Soc., Perkin Trans. 2* **1999**, 2481.
[182] Schach, T.; Papenfuhs, T.; Hackenbruch, J. German Patent 4,0341,109 A1 (1992); *Chem. Abstr.* **1992**, *117*, 26060.
[183] Wang, Z.; Reibenspies, J.; Motekaitis, R. J.; Martell, A. E. *J. Chem. Soc., Dalton Trans.* **1995**, 1511.
[184] Kitai, M.; Katsuro, Y.; Kawamura, S.; Hino, M.; Sato, K. U. S. Patent 4,900,843 (1990); *Chem. Abstr.* **1989**, *111*, 232307.
[185] Venkatraman, S.; Li, C.-J. *Org. Lett.* **1999**, *1*, 1133.
[186a] Mukhopadhyay, S.; Rothenberg, G.; Wiener, H.; Sasson, Y. *Tetrahedron* **1999**, *55*, 14763.
[186b] Mukhopadhyay, S.; Rothenberg, G.; Gitis, D.; Sasson, Y. *J. Org. Chem.* **2000**, *65*, 3107.
[187] Mukhopadhyay, S.; Rothenberg, G.; Sasson, Y. *Adv. Synth. Catal.* **2001**, *343*, 274.
[188] Bamfield, P.; Quan, P. M. U. S. Patent 4,022,795 (1977); *Chem. Abstr.* **1976**, *84*, 164376.
[189] Sato, K.; Takewaki, T.; Katsuro, Y. U. S. Patent 5,095,144 (1992); *Chem. Abstr.* **1991**, *115*, 49107.
[190] Torii, S.; Tanaka, H.; Morisaki, K. *Tetrahedron Lett.* **1985**, *26*, 1655.
[191] Jutand, A.; Négri, S.; Mosleh, A. *J. Chem. Soc., Chem. Commun.* **1992**, 1729.
[192] Jutand, A.; Mosleh, A. *J. Org. Chem.* **1997**, *62*, 261.
[193] Jutand, A.; Mosleh, A. *Synlett* **1993**, 568.
[194] Luo, F.-T.; Jeevanandam, A.; Basu, M. K. *Tetrahedron Lett.* **1998**, *39*, 7939.

[195] Dyker, G.; Kellner, A. *J. Organomet. Chem.* **1998**, *555*, 141.
[196] Hatanaka, Y.; Goda, K.-i.; Okahara, Y.; Hiyama, T. *Tetrahedron* **1994**, *50*, 8301.
[197] Mowery, M. E.; DeShong, P. *J. Org. Chem.* **1999**, *64*, 1684.
[198] Albanese, D.; Landini, D.; Penso, M.; Petricci, S. *Synlett* **1999**, 199.
[199] Amatore, C.; Carré, E.; Jutand, A.; Tanaka, H.; Ren, Q.; Torii, S. *Chem. Eur. J.* **1996**, *2*, 957.
[200] Amatore, C.; Jutand, A. *J. Organomet. Chem.* **1999**, *576*, 254.
[201] Amatore, C.; Jutand, A. *Acc. Chem. Res.* **2000**, *2000*, 314.
[202] Yamamoto, A.; Kayaki, Y.; Nagayama, K.; Shimizu, I. *Synlett* **2000**, 925.
[203] Grushin, V. V.; Alper, H. *Chem. Rev.* **1994**, *94*, 1047.
[204] Jutand, A.; Mosleh, A. *Organometallics* **1995**, *14*, 1810.
[205] Miyano, S.; Tobita, M.; Hashimoto, H. *Bull. Chem. Soc. Jpn.* **1981**, *54*, 3522.
[206] Miyano, S.; Handa, S.; Shimizu, K.; Tagami, K.; Hashimoto, H. *Bull. Chem. Soc. Jpn.* **1984**, *57*, 1943.
[207] Miyano, S.; Fukushima, H.; Handa, S.; Ito, H.; Hashimoto, H. *Bull. Chem. Soc. Jpn.* **1988**, *61*, 3249.
[208] Dai, D.; Martin, O. R. *J. Org. Chem.* **1998**, *63*, 7628.
[209] Rawal, V. H.; Florjancic, A. S.; Singh, S. P. *Tetrahedron Lett.* **1994**, *35*, 8985.
[210] Rosini, C.; Superchi, S.; Bianco, G.; Mecca, T. *Chirality* **2000**, *12*, 256.
[211] Eliel, E. L.; Wilen, S. H. *Stereochemistry of Organic Compounds*; Wiley: New York, 1994, pp. 1119–1190.
[212] Andrus, M. B.; Asgari, D.; Sclafani, J. A. *J. Org. Chem.* **1997**, *62*, 9365.
[213] Miyano, S.; Tobita, M.; Suzuki, S.; Nishikawa, Y.; Hashimoto, H. *Chem. Lett.* **1980**, 1027.
[214] Nelson, T. D.; Meyers, A. I. *Tetrahedron Lett.* **1993**, *34*, 3061.
[215] Nelson, T. D.; Meyers, A. I. *Tetrahedron Lett.* **1994**, *35*, 3259.
[216] Nelson, T. D.; Meyers, A. I. *J. Org. Chem.* **1994**, *59*, 2655.
[217] Nelson, T. D.; Meyers, A. I. *J. Org. Chem.* **1994**, *59*, 7184.
[218] Nelson, T. D.; Meyers, A. I. *J. Org. Chem.* **1994**, *59*, 2577.
[219] Ebbers, E. J.; Ariaans, G. J. A.; Houbiers, J. P. M.; Bruggink, A.; Zwanenburg, B. *Tetrahedron* **1997**, *53*, 9417.
[220] Meyers, A. I.; Price, A. *J. Org. Chem.* **1998**, *63*, 412.
[221] Imai, Y.; Zhang, W.; Kida, T.; Nakatsuji, Y.; Ikeda, I. *Tetrahedron Lett.* **1997**, *38*, 2681.
[222] Tuyet, T. M. T.; Harada, T.; Hashimoto, K.; Hatsuda, M.; Oku, A. *J. Org. Chem.* **2000**, *65*, 1335.
[223] Koike, N.; Hattori, T.; Miyano, S. *Tetrahedron: Asymme*try **1994**, *5*, 1899.
[224] Degnan, A. P.; Meyers, A. I. *J. Am. Chem. Soc.* **1999**, *121*, 2762.
[225] Solladié, G.; Hugelé, P.; Bartsch, R.; Skoulios, A. *Angew. Chem., Int. Ed. Engl.* **1996**, *35*, 1533.
[226] Solladié, G.; Hugelé, P.; Bartsch, R. *J. Org. Chem.* **1998**, *63*, 3895.
[227] Meyers, A. I.; Willemsen, J. J. *Tetrahedron* **1998**, *54*, 10493.
[228] Meyers, A. I.; Willemsen, J. J. *Chem. Commun.* **1997**, 1573.
[229] Meyers, A. I.; Willemsen, J. J. *Tetrahedron Lett.* **1996**, *37*, 791.
[230] Lin, G.-Q.; Zhong, M. *Tetrahedron Lett.* **1996**, *37*, 3015.
[231] Meyers, A. I.; McKennon, M. J. *Tetrahedron Lett.* **1995**, *36*, 5869.
[232] Ried, W.; Freitag, D. *Angew. Chem., Int. Ed. Engl.* **1968**, *7*, 835.
[233] Chaturvedi, V.; Tanaka, S.; Kaeriyama, K. *Macromolecules* **1993**, *26*, 2607.
[234] Speight, J. G.; Kovacic, P.; Koch, F. W. *J. Macromol. Sci. - Rev. M.* **1971**, *6*, 295.
[235] Percec, V.; Okita, S.; Bae, V. *Polym. Bull.* **1992**, *29*, 271.
[236] Gehm, R. *Acta Chem. Scand.* **1951**, *5*, 270.
[237] Claesson, S.; Gehm, R.; Kern, W. *Makromol. Chem.* **1951**, *7*, 46.
[238] Wirth, H. O.; Müller, R.; Kern, W. *Makromol. Chem.* **1964**, *77*, 90.
[239] Braun, D.; Lehmann, P. *Makromol. Chem.* **1976**, *177*, 2221.
[240] Braun, D.; Lehmann, P. *Makromol. Chem.* **1976**, *177*, 1673.
[241] Groenendaal, L.; Peerlings, H. W. I.; van Dongen, J. L. J.; Havinga, E. E.; Vekemans, J. A. J. M.; Meijer, E. W. *Macromolecules* **1995**, *28*, 116.
[242] Pomerantz, M.; Yang, H.; Cheng, Y. *Macromolecules* **1995**, *28*, 5706.
[243] Brockmann, T. W.; Tour, J. M. *J. Am. Chem. Soc.* **1994**, *116*, 7435.
[244] Krigbaum, W. R.; Krause, K. J. *J. Polym. Science: Polym. Chem. Ed.* **1978**, *16*, 3151.
[245] Pomerantz, M.; Cheng, Y.; Kasim, R. K.; Elsenbaumer, R. L. *Synthetic Metals* **1997**, *85*, 1235.

[246] Yamamoto, T.; Morita, A.; Miyazaki, Y.; Maruyama, T.; Wakayama, H.; Zhou, Z.-h.; Nakamura, Y.; Kanbara, T.; Sasaki, S.; Kubota, K. *Macromolecules* **1992**, *25*, 1214.
[247] Kreyenschmidt, M.; Uckert, F.; Müllen, K. *Macromolecules* **1995**, *28*, 4577.
[248] Saito, N.; Kanbara, T.; Sato, T.; Yamamoto, T. *Polym. Bull.* **1993**, *30*, 285.
[249] Miyazaki, Y.; Kanbara, T.; Osakada, K.; Yamamoto, T.; Kubota, K. *Polym. J.* **1994**, *26*, 509.
[250] Yamamoto, T.; Kashiwazaki, A.; Kato, K. *Makromol. Chem.* **1989**, *190*, 1649.
[251] Fauvarque, J.-F.; Petit, M.-A.; Pfluger, F.; Jutand, A.; Chevrot, C.; Troupel, M. *Makromol. Chem., Rapid Commun.* **1983**, *4*, 455.
[252] Percec, V.; Hill, D. H. In *ACS Symposium Series*; Hendrick, J. L., Labadie, J. W., Eds.; American Chemical Society: Washington, DC, 1996; Vol. 624, p. 2.
[253] Colon, I.; Kwiatkowski, G. T. *J. Polym. Science: Part A: Polym. Chem.* **1990**, *28*, 367.
[254] Colon, I. U.S. Patent 4,400,499 (1983); *Chem. Abstr.* **1984**, *100*, 35026.
[255] Wang, Y.; Quirk, R. P. *Macromolecules* **1995**, *28*, 3495.
[256] Chaturvedi, V.; Tanaka, S.; Kaeriyama, K. *J. Chem. Soc., Chem. Commun.* **1992**, 1658.
[257] Ueda, M.; Seino, Y.; Sugiyama, J.-i. *Polym. J.* **1993**, *25*, 1319.
[258] Marrocco III, M. L.; Gagné, R. R.; Trimmer, M. S. U.S. Patent 5,227,457 (1993); *Chem. Abstr.* **1991**, *115*, 160064.
[259] Percec, V.; Okita, S. *J. Polym. Science: Part A: Polym. Chem.* **1993**, *31*, 1087.
[260] Percec, V.; Okita, S. *J. Polym. Science: Part A: Polym. Chem.* **1993**, *31*, 877.
[261] Percec, V.; Okita, S.; Weiss, R. *Macromolecules* **1992**, *25*, 1816.
[262] Percec, V. U.S. Patent 5,241,044 (1993); *Chem. Abstr.* **1994**, *120*, 108108.
[263] Percec, V.; Bae, J.-Y.; Zhao, M.; Hill, D. H. *Macromolecules* **1995**, *28*, 6726.
[264] Stanforth, S. P. *Tetrahedron* **1998**, *54*, 263.
[264a] Hassan, J.; Sévignon, M.; Gozzi, C.; Schulz, E.; Lemaire, M. *Chem. Rev.* **2002**, *102*, 1359.
[265] Bachmann, W. E.; Hoffman, R. A. *Org. React.* **1944**, *2*, 224.
[266] DeTar, D. F. *Org. React.* **1957**, *9*, 409.
[267] Atkinson, E. R.; Lawler, H. J.; Heath, J. C.; Kimball, E. H.; Read, E. R. *J. Am. Chem. Soc.* **1941**, *63*, 730.
[268] Beadle, J. R.; Korzeniowski, S. H.; Rosenberg, D. E.; Garcia-Slanga, B. J.; Gokel, G. W. *J. Org. Chem.* **1984**, *49*, 1594.
[269] Lothrop, W. C. *J. Am. Chem. Soc.* **1942**, *64*, 1698.
[270] Uchiyama, M.; Suzuki, T.; Yamazaki, Y. *Chem. Lett.* **1983**, 1165.
[271] Tamura, Y.; Chun, M.-W.; Inoue, K.; Minamikawa, J. *Synthesis* **1978**, 822.
[272] Kang, S.-K.; Shivkumar, U.; Ahn, C.; Choi, S.-C.; Kim, J.-S. *Synth. Commun.* **1997**, *27*, 1893.
[273] Wolf, W.; Kharasch, N. *J. Org. Chem.* **1965**, *30*, 2493.
[274] Taylor, E. C.; Kienzle, F.; McKillop, A. *J. Am. Chem. Soc.* **1970**, *92*, 6088.
[275] Pummerer, R.; Seligsberger, L. *Chem. Ber.* **1931**, *64*, 2477.
[276] Pummerer, R.; Bittner, K. *Chem. Ber.* **1924**, *57*, 84.
[277] Song, Y.; Gardner, P.; Conrad, H.; Bradshaw, A. M.; White, J. M. *Surf. Science Lett.* **1991**, *248*, L279.
[278] Zhou, X.-L.; White, J. M. *J. Chem. Phys.* **1990**, *92*, 5612.
[279] Bowden, S. T. *J. Chem. Soc.* **1931**, 1111.
[280] Jigajinni, V. B.; Wightman, R. H.; Campbell, M. M. *J. Chem. Soc., Chem. Commun.* **1981**, 87.
[281] Jigajinni, V. B.; Wightman, R. H.; Campbell, M. M. *J. Chem. Res. (M)* **1983**, 1801.
[282] Cameron, D. W.; Feutrill, G. I.; Pannan, L. J. H. *Aust. J. Chem.* **1980**, *33*, 2531.
[283] Cameron, D. W.; Feutrill, G. I.; Pannan, L. J. H.; Raston, C. L.; Skelton, B. W.; White, A. H. *J. Chem. Soc., Perkin Trans. 2* **1981**, 610.
[284] Iranpoor, N.; Shekarriz, M. *J. Chem. Res. (S)* **1999**, 442.
[285] Bell, N. V.; Bowman, W. R.; Coe, P. F.; Turner, A. T.; Whybrow, D. *Tetrahedron Lett.* **1997**, *38*, 2581.
[286] Osborne, A. G.; Glass, K. J.; Staley, M. L. *Tetrahedron Lett.* **1989**, *30*, 3567.
[287] Osborne, A. G.; Clifton, A. A. *Monatsh. Chem.* **1991**, *122*, 529.
[288] Uenishi, J.; Tanaka, T.; Wakabayashi, S.; Oae, S.; Tsukube, H. *Tetrahedron Lett.* **1990**, *31*, 4625.
[289] Kawai, T.; Furukawa, N. *Heterocycles* **1985**, *23*, 177.
[290] Schwartz, E. B.; Knobler, C. B.; Cram, D. J. *J. Am. Chem. Soc.* **1992**, *114*, 10775.
[291] Cade, J. A.; Pilbeam, A. *J. Chem. Soc.* **1964**, 114.

[292] Marcus, E.; Lauer, W. M.; Arnold, R. T. *J. Am. Chem. Soc.* **1958**, *80*, 3742.
[293] Shirley, D. A.; Dean, W. L. *J. Am. Chem. Soc.* **1957**, *79*, 1205.
[294] Jouaiti, A.; Geoffroy, M.; Bernardinelli, G. *Tetrahedron Lett.* **1993**, *34*, 3413.
[295] Gronowitz, S.; Karlsson, H.-O. *Ark. Kemi* **1960**, *17*, 89; *Chem. Abstr.* **1961**, *55*, 144082.
[296] Gronowitz, S.; Hagen, S. *Ark. Kemi* **1967**, *27*, 153; *Chem. Abstr.* **1967**, *67*, 99931.
[297] Ziegler, F. E.; Fowler, K. W.; Kanfer, S. *J. Am. Chem. Soc.* **1976**, *98*, 8282.
[298] Ziegler, F. E.; Chliwner, I.; Fowler, K. W.; Kanfer, S. J.; Kuo, S. J.; Sinha, N. D. *J. Am. Chem. Soc.* **1980**, *102*, 790.
[299] Ziegler, F. E.; Fowler, K. W.; Rodgers, W. B.; Wester, R. T. *Org. Synth. Coll. Vol. 8*, **1993**, 586.
[300] Gies, A.-E.; Pfeffer, M. *J. Org. Chem.* **1999**, *64*, 3650.
[301] Tamao, K.; Kodama, S.; Nakajima, I.; Kumada, M.; Minato, A.; Suzuki, K. *Tetrahedron* **1982**, *38*, 3347.
[302] Khor, E.; Ng, S. C.; Li, H. C.; Chai, S. *Heterocycles* **1991**, *32*, 1805.
[303] Yamamoto, T.; Hayashi, Y.; Yamamoto, A. *Bull. Chem. Soc. Jpn.* **1978**, *51*, 2091.
[304] Amer, A.; Burkhardt, A.; Nkansah, A.; Shabana, R.; Galal, A.; Mark, H. B.; Zimmer, H. *Phosphorus Sulfur Silicon Relat. Elem.* **1989**, *42*, 63.
[305] Tamao, K.; Minato, A.; Miyake, N.; Matsuda, T.; Kiso, Y.; Kumada, M. *Chem. Lett.* **1975**, 133.
[306] Tamao, K.; Yamamoto, H.; Matsumoto, H.; Miyake, N.; Hayashi, T.; Kumada, M. *Tetrahedron Lett.* **1977**, 1389.
[307] Hayashi, T.; Hayashizaki, K.; Kiyoi, T.; Ito, Y. *J. Am. Chem. Soc.* **1988**, *110*, 8153.
[308] Colletti, S. L.; Halterman, R. L. *Tetrahedron Lett.* **1989**, *30*, 3513.
[309] Johnson, D. K.; Ciavarri, J. P.; Ishmael, F. T.; Schillinger, K. J.; van Geel, T. A. P.; Stratton, S. M. *Tetrahedron Lett.* **1995**, *36*, 8565.
[310] Krizewsky, J.; Turner, E. E. *J. Chem. Soc.* **1919**, 559.
[311] Nishiyama, T.; Seshita, T.; Shodai, H.; Aoki, K.; Kameyama, H.; Komura, K. *Chem. Lett.* **1996**, 549.
[312] Taylor, S. K.; Bennett, S. G.; Heinz, K. J.; Lashley, L. K. *J. Org. Chem.* **1981**, *46*, 2194.
[313] Inoue, A.; Kitagawa, K.; Shinokubo, H.; Oshima, K. *Tetrahedron* **2000**, *56*, 9601.
[314] Clayden, J.; Cooney, J. J. A.; Julia, M. *J. Chem. Soc., Perkin Trans. 1* **1995**, 7.
[315] Knochel, P.; Singer, R. D. *Chem. Rev.* **1993**, *93*, 2117.
[316] Rieke, R. D. *Aldrichim. Acta* **2000**, *33*, 52.
[317] Negishi, E.-i.; King, A. O.; Okukado, N. *J. Org. Chem.* **1977**, *42*, 1821.
[318] Dai, C.; Fu, G. C. *J. Am. Chem. Soc.* **2001**, *123*, 2719.
[319] Gosmini, C.; Rollin, Y.; Nédélec, J. Y.; Périchon, J. *J. Org. Chem.* **2000**, *65*, 6024.
[320] Amat, M.; Hadida, S.; Bosh, J. *Tetrahedron Lett.* **1994**, *35*, 793.
[321] Uemura, M.; Nishimura, H.; Kamikawa, K.; Nakayama, K.; Hayashi, Y. *Tetrahedron Lett.* **1994**, *35*, 1909.
[322] Zhu, L.; Wehmeyer, R. M.; Rieke, R. D. *J. Org. Chem.* **1991**, *56*, 1445.
[323] Sibille, S.; Ratovelomanana, V.; Nédélec, J. Y.; Périchon, J. *Synlett* **1993**, 425.
[324] Marshall, J. S. *Chem. Rev.* **2000**, *100*, 3163.
[325] Vicente, J.; Bermúdez, M. D.; Escribano, J. *Organometallics* **1991**, *10*, 3380.
[326] Larock, R. C.; Bernhardt, J. C. *J. Org. Chem.* **1977**, *42*, 1680.
[327] Larock, R. C. *Angew. Chem., Int. Ed. Engl.* **1978**, *17*, 27.
[328] Kretchmer, R. A.; Glowinski, R. *J. Org. Chem.* **1976**, *41*, 2661.
[329] Rosa, P.; Mézailles, N.; Matheny, F.; Le Floch, P. *J. Org. Chem.* **1998**, *63*, 4826.
[330] Nichols, P. J.; Papadopoulos, S.; Raston, C. L. *Chem. Commun.* **2000**, 1227.
[331] Gouda, K.-i.; Hagiwara, E.; Hatanaka, Y.; Hiyama, T. *J. Org. Chem.* **1996**, *61*, 7232.
[332] Hagiwara, E.; Gouda, K.-i.; Hatanaka, Y.; Hiyama, T. *Tetrahedron Lett.* **1997**, *38*, 439.
[333] Farina, V.; Krishnamurthy, V.; Scott, W. J. *Org. React.* **1997**, *50*, 1.
[334] Roshchin, A. I.; Bumagin, N. A.; Beletskaya, I. P. *Tetrahedron Lett.* **1995**, *36*, 125.
[335] Rai, R.; Aubrecht, K. B.; Collum, D. B. *Tetrahedron Lett.* **1995**, *36*, 3111.
[336] Ritter, K. *Synthesis* **1993**, 735.
[337] Kikukawa, K.; Kono, K.; Wada, F.; Matsuda, T. *J. Org. Chem.* **1983**, *48*, 1333.
[338] Shirakawa, E.; Hiyama, T. *J. Organomet. Chem.* **1999**, *576*, 169.
[339] Tour, J. M. *Chem. Rev.* **1996**, *96*, 537.

[340] Larhed, M.; Lindeberg, G.; Hallberg, A. *Tetrahedron Lett.* **1996**, *37*, 8219.
[341] Majeed, A. J.; Antonsen, Ø.; Benneche, T.; Undheim, K. *Tetrahedron* **1989**, *45*, 993.
[342] Iyoda, M.; Kondo, T.; Nakao, K.; Hara, K.; Kuwatani, Y.; Yoshida, M.; Matsuyama, H. *Org. Lett.* **2000**, *2*, 2081.
[343] Kang, S.-K.; Baik, T.-G.; Jiao, X. H.; Lee, Y.-T. *Tetrahedron Lett.* **1999**, *40*, 2383.
[344] Miyaura, N.; Suzuki, A. *Chem. Rev.* **1995**, *95*, 2457.
[345] Suzuki, A. *J. Organomet. Chem.* **1999**, *576*, 147.
[346] Martin, A. R.; Yang, Y. *Acta Chem. Scand.* **1993**, *47*, 221.
[347] Gelman, D.; Schumann, H.; Blum, J. *Tetrahedron Lett.* **2000**, *41*, 7555.
[348] Littke, A. F.; Fu, G. C. *Angew. Chem., Int. Ed. Engl.* **1998**, *37*, 3387.
[349] Wolfe, J. P.; Buchwald, S. L. *Angew. Chem., Int. Ed. Engl.* **1999**, *38*, 2413.
[350] Zapf, A; Ehrentraut, A.; Beller, M. *Angew. Chem., Int. Ed. Engl.* **2000**, *39*, 4153.
[351] Stürmer, R. *Angew. Chem., Int. Ed. Engl.* **1999**, *38*, 3307.
[352] Old, D. W.; Wolfe, J. P.; Buchwald, S. L. *J. Am. Chem. Soc.* **1998**, *120*, 9722.
[353] Zhang, C.; Huang, J.; Trudell, M. L.; Nolan, S. *J. Org. Chem.* **1999**, *64*, 3804.
[354] Bohn, V. P. W.; Gstottmayr, C. W. K.; Weskamp, T.; Herrmann, W. A. *J. Organomet. Chem.* **2000**, *595*, 186.
[355] LeBlond, C. R.; Andrews, A. T.; Sun, Y.; Sowa, J. R. *Org. Lett.* **2001**, *3*, 1555.
[356] Percec, V.; Bae, J.-Y.; Hill, D. H. *J. Org. Chem.* **1995**, *60*, 1060.
[357] Kobayashi, Y.; Mizojiri, R. *Tetrahedron Lett.* **1996**, *37*, 8531.
[358] Wright, S. W.; Hageman, D. L.; McClure, L. D. *J. Org. Chem.* **1994**, *59*, 6095.
[359] Moreno-Mañas, M.; Perez, M.; Pleixats, R. *J. Org. Chem.* **1996**, *61*, 2346.
[360] Andersen, N. G.; Maddaford, S. P.; Keay, B. A. *J. Org. Chem.* **1996**, *61*, 9556.
[361] Kamikawa, K.; Uemura, M. *Synlett* **2000**, 938.
[362] Cammidge, A. N.; Crepy, K. V. L. *Chem. Commun.* **2000**, 1723.
[363] Whiting, D. A. In *Comprehensive Organic Synthesis*; Trost, B. M., Fleming, I., Eds.; Pergamon: Oxford, 1991; Vol. 3, p. 659.
[364] Kozhevnikov, I. V.; Matveev, K. I. *Russ. Chem. Rev.* **1978**, *47*, 649; *Chem. Abstr.* **1978**, *89*, 106962.
[365] Tanaka, M.; Mitsuhashi, H.; Wakamatsu, T. *Tetrahedron Lett.* **1992**, *33*, 4161.
[366] Toda, F.; Tanaka, K.; Iwata, S. *J. Org. Chem.* **1989**, *54*, 3007.
[367] Sartori, G.; Maggi, R.; Bigi, F.; Grandi, M. *J. Org. Chem.* **1993**, *58*, 7271.
[368] Noji, M.; Nakajima, M.; Koga, K. *Tetrahedron Lett.* **1994**, *35*, 7983.
[369] Dhal, R.; Landais, Y.; Lebrun, A.; Lenain, V.; Robin, J.-P. *Tetrahedron* **1994**, *50*, 1153.
[370] Dworkin, A. S.; Poutsma, M. L.; Brynestad, J.; Brown, L. L.; Gilpatrick, L. O.; Smith, G. P. *J. Am. Chem. Soc.* **1979**, *101*, 5299.
[371] Dewar, M. J. S.; Nakaya, T. *J. Am. Chem. Soc.* **1968**, *90*, 7134.
[372] Robin, J.-P.; Landais, Y. *Tetrahedron* **1992**, *48*, 819.
[373] Lipshutz, B. H.; James, B.; Vance, S.; Carrico, I. *Tetrahedron Lett.* **1997**, *38*, 753.
[374] Sakamoto, T.; Yonehara, H.; Pac, C. *J. Org. Chem.* **1994**, *59*, 6859.
[375] Kupchan, S. M.; Liepa, A. J. *J. Am. Chem. Soc.* **1973**, *95*, 4062.
[376] Schwartz, M. A.; Holton, R. A.; Scott, S. W. *J. Am. Chem. Soc.* **1969**, *91*, 2800.
[377] Marin, G. H.; Horak, V. *J. Org. Chem.* **1994**, *59*, 4267.
[378] Brussee, J.; Jansen, A. C. A. *Tetrahedron Lett.* **1983**, *24*, 3261.
[379] Brussee, J.; Groenendijk, J. L. G.; te Koppele, J. M.; Jansen, A. C. A. *Tetrahedron* **1985**, *41*, 3313.
[380] Smrcina, M.; Poláková, J.; Vyskocil, S.; Kocovsk, P. *J. Org. Chem.* **1993**, *58*, 4534.
[381] Yamamoto, K.; Noda, K.; Okamoto, Y. *J. Chem. Soc., Chem. Commun.* **1985**, 1865.
[382] Li, X.; Yang, J.; Kozlowski, M. C. *Org. Lett.* **2001**, *3*, 1137.
[383] Feringa, B.; Wynberg, H. *J. Org. Chem.* **1981**, *46*, 2547.
[384] Osa, T.; Kashiwagi, Y.; Yanagisawa, Y.; Bobbitt, J. M. *J. Chem. Soc., Chem. Commun.* **1994**, 2535.
[385] Unger, M. O.; Fouty, R. A. *J. Org. Chem.* **1969**, *34*, 18.
[386] Clark, F. R. S.; Norman, R. O. C.; Thomas, C. B.; Willson, J. S. *J. Chem. Soc., Perkin Trans. 1* **1974**, 1289.
[387] Iataaki, H.; Yoshimoto, H. *J. Org. Chem.* **1973**, *38*, 76.
[388] Yatsimirsky, A. K.; Deiko, S. A.; Ryabov, A. D. *Tetrahedron* **1983**, *39*, 2381.

[389] Sévignon, M.; Papillon, J.; Schulz, E.; Lemaire, M. *Tetrahedron Lett.* **1999**, *40*, 5873.
[390] McKillop, A.; Elsom, L. F.; Taylor, E. C. *Tetrahedron* **1970**, *26*, 4041.
[391] Tormo, J.; Moreno, F. J.; Ruiz, J.; Fajarí, L.; Juliá, L. *J. Org. Chem.* **1997**, *62*, 878.
[392] Wen, L.-S.; Zawalski, R. C.; Kovacic, P. *J. Org. Chem.* **1978**, *43*, 2435.
[393] Albrecht, M.; Riether, C. *Chem. Ber.* **1996**, *129*, 829.
[394] Sone, T.; Sato, K.; Umetsu, Y.; Yoshino, A.; Takahashi, K. *Bull. Chem. Soc. Jpn.* **1994**, *67*, 2187.
[395] Tanaka, M.; Nakashima, H.; Fujiwara, M.; Ando, H.; Souma, Y. *J. Org. Chem.* **1996**, *61*, 788.
[396] Kaczmarek, L.; Nowak, B.; Zukowski, J.; Borowicz, P.; Sepiol, J.; Grabowska, A. *J. Mol. Struct.* **1991**, *248*, 189.
[397] Vondenhof, M.; Mattay, J. *Chem. Ber.* **1990**, *123*, 2457.
[398] Fort, Y.; Becker, S.; Caubère, P. *Tetrahedron* **1994**, *50*, 11893.
[399] Sone, T.; Umetsu, Y.; Sato, K. *Bull. Chem. Soc. Jpn.* **1991**, *64*, 864.
[400] Fort, Y. *Tetrahedron Lett.* **1995**, *36*, 6051.
[401] Takagi, K.; Hayama, N.; Sasaki, K. *Bull. Chem. Soc. Jpn.* **1984**, *57*, 1887.
[402] Osakada, K.; Sato, R.; Yamamoto, T. *Organometallics* **1994**, *13*, 4645.
[403] Takagi, K.; Hayama, N.; Inokawa, S. *Chem. Lett.* **1979**, 917.
[404] Iyoda, M.; Sato, K.; Oda, M. *Tetrahedron Lett.* **1985**, *26*, 3829.
[405] Budnikova, Y. G.; Kargin, Y. M.; Yanilkin, V. V. *Izv. Akad. Nauk. Ser. Khim.* **1992**, 1674; *Bull. Russ. Acad. Sci. Div. Chem. Sci. (Engl. Transl.)* **1992**, *41*, 1299; *Chem. Abstr.* **1992**, *118*, 80774.
[406] Chen, D.-F.; Zhang, S.-X.; Xie, L.; Xie, J.-X.; Chen, K.; Kashiwada, Y.; Zhou, B.-N.; Wang, P.; Cosentino, L. M.; Lee, K.-H. *Bioorg. Med. Chem.* **1997**, *5*, 1715.
[407] Rodkevich, N. G.; Il'in, A. P. *Russ. J. Org. Chem.* **1996**, *32*, 1706; *Chem. Abstr.* **1997**, *126*, 228479.
[408] Grigg, R.; Stevenson, P.; Worakun, T. *J. Chem. Soc., Chem. Commun.* **1985**, 971.
[409] Fukui, K.; Kitano, H.; Osaka, T.; Inamoto, Y.; Shioji, S. *Nippon Kagaku Zasshi* **1958**, *79*, 1120; *Chem. Abstr.* **1960**, *54*, 5518b.
[410] Hammann, W. C.; Schisla, R. M. *J. Chem. Eng. Data* **1972**, *17*, 112; *Chem. Abstr.* **1972**, *76*, 59106.
[411] Cirigottis, K. A.; Ritchie, E.; Taylor, W. C. *Aust. J. Chem.* **1974**, *27*, 2209.
[412] Mares, F.; Chvalovsky, V. *Collect. Czech. Chem. Commun.* **1967**, *32*, 382; *Chem. Abstr.* **1967**, *66*, 76074.
[413] Ibuki, E.; Ozasa, S.; Murai, K. *Bull. Chem. Soc. Jpn.* **1975**, *48*, 1868.
[414] Mosby, W. L. *J. Org. Chem.* **1957**, *22*, 671.
[415] Baker, W.; Boarland, M. P. V.; McOmie, J. F. W. *J. Chem. Soc.* **1954**, 1476.
[416] Shaw, F. R.; Turner, E. E. *J. Chem. Soc.* **1933**, 135.
[417] Schiemann, G.; Roselius, W. *Chem. Ber.* **1932**, *65*, 737.
[418] Davey, W.; Latter, R. W. *J. Chem. Soc.* **1948**, 264.
[419] Zahn, H.; Zuber, H. *Chem. Ber.* **1953**, *86*, 172.
[420] Cornforth, J.; Ridley, D. D.; Sierakowski, A. F.; Uguen, D.; Wallace, T. W.; Hitchcock, P. B. *J. Chem. Soc., Perkin Trans. 1* **1982**, 2317.
[421] Armarego, W. L. F.; Turner, E. E. *J. Chem. Soc.* **1956**, 1665.
[422] Barber, H. J.; Smiles, S. *J. Chem. Soc.* **1928**, 1141.
[423] Hall, D. M.; Lesslie, M. S.; Turner, E. E. *J. Chem. Soc.* **1950**, 711.
[424] Sakan, T.; Nakazaki, M. *Inst. Polytech. Osaka City Univ.* **1950**, *1*, 23; *Chem. Abstr.* **1952**, *46*, 5036b.
[425] Pettit, M. R.; Tatlow, J. C. *J. Chem. Soc.* **1954**, 1071.
[426] Reisch, H. A.; Enkelmann, V.; Scherf, U. *J. Org. Chem.* **1999**, *64*, 655.
[427] Rapson, W. S.; Shuttleworth, R. G. *J. Chem. Soc.* **1941**, 487.
[428] Bacon, R. G. R.; Lindsay, W. S. *J. Chem. Soc.* **1958**, 1375.
[429] Hurtley, W. R. H. *J. Chem. Soc.* **1929**, 1870.
[430] Brand, K.; Stallmann, O. *J. Prakt. Chem.* **1924**, *107*, 358.
[431] Everitt, P. M.; Hall, D. M.; Turner, E. E. *J. Chem. Soc.* **1956**, 2286.
[432] Bacon, R. G. R.; Lindsay, W. S. *J. Chem. Soc.* **1958**, 1382.
[433] King, F. D.; Walton, D. R. M. *Synthesis* **1976**, 40.
[434] Weitzenböck, R. *Monatsh. Chem.* **1913**, *34*, 193.
[435] Copeland, P. G.; Dean, R. E.; McNeil, D. *J. Chem. Soc.* **1960**, 4522.
[436] Bachmann, W. E.; Clarke, H. T. *J. Am. Chem. Soc.* **1927**, *49*, 2089.

[437] Cade, J. A.; Pilbeam, A. *Tetrahedron* **1964**, *20*, 519.
[438] Ozasa, S.; Fujioka, Y.; Tsukada, M.; Ibuki, E. *Chem. Pharm. Bull.* **1981**, *29*, 344.
[439] Chau, M. M.; Kice, J. L. *J. Org. Chem.* **1977**, *42*, 3265.
[440] Zhang, C.; Wang, Z. Y. *Macromolecules* **1993**, *26*, 3324.
[441] Bachmann, W. E.; Chu, E. J.-H. *J. Am. Chem. Soc.* **1935**, *57*, 1095.
[442] Sadler, A. M.; Powell, G. *J. Am. Chem. Soc.* **1934**, *56*, 2650.
[443] Gray, M.; Chapell, B. J.; Felding, J.; Taylor, N. J.; Snieckus, V. *Synlett* **1998**, 422.
[444] Roling, P. V.; Rausch, M. D. *J. Org. Chem.* **1972**, *37*, 729.
[445] Fuson, R. C.; Hornberger, C. *J. Org. Chem.* **1951**, *16*, 631.
[446] Fuson, R. C.; Kerr, R. O. *J. Org. Chem.* **1954**, *19*, 373.
[447] Ibuki, E.; Ozasa, S.; Fujioka, Y.; Kitamura, H. *Chem. Pharm. Bull.* **1980**, *28*, 1468.
[448] Desponds, O.; Schlosser, M. *J. Organomet. Chem.* **1996**, *507*, 257.
[449] Ozasa, S.; Fujioka, Y.; Fujiwara, M.; Ibuki, E. *Chem. Pharm. Bull.* **1980**, *28*, 3210.
[450] van Alphen, J. *Recl. Trav. Chim. Pays-Bas* **1932**, *51*, 361.
[451] Staab, H. A.; Binnig, F. *Chem. Ber.* **1967**, *100*, 293.
[452] Steinkopf, W.; Jaeger, P. *J. Prakt. Chem.* **1930**, *128*, 63.
[453] Freedman, L. D. *J. Am. Chem. Soc.* **1955**, *77*, 6223.
[454] Wellmar, U.; Hörnfeldt, A.-B.; Gronowitz, S. *J. Heterocycl. Chem.* **1996**, *33*, 409.
[455] Constable, E. C.; Hannon, M. J.; Edwards, A. J.; Raithby, P. R. *J. Chem. Soc., Dalton Trans.* **1994**, 2669.
[456] Baker, W.; Barton, J. W.; McOmie, J. F. W. *J. Chem. Soc.* **1958**, 2658.
[457] Whiting, D. A.; Wood, A. F. *J. Chem. Soc., Perkin Trans. 1* **1980**, 623.
[458] Ozeki, S. *Yakugaku Zasshi* **1965**, *85*, 206; *Chem. Abstr.* **1965**, *63*, 643b.
[459] Longmire, J. M.; Zhu, G.; Zhang, X. *Tetrahedron Lett.* **1997**, *38*, 375.
[460] Carlin, R. B.; Swakon, E. A. *J. Am. Chem. Soc.* **1955**, *77*, 966.
[461] Padmanabhan, S.; Gavaskar, K. V.; Triggle, D. J. *Synth. Commun.* **1996**, *26*, 3109.
[462] Schach, T.; Papenfuhs, T.; Hackenbruch, J. U. S. Patent 5,451,703 (1995); *Chem. Abstr.* **1992**, *117*, 26060.
[463] Kageyama, H.; Furusawa, O.; Kimura, Y. *Chem. Express* **1991**, *6*, 229; *Chem. Abstr.* **1991**, *114*, 228479w.
[464] Meyer, R.; Meyer, W.; Taeger, K. *Chem. Ber.* **1920**, *53*, 2034.
[465] Ueda, M.; Ito, T. *Polym. J.* **1991**, *23*, 297.
[466] Schreiner, E. *J. Prakt. Chem.* **1910**, *81*, 422.
[467] Finger, H.; Schott, W. *J. Prakt. Chem.* **1927**, *115*, 281.
[468] Kageyama, H.; Furusawa, O.; Kimura, Y. *Chem. Express* **1990**, *5*, 645; *Chem. Abstr.* **1991**, *114*, 228479.
[469] Fields, E. K.; Meyerson, S. *J. Org. Chem.* **1978**, *43*, 4705.
[470] Boedtker, M. E. *Bull. Soc. Chim. Belg.* **1929**, *45*, 645.
[471] Faid, K.; Siove, A.; Chevrot, C.; Riou, M. T.; Froyer, G. *J. Chim. Phys.* **1992**, *89*, 1305.
[472] Wibaut, J. P.; Overhoff, J.; Gratama, K. *Recl. Trav. Chim. Pays-Bays* **1940**, *59*, 298.
[473] Novikov, A. N.; Khalimova, T. A. *Tr. Tomskogo Gos. Univ., Ser. Khim.* **1964**, *170*, 45; *Chem. Abstr.* **1965**, *63*, 3124.
[474] Harley-Mason, J.; Mann, F. G. *J. Chem. Soc.* **1940**, 1379.
[475] Sybert, P. D.; Beever, W. H.; Stille, J. K. *Macromolecules* **1981**, *14*, 493.
[476] Balaban, A. T.; Birladeanu, L.; Bally, I.; Frangopol, P. T.; Mocanu, M.; Simon, Z. *Tetrahedron* **1963**, *19*, 2199.
[477] Dehne, H.; Zahnow, R.; Steinhagen, H. G. *Z. Chem.* **1971**, *11*, 305.
[478] Chaikovskii, V. K.; Novikov, A. N. *Zh. Org. Khim.* **1985**, *21*, 909; *J. Org. Chem. USSR (Engl. Translation)* **1985**, *21*, 827; *Chem. Abstr.* **1986**, *104*, 33809.
[479] Reiser, A.; Leyshon, L. J.; Saunders, D.; Mijovic, M. V.; Bright, A.; Bogie, J. *J. Am. Chem. Soc.* **1972**, *94*, 2414.
[480] Gross, U.; Kaufmann, D. *Chem. Ber.* **1987**, *120*, 991.
[481] Leupold, I.; Musso, H. *Liebigs Ann. Chem.* **1971**, *746*, 134.
[482] Newman, M. S.; Wiseman, E. H. *J. Org. Chem.* **1961**, *26*, 3208.

[483] Lounasmaa, M. *Acta Chem. Scand.* **1968**, *22*, 2388.
[484] Horner, L.; Weber, K.-H. *Chem. Ber.* **1963**, *96*, 1568.
[485] Pascal, R. A.; Ho, D. M. *Tetrahedron Lett.* **1992**, *33*, 13.
[486] Ozasa, S.; Fujioka, Y.; Kikutake, J.-I.; Ibuki, E. *Chem. Pharm. Bull.* **1983**, *31*, 1572.
[487] Litvinenko, L. M.; Grekov, A. P.; Verkhovod, N. N.; Dzyuba, V. P. *Zh. Obshch. Khim.* **1956**, *26*, 2524; *J. Gen. Chem. U.S.S.R.* **1956**, *26*, 2817; *Chem. Abstr.* **1957**, *51*, 25396.
[488] Shaw, F. R.; Turner, E. E. *J. Chem. Soc.* **1932**, 509.
[489] Corbett, J. F.; Holt, P. F. *J. Chem. Soc.* **1961**, 5029.
[490] Yamato, T.; Hideshima, C.; Suehiro, K.; Tashiro, M.; Prakash, G. K. S.; Olah, G. A. *J. Org. Chem.* **1991**, *56*, 6248.
[491] Carlin, R. B.; Foltz, G. E. *J. Am. Chem. Soc.* **1956**, *78*, 1997.
[492] Ross, S. D.; Kuntz, I. *J. Am. Chem. Soc.* **1952**, *74*, 1297.
[493] Bradsher, C. K.; Bond, J. B. *J. Am. Chem. Soc.* **1949**, *71*, 2659.
[494] Hata, K.; Tatematsu, K.; Kubota, B. *Bull. Chem. Soc. Jpn.* **1935**, *10*, 425.
[495] Carlin, R. B.; Odioso, R. C. *J. Am. Chem. Soc.* **1954**, *76*, 2345.
[496] Pettit, M. R.; Tatlow, J. C. *J. Chem. Soc.* **1951**, 3459.
[497] Wolf, C.; König, W. A.; Roussel, C. *Liebigs Ann. Chem.* **1995**, 781.
[498] Bloomfield, C.; Manglik, A. K.; Moodie, R. B.; Schofield, K.; Tobin, G. D. *J. Chem. Soc., Perkin Trans. 2* **1983**, 75.
[499] Litvinenko, L. M.; Grekov, A. P.; Shapoval, L. D. *Zh. Obshch. Khim.* **1957**, *22*, 3115; *J. Gen. Chem. U.S.S.R.* **1957**, *27*, 3154; *Chem. Abstr.* **1957**, *51*, 25396.
[500] Yamashiro, S. *Bull. Chem. Soc. Jpn.* **1942**, *17*, 10.
[501] Ling, C. C. K.; Harris, M. M. *J. Chem. Soc.* **1964**, 1825.
[502] Kempter, F. E.; Castle, R. N. *J. Heterocycl. Chem.* **1969**, *6*, 523.
[503] Osawa, Y. *Nippon Kagaku Zasshi* **1963**, *84*, 140; *Chem. Abstr.* **1963**, *59*, 13860f.
[504] Wolf, C.; Hochmuth, D. H.; König, W. A.; Roussel, C. *Liebigs Ann. Chem.* **1996**, 357.
[505] Pan, H.-L.; Fletcher, T. L. *J. Med. Chem.* **1970**, *13*, 567.
[506] Truce, W. E.; Emrick, D. D. *J. Am. Chem. Soc.* **1956**, *78*, 6130.
[507] Späth, E.; Gibian, K. *Monatsh. Chem.* **1930**, *55*, 342.
[508] Searle, N. E.; Adams, R. *J. Am. Chem. Soc.* **1933**, *55*, 1649.
[509] Bunton, C. A.; Kenner, G. W.; Robinson, M. J. T.; Webster, B. R. *Tetrahedron* **1963**, *19*, 1001.
[510] Rizzacasa, M. A.; Sargent, M. V. *Aust. J. Chem.* **1988**, *41*, 1087.
[511] Govindachari, T. R.; Viswanathan, N.; Ravindranath, K. R.; Anjaneyulu, B. *Indian J. Chem.* **1973**, *11*, 1081; *Chem. Abstr.* **1974**, *80*, 108235.
[512] Lesslie, M. S.; Turner, E. E. *J. Chem. Soc.* **1932**, 2021.
[513] Simpson, J. E.; Daub, G. H.; Hayes, F. N. *J. Org. Chem.* **1973**, *38*, 4428.
[514] Sako, S.-i. *Bull. Chem. Soc. Jpn.* **1935**, *10*, 593.
[515] Taber, R. L.; Daub, G. H.; Hayes, F. N.; Ott, D. G. *J. Heterocycl. Chem.* **1965**, *2*, 181.
[516] Müller, E.; Hertel, E. *Liebigs Ann. Chem.* **1944**, *555*, 157.
[517] Kern, W.; Gruber, W.; Wirth, H. O. *Makromol. Chem.* **1960**, *37*, 198.
[518] Rieger, M.; Westheimer, F. H. *J. Am. Chem. Soc.* **1950**, *72*, 28.
[519] Pummerer, R.; Puttfarcken, H.; Schopflocher, P. *Chem. Ber.* **1925**, *58*, 1808.
[520] Sato, T. *Bull. Chem. Soc. Jpn.* **1960**, *33*, 501.
[521] Abbaszadeh, M. R.; Bowden, K. *J. Chem. Soc., Perkin Trans. 2* **1990**, 2081.
[522] Brand, K.; Groebe, W. *J. Prakt. Chem.* **1924**, *108*, 1.
[523] Pachaly, P.; Schäfer, M. *Arch. Pharm. (Weinheim)* **1989**, *322*, 483.
[524] Ross, S. D.; Markarian, M.; Schwarz, M. *J. Am. Chem. Soc.* **1953**, *75*, 4967.
[525] Pufahl, F. *Chem. Ber.* **1929**, *62*, 2817.
[526] Brockmann, H.; Vorbrüggen, H. *Chem. Ber.* **1962**, *95*, 810.
[527] Rao, K. V. J.; Row, R. *J. Org. Chem.* **1960**, *25*, 981.
[528] Kenner, J.; Witham, E. *J. Chem. Soc.* **1913**, 232.
[529] Tashiro, M.; Yamato, T. *J. Org. Chem.* **1979**, *44*, 3037.
[530] Runeberg, J. *Acta Chem. Scand.* **1958**, *12*, 188.
[531] Fujita, E.; Fuji, K.; Tanaka, K. *Tetrahedron Lett.* **1968**, 5905.

[532] Lesslie, M. S.; Mayer, U. J. H. *J. Chem. Soc.* **1961**, 611.
[533] Blatchly, J. M.; McOmie, J. F. W.; Watts, M. L. *J. Chem. Soc.* **1962**, 5085.
[534] Stetter, H.; Schwarz, M. *Chem. Ber.* **1957**, *90*, 1349.
[535] Staab, H. A.; Höne, M.; Krieger, C. *Tetrahedron Lett.* **1988**, *29*, 1905.
[536] Iqbal, K.; Wilson, R. C. *J. Chem. Soc. (C)* **1967**, 1690.
[537] Waller, S. C.; Mash, E. A. *Org. Prep. Proced. Int.* **1997**, *29*, 679.
[538] Field, L. D.; Skelton, B. W.; Sternhell, S.; White, A. H. *Aust. J. Chem.* **1985**, *38*, 391.
[539] Wittig, G.; Stichnoth, O. *Chem. Ber.* **1935**, *68*, 928.
[540] Mascarelli, L.; Longo, B. *Gazz. Chim. Ital.* **1938**, *68*, 121; *Chem. Abstr.* **1938**, *32*, 44873.
[541] Dethloff, W.; Mix, H. *Chem. Ber.* **1949**, *82*, 534.
[542] Schmid, R.; Cereghetti, M.; Heiser, B.; Schönholzer, P.; Hansen, H.-J. *Helv. Chim. Acta* **1988**, *71*, 897.
[543] Cereghetti, M.; Schmid, R.; Schönholzer, P.; Rageot, A. *Tetrahedron Lett.* **1996**, *37*, 5343.
[544] Adams, R.; Finger, G. C. *J. Am. Chem. Soc.* **1939**, *61*, 2828.
[545] Becker, B. C.; Adams, R. *J. Am. Chem. Soc.* **1932**, *54*, 2973.
[546] Iffland, D. C.; Siegel, H. *J. Am. Chem. Soc.* **1958**, *80*, 1947.
[547] Ingersoll, A. W.; Little, J. R. *J. Am. Chem. Soc.* **1934**, *56*, 2123.
[548] Seno, K.; Hagishita, S.; Sato, T.; Kuriyama, K. *J. Chem. Soc., Perkin Trans. 1* **1984**, 2013.
[549] Díaz, E.; Guzmán, A.; Cruz, M.; Mares, J.; Ramírez, D. J.; Joseph-Nathan, P. *Org. Magn. Reson.* **1980**, *13*, 180.
[550] Kenner, J.; Stubbings, W. V. *J. Chem. Soc.* **1921**, 593.
[551] Stanley, W. M.; McMahon, E.; Adams, R. *J. Am. Chem. Soc.* **1933**, *55*, 706.
[552] Carlin, R. B. *J. Am. Chem. Soc.* **1945**, *67*, 928.
[553] Adams, R.; Baker, B. R. *J. Am. Chem. Soc.* **1941**, *63*, 535.
[554] VanArendonk, A. M.; Cupery, M. E.; Adams, R. *J. Am. Chem. Soc.* **1933**, *55*, 4225.
[555] Lettré, H.; Jahn, A. *Chem. Ber.* **1952**, *85*, 346.
[556] Brune, H.-A.; Lerche, J.; Schmidtberg, G.; Baur, A. *J. Organometallic Chem.* **1993**, *450*, 269.
[557] Kuhn, R.; Albrecht, O. *Liebigs Ann. Chem.* **1927**, *455*, 272.
[558] Wittig, G.; Zimmermann, H. *Chem. Ber.* **1953**, *86*, 629.
[559] Kanoh, S.; Muramoto, H.; Kobayashi, N.; Motoi, M.; Suda, H. *Bull. Chem. Soc. Jpn.* **1987**, *60*, 3659.
[560] Bell, F. *J. Chem. Soc.* **1934**, 835.
[561] Ames, D. E.; Hansen, K. J.; Griffiths, N. D. *J. Chem. Soc., Perkin Trans. 1* **1973**, 2818.
[562] Sako, S.-i. *Bull. Chem. Soc. Jpn.* **1934**, *9*, 55.
[563] Sako, S.-i. *Bull. Chem. Soc. Jpn.* **1936**, *11*, 144.
[564] Jendralla, H.; Li, C. H.; Paulus, E. *Tetrahedron: Asymmetry* **1994**, *5*, 1297.
[565] Schmid, R.; Foricher, J.; Cereghetti, M.; Schönholzer, P. *Helv. Chim. Acta* **1991**, *74*, 370.
[566] Patterson, W. I.; Adams, R. *J. Am. Chem. Soc.* **1935**, *57*, 762.
[567] Beckwith, A. L. J.; Waters, W. A. *J. Chem. Soc.* **1957**, 1665.
[568] Dallacker, F.; Adolphen, G. *Liebigs Ann. Chem.* **1966**, *694*, 110.
[569] Crossley, A. W.; Hampshire, C. H. *J. Chem. Soc.* **1911**, 721.
[570] Scholl, R.; Liese, K.; Michelson, K.; Grunewald, E. *Chem. Ber.* **1910**, *43*, 512.
[571] Lam, H.; Marcuccio, S. M.; Svirskaya, P. I.; Greenberg, S.; Lever, A. B. P.; Leznoff, C. C.; Cerny, R. L. *Can. J. Chem.* **1989**, *67*, 1087.
[572] Whaley, W. M.; White, C. *J. Org. Chem.* **1953**, *18*, 184.
[573] Ritchie, E. *J. Proc. Roy. N. S. Wales* **1945**, *78*, 134; *Chem. Abstr.* **1946**, *40*, 5229.
[574] Joulié, L. F.; Schatz, E.; Ward, M. D.; Weber, F.; Yellowlees, L. J. *J. Chem. Soc., Dalton Trans.* **1994**, 799.
[575] Gardent, J. *Bull. Soc. Chim. Fr.* **1962**, 1049.
[576] Ding, M.; Wang, Z.; Yang, Z.; Zhang, J. U.S. Patent 5,081,281 (1992); *Chem. Abstr.* **1990**, *113*, 152043.
[577] Sharma, V.; Bachand, B.; Simard, M.; Wuest, J. D. *J. Org. Chem.* **1994**, *59*, 7785.
[578] Boden, N.; Bushby, R. J.; Cammidge, A. N. *J. Am. Chem. Soc.* **1995**, *117*, 924.
[579] Britton, E. C.; Livak, J. E. U.S. Patent 2,260,739 (1941); *Chem. Abstr.* **1942**, *36*, 5487.
[580] Hung, J.; Werbel, L. M. *Eur. J. Med. Chem.* **1983**, *18*, 61.

[581] Shen, X.; Dong, R. Y.; Boden, N.; Bushby, R. J.; Martin, P. S.; Wood, A. *J. Chem. Phys.* **1998**, *108*, 4324.
[582] Case, F. H. *J. Am. Chem. Soc.* **1942**, *64*, 1848.
[583] Bräunling, H.; Binnig, F.; Staab, H. A. *Chem. Ber.* **1967**, *100*, 880.
[584] McAlister, F. B.; Kenner, J. *J. Chem. Soc.* **1928**, 1913.
[585] Kern, W.; Ebersbach, H. W.; Ziegler, I. *Makromol. Chem.* **1959**, *31*, 154.
[586] Riedl, W.; Imhof, W. *Liebigs Ann. Chem.* **1955**, *597*, 153.
[587] Beley, M.; Chodorowski, S.; Collin, J.-P.; Sauvage, J.-P. *Tetrahedron Lett.* **1993**, *34*, 2933.
[588] Ozasa, S.; Fujioka, Y.; Hashino, H.; Kimura, N.; Ibuki, E. *Chem. Pharm. Bull.* **1983**, *31*, 2313.
[589] Carruthers, W.; Douglas, A. G. *J. Chem. Soc.* **1959**, 2813.
[590] Barnes, R. A.; Faessinger, R. W. *J. Am. Chem. Soc.* **1961**, *26*, 4544.
[591] Cumming, W. M.; Howie, G. *J. Chem. Soc.* **1931**, 3176.
[592] Tsuji, N. *Tetrahedron* **1968**, *24*, 1765.
[593] ApSimon, J. W.; Creasey, N. G.; Marlow, W.; Sim, K. Y.; Whalley, W. B. *J. Chem. Soc.* **1965**, 4156.
[594] Cornforth, J.; Sierakowski, A. F.; Wallace, T. W. *J. Chem. Soc., Perkin Trans. 1* **1982**, 2299.
[595] Carroll, A. R.; Read, R. W.; Taylor, W. C. *Aust. J. Chem.* **1994**, *47*, 1579.
[596] Farrar, J. M.; Sienkowska, M.; Kaszynski, P. *Synth. Commun.* **2000**, *30*, 4039.
[597] Murphy, D. B.; Schwartz, F. R.; Picard, J. P.; Kaufman, J. V. R. *J. Am. Chem. Soc.* **1953**, *75*, 4289.
[598] Case, F. H.; Schock, R. U. *J. Am. Chem. Soc.* **1943**, *65*, 2086.
[599] Tomita, M.; Kikuchi, T.; Bessho, K.; Hori, T.; Inubushi, Y. *Chem. Pharm. Bull.* **1963**, *11*, 1484.
[600] Nilsson, M. *Acta Chem. Scand.* **1958**, *12*, 1830.
[601] Omote, Y.; Fujinuma, Y.; Sugiyama, N. *Bull. Chem. Soc. Jpn.* **1971**, *44*, 572.
[602] Faber, A. C.; Nauta, W. T. *Rec. Trav. Chim. Pays-Bas* **1943**, *62*, 469.
[603] Hine, J.; Hahn, S.; Miles, D. E.; Ahn, K. *J. Org. Chem.* **1985**, *50*, 5092.
[604] Kondo, K.; Takahashi, M.; Ohmizu, H.; Matsumoto, M.; Taguchi, I.; Iwasaki, T. *Chem. Pharm. Bull.* **1994**, *42*, 62.
[605] Chapman, R. F.; Swan, G. A. *J. Chem. Soc. (C)* **1970**, 865.
[606] Gilman, H.; Thirtle, J. R. *J. Am. Chem. Soc.* **1944**, *66*, 858.
[607] Miyazaki, T.; Mihashi, S.; Okabayashi, K. *Chem. Pharm. Bull.* **1964**, *12*, 1236.
[608] Kobayashi, S.; Azekawa, M.; Taoka, M. *Chem. Pharm. Bull.* **1969**, *17*, 1279.
[609] Scarpati, M. L.; Bianco, A.; Lo Scalzo, R. *Synth. Commun.* **1991**, *21*, 849.
[610] Brown, J. M.; Woodward, S. *J. Org. Chem.* **1991**, *56*, 6803.
[611] Carter, R. E.; Dahlgren, L. *Arkiv Kemi* **1967**, *27*, 257; *Chem. Abstr.* **1968**, *68*, 12646.
[612] Baker, W.; Barton, J. W.; McOmie, J. F. W.; Penneck, R. J.; Watts, M. L. *J. Chem. Soc.* **1961**, 3986.
[613] Hughes, G. K.; Lions, F.; Maunsell, J. J.; Wright, L. E. A. *J. Proc. Soc. N.S. Wales* **1938**, *71*, 428; *Chem. Abstr.* **1939**, *33*, 613.
[614] Cromartie, R. I. T.; Harley-Mason, J.; Wannigama, D. G. P. *J. Chem. Soc.* **1958**, 1982.
[615] Kobayashi, S.; Azekawa, M. *Tokushima Daigaku Yakugakubu Kenkyu Nenpo* **1969**, *18*, 11; *Chem. Abstr.* **1970**, *73*, 98558t.
[616] Ward, E. R.; Pearson, B. D. *J. Chem. Soc.* **1959**, 1676.
[617] Hewgill, F. R.; Slamet, R.; Stewart, J. M. *J. Chem. Soc., Perkin Trans. 1* **1991**, 3033.
[618] Bowman, D. F.; Hewgill, F. R.; Kennedy, B. R. *J. Chem. Soc. (C)* **1966**, 2274.
[619] Hein, D. W.; Radkowski, S. J. U.S. Patent 3,402,202 (1968); *Chem. Abstr.* **1969**, *70*, 47077.
[620] van Duin, C. F. *Recl. Trav. Chim. Pays-Bas* **1920**, *39*, 685.
[621] Bourdon, J.; Calvin, M. *J. Org. Chem.* **1957**, *22*, 101.
[622] Carlin, R. B.; Heininger, S. A. *J. Am. Chem. Soc.* **1955**, *77*, 2272.
[623] Corbett, J. F.; Holt, P. F. *J. Chem. Soc.* **1961**, 4261.
[624] Castle, R. N.; Guither, W. D.; Hilbert, P.; Kempter, F. E.; Patel, N. R. *J. Heterocycl. Chem.* **1969**, *6*, 533.
[625] Goldschmidt, S.; Suchanek, L. *Chem. Ber.* **1957**, *90*, 19.
[626] Theilacker, W.; Baxmann, F. *Liebigs Ann. Chem.* **1953**, *581*, 117.
[627] Musso, H.; Steckelberg, W. *Liebigs Ann. Chem.* **1966**, *693*, 187.
[628] Müller, E.; Tietz, E. *Chem. Ber.* **1941**, *74*, 807.
[629] Ullmann, F.; Engi, G.; Wosnessensky, N.; Kuhn, E.; Herre, E. *Liebigs Ann. Chem.* **1909**, *366*, 79.

[630] Posternak, T.; Ruelius, H. W.; Teherniak, J. *Helv. Chim. Acta* **1943**, *26*, 2031.
[631] Inubushi, Y.; Nomura, K. *Yakugaku Zasshi* **1961**, *81*, 7; *Chem. Abstr.* **1961**, *55*, 15493.
[632] Yang, K.; Lemieux, R. P. *Mol. Cryst. Liq. Cryst.* **1995**, *260*, 247.
[633] Nomura, Y.; Takeuchi, Y. *J. Chem. Soc. (B)* **1970**, 956.
[634] Musso, H. *Chem. Ber.* **1958**, *91*, 349.
[635] Riedl, W. *Liebigs Ann. Chem.* **1955**, *597*, 148.
[636] Dallacker, F.; Leidig, H. *Chem. Ber.* **1979**, *112*, 2672.
[637] Mix, H. *Liebigs Ann. Chem.* **1955**, *592*, 146.
[638] Williams, V. E.; Lemieux, R. P. *Chem. Commun.* **1996**, 2259.
[639] Ridley, D. D.; Ritchie, E.; Taylor, W. C. *Aust. J. Chem.* **1970**, *23*, 147.
[640] Insole, J. M. *J. Chem. Res. (M)* **1990**, 2831.
[641] Wünsche, C.; Sachs, A.; Mayer, W. *Tetrahedron* **1969**, *25*, 73.
[642] Shibata, S. *Acta Phytochim. (Japan)* **1944**, *14*, 9; *Chem. Abstr.* **1951**, *45*, 7100.
[643] Müller, E.; Neuhoff, H. *Chem. Ber.* **1939**, *72*, 2063.
[644] Chester, D. O.; Elix, J. A. *Aust. J. Chem.* **1981**, *34*, 1501.
[645] Armarego, W. L. F.; Turner, E. E. *J. Chem. Soc.* **1956**, 3668.
[646] Chao, C.; Zhang, P. *Tetrahedron Lett.* **1988**, *29*, 225.
[647] Fujioka, Y.; Ozasa, S.; Sato, K.; Ibuki, E. *Chem. Pharm. Bull.* **1985**, *33*, 22.
[648] Becker, A.; Ewenson, A. A.; Croitoru, B. Eur. Patent Appl. EP 514821 (1992); *Chem. Abstr.* **1993**, *118*, 61912.
[649] Belf, L. J.; Buxton, M. W.; Tilney-Bassett, J. F. *Tetrahedron* **1967**, *23*, 4719.
[650] Stjernstrom, N. E. *Arkiv Kemi* **1963**, *21*, 73; *Chem. Abstr.* **1963**, *59*, 41523.
[651] Lawson, D. W.; McOmie, J. F. W.; West, D. E. *J. Chem. Soc. (C)* **1968**, 2414.
[652] Baker, W.; Miles, D. *J. Chem. Soc.* **1955**, 2089.
[653] Cherkaoui, M. Z.; Scherowsky, G. *New J. Chem.* **1997**, *21*, 1203.
[654] Baker, W.; McLean, N. J.; McOmie, J. F. W. *J. Chem. Soc.* **1963**, 922.
[655] Chen, D.-F.; Zhang, S.-X.; Xie, L.; Xie, J.-X.; Chen, K.; Kashiwada, Y.; Zhou, B.-N.; Wang, P.; Cosentino, L. M.; Lee, K.-H. *Bioorg. Med. Chem.* **1997**, *5*, 1715.
[656] Bock, L. H.; Moyer, W. W.; Adams, R. *J. Am. Chem. Soc.* **1930**, *52*, 2054.
[657] Dacons, J. C.; Adolph, H. G.; Kamlet, M. J. *Tetrahedron* **1963**, *19*, 791.
[658] Oesterling, R. E.; Dacons, J. C.; Kaplan, L. A. U.S. Patent 3,404,184 (1968); *Chem. Abstr.* **1969**, *70*, 37444.
[659] Bellamy, A. J.; Hudson, P. N. *J. Chem. Res. (M)* **1996**, 959.
[660] Brown, E.; Robin, J.-P. *Tetrahedron Lett.* **1977**, 2015.
[661] Brown, E.; Robin, J.-P.; Dhal, R. *Tetrahedron* **1982**, *38*, 2569.
[662] Moyer, W. W.; Adams, R. *J. Am. Chem. Soc.* **1929**, *51*, 630.
[663] Kanojia, R. M.; Ohemeng, K. A.; Schwender, C. F.; Barrett, J. F. *Tetrahedron Lett.* **1995**, *36*, 8553.
[664] Wu, W. L.; Chen, S. E.; Chang, W. L.; Chen, C. F.; Lee, A. R. *Eur. J. Med. Chem.* **1992**, *27*, 353.
[665] Giles, R. G. F.; Sargent, M. V. *Aust. J. Chem.* **1986**, *39*, 2177.
[666] Hathway, D. E. *J. Chem. Soc.* **1957**, 519.
[667] Grimshaw, J.; Haworth, R. D. *J. Chem. Soc.* **1956**, 4225.
[668] Hauser, F. M.; Gauuan, J. F. *Org. Lett.* **1999**, *1*, 671.
[669] Fischer, E.; Hess, H.; Lorenz, T.; Musso, H.; Rossnagel, I. *Chem. Ber.* **1991**, *124*, 783.
[670] Ragan, M. A. *Can. J. Chem.* **1985**, *63*, 294.
[671] Elix, J. A.; Jayanthi, V. K.; Jones, A. J.; Lennard, C. J. *Aust. J. Chem.* **1984**, *37*, 1531.
[672] Binns, F.; Suschitzky, H. *J. Chem. Soc. (C)* **1971**, 1913.
[673] Bruce, J. M.; Sutcliffe, F. K. *J. Chem. Soc.* **1956**, 3820.
[674] Kleiderer, E. C.; Adams, R. *J. Am. Chem. Soc.* **1931**, *53*, 1575.
[675] Kleiderer, E. C.; Adams, R. *J. Am. Chem. Soc.* **1933**, *55*, 716.
[676] Kalamar, J.; Steiner, E.; Charollais, E.; Posternak, T. *Helv. Chim. Acta* **1974**, *57*, 2368.
[677] Wünsche, C.; Sachs, A.; Einwiller, A.; Mayer, W. *Tetrahedron* **1968**, *24*, 3407.
[678] Birchall, J. M.; Hazard, R.; Haszeldine, R. N.; Wakalski, W. W. *J. Chem. Soc. (C)* **1967**, 47.
[679] Pummer, W. J.; Wall, L. A. U. S. Patent 3,046,313 (1962); *Chem. Abstr.* **1962**, *57*, 75681.
[680] Nield, E.; Stephens, R.; Tatlow, J. C. *J. Chem. Soc.* **1959**, 166.

[681] Pummer, W. J.; Wall, L. A. *J. Res. NBS. A. Phys. Ch.* **1959**, *63A*, 167; *Chem. Abstr.* **1960**, *54*, 56144.
[682] Thrower, J.; White, M. A. *Polym. Prepr., Am. Chem. Soc., Div. Polym. Chem.* **1966**, *7*, 1077; *Chem. Abstr.* **1967**, *66*, 29173.
[683] Yakobson, G. G.; Shteingarts, V. D.; Miroshnikov, A. I.; Vorozhtsov, N. N., Jr. *Dokl. Akad. Nauk SSSR* **1964**, *159*, 1109; *Dokl. Acad. Nauk. SSSR (Engl. Translation)* **1964**, *159*, 1347; *Chem. Abstr.* **1965**, *62*, 51263.
[684] Brooke, G. M.; Chambers, R. D.; Heyes, J.; Musgrave, W. K. R. *J. Chem. Soc.* **1964**, 729.
[685] Imperial Smelting Corp. (N.S.C) Ltd. Belg. Patent 659,239 (1965); *Chem. Abstr.* **1966**, *64*, 15792h.
[686] Osina, O. I.; Shteingarts, V. D. *J. Org. Chem. U.S.S.R.* **1974**, *10*, 329; *Chem. Abstr.* **1974**, *80*, 120611.
[687] Hodgson, H. H.; Crook, J. H. *J. Chem. Soc.* **1937**, 571.
[688] Hodgson, H. H.; Crook, J. H. *J. Chem. Soc.* **1937**, 571.
[689] Schoepfle, C. S. *J. Am. Chem. Soc.* **1923**, *45*, 1566.
[690] Dixon, W.; Harris, M. M.; Mazengo, R. Z. *J. Chem. Soc. (B)* **1971**, 775.
[691] Edwards, J. D.; Cashaw, J. L. *J. Am. Chem. Soc.* **1954**, *76*, 6141.
[692] Chudozilov, L. K. *Chem. Listy* **1925**, *19*, 187; *Chem. Abstr.* **1925**, *19*, 3268.
[693] Escudero, S.; Pérez, D.; Guitián, E.; Castedo, L. *Tetrahedron Lett.* **1997**, *38*, 5375.
[694] Brass, K.; Patzelt, R. *Chem. Ber.* **1937**, *70*, 1349.
[695] Venkatraman, S.; Li, C.-J. *Tetrahedron Lett.* **2000**, *41*, 4831.
[696] Fritsch, R.; Hartmann, E.; Andert, D.; Mannschreck, A. *Chem. Ber.* **1992**, *125*, 849.
[697] Bacon, R. G. R.; Bankhead, R. *J. Chem. Soc.* **1963**, 839.
[698] Martin, R. H. *J. Chem. Soc.* **1941**, 679.
[699] Bergmann, E. D.; Szmuszkovicz, J. *J. Am. Chem. Soc.* **1951**, *73*, 5153.
[700] Rosini, C.; Tanturli, R.; Pertici, P.; Salvadori, P. *Tetrahedron: Asymmetry* **1996**, *7*, 2971.
[701] Hall, D. M.; Turner, E. E. *J. Chem. Soc.* **1955**, 1242.
[702] Mazaleyrat, J.-P. *Tetrahedron: Asymmetry* **1997**, *8*, 2709.
[703] Kuhn, R.; Albrecht, O. *Liebigs Ann. Chem.* **1928**, *465*, 282.
[704] Weber, E.; Csöregh, I.; Stensland, B.; Czugler, M. *J. Am. Chem. Soc.* **1984**, *106*, 3297.
[705] Colletti, S. L.; Halterman, R. L. *Organometallics* **1991**, *10*, 3438.
[706] Kim, J.-I.; Schuster, G. B. *J. Am. Chem. Soc.* **1992**, *114*, 9309.
[707] Armarego, W. L. F.; Turner, E. E. *J. Chem. Soc.* **1957**, 13.
[708] Seer, C.; Scholl, R. *Liebigs Ann. Chem.* **1913**, *398*, 82.
[709] Hall, D. M.; Ridgwell, S.; Turner, E. E. *J. Chem. Soc.* **1954**, 2498.
[710] Fierz-David, H. E.; Blangey, L.; Dübendorfer, H. *Helv. Chim. Acta* **1946**, *29*, 1661.
[711] Puts, R. D.; Chao, J.; Sogah, D. Y. *Synthesis* **1997**, 431.
[712] Morishita, E.; Shibata, S. *Chem. Pharm. Bull.* **1967**, *15*, 1765.
[713] Cameron, D. W.; Feutrill, G. I.; Pannan, L. J. H. *Tetrahedron Lett.* **1980**, *21*, 1385.
[714] Charmant, J. P. H.; Fallis, I. A.; Hunt, N. J.; Lloyd-Jones, G. C.; Murray, M.; Nowak, T. *J. Chem. Soc., Dalton Trans.* **2000**, *11*, 1723.
[715] Pracejus, H. *Liebigs Ann. Chem.* **1956**, *601*, 61.
[716] Judice, J. K.; Keipert, S. J.; Cram, D. J. *J. Chem. Soc., Chem. Commun.* **1993**, 1323.
[717] Drefahl, G.; Winnefeld, K. *J. Prakt. Chem.* **1965**, *28*, 242.
[718] Curtis, R. F.; Viswanath, G. *J. Chem. Soc.* **1959**, 1670.
[719] Bentley, K. W. *J. Chem. Soc.* **1955**, 2398.
[720] Vesel, V. *Chem. Ber.* **1905**, *38*, 136.
[721] Clemo, G. R.; Cockburn, J. G.; Spence, R. *J. Chem. Soc.* **1931**, 1265.
[722] Armarego, W. L. F. *J. Chem. Soc.* **1960**, 433.
[723] Loder, J. W.; Mongolsuk, S.; Robertson, A.; Whalley, W. B. *J. Chem. Soc.* **1957**, 2233.
[724] Morita, T.; Takase, K. *Bull. Chem. Soc. Jpn.* **1982**, *55*, 1144.
[725] Meyer, A.; Schlögl, K.; Keller, W.; Kratky, C. *Monatsh. Chem.* **1989**, *120*, 453.
[726] Kelly, T. R.; Garcia, A.; Lang, F.; Walsh, J. J.; Bhaskar, K. V.; Boyd, M. R.; Götz, R.; Keller, P. A.; Walter, R.; Bringmann, G. *Tetrahedron Lett.* **1994**, *35*, 7621.
[727] Barnett, M. D.; Daub, G. H.; Hayes, F. N.; Ott, D. G. *J. Am. Chem. Soc.* **1959**, *81*, 4583.
[728] Suzuki, K.; Weisburger, E. K.; Weisburger, J. H. *J. Org. Chem.* **1961**, *26*, 2236.

[729] Wirth, H. O.; Gönner, K. H.; Stück, R.; Kern, W. *Makromol. Chem.* **1963**, *63*, 30.
[730] Hughes, A. N.; Prankprakma, V. *Tetrahedron* **1966**, *22*, 2053.
[731] Zhou, Z.-h.; Yamamoto, T. *J. Organomet. Chem.* **1991**, *414*, 119.
[732] Staab, H. A.; Bräunling, H. *Tetrahedron Lett.* **1965**, 45.
[733] Wirth, H. O.; Gönner, K. H.; Kern, W. *Makromol. Chem.* **1963**, *63*, 53.
[734] Zinke, A.; Ziegler, E. *Chem. Ber.* **1941**, *74*, 115.
[735] De Ridder, R.; Martin, R. H. *Bull. Soc. Chim. Belg.* **1960**, *69*, 534.
[736] Yamamoto, K.; Kitsuki, T.; Okamoto, Y. *Bull. Chem. Soc. Jpn.* **1986**, *59*, 1269.
[737] Diwu, Z.; Lown, J. W. *Tetrahedron Lett.* **1992**, *48*, 45.
[738] Ammerer, L.; Zinke, A. *Monatsh. Chem.* **1953**, *84*, 25.
[739] Nagai, Y.; Gotoh, N.; Ogawa, S. *Yuki Gosei Kagaku Kyokai Shi* **1970**, *28*, 930; *Chem. Abstr.* **1971**, *74*, 43506q.
[740] Nagai, Y.; Nagasawa, K. *Kogyo Kagaku Zasshi* **1966**, *69*, 666; *Chem. Abstr.* **1967**, *66*, 37696.
[741] Mitchell, R. H.; Chaudhary, M.; Dingle, T. W.; Williams, R. V. *J. Am. Chem. Soc.* **1984**, *106*, 7776.
[742] Gotoh, N.; Koga, Y. *Seisan-Kenkyu* **1967**, *19*, 175; *Chem. Abstr.* **1968**, *69*, 58980.
[743] Debad, J. D.; Morris, J. C.; Magnus, P.; Bard, A. J. *J. Org. Chem.* **1997**, *62*, 530.
[744] Kreyenschmidt, M.; Baumgarten, M.; Tyutyulkov, N.; Mullen, K. *Angew. Chem., Int. Ed. Engl.* **1994**, *33*, 1957.
[745] Quante, H.; Mullen, K. *Angew. Chem., Int. Ed. Engl.* **1995**, *34*, 1323.
[746] Ito, S.; Herwig, P. T.; Böhme, T.; Rabe, J. P.; Rettig, W.; Müllen, K. *J. Am. Chem. Soc.* **2000**, *122*, 7698.
[747] Schwenk, E.; Waldmann, H. *J. Prakt. Chem.* **1931**, *130*, 79.
[748] Ullmann, F.; Eiser, O. *Chem. Ber.* **1916**, *49*, 2154.
[749] Eckert, A.; Tomaschek, R. *Monatsh. Chem.* **1918**, *39*, 839.
[750] Sauvage, G. *Ann. Chim. (Paris)* **1947**, *2*, 844.
[751] Minaev, V. *Chem.-Ztg.* **1913**, *36*, 199; *Chem. Abstr.* **1913**, *7*, 1481.
[752] Scholl, R.; Mansfeld, J. *Chem. Ber.* **1910**, *43*, 1734.
[753] Ookubo, S.; Ookuma, T.; Ito, N. Japanese Patent JP 09176108 (1997); *Chem. Abstr.* **1997**, *127*, 121572.
[754] Bell, F.; Waring, D. H. *J. Chem. Soc.* **1949**, 267.
[755] Scholl, R. *Chem. Ber.* **1907**, *40*, 1691.
[756] Kuhn, R.; Albrecht, O. *Liebigs Ann. Chem.* **1928**, *464*, 91.
[757] Ruggli, P.; Merz, E. *Helv. Chim. Acta* **1929**, *12*, 71.
[758] Ullmann, F.; Minajeff, W. *Chem. Ber.* **1912**, *45*, 687.
[759] Hardacre, R. W.; Perkin, A. G. *J. Chem. Soc.* **1929**, 180.
[760] Benesch, E. *Monatsh. Chem.* **1911**, *32*, 447.
[761] Brockmann, H.; von Falkenhausen, E. H. F.; Neeff, R.; Dorlars, A.; Budde, G. *Chem. Ber.* **1951**, *84*, 865.
[762] Scholl, R.; Potschiwauscheg, J.; Lenko, J. *Monatsh. Chem.* **1911**, *32*, 687.
[763] Stanley, W. M.; Adams, R. *J. Am. Chem. Soc.* **1931**, *53*, 2364.
[764] Shibata, S.; Tanaka, O.; Kitagawa, I. *Chem. Pharm. Bull.* **1955**, *3*, 278.
[765] Brockmann, H.; Dolars, A. *Chem. Ber.* **1952**, *85*, 1168.
[766] Seer, C.; Karl, E. *Monatsh. Chem.* **1913**, *34*, 631.
[767] Scholl, R. *Chem. Ber.* **1910**, *43*, 346.
[768] Scholl, R.; Meyer, H. K. *Chem. Ber.* **1935**, *68*, 1307.
[769] Ullmann, F.; Junghans, W. *Liebigs Ann. Chem.* **1913**, *399*, 330.
[770] Perkin, A. G.; Haddock, N. H. *J. Chem. Soc.* **1933**, 1512.
[771] CIBA, Ltd. British Patent 889,746 (1962); *Chem. Abstr.* **1962**, *57*, 7202a.
[772] Brockmann, H.; Neeff, R.; Mühlmann, E. *Chem. Ber.* **1950**, *83*, 467.
[773] Brockmann, H.; Muxfeldt, H. German Patent 956,307 (1959); *Chem. Abstr.* **1959**, *53*, 6193b.
[774] Brockmann, H.; Kluge, F. *Naturwissenschaften* **1951**, *38*, 141.
[775] Brockmann, H.; Kluge, F.; Muxfeldt, H. *Chem. Ber.* **1957**, *90*, 2302.
[776] Iio, H.; Zenfuku, K.; Tokoroyama, T. *Tetrahedron Lett.* **1995**, *36*, 5921.
[777] Minaeff, W. J.; Ripper, K. *Monatsh. Chem.* **1921**, *42*, 73.

[778] Scholl, R.; Neovius, W. *Chem. Ber.* **1911**, *44*, 1075.
[779] Scholl, R.; Müller, E. J.; Böttger, O. *Chem. Ber.* **1935**, *68*, 45.
[780] Huebner, C. F.; Link, K. P. *J. Am. Chem. Soc.* **1945**, *67*, 99.
[781] Lele, S. S.; Patel, M. G.; Sethna, S. *J. Chem. Soc.* **1961**, 969.
[782] Shah, M. V. *Current Sci. (India)* **1962**, *31*, 57; *Chem. Abstr.* **1963**, *58*, 497h.
[783] Cairns, H.; Fitzmaurice, C.; Hunter, D.; Johnson, P. B.; King, J.; Lee, T. B.; Lord, G. H.; Minshull, R.; Cox, J. S. G. *J. Med. Chem.* **1972**, *15*, 583.
[784] Zhang, F.-J.; Lin, G.-Q.; Huang, Q.-C. *J. Org. Chem.* **1995**, *60*, 6427.
[785] Mugnier, Y.; Laviron, E. *J. Chem. Soc., Perkin Trans. 2* **1979**, 1264.
[786] Ward, E. R.; Marriott, J. E. *J. Chem. Soc.* **1963**, 4999.
[787] Ward, E. R.; Pearson, B. D. *J. Chem. Soc.* **1961**, 515.
[788] Staab, H. A.; Bräunling, H.; Schneider, K. *Chem. Ber.* **1968**, *101*, 879.
[789] Chambers, R. D.; Cunningham, J. A.; Spring, D. J. *Tetrahedron* **1968**, *24*, 3997.
[790] Farrell, P. G.; Moskowitz, D.; Terrier, F. *Synth. Commun.* **1993**, *23*, 231.
[791] Newman, M. S.; Cella, J. A. *J. Org. Chem.* **1974**, *39*, 2084.
[792] Delogu, G.; Fabbri, D. *Tetrahedron: Asymmetry* **1997**, *8*, 759.
[793] Karimipour, M.; Semones, A. M.; Asleson, G. L.; Heldrich, F. J. *Synlett* **1990**, 525.
[794] Takahashi, M.; Ogiku, T.; Okamura, K.; Da-te, T.; Ohmizu, H.; Kondo, K.; Iwasaki, T. *J. Chem. Soc., Perkin Trans. 1* **1993**, 1473.
[795] Brandt, S.; Marfat, A.; Helquist, P. *Tetrahedron Lett.* **1979**, 2193.
[796] Horner, L.; Weber, K.-H. *Chem. Ber.* **1967**, *100*, 2842.
[797] Kajigaeshi, S.; Kadowaki, T.; Nishida, A.; Fujisaki, S. *Bull. Chem. Soc. Jpn.* **1986**, *59*, 97.
[798] Paul, G. C.; Gajewski, J. J. *Org. Prep. Proced. Int.* **1998**, *30*, 222.
[799] Seki, M.; Furutani, T.; Hatsuda, M.; Imashiro, R. *Tetrahedron Lett.* **2000**, *41*, 2149.
[800] Carruthers, W.; Coggins, P.; Weston, J. B. *J. Chem. Soc., Perkin Trans. 1* **1991**, 611.
[801] Miyano, S.; Tobita, M.; Nawa, M.; Sato, S.; Hashimoto, H. *J. Chem. Soc., Chem. Commun.* **1980**, 1233.
[802] Miyano, S.; Handa, S.; Tobita, M.; Hashimoto, H. *Bull. Chem. Soc. Jpn.* **1986**, *59*, 235.
[803] Miyano, S.; Shimizu, K.; Sato, S.; Hashimoto, H. *Bull. Chem. Soc. Jpn.* **1985**, *58*, 1345.
[804] Takahashi, M.; Moritani, Y.; Ogiku, T.; Ohmizu, H.; Kondo, K.; Iwasaki, T. *Tetrahedron Lett.* **1992**, *33*, 5103.
[805] Takahashi, M.; Kuroda, T.; Ogiku, T.; Ohmizu, H.; Kondo, K.; Iwasaki, T. *Tetrahedron Lett.* **1991**, *32*, 6919.
[806] Mohamed, S. E. N.; Whiting, D. A. *J. Chem. Soc., Perkin Trans. 1* **1983**, 2577.
[807] Nicolaou, K. C.; Chu, X.-J.; Ramanjulu, J. M.; Natarajan, S.; Bräse, S.; Rübsam, F.; Boddy, C. N. C. *Angew. Chem., Int. Ed. Engl.* **1997**, *36*, 1539.
[808] Fukuyama, Y.; Yaso, H.; Nakamura, K.; Kodama, M. *Tetrahedron Lett.* **1999**, *40*, 105.
[809] Whiting, D. A.; Wood, A. F. *Tetrahedron Lett.* **1978**, 2335.
[810] Hart, D. J.; Hong, W.-P.; Hsu, L.-Y. *J. Org. Chem.* **1987**, *52*, 4665.
[811] Lothrup, W. C. *J. Am. Chem. Soc.* **1941**, *63*, 1187.
[812] Proctor, C. J.; Kralj, B.; Larka, E. A.; Porter, C. J.; Maquestiau, A.; Beynon, J. H. *Org. Mass Spectrom.* **1981**, *16*, 312.
[813] Constantine, P. R.; Hall, G. E.; Harrison, C. R.; McOmie, J. F. W.; Searle, R. J. G. *J. Chem. Soc. (C)* **1966**, 1767.
[814] Cohen, S. C.; Massey, A. G. *Tetrahedron Lett.* **1966**, 4393.
[815] Gribble, G. W.; Douglas, J. R. *J. Am. Chem. Soc.* **1970**, *92*, 5764.
[816] Hine, J.; Ahn, K. *J. Org. Chem.* **1987**, *52*, 2089.
[817] Barton, J. W. *J. Chem. Soc.* **1964**, 5161.
[818] Cava, M. P.; Stucker, J. F. *J. Am. Chem. Soc.* **1955**, *77*, 6022.
[819] Kelly, T. R.; Meghani, P.; Ekkundi, V. S. *Tetrahedron Lett.* **1990**, *31*, 3381.
[820] Webb, J. L. A. *J. Org. Chem.* **1953**, *18*, 1413.
[821] Grigg, R.; Johnson, A. W. *J. Chem. Soc.* **1964**, 3315.
[822] Grigg, R.; Knight, J. A.; Sargent, M. V. *J. Chem. Soc. (C)* **1966**, 976.
[823] Grigg, R.; Johnson, A. W.; Wasley, J. W. F. *J. Chem. Soc.* **1963**, 359.

[824] Sessler, J. L.; Cyr, M. J.; Lynch, V.; McGhee, E.; Ibers, J. A. *J. Am. Chem. Soc.* **1990**, *112*, 2810.
[825] Sessler, J. L.; Cyr, M.; Burrell, A. K. *Tetrahedron* **1992**, *48*, 9661.
[826] Guilard, R.; Aukauloo, M. A.; Tardieux, C.; Vogel, E. *Synthesis* **1995**, 1480.
[827] Vogel, E.; Koch, P.; Hou, X.-L.; Lex, J.; Lausmann, M.; Kisters, M.; Aukauloo, M. A.; Richard, P.; Guilard, R. *Angew. Chem., Int. Ed. Engl.* **1993**, *32*, 1600.
[828] Webb, J. L. A.; Threlkeld, R. R. *J. Org. Chem.* **1953**, *18*, 1406.
[829] Vogel, E.; Balci, M.; Pramod, K.; Koch, P.; Lex, J.; Ermer, O. *Angew. Chem., Int. Ed. Engl.* **1987**, *26*, 928.
[830] Richert, C.; Wessels, J. M.; Müller, M.; Kisters, M.; Benninghaus, T.; Goetz, A. E. *J. Med. Chem.* **1994**, *37*, 2797.
[831] Nonell, S.; Bou, N.; Borrell, J. I.; Teixidó, J.; Villanueva, A.; Juarranz, A.; Cañete, M. *Tetrahedron Lett.* **1995**, *36*, 3405.
[832] Bauer, V. J.; Clive, D. L. J.; Dolphin, D.; Paine III, J. B.; Harris, F. L.; King, M. M.; Loder, J.; Wang, S.-W. C.; Woodward, R. B. *J. Am. Chem. Soc.* **1983**, *105*, 6429.
[833] Ikeda, H.; Sessler, J. L. *J. Org. Chem.* **1993**, *58*, 2340.
[834] Vanderesse, R.; Lourak, M.; Fort, Y.; Caubère, P. *Tetrahedron Lett.* **1986**, *27*, 5483.
[835] Karminski, W.; Kulichi, Z. *Chem. Stowana, Ser. A.* **1965**, *9*, 129; *Chem. Abstr.* **1965**, *63*, 18018.
[836] Wibaut, J. P.; Overhoff, J. *Recl. Trav. Chim. Pays-Bas* **1928**, *47*, 761.
[837] Geissman, T. A.; Schlatter, M. J.; Webb, I. D.; Roberts, J. D. *J. Org. Chem.* **1946**, *11*, 741.
[838] Burstall, F. H. *J. Chem. Soc.* **1938**, 1662.
[839] Tiecco, M. *Bull. Soc. Chim. Belg.* **1986**, *95*, 1009.
[840] Sakamoto, T.; Arakida, H.; Edo, K.; Yamanaka, H. *Chem. Pharm. Bull.* **1982**, *30*, 3647.
[841] Dehmlow, E. V.; Schulz, H.-J. *J. Chem. Res. (M)* **1987**, 2951.
[842] Vekemans, J. A. J. M.; Groenendaal, L.; Palmans, A. R. A.; Delnoye, D. A. P.; van Mullekom, H. A. M.; Meijer, E. W. *Bull. Soc. Chim. Belg.* **1996**, *105*, 659.
[843] Matsuda, K.; Yanagisawa, I.; Isomura, Y.; Mase, T.; Shibanuma, T. *Synth. Commun.* **1997**, *27*, 2393.
[844] Etienne, A.; Izoret, G. French Patent 1,369,401 (1965); *Chem. Abstr.* **1965**, *62*, 570.
[845] Palmans, A. R. A.; Vekemans, J. A. J. M.; Meijer, E. W. *Recl. Trav. Chim. Pays-Bas* **1995**, *114*, 277.
[846] Case, F. H. *J. Am. Chem. Soc.* **1946**, *68*, 2574.
[847] Nakamaru, K. *Bull. Chem. Soc. Jpn.* **1982**, *55*, 2697.
[848] Chan, K. S.; Tse, A. K.-S. *Synth. Commun.* **1993**, *23*, 1929.
[849] Munavalli, S.; Rossman, D. I.; Szafraniec, L. L.; Beaudry, W. T.; Rohrbaugh, D. K.; Ferguson, C. P.; Grätzel, M. *J. Fluorine Chem.* **1995**, *73*, 1.
[850] Baxter, P. N. W.; Connor, J. A.; Povey, D. C.; Wallis, J. D. *J. Chem. Soc., Chem. Commun.* **1991**, 1135.
[851] Hasseberg, H.-A.; Gerlach, H. *Helv. Chim. Acta* **1988**, *71*, 957.
[852] Dehmlow, E. V.; Sleegers, A. *Liebigs Ann. Chem.* **1992**, 953.
[853] Case, F. H.; Kasper, T. J. *J. Am. Chem. Soc.* **1956**, *78*, 5842.
[854] Chambron, J.-C.; Sauvage, J.-P. *Tetrahedron* **1987**, *43*, 895.
[855] Janiak, C.; Deblon, S.; Wu, H.-P. *Synth. Commun.* **1999**, *29*, 3341.
[856] Janiak, C.; Deblon, S.; Wu, H.-P.; Kolm, M. J.; Klüfers, P.; Piotrowski, H.; Mayer, P. *Eur. J. Inorg. Chem.* **1999**, 1507.
[857] Schmidt, B.; Neitemeier, V. *Synthesis* **1998**, 42.
[858] Zhang, B.; Breslow, R. *J. Am. Chem. Soc.* **1997**, *119*, 1676.
[859] Vögtle, F.; Hochberg, R.; Kochendörfer, F.; Windscheif, P.-M.; Volkmann, M.; Jansen, M. *Chem. Ber.* **1990**, *123*, 2181.
[860] Rode, T.; Breitmaier, E. *Synthesis* **1987**, 574.
[861] Maruyama, T.; Yamamoto, T. *Inorg. Chim. Acta* **1995**, *238*, 9.
[862] Cassol, T. M.; Demnitz, F. W. J.; Navarro, M.; Neves, E. A. d. *Tetrahedron Lett.* **2000**, *41*, 8203.
[863] Newkome, G. R.; Puckett, W. E.; Kiefer, G. E.; Gupta, V. K.; Xia, Y.; Coreil, M.; Hackney, M. A. *J. Org. Chem.* **1982**, *47*, 4116.
[864] Newkome, G. R.; Pantaleo, D. C.; Puckett, W. E.; Ziefle, P. L.; Deutsch, W. A. *J. Inorg. Nucl. Chem.* **1981**, *43*, 1529.
[865] Tiecco, M.; Tingoli, M.; Testaferri, L.; Bartoli, D.; Chianelli, D. *Tetrahedron* **1989**, *45*, 2857.
[866] Falk, H.; Suste, A. *Monatsh. Chem.* **1993**, *124*, 881.

[867] Bolm, C.; Ewald, M.; Felder, M.; Schlingloff, G. *Chem. Ber.* **1992**, *125*, 1169.
[868] Bolm, C.; Zehnder, M.; Bur, D. *Angew. Chem., Int. Ed. Engl.* **1990**, *29*, 205.
[869] Constable, E. C.; Elder, S. M.; Healy, J.; Tocher, D. A. *J. Chem. Soc., Dalton Trans.* **1990**, 1669.
[870] Constable, E. C.; Elder, S. M.; Hannon, M. J.; Martin, A.; Raithby, P. R.; Tocher, D. A. *J. Chem. Soc., Dalton Trans.* **1996**, 2423.
[871] Rodríguez-Ubis, J. C.; Sedano, R.; Barroso, G.; Juanes, O.; Brunet, E. *Helv. Chim. Acta* **1997**, *80*, 86.
[872] Chelucci, G.; Falorni, M.; Giacomelli, G. *Tetrahedron* **1992**, *48*, 3653.
[873] Chotalia, R.; Constable, E. C.; Neuburger, M.; Smith, D. R.; Zehnder, M. *J. Chem. Soc., Dalton Trans.* **1996**, 4207.
[874] Peterson, M. A.; Dalley, N. K. *Synth. Commun.* **1996**, *26*, 2223.
[875] Funeriu, D.-P.; He, Y.-B.; Bister, H.-J.; Lehn, J.-M. *Bull. Soc. Chim. Fr.* **1996**, *133*, 673.
[876] Dehmlow, E. V.; Schulz, H.-J. *Liebigs Ann. Chem.* **1987**, 857.
[877] Dehmlow, E. V.; Schulz, H.-J. *Tetrahedron Lett.* **1985**, *26*, 4903.
[878] Tiecco, M.; Tingoli, M.; Testaferri, L.; Chianelli, D.; Wenkert, E. *Experientia* **1987**, *43*, 462.
[879] Godard, A.; Marsais, F.; Plé, N.; Trécourt, F.; Turck, A.; Quéguiner, G. *Heterocycles* **1995**, *40*, 1055.
[880] Trécourt, F.; Mallet, M.; Mongin, O.; Gevais, B.; Quéguiner, G. *Tetrahedron* **1993**, *49*, 8373.
[881] Mongin, O.; Rocca, P.; Thomas-dit-Dumont, L.; Trécourt, F.; Marsais, F.; Godard, A.; Quéguiner, G. *J. Chem. Soc., Perkin Trans. 1* **1995**, 2503.
[882] Lehn, J.-M.; Sauvage, J.-P.; Simon, J.; Ziessel, R.; Piccinni-Leopardi, C.; Germain, G.; Declercq, J.-P.; Van Meerssche, M. *Nouv. J. Chim.* **1983**, *7*, 413; *Chem. Abstr.* **1984**, *100*, 16686.
[883] Bolm, C.; Ewald, M.; Zehnder, M.; Neuburger, M. A. *Chem. Ber.* **1992**, *125*, 453.
[884] Ito, K.; Katsuki, T. *Tetrahedron Lett.* **1993**, *34*, 2661.
[885] Malkov, A. V.; Bella, M.; Langer, V.; Kocovsk, P. *Org. Lett.* **2000**, *2*, 3047.
[886] Ito, K.; Yoshitake, M.; Katsuki, T. *Heterocycles* **1996**, *42*, 305.
[887] Ito, K.; Katsuki, T. *Chem. Lett.* **1994**, 1857.
[888] Mack, A. G.; Suschitzky, H.; Wakefield, B. J. *J. Chem. Soc., Perkin Trans. 1* **1980**, 1682.
[889] Goshaev, M.; Otroshchenko, O. S.; Sadyko, A. S. *Tr. Samarkand. Gos. Univ.* **1969**, *167*, 95; *Chem. Abstr.* **1971**, *74*, 53433.
[890] Plaquevent, J.-C.; Chichaoui, I. *Bull. Soc. Chim. Fr.* **1996**, *133*, 369.
[891] Frank, R. L.; Crawford, J. V. *Bull. Soc. Chim. Fr.* **1958**, 419.
[892] Moran, D. B.; Morton, G. O.; Albright, J. D. *J. Heterocycl. Chem.* **1986**, *23*, 1071.
[893] Jones, W. D.; Jenkins, G. L.; Christian, J. E. *J. Am. Pharm. Assoc.* **1949**, *38*, 70.
[894] Pai, C.-C.; Lin, C.-W.; Lin, C.-C.; Chen, C.-C.; Chan, A. S. C.; Wong, W. T. *J. Am. Chem. Soc.* **2000**, *122*, 11513.
[895] Chambers, R. D.; Hutchinson, J.; Musgrave, W. K. R. *J. Chem. Soc.* **1965**, 5040.
[896] Banks, R. E.; Haszeldine, R. N.; Phillips, E.; Young, I. M. *J. Chem. Soc. (C)* **1967**, 2091.
[897] Banks, R. E.; Haszeldine, R. N.; Phillips, E. *J. Fluorine Chem.* **1977**, *9*, 243.
[898] Collins, I.; Roberts, S. M.; Suschitzky, H. *J. Chem. Soc. (C)* **1971**, 167.
[899] Profft, E.; Richter, H. *J. Prakt. Chem.* **1959**, *9*, 164.
[900] de Geest, D. J.; Steel, P. J. *Inorg. Chem. Commun.* **1998**, *1*, 358.
[901] Constable, E. C.; Ward, M. D. *J. Chem. Soc., Dalton Trans.* **1990**, 1405.
[902] Constable, E. C.; Thompson, A. M. W. C. *J. Chem. Soc., Dalton Trans.* **1992**, 3467.
[903] Bly, D. D.; Mellon, M. G. *J. Org. Chem.* **1962**, *27*, 2945.
[904] Musgrave, T. R.; Wescott, P. A. *Org. Synth.* **1972**, *52*, 1799.
[905] Petty, R. H.; Welch, B. R.; Wilson, L. J.; Bottomley, L. A.; Kadish, K. M. *J. Am. Chem. Soc.* **1980**, *102*, 611.
[906] Mukkala, V.-M.; Sund, C.; Kwiatkowski, M.; Pasanen, P.; Högberg, M.; Kankare, J.; Takalo, H. *Helv. Chim. Acta* **1992**, *75*, 1621.
[907] Bly, D. D. *J. Org. Chem.* **1964**, *29*, 943.
[908] Crossley, M. J.; Gorjian, S.; Sternhell, S.; Tansey, K. M. *Aust. J. Chem.* **1994**, *47*, 723.
[909] Lafferty, J. J.; Case, F. H. *J. Org. Chem.* **1967**, *32*, 1591.
[910] Yanai, K.; Naito, T. *Yakugaku Zasshi* **1941**, *61*, 99; *Chem. Abstr.* **1942**, *36*, 479.
[911] Banks, R. E.; Field, D. S.; Haszeldine, R. N. *J. Chem. Soc. (C)* **1969**, 1866.
[912] Caton, M. P. L.; Hurst, D. T.; McOmie, J. F. W.; Hunt, R. R. *J. Chem. Soc. (C)* **1967**, 1204.

[913] Papet, A.-L.; Marsura, A. *Synthesis* **1993**, 478.
[914] Papet, A. L.; Marsura, A.; Ghermani, N.; Lecomte, C.; Friant, P.; Rivail, J. L. *New J. Chem.* **1993**, *17*, 181.
[915] Yanai, M.; Naito, T. *J. Pharm. Soc. Jpn.* **1941**, *61*, 99; *Chem. Abstr.* **1942**, *36*, 479.
[916] Breckenridge, J. G. *Can. J. Res. B* **1950**, *28*, 593; *Chem. 'Abstr.* **1951**, *45*, 41479.
[917] Case, F. H.; Maerker, G. *J. Am. Chem. Soc.* **1953**, *75*, 4920.
[918] Nakano, S. *Yakugaku Zasshi* **1959**, *79*, 314; *Chem. Abstr.* **1959**, *53*, 16134a.
[919] Case, F. H.; Lafferty, J. J. *J. Org. Chem.* **1958**, *23*, 1375.
[920] Yamanaka, H.; An-naka, M.; Kondo, Y.; Sakamoto, T. *Chem. Pharm. Bull.* **1985**, *33*, 4309.
[921] Slany, M.; Stang, P. J. *Synthesis* **1996**, 1019.
[922] Lee, V. J.; Wang, Y.; Taran, C.; Marocco, M. L. U. S. Patent 5,532,374 (1996); *Chem. Abstr.* **1996**, *125*, 143513.
[923] Ueda, K. *Yakugaku Zasshi* **1940**, *60*, 536; *Chem. Abstr.* **1941**, *35*, 1791.
[924] Fujii, M.; Honda, A. *J. Heterocycl. Chem.* **1992**, *29*, 931.
[925] Case, F. H. *J. Org. Chem.* **1952**, *17*, 471.
[926] Dai, L.-x.; Zhou, Z.-h.; Zhang, Y.-z.; Ni, C.-z.; Zhang, Z.-m.; Zhou, Y.-f. *J. Chem. Soc., Chem. Commun.* **1987**, 1760.
[927] Hirao, K.-i.; Tsuchiya, R.; Yano, Y.; Tsue, H. *Heterocycles* **1996**, *42*, 415.
[928] Chelucci, G.; Cabras, M. A.; Saba, A.; Sechi, A. *Tetrahedron: Asymmetry* **1996**, *7*, 1027.
[929] Ford, A.; Sinn, E.; Woodward, S. *J. Chem. Soc., Perkin Trans. 1* **1997**, 927.
[930] Kitamura, C.; Yamamoto, S.; Ouchi, M.; Yoneda, A. *J. Chem. Res. (S)* **2000**, 46.
[931] Rinkes, I. J. *Recl. Trav. Chim. Pays-Bas* **1931**, *50*, 981.
[932] Märkl, G.; Knott, T.; Kreitmeier, P.; Burgemeister, T.; Kastner, F. *Tetrahedron* **1996**, *52*, 11763.
[933] Wirth, H. O.; Waese, G.; Kern, W. *Makromol. Chem.* **1965**, *86*, 139.
[934] Gilman, H.; Weipert, E. A.; Dietrich, J. J.; Hayes, F. N. *J. Org. Chem.* **1958**, *23*, 361.
[935] Sease, J. W.; Zechmeister, L. *J. Am. Chem. Soc.* **1947**, *69*, 270.
[936] Wynberg, H.; Logothetis, A. *J. Am. Chem. Soc.* **1956**, *78*, 1958.
[937] Jean, G. N.; Nord, F. F. *J. Org. Chem.* **1955**, *20*, 1363.
[938] Lipkin, A. E. *Zh. Obsh. Khim.* **1963**, *33*, 196; *J. Gen. Chem. U.S.S.R.* **1963**, *33*, 188; *Chem. Abstr.* **1963**, *59*, 75160.
[939] Hassan, J.; Lavenot, L.; Gozzi, C.; Lemaire, M. *Tetrahedron Lett.* **1999**, *40*, 857.
[940] Yoshida, S.; Fujii, M.; Aso, Y.; Otsubo, T.; Ogura, F. *J. Org. Chem.* **1994**, *59*, 3077.
[941] Wei, Y.; Wang, B.; Wang, W.; Tian, J. *Tetrahedron Lett.* **1995**, *36*, 665.
[942] Uhlenbroek, J. H.; Bijloo, J. D. *Recl. Trav. Chim Pays-Bas* **1960**, *79*, 1181.
[943] Steinkopf, W.; Leitsmann, R.; Müller, A. H.; Wilhelm, H. *Liebigs Ann. Chem.* **1939**, *541*, 260.
[944] Owen, L. J.; Nord, F. F. *J. Org. Chem.* **1951**, *16*, 1864.
[945] Owen, L. J.; Nord, F. F. *Nature* **1951**, *167*, 1035.
[946] Steinkopf, W.; Merckoll, A.; Strauch, H. *Liebigs Ann. Chem.* **1940**, *545*, 45.
[947] Sy, M.; Buu-Hoï, N. P.; Xuong, N. D. *J. Chem. Soc.* **1954**, 1975.
[948] Steinkopf, W.; Hanske, W. *Liebigs Ann. Chem.* **1939**, *541*, 238.
[949] Nakayama, J.; Konishi, T.; Murabayashi, S.; Hoshino, M. *Heterocycles* **1987**, *26*, 1793.
[950] Yui, K.; Aso, Y.; Otsubo, T.; Ogura, F. *Bull. Chem. Soc. Jpn.* **1989**, *62*, 1539.
[951] Steinkopf, W.; Leitsmann, R.; Hofmann, K. H. *Liebigs Ann. Chem.* **1941**, *546*, 180.
[952] Kuroda, M.; Nakayama, J.; Hoshino, M.; Furusho, N.; Ohba, S. *Tetrahedron Lett.* **1994**, *35*, 3957.
[953] Minnis, W. *J. Am. Chem. Soc.* **1929**, *51*, 2143.
[954] Bäuerle, P.; Pfau, F.; Schlupp, H.; Würthner, F.; Gaudl, K.-U.; Caro, M. B.; Fisher, P. *J. Chem. Soc., Perkin Trans. 2* **1993**, 489.
[955] Parakka, J. P.; Cava, M. P. *Tetrahedron* **1995**, *51*, 2229.
[956] Shepherd, M. K. *J. Chem. Soc., Chem. Commun.* **1985**, 880.
[957] Gronowitz, S.; Dahlgren, K. *Arkiv Kemi* **1963**, *21*, 201; *Chem. Abstr.* **1963**, *59*, 69005.
[958] Håkansson, R.; Wiklund, E. *Arkiv Kemi* **1969**, *31*, 101; *Chem. Abstr.* **1969**, *71*, 12917.
[959] Steinkopf, W.; Petersdorff, H.-J. *Liebigs Ann. Chem.* **1940**, *543*, 119.
[960] Schuetz, R. D.; Ciporin, L. *J. Org. Chem.* **1958**, *23*, 206.
[961] Gilman, H.; Wilder, G. R. *J. Org. Chem.* **1957**, *22*, 523.

962. Shabana, R.; Galal, A.; Mark, H. B.; Zimmer, H.; Gronowitz, S.; Hörnfeldt, A.-B. *Phosphorus Sulfur Silicon Relat. Elem.* **1990**, *48*, 239.
963. Chierici, L.; Dell'Erba, C.; Guareschi, A.; Spinelli, D. *Ric. Sci., Rend., Sez. A* **1965**, *8*, 1537; *Chem. Abstr.* **1966**, *65*, 2203h.
964. Dell'Erba, C.; Spinelli, D.; Garbarino, G.; Leandri, G. *J. Heterocycl. Chem.* **1968**, *5*, 45.
965. Perevalova, E. G.; Nesmeyanova, O. A. *Dokl. Akad. Nauk SSSR* **1960**, *132*, 1093; *Proc. Acad. Sci. USSR (English Translation)* **1960**, *132*, 673; *Chem. Abstr.* **1960**, *54*, 110429.
966. Neuse, E. W. *J. Macromol. Sci.; Chem.* **1981**, *A16*, 3.
967. Apen, P. G.; Rasmussen, P. G. *J. Am. Chem. Soc.* **1991**, *113*, 6178.
968. Trauner, H.; Le Floch, P.; Lefour, J.-M.; Ricard, L.; Mathey, F. *Synthesis* **1995**, 717.
969. Bergman, J.; Eklund, N. *Tetrahedron* **1980**, *36*, 1439.
970. Lehmstedt, K.; Hundertmark, H. *Chem. Ber.* **1929**, *62*, 1065.
971. Ullmann, F.; von Glenck, O. *Chem. Ber.* **1916**, *49*, 2487.
972. Jordens, P.; Rawson, G.; Wynberg, H. *J. Chem. Soc. (C)* **1970**, 273.
973. Lucas, P.; Mehdi, N. E.; Ho, H. A.; Bélanger, D.; Breau, L. *Synthesis* **2000**, 1253.
974. Wiersema, A.; Gronowitz, S. *Acta Chem. Scand.* **1970**, *24*, 2593.
975. Yamamoto, K.; Tateishi, H.; Watanabe, K.; Adachi, T.; Matsubara, H.; Ueda, T.; Yoshida, T. *J. Chem. Soc., Chem. Commun.* **1995**, 1637.
976. Falk, H.; Chen, Q.-Q. *Monatsh. Chem.* **1996**, *127*, 69.
977. Kelly, T. R.; Lee, Y.-J.; Mears, R. J. *J. Org. Chem.* **1997**, *62*, 2774.
978. Takahashi, M.; Kuroda, T.; Ogiku, T.; Ohmizu, H.; Kondo, K.; Iwasaki, T. *Heterocycles* **1993**, *36*, 1867.
979. Takahashi, M.; Kuroda, T.; Ogiku, T.; Ohmizu, H.; Kondo, K.; Iwasaki, T. *Heterocycles* **1992**, *34*, 2061.
980. Kizu, K.; Maruyama, T.; Yamamoto, T. *Polym. J.* **1995**, *27*, 205.
981. Yamamoto, T.; Ito, T.; Kubota, K. *Chem. Lett.* **1988**, 153.
982. Brockmann, T. W.; Tour, J. M. *J. Am. Chem. Soc.* **1995**, *117*, 4437.
983. Yamamoto, T.; Zhou, Z.-h.; Kanbara, T.; Maruyama, T. *Chem. Lett.* **1990**, 223.

CUMULATIVE CHAPTER TITLES BY VOLUME

Volume 1 (1942)

1. **The Reformatsky Reaction**: Ralph L. Shriner

2. **The Arndt-Eistert Reaction**: W. E. Bachmann and W. S. Struve

3. **Chloromethylation of Aromatic Compounds**: Reynold C. Fuson and C. H. McKeever

4. **The Amination of Heterocyclic Bases by Alkali Amides**: Marlin T. Leffler

5. **The Bucherer Reaction**: Nathan L. Drake

6. **The Elbs Reaction**: Louis F. Fieser

7. **The Clemmensen Reduction**: Elmore L. Martin

8. **The Perkin Reaction and Related Reactions**: John R. Johnson

9. **The Acetoacetic Ester Condensation and Certain Related Reactions**: Charles R. Hauser and Boyd E. Hudson, Jr.

10. **The Mannich Reaction**: F. F. Blicke

11. **The Fries Reaction**: A. H. Blatt

12. **The Jacobson Reaction**: Lee Irvin Smith

Volume 2 (1944)

1. **The Claisen Rearrangement**: D. Stanley Tarbell

2. **The Preparation of Aliphatic Fluorine Compounds**: Albert L. Henne

3. **The Cannizzaro Reaction**: T. A. Geissman

4. **The Formation of Cyclic Ketones by Intramolecular Acylation**: William S. Johnson

5. **Reduction with Aluminum Alkoxides (The Meerwein-Ponndorf-Verley Reduction)**: A. L. Wilds

557

6. **The Preparation of Unsymmetrical Biaryls by the Diazo Reaction and the Nitrosoacetylamine Reaction:** Werner E. Bachmann and Roger A. Hoffman

7. **Replacement of the Aromatic Primary Amino Group by Hydrogen:** Nathan Kornblum

8. **Periodic Acid Oxidation:** Ernest L. Jackson

9. **The Resolution of Alcohols:** A. W. Ingersoll

10. **The Preparation of Aromatic Arsonic and Arsinic Acids by the Bart, Béchamp, and Rosenmund Reactions:** Cliff S. Hamilton and Jack F. Morgan

Volume 3 (1946)

1. **The Alkylation of Aromatic Compounds by the Friedel-Crafts Method:** Charles C. Price

2. **The Willgerodt Reaction:** Marvin Carmack and M. A. Spielman

3. **Preparation of Ketenes and Ketene Dimers:** W. E. Hanford and John C. Sauer

4. **Direct Sulfonation of Aromatic Hydrocarbons and Their Halogen Derivatives:** C. M. Suter and Arthur W. Weston

5. **Azlactones:** H. E. Carter

6. **Substitution and Addition Reactions of Thiocyanogen:** John L. Wood

7. **The Hofmann Reaction:** Everett L. Wallis and John F. Lane

8. **The Schmidt Reaction:** Hans Wolff

9. **The Curtius Reaction:** Peter A. S. Smith

Volume 4 (1948)

1. **The Diels-Alder Reaction with Maleic Anhydride:** Milton C. Kloetzel

2. **The Diels-Alder Reaction: Ethylenic and Acetylenic Dienophiles:** H. L. Holmes

3. **The Preparation of Amines by Reductive Alkylation:** William S. Emerson

4. **The Acyloins:** S. M. McElvain

5. **The Synthesis of Benzoins:** Walter S. Ide and Johannes S. Buck

6. **Synthesis of Benzoquinones by Oxidation:** James Cason

7. **The Rosenmund Reduction of Acid Chlorides to Aldehydes:** Erich Mosettig and Ralph Mozingo

8. **The Wolff-Kishner Reduction:** David Todd

Volume 5 (1949)

1. **The Synthesis of Acetylenes**: Thomas L. Jacobs

2. **Cyanoethylation**: Herman L. Bruson

3. **The Diels-Alder Reaction: Quinones and Other Cyclenones**: Lewis L. Butz and Anton W. Rytina

4. **Preparation of Aromatic Fluorine Compounds from Diazonium Fluoborates: The Schiemann Reaction**: Arthur Roe

5. **The Friedel and Crafts Reaction with Aliphatic Dibasic Acid Anhydrides**: Ernst Berliner

6. **The Gattermann-Koch Reaction**: Nathan N. Crounse

7. **The Leuckart Reaction**: Maurice L. Moore

8. **Selenium Dioxide Oxidation**: Norman Rabjohn

9. **The Hoesch Synthesis**: Paul E. Spoerri and Adrien S. DuBois

10. **The Darzens Glycidic Ester Condensation**: Melvin S. Newman and Barney J. Magerlein

Volume 6 (1951)

1. **The Stobbe Condensation**: William S. Johnson and Guido H. Daub

2. **The Preparation of 3,4-Dihydroisoquinolines and Related Compounds by the Bischler-Napieralski Reaction**: Wilson M. Whaley and Tutucorin R. Govindachari

3. **The Pictet-Spengler Synthesis of Tetrahydroisoquinolines and Related Compounds**: Wilson M. Whaley and Tutucorin R. Govindachari

4. **The Synthesis of Isoquinolines by the Pomeranz-Fritsch Reaction**: Walter J. Gensler

5. **The Oppenauer Oxidation**: Carl Djerassi

6. **The Synthesis of Phosphonic and Phosphinic Acids**: Gennady M. Kosolapoff

7. **The Halogen-Metal Interconversion Reaction with Organolithium Compounds**: Reuben G. Jones and Henry Gilman

8. **The Preparation of Thiazoles**: Richard H. Wiley, D. C. England, and Lyell C. Behr

9. **The Preparation of Thiophenes and Tetrahydrothiophenes**: Donald E. Wolf and Karl Folkers

10. **Reductions by Lithium Aluminum Hydride**: Weldon G. Brown

Volume 7 (1953)

1. **The Pechmann Reaction**: Suresh Sethna and Ragini Phadke

2. **The Skraup Synthesis of Quinolines**: R. H. F. Manske and Marshall Kulka

3. **Carbon-Carbon Alkylations with Amines and Ammonium Salts**: James H. Brewster and Ernest L. Eliel

4. **The von Braun Cyanogen Bromide Reaction**: Howard A. Hageman

5. **Hydrogenolysis of Benzyl Groups Attached to Oxygen, Nitrogen, or Sulfur**: Walter H. Hartung and Robert Simonoff

6. **The Nitrosation of Aliphatic Carbon Atoms**: Oscar Touster

7. **Epoxidation and Hydroxylation of Ethylenic Compounds with Organic Peracids**: Daniel Swern

Volume 8 (1954)

1. **Catalytic Hydrogenation of Esters to Alcohols**: Homer Adkins

2. **The Synthesis of Ketones from Acid Halides and Organometallic Compounds of Magnesium, Zinc, and Cadmium**: David A. Shirley

3. **The Acylation of Ketones to Form β-Diketones or β-Keto Aldehydes**: Charles R. Hauser, Frederic W. Swamer, and Joe T. Adams

4. **The Sommelet Reaction**: S. J. Angyal

5. **The Synthesis of Aldehydes from Carboxylic Acids**: Erich Mosettig

6. **The Metalation Reaction with Organolithium Compounds**: Henry Gilman and John W. Morton, Jr.

7. **β-Lactones**: Harold E. Zaugg

8. **The Reaction of Diazomethane and Its Derivatives with Aldehydes and Ketones**: C. David Gutsche

Volume 9 (1957)

1. **The Cleavage of Non-enolizable Ketones with Sodium Amide**: K. E. Hamlin and Arthur W. Weston

2. **The Gattermann Synthesis of Aldehydes**: William E. Truce

3. **The Baeyer-Villiger Oxidation of Aldehydes and Ketones**: C. H. Hassall

4. **The Alkylation of Esters and Nitriles**: Arthur C. Cope, H. L. Holmes, and Herbert O. House

5. **The Reaction of Halogens with Silver Salts of Carboxylic Acids**: C. V. Wilson

6. **The Synthesis of β-Lactams**: John C. Sheehan and Elias J. Corey

7. **The Pschorr Synthesis and Related Diazonium Ring Closure Reactions**: DeLos F. DeTar

Volume 10 (1959)

1. **The Coupling of Diazonium Salts with Aliphatic Carbon Atoms**: Stanley J. Parmerter

2. **The Japp-Klingemann Reaction**: Robert R. Phillips

3. **The Michael Reaction**: Ernst D. Bergmann, David Ginsburg, and Raphael Pappo

Volume 11 (1960)

1. **The Beckmann Rearrangement**: L. Guy Donaruma and Walter Z. Heldt

2. **The Demjanov and Tiffeneau-Demjanov Ring Expansions**: Peter A. S. Smith and Donald R. Baer

3. **Arylation of Unsaturated Compounds by Diazonium Salts**: Christian S. Rondestvedt, Jr.

4. **The Favorskii Rearrangement of Haloketones**: Andrew S. Kende

5. **Olefins from Amines: The Hofmann Elimination Reaction and Amine Oxide Pyrolysis**: Arthur C. Cope and Elmer R. Trumbull

Volume 12 (1962)

1. **Cyclobutane Derivatives from Thermal Cycloaddition Reactions**: John D. Roberts and Clay M. Sharts

2. **The Preparation of Olefins by the Pyrolysis of Xanthates. The Chugaev Reaction**: Harold R. Nace

3. **The Synthesis of Aliphatic and Alicyclic Nitro Compounds**: Nathan Kornblum

4. **Synthesis of Peptides with Mixed Anhydrides**: Noel F. Albertson

5. **Desulfurization with Raney Nickel**: George R. Pettit and Eugene E. van Tamelen

Volume 13 (1963)

1. **Hydration of Olefins, Dienes, and Acetylenes via Hydroboration**: George Zweifel and Herbert C. Brown

2. **Halocyclopropanes from Halocarbenes**: William E. Parham and Edward E. Schweizer

3. **Free Radical Addition to Olefins to Form Carbon-Carbon Bonds**: Cheves Walling and Earl S. Huyser

4. **Formation of Carbon-Heteroatom Bonds by Free Radical Chain Additions to Carbon-Carbon Multiple Bonds**: F. W. Stacey and J. F. Harris, Jr.

Volume 14 (1965)

1. **The Chapman Rearrangement**: J. W. Schulenberg and S. Archer

2. **α-Amidoalkylations at Carbon**: Harold E. Zaugg and William B. Martin

3. **The Wittig Reaction**: Adalbert Maercker

Volume 15 (1967)

1. **The Dieckmann Condensation**: John P. Schaefer and Jordan J. Bloomfield

2. **The Knoevenagel Condensation**: G. Jones

Volume 16 (1968)

1. **The Aldol Condensation**: Arnold T. Nielsen and William J. Houlihan

Volume 17 (1969)

1. **The Synthesis of Substituted Ferrocenes and Other π-Cyclopentadienyl-Transition Metal Compounds**: Donald E. Bublitz and Kenneth L. Rinehart, Jr.

2. **The γ-Alkylation and γ-Arylation of Dianions of β-Dicarbonyl Compounds**: Thomas M. Harris and Constance M. Harris

3. **The Ritter Reaction**: L. I. Krimen and Donald J. Cota

Volume 18 (1970)

1. **Preparation of Ketones from the Reaction of Organolithium Reagents with Carboxylic Acids**: Margaret J. Jorgenson

2. **The Smiles and Related Rearrangements of Aromatic Systems**: W. E. Truce, Eunice M. Kreider, and William W. Brand

3. **The Reactions of Diazoacetic Esters with Alkenes, Alkynes, Heterocyclic, and Aromatic Compounds**: Vinod Dave and E. W. Warnhoff

4. **The Base-Promoted Rearrangements of Quaternary Ammonium Salts**: Stanley H. Pine

Volume 19 (1972)

1. **Conjugate Addition Reactions of Organocopper Reagents**: Gary H. Posner

2. **Formation of Carbon-Carbon Bonds via π-Allylnickel Compounds**: Martin F. Semmelhack

3. **The Thiele-Winter Acetoxylation of Quinones**: J. F. W. McOmie and J. M. Blatchly

4. **Oxidative Decarboxylation of Acids by Lead Tetraacetate**: Roger A. Sheldon and Jay K. Kochi

Volume 20 (1973)

1. **Cyclopropanes from Unsaturated Compounds, Methylene Iodide, and Zinc-Copper Couple**: H. E. Simmons, T. L. Cairns, Susan A. Vladuchick, and Connie M. Hoiness

2. **Sensitized Photooxygenation of Olefins**: R. W. Denny and A. Nickon

3. **The Synthesis of 5-Hydroxyindoles by the Nenitzescu Reaction**: George R. Allen, Jr.

4. **The Zinin Reaction of Nitroarenes**: H. K. Porter

Volume 21 (1974)

1. **Fluorination with Sulfur Tetrafluoride**: G. A. Boswell, Jr., W. C. Ripka, R. M. Scribner, and C. W. Tullock

2. **Modern Methods to Prepare Monofluoroaliphatic Compounds**: Clay M. Sharts and William A. Sheppard

Volume 22 (1975)

1. **The Claisen and Cope Rearrangements**: Sara Jane Rhoads and N. Rebecca Raulins

2. **Substitution Reactions Using Organocopper Reagents**: Gary H. Posner

3. **Clemmensen Reduction of Ketones in Anhydrous Organic Solvents**: E. Vedejs

4. **The Reformatsky Reaction**: Michael W. Rathke

Volume 23 (1976)

1. **Reduction and Related Reactions of α,β-Unsaturated Compounds with Metals in Liquid Ammonia**: Drury Caine

2. **The Acyloin Condensation**: Jordan J. Bloomfield, Dennis C. Owsley, and Janice M. Nelke

3. **Alkenes from Tosylhydrazones**: Robert H. Shapiro

Volume 24 (1976)

1. **Homogeneous Hydrogenation Catalysts in Organic Solvents**: Arthur J. Birch and David H. Williamson

2. **Ester Cleavages via S_N2-Type Dealkylation**: John E. McMurry

3. **Arylation of Unsaturated Compounds by Diazonium Salts (The Meerwein Arylation Reaction)**: Christian S. Rondestvedt, Jr.

4. **Selenium Dioxide Oxidation**: Norman Rabjohn

Volume 25 (1977)

1. **The Ramberg-Bäcklund Rearrangement**: Leo A. Paquette

2. **Synthetic Applications of Phosphoryl-Stabilized Anions**: William S. Wadsworth, Jr.

3. **Hydrocyanation of Conjugated Carbonyl Compounds**: Wataru Nagata and Mitsuru Yoshioka

Volume 26 (1979)

1. **Heteroatom-Facilitated Lithiations**: Heinz W. Gschwend and Herman R. Rodriguez

2. **Intramolecular Reactions of Diazocarbonyl Compounds**: Steven D. Burke and Paul A. Grieco

Volume 27 (1982)

1. **Allylic and Benzylic Carbanions Substituted by Heteroatoms**: Jean-François Biellmann and Jean-Bernard Ducep

2. **Palladium-Catalyzed Vinylation of Organic Halides**: Richard F. Heck

Volume 28 (1982)

1. **The Reimer-Tiemann Reaction**: Hans Wynberg and Egbert W. Meijer

2. **The Friedländer Synthesis of Quinolines**: Chia-Chung Cheng and Shou-Jen Yan

3. **The Directed Aldol Reaction**: Teruaki Mukaiyama

Volume 29 (1983)

1. **Replacement of Alcoholic Hydroxy Groups by Halogens and Other Nucleophiles via Oxyphosphonium Intermediates**: Bertrand R. Castro

2. **Reductive Dehalogenation of Polyhalo Ketones with Low-Valent Metals and Related Reducing Agents**: Ryoji Noyori and Yoshihiro Hayakawa

3. **Base-Promoted Isomerizations of Epoxides**: Jack K. Crandall and Marcel Apparu

Volume 30 (1984)

1. **Photocyclization of Stilbenes and Related Molecules**: Frank B. Mallory and Clelia W. Mallory

2. **Olefin Synthesis via Deoxygenation of Vicinal Diols**: Eric Block

Volume 31 (1984)

1. **Addition and Substitution Reactions of Nitrile-Stabilized Carbanions**: Siméon Arseniyadis, Keith S. Kyler, and David S. Watt

Volume 32 (1984)

1. **The Intramolecular Diels-Alder Reaction**: Engelbert Ciganek

2. **Synthesis Using Alkyne-Derived Alkenyl- and Alkynylaluminum Compounds**: George Zweifel and Joseph A. Miller

Volume 33 (1985)

1. **Formation of Carbon-Carbon and Carbon-Heteroatom Bonds via Organoboranes and Organoborates**: Ei-Ichi Negishi and Michael J. Idacavage

2. **The Vinylcyclopropane-Cyclopentene Rearrangement**: Tomáš Hudlický, Toni M. Kutchan, and Saiyid M. Naqvi

Volume 34 (1985)

1. **Reductions by Metal Alkoxyaluminum Hydrides**: Jaroslav Málek

2. **Fluorination by Sulfur Tetrafluoride**: Chia-Lin J. Wang

Volume 35 (1988)

1. **The Beckmann Reactions: Rearrangements, Elimination-Additions, Fragmentations, and Rearrangement-Cyclizations**: Robert E. Gawley

2. **The Persulfate Oxidation of Phenols and Arylamines (The Elbs and the Boyland-Sims Oxidations)**: E. J. Behrman

3. **Fluorination with Diethylaminosulfur Trifluoride and Related Aminofluorosulfuranes**: Miloš Hudlický

Volume 36 (1988)

1. **The [3 + 2] Nitrone-Olefin Cycloaddition Reaction**: Pat N. Confalone and Edward M. Huie

2. **Phosphorus Addition at sp^2 Carbon**: Robert Engel

3. **Reduction by Metal Alkoxyaluminum Hydrides. Part II. Carboxylic Acids and Derivatives, Nitrogen Compounds, and Sulfur Compounds**: Jaroslav Málek

Volume 37 (1989)

1. **Chiral Synthons by Ester Hydrolysis Catalyzed by Pig Liver Esterase**: Masaji Ohno and Masami Otsuka

2. **The Electrophilic Substitution of Allylsilanes and Vinylsilanes**: Ian Fleming, Jacques Dunoguès, and Roger Smithers

Volume 38 (1990)

1. **The Peterson Olefination Reaction**: David J. Ager

2. **Tandem Vicinal Difunctionalization: β-Addition to α,β-Unsaturated Carbonyl Substrates Followed by α-Functionalization**: Marc J. Chapdelaine and Martin Hulce

3. **The Nef Reaction**: Harold W. Pinnick

Volume 39 (1990)

1. **Lithioalkenes from Arenesulfonylhydrazones**: A. Richard Chamberlin and Steven H. Bloom

2. **The Polonovski Reaction**: David Grierson

3. **Oxidation of Alcohols to Carbonyl Compounds via Alkoxysulfonium Ylides: The Moffatt, Swern, and Related Oxidations**: Thomas T. Tidwell

Volume 40 (1991)

1. **The Pauson-Khand Cycloaddition Reaction for Synthesis of Cyclopentenones**: Neil E. Schore

2. **Reduction with Diimide**: Daniel J. Pasto and Richard T. Taylor

3. **The Pummerer Reaction of Sulfinyl Compounds**: Ottorino DeLucchi, Umberto Miotti, and Giorgio Modena

4. **The Catalyzed Nucleophilic Addition of Aldehydes to Electrophilic Double Bonds**: Hermann Stetter and Heinrich Kuhlmann

Volume 41 (1992)

1. **Divinylcyclopropane-Cycloheptadiene Rearrangement**: Tomáš Hudlický, Rulin Fan, Josephine W. Reed, and Kumar G. Gadamasetti

2. **Organocopper Reagents: Substitution, Conjugate Addition, Carbo/Metallocupration, and Other Reactions**: Bruce H. Lipshutz and Saumitra Sengupta

Volume 42 (1992)

1. **The Birch Reduction of Aromatic Compounds**: Peter W. Rabideau and Zbigniew Marcinow

2. **The Mitsunobu Reaction**: David L. Hughes

Volume 43 (1993)

1. **Carbonyl Methylenation and Alkylidenation Using Titanium-Based Reagents**: Stanley H. Pine

2. **Anion-Assisted Sigmatropic Rearrangements**: Stephen R. Wilson

3. **The Baeyer-Villiger Oxidation of Ketones and Aldehydes**: Grant R. Krow

Volume 44 (1993)

1. **Preparation of α,β-Unsaturated Carbonyl Compounds and Nitriles by Selenoxide Elimination**: Hans J. Reich and Susan Wollowitz

2. **Enone Olefin [2 + 2] Photochemical Cyclizations**: Michael T. Crimmins and Tracy L. Reinhold

Volume 45 (1994)

1. **The Nazarov Cyclization**: Karl L. Habermas, Scott E. Denmark, and Todd K. Jones

2. **Ketene Cycloadditions**: John A. Hyatt and Peter W. Raynolds

Volume 46 (1994)

1. **Tin(II) Enolates in the Aldol, Michael, and Related Reactions**: Teruaki Mukaiyama and Shū Kobayashi

2. **The [2,3]-Wittig Reaction**: Takeshi Nakai and Koichi Mikami

3. **Reductions with Samarium(II) Iodide**: Gary A. Molander

Volume 47 (1995)

1. **Lateral Lithiation Reactions Promoted by Heteroatomic Substituents**: Robin D. Clark and Alam Jahangir

2. **The Intramolecular Michael Reaction**:   R. Daniel Little, Mohammad R. Masjedizadeh, Olof Wallquist (in part), and Jim I. McLoughlin (in part)

Volume 48 (1996)

1. **Asymmetric Epoxidation of Allylic Alcohols: The Katsuki–Sharpless Epoxidation Reaction**:   Tsutomu Katsuki and Victor S. Martin

2. **Radical Cyclization Reactions**:   B. Giese, B. Kopping, T. Göbel, J. Dickhaut, G. Thoma, K. J. Kulicke, and F. Trach

Volume 49 (1997)

1. **The Vilsmeier Reaction of Fully Conjugated Carbocycles and Heterocycles**: Gurnos Jones and Stephen P. Stanforth

2. **[6 + 4] Cycloaddition Reactions**:   James H. Rigby

3. **Carbon–Carbon Bond-Forming Reactions Promoted by Trivalent Manganese**: Gagik G. Melikyan

Volume 50 (1997)

1. **The Stille Reaction**:   Vittorio Farina, Venkat Krishnamurthy, and William J. Scott

Volume 51 (1997)

1. **Asymmetric Aldol Reactions Using Boron Enolates**:   Cameron J. Cowden and Ian Paterson

2. **The Catalyzed α-Hydroxylation and α-Aminoalkylation of Activated Olefins (The Morita–Baylis–Hillman Reaction)**:   Engelbert Ciganek

3. **[4 + 3] Cycloaddition Reactions**:   James H. Rigby and F. Christopher Pigge

Volume 52 (1998)

1. **The Retro–Diels–Alder Reaction. Part I. C—C Dienophiles**:   Bruce Rickborn

2. **Enantioselective Reduction of Ketones**:   Shinichi Itsuno

Volume 53 (1998)

1. **The Oxidation of Alcohols by Modified Oxochromium(VI)-Amine Reagents**: Frederick A. Luzzio

2. **The Retro–Diels–Alder Reaction. Part II. Dienophiles with One or More Heteroatoms**: Bruce Rickborn

Volume 54 (1999)

1. **Aromatic Substitution by the $S_{RN}1$ Reaction**:   Roberto Rossi, Adriana B. Pierini, and Ana N. Santiago

2. **Oxidation of Carbonyl Compounds with Organohypervalent Iodine Reagents**: Robert M. Moriarty and Om Prakash

Volume 55 (1999)

1. **Synthesis of Nucleosides**: Helmut Vorbrüggen and Carmen Ruh-Pohlenz

Volume 56 (2000)

1. **The Hydroformylation Reaction**: Iwao Ojima, Chung-Ying Tsai, Maria Tzamarioudaki, and Dominique Bonafoux

2. **The Vilsmeier Reaction. 2. Reactions with Compounds Other Than Fully Conjugated Carbocycles and Heterocycles**: Gurnos Jones and Stephen P. Stanforth

Volume 57 (2001)

1. **Intermolecular Metal-Catalyzed Carbenoid Cyclopropanations**: Huw M. L. Davies and Evan G. Antoulinakis

2. **Oxidation of Phenolic Compounds with Organohypervalent Iodine Reagents**: Robert M. Moriarty and Om Prakash

3. **Synthetic Uses of Tosylmethyl Isocyanide (TosMIC)**: Daan van Leusen and Albert M. van Leusen

Volume 58 (2001)

1. **Simmons-Smith Cyclopropanation Reaction:** André B. Charette and André Beauchemin

2. **Preparation and Applications of Functionalized Organozinc Compounds:** Paul Knochel, Nicolas Millot, Alain L. Rodriguez, and Charles E. Tucker

Volume 59 (2002)

1. **Reductive Aminations of Carbonyl Compounds with Borohydride and Borane Reducing Agents:** Ellen W. Baxter and Allen B. Reitz

Volume 60 (2002)

1. **Epoxide Migration (Payne Rearrangement) and Related Reactions:** Robert M. Hanson

2. **The Intramolecular Heck Reaction:** J. T. Link

Volume 61 (2002)

1. **[3 + 2] Cycloaddition of Trimethylenemethane and its Synthetic Equivalents:** Shigeru Yamago and Eiichi Nakamura

2. **Dioxirane Epoxidation of Alkenes:** Waldemar Adam, Chantu R. Saha-Möller, and Cong-Gui Zhao

Volume 62 (2003)

1. **The α-Hydroxylation of Enolates and Silyl Enol Ethers**: Bang-Chi Chen, Ping Zhou, Franklin A. Davis, and Engelbert Ciganek

2. **The Ramberg-Bäcklund Reaction**:   Richard J. K. Taylor and Guy Casy

3. **The α-Hydroxy Ketone (α-Ketol) And Related Rearrangements**:   Leo A. Paquette and John E. Hofferberth

4. **Transformation Of Glycals Into 2,3-Unsaturated Glycosyl Derivatives**:   Robert J. Ferrier and Oleg A. Zubkov

AUTHOR INDEX, VOLUMES 1–63

Volume number only is designated in this index.

Adam, Waldemar, 61
Adams, Joe T., 8
Adkins, Homer, 8
Ager, David J., 38
Albertson, Noel F., 12
Allen, George R., Jr., 20
Angyal, S. J., 8
Antoulinkis, Evan G., 57
Apparu, Marcel, 29
Archer, S., 14
Arseniyadis, Siméon, 31

Bachmann, W. E., 1, 2
Baer, Donald R., 11
Baxter, Ellen W., 59
Beauchemin, André, 58
Behr, Lyell C., 6
Behrman, E. J., 35
Bergmann, Ernst D., 10
Berliner, Ernst, 5
Biellmann, Jean-François, 27
Birch, Arthur J., 24
Blatchly, J. M., 19
Blatt, A. H., 1
Blicke, F. F., 1
Block, Eric, 30
Bloom, Steven H., 39
Bloomfield, Jordan J., 15, 23
Bonafoux, Dominique, 56
Boswell, G. A., Jr., 21
Brand, William W., 18
Brewster, James H., 7
Brown, Herbert C., 13
Brown, Weldon G., 6
Bruson, Herman Alexander, 5
Bublitz, Donald E., 17
Buck, Johannes S., 4
Burke, Steven D., 26
Butz, Lewis W., 5

Caine, Drury, 23
Cairns, Theodore L., 20
Carmack, Marvin, 3
Carter, H. E., 3

Cason, James, 4
Castro, Bertrand R., 29
Casy, Guy, 62
Chamberlin, A. Richard, 39
Chapdelaine, Marc J., 38
Charette, André B., 58
Chen, Bang-Chi, 62
Cheng, Chia-Chung, 28
Ciganek, Engelbert, 32, 51, 62
Clark, Robin D., 47
Confalone, Pat N., 36
Cope, Arthur C., 9, 11
Corey, Elias J., 9
Cota, Donald J., 17
Cowden, Cameron J., 51
Crandall, Jack K., 29
Crimmins, Michael T., 44
Crouch, R. David, 63
Crounse, Nathan N., 5

Daub, Guido H., 6
Dave, Vinod, 18
Davies, Huw M. L., 57
Davis, Franklin A., 62
Denmark, Scott E., 45
Denny, R. W., 20
DeLucchi, Ottorino, 40
DeTar, DeLos F., 9
Dickhaut, J., 48
Djerassi, Carl, 6
Donaruma, L. Guy, 11
Drake, Nathan L., 1
DuBois, Adrien S., 5
Ducep, Jean-Bernard, 27
Dunoguès, Jacques, 37

Eliel, Ernest L., 7
Emerson, William S., 4
Engel, Robert, 36
England, D. C., 6

Fan, Rulin, 41
Farina, Vittorio, 50
Ferrier, Robert J., 62

Fieser, Louis F., 1
Fleming, Ian, 37
Folkers, Karl, 6
Fuson, Reynold C., 1

Gadamasetti, Kumar G., 41
Gawley, Robert E., 35
Geissman, T. A., 2
Gensler, Walter J., 6
Giese, B., 48
Gilman, Henry, 6, 8
Ginsburg, David, 10
Göbel, T., 48
Govindachari, Tuticorin R., 6
Grieco, Paul A., 26
Grierson, David, 39
Gschwend, Heinz W., 26
Gutsche, C. David, 8

Habermas, Karl L., 45
Hageman, Howard A., 7
Hamilton, Cliff S., 2
Hamlin, K. E., 9
Hanford, W. E., 3
Hanson, Robert M., 60
Harris, Constance M., 17
Harris, J. F., Jr., 13
Harris, Thomas M., 17
Hartung, Walter H., 7
Hassall, C. H., 9
Hauser, Charles R., 1, 8
Hayakawa, Yoshihiro, 29
Heck, Richard F., 27
Heldt, Walter Z., 11
Henne, Albert L., 2
Hofferberth, John E., 62
Hoffman, Roger A., 2
Hoiness, Connie M., 20
Holmes, H. L., 4, 9
Houlihan, William J., 16
House, Herbert O., 9
Hudlický, Miloš, 35
Hudlický, Tomáš, 33, 41
Hudson, Boyd E., Jr., 1
Hughes, David L., 42
Huie, E. M., 36
Hulce, Martin, 38
Huyser, Earl S., 13
Hyatt, John A., 45

Idacavage, Michael J., 33
Ide, Walter S., 4
Ingersoll, A. W., 2
Itsuno, Shinichi, 52

Jackson, Ernest L., 2
Jacobs, Thomas L., 5
Jahangir, Alam, 47
Johnson, John R., 1
Johnson, Roy A., 63
Johnson, William S., 2, 6
Jones, Gurnos, 15, 49, 56
Jones, Reuben G., 6
Jones, Todd K., 45
Jorgenson, Margaret J., 18

Kappe, C. Oliver, 63
Katsuki, Tsutomu, 48
Kende, Andrew S., 11
Kloetzel, Milton C., 4
Knochel, Paul, 58
Kobayashi, Shū, 46
Kochi, Jay K., 19
Kopping, B., 48
Kornblum, Nathan, 2, 12
Kosolapoff, Gennady M., 6
Kreider, Eunice M., 18
Krimen, L. I., 17
Krishnamurthy, Venkat, 50
Krow, Grant R., 43
Kuhlmann, Heinrich, 40
Kulicke, K. J., 48
Kulka, Marshall, 7
Kutchan, Toni M., 33
Kyler, Keith S., 31

Lane, John F., 3
Leffler, Marlin T., 1
Link, J. T., 60
Little, R. Daniel, 47
Lipshutz, Bruce H., 41
Luzzio, Frederick A., 53

McElvain, S. M., 4
McKeever, C. H., 1
McLoughlin, Jim I., 47
McMurry, John E., 24
McOmie, J. F. W., 19
Maercker, Adalbert, 14
Magerlein, Barney J., 5
Málek, Jaroslav, 34, 36
Mallory, Clelia W., 30
Mallory, Frank B., 30
Manske, Richard H. F., 7
Marcinow, Zbigniew, 42
Martin, Elmore L., 1
Martin, Victor S., 48
Martin, William B., 14
Masjedizadeh, Mohammad R., 47

Meijer, Egbert W., 28
Melikyan, Gagik G., 49
Mikami, Koichi, 46
Miller, Joseph A., 32
Millot, Nicolas, 58
Miotti, Umberto, 40
Modena, Giorgio, 40
Molander, Gary, 46
Moore, Maurice L., 5
Morgan, Jack F., 2
Moriarty, Robert M., 54, 57
Morton, John W., Jr., 8
Mosettig, Erich, 4, 8
Mozingo, Ralph, 4
Mukaiyama, Teruaki, 28, 46

Nace, Harold R., 12
Nagata, Wataru, 25
Nakai, Takeshi, 46
Nakamura, Eiichi, 61
Naqvi, Saiyid M., 33
Negishi, Ei-Ichi, 33
Nelke, Janice M., 23
Nelson, Todd D., 63
Newman, Melvin S., 5
Nickon, A., 20
Nielsen, Arnold T., 16
Noyori, Ryoji, 29

Ohno, Masaji, 37
Ojima, Iwao, 56
Otsuka, Masami, 37
Owsley, Dennis C., 23

Pappo, Raphael, 10
Paquette, Leo A., 25, 62
Parham, William E., 13
Parmerter, Stanley M., 10
Pasto, Daniel J., 40
Paterson, Ian, 51
Pettit, George R., 12
Phadke, Ragini, 7
Phillips, Robert R., 10
Pierini, Adriana B., 54
Pigge, F. Christopher, 51
Pine, Stanley H., 18, 43
Pinnick, Harold W., 38
Porter, H. K., 20
Posner, Gary H., 19, 22
Prakash, Om, 54, 57
Price, Charles C., 3

Rabideau, Peter W., 42
Rabjohn, Norman, 5, 24
Rathke, Michael W., 22

Raulins, N. Rebecca, 22
Raynolds, Peter W., 45
Reed, Josephine W., 41
Reich, Hans J., 44
Reinhold, Tracy L., 44
Reitz, Allen B., 59
Rhoads, Sara Jane, 22
Rickborn, Bruce, 52, 53
Rigby, James H., 49, 51
Rinehart, Kenneth L., Jr., 17
Ripka, W. C., 21
Roberts, John D., 12
Rodriguez, Alain L., 58
Rodriguez, Herman R., 26
Roe, Arthur, 5
Rondestvedt, Christian S., Jr., 11, 24
Rossi, Roberto A., 54
Ruh-Pohlenz, Carmen, 55
Rytina, Anton W., 5

Saha-Möller, Chantu R., 61
Santiago, Ana N., 54
Sauer, John C., 3
Schaefer, John P., 15
Schore, Neil E., 40
Schulenberg, J. W., 14
Schweizer, Edward E., 13
Scott, William J., 50
Scribner, R. M., 21
Semmelhack, Martin F., 19
Sengupta, Saumitra, 41
Sethna, Suresh, 7
Shapiro, Robert H., 23
Sharts, Clay M., 12, 21
Sheehan, John C., 9
Sheldon, Roger A., 19
Sheppard, W. A., 21
Shirley, David A., 8
Shriner, Ralph L., 1
Simmons, Howard E., 20
Simonoff, Robert, 7
Smith, Lee Irvin, 1
Smith, Peter A. S., 3, 11
Smithers, Roger, 37
Spielman, M. A., 3
Spoerri, Paul E., 5
Stacey, F. W., 13
Stadler, Alexander, 63
Stanforth, Stephen P., 49, 56
Stetter, Hermann, 40
Struve, W. S., 1
Suter, C. M., 3
Swamer, Frederic W., 8
Swern, Daniel, 7

Tarbell, D. Stanley, 2
Taylor, Richard J. K., 62
Taylor, Richard T., 40
Thoma, G., 48
Tidwell, Thomas T., 39
Todd, David, 4
Touster, Oscar, 7
Trach, F., 48
Truce, William E., 9, 18
Trumbull, Elmer R., 11
Tsai, Chung-Ying, 56
Tucker, Charles, E., 58
Tullock, C. W., 21
Tzamarioudaki, Maria, 56

van Leusen, Albert M., 57
van Leusen, Daan, 57
van Tamelen, Eugene E., 12
Vedejs, E., 22
Vladuchick, Susan A., 20
Vorbrüggen, Helmut, 55

Wadsworth, William S., Jr., 25
Walling, Cheves, 13
Wallis, Everett S., 3
Wallquist, Olof, 47

Wang, Chia-Lin L., 34
Warnhoff, E. W., 18
Watt, David S., 31
Weston, Arthur W., 3, 9
Whaley, Wilson M., 6
Wilds, A. L., 2
Wiley, Richard H., 6
Williamson, David H., 24
Wilson, C. V., 9
Wilson, Stephen R., 43
Wolf, Donald E., 6
Wolff, Hans, 3
Wollowitz, Susan, 44
Wood, John L., 3
Wynberg, Hans, 28

Yamago, Shigeru, 61
Yan, Shou-Jen, 28
Yoshioka, Mitsuru, 25

Zaugg, Harold E., 8, 14
Zhao, Cong-Gui, 61
Zhou, Ping, 62
Zubkov, Oleg A., 62
Zweifel, George, 13, 32

CHAPTER AND TOPIC INDEX, VOLUMES 1–63

Many chapters contain brief discussions of reactions and comparisons of alternative synthetic methods related to the reaction that is the subject of the chapter. These related reactions and alternative methods are not usually listed in this index. In this index, the volume number is in **boldface**, the chapter number is in ordinary type.

Acetoacetic ester condensation, **1**, 9
Acetylenes, synthesis of, **5**, 1; **23**, 3; **32**, 2
Acid halides:
   reactions with esters, **1**, 9
   reactions with organometallic compounds, **8**, 2
α-Acylamino acid mixed anhydrides, **12**, 4
α-Acylamino acids, azlactonization of, **3**, 5
Acylation:
   of esters with acid chlorides, **1**, 9
   intramolecular, to form cyclic ketones, **2**, 4; **23**, 2
   of ketones to form diketones, **8**, 3
Acyl fluorides, synthesis of, **21**, 1; **34**, 2; **35**, 3
Acyl hypohalites, reactions of, **9**, 5
Acyloins, **4**, 4; **15**, 1; **23**, 2
Alcohols:
   conversion to fluorides, **21**, 1, 2; **34**, 2; **35**, 3
   conversion to olefins, **12**, 2
   oxidation of, **6**, 5; **39**, 3; **53**, 1
   replacement of hydroxy group by nucleophiles, **29**, 1; **42**, 2
   resolution of, **2**, 9
Alcohols, synthesis:
   by base-promoted isomerization of epoxides, **29**, 3
   by hydroboration, **13**, 1
   by hydroxylation of ethylenic compounds, **7**, 7
   from organoboranes, **33**, 1
   by reduction, **6**, 10; **8**, 1
Aldehydes, catalyzed addition to double bonds, **40**, 4
Aldehydes, synthesis of, **4**, 7; **5**, 10; **8**, 4, 5; **9**, 2; **33**, 1
Aldol condensation, **16**
   directed, **28**, 3
   with boron enolates, **51**, 1
Aliphatic fluorides, **2**, 2; **21**, 1, 2; **34**, 2; **35**, 3
Alkenes:
   arylation of, **11**, 3; **24**, 3; **27**, 2
   cyclopropanes from, **20**, 1
   cyclization in intramolecular Heck reactions **60**, 2

dioxirane epoxidation of, **61**, 2
epoxidation and hydroxylation of, **7**, 7
free-radical additions to, **13**, 3, 4
hydroboration of, **13**, 1
hydrogenation with homogeneous catalysts, **24**, 1
reactions with diazoacetic esters, **18**, 3
reactions with nitrones, **36**, 1
reduction by alkoxyaluminum hydrides, **34**, 1
Alkenes, synthesis:
   from amines, **11**, 5
   from aryl and vinyl halides, **27**, 2
   by Bamford–Stevens reaction, **23**, 3
   by Claisen and Cope rearrangements, **22**, 1
   by dehydrocyanation of nitriles, **31**
   by deoxygenation of vicinal diols, **30**, 2
   from α-halosulfones, **25**, 1; **62**, 2
   by palladium-catalyzed vinylation, **27**, 2
   from phosphoryl-stabilized anions, **25**, 2
   by pyrolysis of xanthates, **12**, 2
   from silicon-stabilized anions, **38**, 1
   from tosylhydrazones, **23**, 3; **39**, 1
   by Wittig reaction, **14**, 3
Alkene reduction by diimide, **40**, 2
Alkenyl- and alkynylaluminum reagents, **32**, 2
Alkenyllithiums, formation of **39**, 1
Alkoxyaluminum hydride reductions, **34**, 1; **36**, 3
Alkoxyphosphonium cations, nucleophilic displacements on, **29**, 1
Alkylation:
   of allylic and benzylic carbanions, **27**, 1
   with amines and ammonium salts, **7**, 3
   of aromatic compounds, **3**, 1
   of esters and nitriles, **9**, 4
   γ-, of dianions of β-dicarbonyl compounds, **17**, 2
   of metallic acetylides, **5**, 1
   of nitrile-stabilized carbanions, **31**
   with organopalladium complexes, **27**, 2
Alkylidenation by titanium-based reagents, **43**, 1
Alkylidenesuccinic acids, synthesis and reactions of, **6**, 1

Alkylidene triphenylphosphoranes, synthesis and reactions of, **14**, 3
Allenylsilanes, electrophilic substitution reactions of, **37**, 2
Allylic alcohols, synthesis:
　from epoxides, **29**, 3
　by Wittig rearrangement, **46**, 2
Allylic and benzylic carbanions, heteroatom-substituted, **27**, 1
Allylic hydroperoxides, in photooxygenations, **20**, 2
Allylic rearrangements, transformation of glycols into 2,3-unsaturated glycosyl derivatives, **62**, 4
π-Allylnickel complexes, **19**, 2
Allylphenols, synthesis by Claisen rearrangement, **2**, 1; **22**, 1
Allylsilanes, electrophilic substitution reactions of, **37**, 2
Aluminum alkoxides:
　in Meerwein–Ponndorf–Verley reduction, **2**, 5
　in Oppenauer oxidation, **6**, 5
Amide formation by oxime rearrangement, **35**, 1
α-Amidoalkylations at carbon, **14**, 2
Amination:
　of heterocyclic bases by alkali amides, **1**, 4
　of hydroxy compounds by Bucherer reaction, **1**, 5
Amine oxides:
　Polonovski reaction of, **39**, 2
　pyrolysis of, **11**, 5
Amines:
　synthesis from organoboranes, **33**, 1
　synthesis by reductive alkylation, **4**, 3; **5**, 7
　synthesis by Zinin reaction, **20**, 4
　reactions with cyanogen bromide, **7**, 4
α-Aminoalkylation of activated olefins, **51**, 2
Aminophenols from anilines, **35**, 2
Anhydrides of aliphatic dibasic acids, Friedel–Crafts reaction with, **5**, 5
Anion-assisted sigmatropic rearrangements, **43**, 2
Anthracene homologs, synthesis of, **1**, 6
Anti-Markownikoff hydration of alkenes, **13**, 1
π-Arenechromium tricarbonyls, reaction with nitrile-stabilized carbanions, **31**
Arndt–Eistert reaction, **1**, 2
Aromatic aldehydes, synthesis of, **5**, 6; **28**, 1
Aromatic compounds, chloromethylation of, **1**, 3
Aromatic fluorides, synthesis of, **5**, 4
Aromatic hydrocarbons, synthesis of, **1**, 6; **30**, 1
Aromatic substitution by the $S_{RN}1$ reaction, **54**, 1
Arsinic acids, **2**, 10
Arsonic acids, **2**, 10
Arylacetic acids, synthesis of, **1**, 2; **22**, 4

β-Arylacrylic acids, synthesis of, **1**, 8
Arylamines, synthesis and reactions of, **1**, 5
Arylation:
　by aryl halides, **27**, 2
　by diazonium salts, **11**, 3; **24**, 3
　γ-, of dianions of β-dicarbonyl compounds, **17**, 2
　of nitrile-stabilized carbanions, **31**
　of alkenes, **11**, 3; **24**, 3; **27**, 2
Arylglyoxals, condensation with aromatic hydrocarbons, **4**, 5
Arylsulfonic acids, synthesis of, **3**, 4
Aryl halides, homocoupling of, **63**, 3
Aryl thiocyanates, **3**, 6
Asymmetric aldol reactions using boron enolates, **51**, 1
Asymmetric cyclopropanation, **57**, 1
Asymmetric epoxidation, **48**, 1; **61**, 2
Atom transfer preparation of radicals, **48**, 2
Aza-Payne rearrangements, **60**, 1
Azaphenanthrenes, synthesis by photocyclization, **30**, 1
Azides, synthesis and rearrangement of, **3**, 9
Azlactones, **3**, 5

Baeyer–Villiger reaction, **9**, 3; **43**, 3
Bamford–Stevens reaction, **23**, 3
Barbier reaction, **58**, 2
Bart reaction, **2**, 10
Barton fragmentation reaction, **48**, 2
Béchamp reaction, **2**, 10
Beckmann rearrangement, **11**, 1; **35**, 1
Benzils, reduction of, **4**, 5
Benzoin condensation, **4**, 5
Benzoquinones:
　acetoxylation of, **19**, 3
　in Nenitzescu reaction, **20**, 3
　synthesis of, **4**, 6
Benzylic carbanions, **27**, 1
Biaryls, synthesis of, **2**, 6; **63**, 3
Bicyclobutanes, from cyclopropenes, **18**, 3
Biginelli dihydropyrimidine synthesis, **63**, 1
Birch reaction, **23**, 1; **42**, 1
Bischler–Napieralski reaction, **6**, 2
Bis(chloromethyl) ether, **1**, 3; **19**, *warning*
Borane reduction, chiral, **52**, 2
Borohydride reduction, chiral, **52**, 2
　in reductive aminations, **59**, 1
Boron enolates, **51**, 1
Boyland–Sims oxidation, **35**, 2
Bucherer reaction, **1**, 5

Cannizzaro reaction, **2**, 3
Carbenes, **13**, 2; **26**, 2; **28**, 1

Carbenoid cyclopropanation reactions, **57**, 1; **58**, 1
Carbohydrates, deoxy, synthesis of, **30**, 2
Carbo/metallocupration, **41**, 2
Carbon–carbon bond formation:
  by acetoacetic ester condensation, **1**, 9
  by acyloin condensation, **23**, 2
  by aldol condensation, **16**; **28**, 3; **46**, 1
  by alkylation with amines and ammonium salts, **7**, 3
  by γ-alkylation and arylation, **17**, 2
  by allylic and benzylic carbanions, **27**, 1
  by amidoalkylation, **14**, 2
  by Cannizzaro reaction, **2**, 3
  by Claisen rearrangement, **2**, 1; **22**, 1
  by Cope rearrangement, **22**, 1
  by cyclopropanation reaction, **13**, 2; **20**, 1
  by Darzens condensation, **5**, 10
  by diazonium salt coupling, **10**, 1; **11**, 3; **24**, 3
  by Dieckmann condensation, **15**, 1
  by Diels–Alder reaction, **4**, 1, 2; **5**, 3; **32**, 1
  by free-radical additions to alkenes, **13**, 3
  by Friedel–Crafts reaction, **3**, 1; **5**, 5
  by Knoevenagel condensation, **15**, 2
  by Mannich reaction, **1**, 10; **7**, 3
  by Michael addition, **10**, 3
  by nitrile-stabilized carbanions, **31**
  by organoboranes and organoborates, **33**, 1
  by organocopper reagents, **19**, 1; **38**, 2; **41**, 2
  by organopalladium complexes, **27**, 2
  by organozinc reagents, **20**, 1
  by rearrangement of α-halosulfones, **25**, 1; **62**, 2
  by Reformatsky reaction, **1**, 1; **28**, 3
  by trivalent manganese, **49**, 3
  by Vilsmeier reaction, **49**, 1; **56**, 2
  by vinylcyclopropane-cyclopentene rearrangement, **33**, 2
Carbon–halogen bond formation, by replacement of hydroxy groups, **29**, 1
Carbon–heteroatom bond formation:
  by free-radical chain additions to carbon–carbon multiple bonds, **13**, 4
  by organoboranes and organoborates, **33**, 1
Carbon–nitrogen bond formation, by reductive amination, **59**, 1
Carbon–phosphorus bond formation, **36**, 2
Carbonyl compounds, α,β-unsaturated:
  formation by selenoxide elimination, **44**, 1
  vicinal difunctionalization of, **38**, 2
Carbonyl compounds, from nitro compounds, **38**, 3
  oxidation with hypervalent iodine reagents, **54**, 2
  reductive amination of, **59**, 1

Carbonylation as part of intramolecular Heck reaction, **60**, 2
Carboxylic acid derivatives, conversion to fluorides, **21**, 1, 2; **34**, 2; **35**, 3
Carboxylic acids:
  reaction with organolithium reagents, **18**, 1
  synthesis from organoboranes, **33**, 1
Chapman rearrangement, **14**, 1; **18**, 2
Chloromethylation of aromatic compounds, **2**, 3; **9**, *warning*
Cholanthrenes, synthesis of, **1**, 6
Chugaev reaction, **12**, 2
Claisen condensation, **1**, 8
Claisen rearrangement, **2**, 1; **22**, 1
Cleavage:
  of benzyl–oxygen, benzyl–nitrogen, and benzyl–sulfur bonds, **7**, 5
  of carbon–carbon bonds by periodic acid, **2**, 8
  of esters via S_N2-type dealkylation, **24**, 2
  of non-enolizable ketones with sodium amide, **9**, 1
  in sensitized photooxidation, **20**, 2
Clemmensen reduction, **1**, 7; **22**, 3
Collins reagent, **53**, 1
Condensation:
  acetoacetic ester, **1**, 9
  acyloin, **4**, 4; **23**, 2
  aldol, **16**
  benzoin, **4**, 5
  Biginelli, **63**, 1
  Claisen, **1**, 8
  Darzens, **5**, 10; **31**
  Dieckmann, **1**, 9; **6**, 9; **15**, 1
  directed aldol, **28**, 3
  Knoevenagel, **1**, 8; **15**, 2
  Stobbe, **6**, 1
  Thorpe–Ziegler, **15**, 1; **31**
Conjugate addition:
  of hydrogen cyanide, **25**, 3
  of organocopper reagents, **19**, 1; **41**, 2
Cope rearrangement, **22**, 1; **41**, 1; **43**, 2
Copper–Grignard complexes, conjugate additions of, **19**, 1; **41**, 2
Corey–Winter reaction, **30**, 2
Coumarins, synthesis of, **7**, 1; **20**, 3
Coupling reaction of organostannanes, **50**, 1
Cuprate reagents, **19**, 1; **38**, 2; **41**, 2
Curtius rearrangement, **3**, 7, 9
Cyanoborohydride, in reductive aminations, **59**, 1
Cyanoethylation, **5**, 2
Cyanogen bromide, reactions with tertiary amines, **7**, 4
Cyclic ketones, formation by intramolecular acylation, **2**, 4; **23**, 2

Cyclization:
  of alkyl dihalides, **19**, 2
  of aryl-substituted aliphatic acids, acid chlorides, and anhydrides, **2**, 4; **23**, 2
  of α-carbonyl carbenes and carbenoids, **26**, 2
  cycloheptenones from α-bromoketones, **29**, 2
Cyclization (*Continued*)
  of diesters and dinitriles, **15**, 1
  Fischer indole, **10**, 2
  intramolecular by acylation, **2**, 4
  intramolecular by acyloin condensation, **4**, 4
  intramolecular by Diels–Alder reaction, **32**, 1
  intramolecular by Heck reaction, **60**, 2
  intramolecular by Michael reaction, **47**, 2
  Nazarov, **45**, 1
  by radical reactions, **48**, 2
  of stilbenes, **30**, 1
  tandem cyclization by Heck reaction, **60**, 2
Cycloaddition reactions:
  of cyclenones and quinones, **5**, 3
  cyclobutanes, synthesis of, **12**, 1; **44**, 2
  Diels–Alder, acetylenes and alkenes, **4**, 2
  Diels–Alder, intramolecular, **32**, 1
  Diels–Alder, maleic anhydride, **4**, 1
  [4 + 3], **51**, 3
  of enones, **44**, 2
  of ketenes, **45**, 2
  of nitrones and alkenes, **36**, 1
  Pauson–Khand, **40**, 1
  photochemical, **44**, 2
  retro Diels–Alder reaction, **52**, 1; **53**, 2
  [6 + 4], **49**, 2
  [3 + 2], **61**, 1
Cyclobutanes, synthesis:
  from nitrile-stabilized carbanions, **31**
  by thermal cycloaddition reactions, **12**, 1
Cycloheptadienes, from:
  divinylcyclopropanes, **41**, 1
  polyhalo ketones, **29**, 2
π-Cyclopentadienyl transition metal carbonyls, **17**, 1
Cyclopentenones:
  annulation, **45**, 1
  synthesis, **40**, 1; **45**, 1
Cyclopropane carboxylates, from diazoacetic esters, **18**, 3
Cyclopropanes:
  from α-diazocarbonyl compounds, **26**, 2
  from metal-catalyzed decomposition of diazo compounds, **57**, 1
  from nitrile-stabilized carbanions, **31**
  from tosylhydrazones, **23**, 3
  from unsaturated compounds, methylene iodide, and zinc-copper couple, **20**, 1; **58**, 1; **58**, 2

Cyclopropenes, synthesis of, **18**, 3

Darzens glycidic ester condensation, **5**, 10; **31**
DAST, **34**, 2; **35**, 3
Deamination of aromatic primary amines, **2**, 7
Debenzylation, **7**, 5; **18**, 4
Decarboxylation of acids, **9**, 5; **19**, 4
Dehalogenation of α-haloacyl halides, **3**, 3
Dehydrogenation:
  in synthesis of ketenes, **3**, 3
  in synthesis of acetylenes, **5**, 1
Demjanov reaction, **11**, 2
Deoxygenation of vicinal diols, **30**, 2
Desoxybenzoins, conversion to benzoins, **4**, 5
Dess-Martin oxidation, **53**, 1
Desulfurization:
  of α-(alkylthio)nitriles, **31**
  in alkene synthesis, **30**, 2
  with Raney nickel, **12**, 5
Diazo compounds, carbenoids derived from, **57**, 1
Diazoacetic esters, reactions with alkenes, alkynes, heterocyclic and aromatic compounds, **18**, 3; **26**, 2
α-Diazocarbonyl compounds, insertion and addition reactions, **26**, 2
Diazomethane:
  in Arndt–Eistert reaction, **1**, 2
  reactions with aldehydes and ketones, **8**, 8
Diazonium fluoroborates, synthesis and decomposition, **5**, 4
Diazonium salts:
  coupling with aliphatic compounds, **10**, 1, 2
  in deamination of aromatic primary amines, **2**, 7
  in Meerwein arylation reaction, **11**, 3; **24**, 3
  in ring closure reactions, **9**, 7
  in synthesis of biaryls and aryl quinones, **2**, 6
Dieckmann condensation, **1**, 9; **15**, 1
  for synthesis of tetrahydrothiophenes, **6**, 9
Diels–Alder reaction:
  intramolecular, **32**, 1
  retro–Diels–Alder reaction, **52**, 1; **53**, 2
  with alkynyl and alkenyl dienophiles, **4**, 2
  with cyclenones and quinones, **5**, 3
  with maleic anhydride, **4**, 1
Dihydrodiols, **63**, 2
Dihydropyrimidine synthesis, **63**, 1
Diimide, **40**, 2
Diketones:
  pyrolysis of diaryl, **1**, 6
  reduction by acid in organic solvents, **22**, 3
  synthesis by acylation of ketones, **8**, 3
  synthesis by alkylation of β-diketone anions, **17**, 2
Dimethyl sulfide, in oxidation reactions, **39**, 3

Dimethyl sulfoxide, in oxidation reactions, **39**, 3
Diols:
  deoxygenation of, **30**, 2
  oxidation of, **2**, 8
Dioxetanes, **20**, 2
Dioxiranes, **61**, 2
Dioxygenases, **63**, 2
Divinyl-aziridines, -cyclopropanes, -oxiranes, and -thiiranes, rearrangements of, **41**, 1
Doebner reaction, **1**, 8

Eastwood reaction, **30**, 2
Elbs reaction, **1**, 6; **35**, 2
Enamines, reaction with quinones, **20**, 3
Ene reaction, in photosensitized oxygenation, **20**, 2
Enolates:
  α-Hydroxylation of, **62**, 1
  in directed aldol reactions, **28**, 3; **46**, 1; **51**, 1
Enone cycloadditions, **44**, 2
Enzymatic reduction, **52**, 2
Enzymatic resolution, **37**, 1
Epoxidation:
  of alkenes, **61**, 2
  of allylic alcohols, **48**, 1
  with organic peracids, **7**, 7
Epoxide isomerizations, **29**, 3
Epoxide:
  formation, **61**, 2
  migration, **60**, 1
Esters:
  acylation with acid chlorides, **1**, 9
  alkylation of, **9**, 4
  alkylidenation of, **43**, 1
  cleavage via S_N2-type dealkylation, **24**, 2
  dimerization, **23**, 2
  glycidic, synthesis of, **5**, 10
  hydrolysis, catalyzed by pig liver esterase, **37**, 1
  β-hydroxy, synthesis of, **1**, 1; **22**, 4
  β-keto, synthesis of, **15**, 1
  reaction with organolithium reagents, **18**, 1
  reduction of, **8**, 1
  synthesis from diazoacetic esters, **18**, 3
  synthesis by Mitsunobu reaction, **42**, 2
Ethers, synthesis by Mitsunobu reaction, **42**, 2
Exhaustive methylation, Hofmann, **11**, 5

Favorskii rearrangement, **11**, 4
Ferrocenes, **17**, 1
Fischer indole cyclization, **10**, 2
Fluorination of aliphatic compounds, **2**, 2; **21**, 1, 2; **34**, 2; **35**, 3
Fluorination by DAST, **35**, 3

Fluorination by sulfur tetrafluoride, **21**, 1; **34**, 2
Formylation:
  by hydroformulation, **56**, 1
  of alkylphenols, **28**, 1
  of aromatic hydrocarbons, **5**, 6
  of aromatic compounds, **49**, 1
  of nonaromatic compounds, **56**, 2
Free radical additions:
  to alkenes and alkynes to form carbon–heteroatom bonds, **13**, 4
  to alkenes to form carbon-carbon bonds, **13**, 3
Friedel-Crafts catalysts, in nucleoside synthesis, **55**, 1
Friedel–Crafts reaction, **2**, 4; **3**, 1; **5**, 5; **18**, 1
Friedländer synthesis of quinolines, **28**, 2
Fries reaction, **1**, 11

Gattermann aldehyde synthesis, **9**, 2
Gattermann–Koch reaction, **5**, 6
Germanes, addition to alkenes and alkynes, **13**, 4
Glycals, transformation in glycosyl derivatives, **62**, 4
Glycidic esters, synthesis and reactions of, **5**, 10
Gomberg–Bachmann reaction, **2**, 6; **9**, 7
Grundmann synthesis of aldehydes, **8**, 5

Halides, displacement reactions of, **22**, 2; **27**, 2
Halide-metal exchange, **58**, 2
Halides, synthesis:
  from alcohols, **34**, 2
  by chloromethylation, **1**, 3
  from organoboranes, **33**, 1
  from primary and secondary alcohols, **29**, 1
Haller–Bauer reaction, **9**, 1
Halocarbenes, synthesis and reactions of, **13**, 2
Halocyclopropanes, reactions of, **13**, 2
Halogen-metal interconversion reactions, **6**, 7
α-Haloketones, rearrangement of, **11**, 4
α-Halosulfones, synthesis and reactions of, **25**, 1; **62**, 2
Heck reaction, intramolecular, **60**, 2
Helicenes, synthesis by photocyclization, **30**, 1
Heterocyclic aromatic systems, lithiation of, **26**, 1
Heterocyclic bases, amination of, **1**, 4
  in nucleosides, **55**, 1
Heterodienophiles, **53**, 2
Hilbert-Johnson method, **55**, 1
Hoesch reaction, **5**, 9
Hofmann elimination reaction, **11**, 5; **18**, 4
Hofmann reaction of amides, **3**, 7, 9
Homocouplings mediated by Cu, Ni, and Pd, **63**, 3
Homogeneous hydrogenation catalysts, **24**, 1
Hunsdiecker reaction, **9**, 5; **19**, 4
Hydration of alkenes, dienes, and alkynes, **13**, 1

Hydrazoic acid, reactions and generation of, **3**, 8
Hydroboration, **13**, 1
Hydrocyanation of conjugated carbonyl compounds, **25**, 3
Hydroformylation, **56**, 1
Hydrogenation catalysts, homogeneous, **24**, 1
Hydrogenation of esters, with copper chromite and Raney nickel, **8**, 1
Hydrohalogenation, **13**, 4
Hydroxyaldehydes, aromatic, **28**, 1
α-Hydroxyalkylation of activated olefins, **51**, 2
α-Hydroxyketones:
  rearrangement, **62**, 3
  synthesis of, **23**, 2
Hydroxylation:
  of enolates, **62**, 1
  of ethylenic compounds with organic peracids, **7**, 7
Hypervalent iodine reagents, **54**, 2; **57**, 2

Imidates, rearrangement of, **14**, 1
Iminium ions, **39**, 2
Indoles, by Nenitzescu reaction, **20**, 3
  via reaction with TosMIC, **57**, 3
Isocyanides, sulfonylmethyl, reactions of, **57**, 3
Isoquinolines, synthesis of, **6**, 2, 3, 4; **20**, 3

Jacobsen reaction, **1**, 12
Japp–Klingemann reaction, **10**, 2

Katsuki–Sharpless epoxidation, **48**, 1
Ketene cycloadditions, **45**, 2
Ketenes and ketene dimers, synthesis of, **3**, 3; **45**, 2
α-Ketol rearrangement, **62**, 3
Ketones:
  acylation of, **8**, 3
  alkylidenation of, **43**, 1
  Baeyer–Villiger oxidation of, **9**, 3; **43**, 3
  cleavage of non-enolizable, **9**, 1
  comparison of synthetic methods, **18**, 1
  conversion to amides, **3**, 8; **11**, 1
  conversion to fluorides, **34**, 2; **35**, 3
  cyclic, synthesis of, **2**, 4; **23**, 2
  cyclization of divinyl ketones, **45**, 1
  synthesis from acid chlorides and organometallic compounds, **8**, 2; **18**, 1
  synthesis from organoboranes, **33**, 1
  synthesis from α,β-unsaturated carbonyl compounds and metals in liquid ammonia, **23**, 1
  reaction with diazomethane, **8**, 8
  reduction to aliphatic compounds, **4**, 8
  reduction by alkoxyaluminum hydrides, **34**, 1
  reduction in anhydrous organic solvents, **22**, 3
  synthesis from organolithium reagents and carboxylic acids, **18**, 1
  synthesis by oxidation of alcohols, **6**, 5; **39**, 3
Kindler modification of Willgerodt reaction, **3**, 2
Knoevenagel condensation, **1**, 8; **15**, 2; **57**, 3
Koch–Haaf reaction, **17**, 3
Kornblum oxidation, **39**, 3
Kostaneki synthesis of chromanes, flavones, and isoflavones, **8**, 3

β-Lactams, synthesis of, **9**, 6; **26**, 2
β-Lactones, synthesis and reactions of, **8**, 7
Leuckart reaction, **5**, 7
Lithiation:
  of allylic and benzylic systems, **27**, 1
  by halogen-metal exchange, **6**, 7
  heteroatom facilitated, **26**, 1; **47**, 1
  of heterocyclic and olefinic compounds, **26**, 1
Lithioorganocuprates, **19**, 1; **22**, 2; **41**, 2
Lithium aluminum hydride reductions, **6**, 2
  chirally modified, **52**, 2
Lossen rearrangement, **3**, 7, 9

Mannich reaction, **1**, 10; **7**, 3
Meerwein arylation reaction, **11**, 3; **24**, 3
Meerwein–Ponndorf–Verley reduction, **2**, 5
Mercury hydride method to prepare radicals, **48**, 2
Metalations with organolithium compounds, **8**, 6; **26**, 1; **27**, 1
Methylenation of carbonyl groups, **43**, 1
Methylenecyclopropane, in cycloaddition reactions, **61**, 1
Methylene-transfer reactions, **18**, 3; **20**, 1; **58**, 1
Michael reaction, **10**, 3; **15**, 1, 2; **19**, 1; **20**, 3; **46**, 1; **47**, 2
Microbiological oxygenations, **63**, 2
Mitsunobu reaction, **42**, 2
Moffatt oxidation, **39**, 3; **53**, 1
Morita–Baylis–Hillman reaction, **51**, 2

Nazarov cyclization, **45**, 1
Nef reaction, **38**, 3
Nenitzescu reaction, **20**, 3
Nitriles:
  formation from oximes, **35**, 2
  synthesis from organoboranes, **33**, 1
  α,β-unsaturated:
    by elimination of selenoxides, **44**, 1
Nitrile-stabilized carbanions:
  alkylation and arylation of, **31**
Nitroamines, **20**, 4

Nitro compounds, conversion to carbonyl
   compounds, **38**, 3
Nitro compounds, synthesis of, **12**, 3
Nitrone-olefin cycloadditions, **36**, 1
Nitrosation, **2**, 6; **7**, 6
Nucleosides, synthesis of, **55**, 1

Olefins, hydroformylation of, **56**, 1
Oligomerization of 1,3-dienes, **19**, 2
Oppenauer oxidation, **6**, 5
Organoboranes:
   formation of carbon–carbon and carbon–
      heteroatom bonds from, **33**, 1
   isomerization and oxidation of, **13**, 1
   reaction with anions of α-chloronitriles, **31**
Organohypervalent iodine reagents, **54**, 2; **57**, 2
Organometallic compounds:
   of aluminum, **25**, 3
   of copper, **19**, 1; **22**, 2; **38**, 2; **41**, 2
   of lithium, **6**, 7; **8**, 6; **18**, 1; **27**, 1
   of magnesium, zinc, and cadmium, **8**, 2;
   of palladium, **27**, 2
   of tin, **50**, 1
   of zinc, **1**, 1; **20**, 1; **22**, 4; **58**, 2
Oxidation:
   by dioxiranes, **61**, 2
   of alcohols and polyhydroxy compounds, **6**, 5;
      39, 3; **53**, 1
   of aldehydes and ketones, Baeyer–Villiger
      reaction, **9**, 3; **43**, 3
   of amines, phenols, aminophenols, diamines,
      hydroquinones, and halophenols, **4**, 6; **35**, 2
   of enolates and silyl enol ethers, **62**, 1
   of α-glycols, α-amino alcohols, and
      polyhydroxy compounds by periodic acid,
      2, 8
   with hypervalent iodine reagents, **54**, 2
   of organoboranes, **13**, 1
   of phenolic compounds, **57**, 2
   with peracids, **7**, 7
   by photooxygenation, **20**, 2
   with selenium dioxide, **5**, 8; **24**, 4
Oxidative decarboxylation, **19**, 4
Oximes, formation by nitrosation, **7**, 6
Oxochromium(VI)-amine complexes, **53**, 1
Oxo process, **56**, 1
Oxygenation of arenes by dioxygenases, **63**, 2

Palladium-catalyzed vinylic substitution, **27**, 2
Palladium-catalyzed coupling of organostannane,
   50, 1
Palladium intermediates in Heck reactions, **60**, 2
Pauson–Khand reaction to prepare
   cyclopentenones, **40**, 1

Payne rearrangement, **60**, 1
Pechmann reaction, **7**, 1
Peptides, synthesis of, **3**, 5; **12**, 4
Peracids, epoxidation and hydroxylation with,
   7, 7
   in Baeyer–Villiger oxidation, **9**, 3; **43**, 3
Periodic acid oxidation, **2**, 8
Perkin reaction, **1**, 8
Persulfate oxidation, **35**, 2
Peterson olefination, **38**, 1
Phenanthrenes, synthesis by photocyclization,
   30, 1
Phenols, dihydric from phenols, **35**, 2
   oxidation of **57**, 2
Phosphinic acids, synthesis of, **6**, 6
Phosphonic acids, synthesis of, **6**, 6
Phosphonium salts:
   halide synthesis, use in, **29**, 1
   synthesis and reactions of, **14**, 3
Phosphorus compounds, addition to carbonyl
   group, **6**, 6; **14**, 3; **25**, 2; **36**, 2
   addition reactions at imine carbon, **36**, 2
Phosphoryl-stabilized anions, **25**, 2
Photochemical cycloadditions, **44**, 2
Photocyclization of stilbenes, **30**, 1
Photooxygenation of olefins, **20**, 2
Photosensitizers, **20**, 2
Pictet–Spengler reaction, **6**, 3
Pig liver esterase, **37**, 1
Polonovski reaction, **39**, 2
Polyalkylbenzenes, in Jacobsen reaction, **1**, 12
Polycyclic aromatic compounds, synthesis by
   photocyclization of stilbenes, **30**, 1
Polyhalo ketones, reductive dehalogenation of,
   29, 2
Pomeranz–Fritsch reaction, **6**, 4
Prévost reaction, **9**, 5
Pschorr synthesis, **2**, 6; **9**, 7
Pummerer reaction, **40**, 3
Pyrazolines, intermediates in diazoacetic ester
   reactions, **18**, 3
Pyridinium chlorochromate, **53**, 1
Pyrolysis:
   of amine oxides, phosphates, and acyl
      derivatives, **11**, 5
   of ketones and diketones, **1**, 6
   for synthesis of ketenes, **3**, 3
   of xanthates, **12**, 2

Quaternary ammonium salts, rearrangements of,
   18, 4
Quinolines, synthesis of:
   by Friedländer synthesis, **28**, 2
   by Skraup synthesis, **7**, 2

Quinones:
    acetoxylation of, **19**, 3
    diene additions to, **5**, 3
    synthesis of, **4**, 6
    in synthesis of 5-hydroxyindoles, **20**, 3

Radical formation and cyclization, **48**, 2
Ramberg–Bäcklund rearrangement, **25**, 1; **62**, 2
Rearrangements:
    anion-assisted sigmatropic, **43**, 2
    Beckmann, **11**, 1; **35**, 1
    Chapman, **14**, 1; **18**, 2
    Claisen, **2**, 1; **22**, 1
    Cope, **22**, 1; **41**, 1, **43**, 2
    Curtius, **3**, 7, 9
    divinylcyclopropane, **41**, 1
    Favorskii, **11**, 4
    Lossen, **3**, 7, 9
    Ramberg–Bäcklund, **25**, 1; **62**, 2
    Smiles, **18**, 2
    Sommelet–Hauser, **18**, 4
    Stevens, **18**, 4
    [2,3] Wittig, **46**, 2
    vinylcyclopropane-cyclopentene, **33**, 2
Reduction:
    of acid chlorides to aldehydes, **4**, 7; **8**, 5
    of aromatic compounds, **42**, 1
    of benzils, **4**, 5
    of ketones, enantioselective, **52**, 2
    Clemmensen, **1**, 7; **22**, 3
    desulfurization, **12**, 5
    with diimide, **40**, 2
    by dissolving metal, **42**, 1
    by homogeneous hydrogenation catalysts, **24**, 1
    by hydrogenation of esters with copper chromite and Raney nickel, **8**, 1
    hydrogenolysis of benzyl groups, **7**, 5
    by lithium aluminum hydride, **6**, 10
    by Meerwein–Ponndorf–Verley reaction, **2**, 5
    chiral, **52**, 2
    by metal alkoxyaluminum hydrides, **34**, 1; **36**, 3
    of mono- and polynitroarenes, **20**, 4
    of olefins by diimide, **40**, 2
    of α,β-unsaturated carbonyl compounds, **23**, 1
    by samarium(II) iodide, **46**, 3
    by Wolff–Kishner reaction, **4**, 8
Reductive alkylation, synthesis of amines, **4**, 3; **5**, 7
Reductive amination of carbonyl compounds, **59**, 1
Reductive cyanation, **57**, 3
Reductive desulfurization of thiol esters, **8**, 5
Reformatsky reaction, **1**, 1; **22**, 4

Reimer–Tiemann reaction, **13**, 2; **28**, 1
Resolution of alcohols, **2**, 9
Retro–Diels–Alder reactions, **52**, 1; **53**, 2
Ritter reaction, **17**, 3
Rosenmund reaction for synthesis of arsonic acids, **2**, 10
Rosenmund reduction, **4**, 7

Samarium(II) iodide, **46**, 3
Sandmeyer reaction, **2**, 7
Schiemann reaction, **5**, 4
Schmidt reaction, **3**, 8, 9
Selenium dioxide oxidation, **5**, 8; **24**, 4
Seleno–Pummerer reaction, **40**, 3
Selenoxide elimination, **44**, 1
Shapiro reaction, **23**, 3; **39**, 1
Silanes:
    addition to olefins and acetylenes, **13**, 4
    electrophilic substitution reactions, **37**, 2
Sila–Pummerer reaction, **40**, 3
Silyl carbanions, **38**, 1
Silyl enol ether, α-hydroxylation, **62**, 1
Simmons–Smith reaction, **20**, 1; **58**, 1
Simonini reaction, **9**, 5
Singlet oxygen, **20**, 2
Skraup synthesis, **7**, 2; **28**, 2
Smiles rearrangement, **18**, 2
Sommelet–Hauser rearrangement, **18**, 4
Sommelet reaction, **8**, 4
$S_{RN}1$ reactions of aromatic systems, **54**, 1
Stevens rearrangement, **18**, 4
Stetter reaction of aldehydes with olefins, **40**, 4
Stilbenes, photocyclization of, **30**, 1
Stille reaction, **50**, 1
Stobbe condensation, **6**, 1
Substitution reactions using organocopper reagents, **22**, 2; **41**, 2
Sulfide reduction of nitroarenes, **20**, 4
Sulfonation of aromatic hydrocarbons and aryl halides, **3**, 4
Swern oxidation, **39**, 3; **53**, 1

Tetrahydroisoquinolines, synthesis of, **6**, 3
Tetrahydrothiophenes, synthesis of, **6**, 9
Thia-Payne rearrangement, **60**, 1
Thiazoles, synthesis of, **6**, 8
Thiele–Winter acetoxylation of quinones, **19**, 3
Thiocarbonates, synthesis of, **17**, 3
Thiocyanation of aromatic amines, phenols, and polynuclear hydrocarbons, **3**, 6
Thiophenes, synthesis of, **6**, 9
Thorpe–Ziegler condensation, **15**, 1; **31**
Tiemann reaction, **3**, 9
Tiffeneau–Demjanov reaction, **11**, 2

Tin(II) enolates, **46**, 1
Tin hydride method to prepare radicals, **48**, 2
Tipson–Cohen reaction, **30**, 2
Tosylhydrazones, **23**, 3; **39**, 1
Tosylmethyl isocyanide (TosMIC), **57**, 3
Transmetallation reactions, **58**, 2
Trimethylenemethane, [3 + 2] cycloaddition of, **61**, 1

Ullmann reaction:
  homocoupling mediated by Cu, Ni, and Pd, **63**, 3
  in synthesis of diphenylamines, **14**, 1
  in synthesis of unsymmetrical biaryls, **2**, 6
Unsaturated compounds, synthesis with alkenyl- and alkynylaluminum reagents, **32**, 2

Vilsmeier reaction, **49**, 1; **56**, 2
Vinylcyclopropanes, rearrangement to cyclopentenes, **33**, 2
Vinyllithiums, from sulfonylhydrazones, **39**, 1
Vinylsilanes, electrophilic substitution reactions of, **37**, 2
Vinyl substitution, catalyzed by palladium complexes, **27**, 2
von Braun cyanogen bromide reaction, **7**, 4
Vorbrüggen reaction, **55**, 1

Willgerodt reaction, **3**, 2
Wittig reaction, **14**, 3; **31**
[2,3]-Wittig rearrangement, **46**, 2
Wolff–Kishner reaction, **4**, 8

Xanthates, synthesis and pyrolysis of, **12**, 2

Ylides:
  in Stevens rearrangement, **18**, 4
  in Wittig reaction, structure and properties, **14**, 3

Zinc–copper couple, **20**, 1; **58**, 1, 2
Zinin reduction of nitroarenes, **20**, 4

William F. Maag Library
Youngstown State University